William Richards

A Practical Treatise on the Manufacture and Distribution of Coal Gas

William Richards

A Practical Treatise on the Manufacture and Distribution of Coal Gas

ISBN/EAN: 9783337059927

Printed in Europe, USA, Canada, Australia, Japan

Cover: Foto ©berggeist007 / pixelio.de

More available books at **www.hansebooks.com**

A PRACTICAL TREATISE

ON

THE MANUFACTURE AND DISTRIBUTION

OF

COAL GAS.

BY

WILLIAM RICHARDS, C.E.

WITH NUMEROUS PLATES AND ILLUSTRATIONS.

LONDON:
E. & F. N. SPON, 46, CHARING CROSS.
NEW YORK: 446, BROOME STREET.
1877.

PREFACE.

THIS work is presented to the Public with the view of supplying a long experienced want, and will undoubtedly be acceptable to engineers, to managers of gasworks of every description, and to all interested in the art of gas lighting.

A considerable degree of detail has been entered into in the work, alike in the description of the various apparatus as well as in the modes of operation, which has never been attempted in any production of the kind hitherto published; and whilst a clear and comprehensive description is given of the machinery and processes employed at the present day for the economical manufacture of gas, together with its proper distribution and consumption, at the same time no extraneous matter has been admitted, and all abstruse calculations avoided; and it is confidently assumed that the work will be appreciated by those for whom it is intended.

In producing this work I have to express my great obligations to Mr. Lewis Thompson for his valuable assistance and contributions, which, like all his exertions in the art of gas lighting, have been labours of love devoid of all pecuniary considerations. My thanks are also due to Messrs. H. Brothers, F. W. Hartley, Hislop, J. Paterson, Wyatt, and other gentlemen mentioned, to whom I am indebted for much valuable information, the sources of which are invariably acknowledged.

THE AUTHOR.

CONTENTS.

	PAGE

Ponderability of air ascertained.—Gas discovered and named.—Inflammability of gases determined.—Phlogistic theory of combustion.—Oxygen, hydrogen, carbonic acid, and the composition of water discovered 1

First application of coal gas for illumination.—Lebon's patent.—Winsor's efforts to establish a gas company.—Establishment of the London and Westminster Gas Company.—Companies established in various parts of the kingdom.—The London Portable Gas Company.—Proposed oil-gas company.—Gas lighting introduced to the Continent and United States.—Competition amongst gas companies.—Various inventions connected with gas lighting.—Beneficial results derived from the use of gas .. 11

Chemistry of gas manufacture.—Observations on the composition of coal.—Hydraulic main a source of purification.—Superiority of wet-lime purification.—Purification by the bisulphide of calcium.—A metalloid suggested as a means of purification.—Ordinary method of purifying gas.—W. R. Phillips's method of purification 39

Coal.—Its origin.—Derivation of name.—The earliest mention of pit coal.—Coal used by the Romans.—Became a commercial product.—Various patents for employing coal.—Different descriptions of coal.—Analyses of coal.—Specific gravity no criterion of the value of coal.—Lignites.—Pitch.—Chemical evidence 53

Carbonization.—Value of coal for gas.—Different temperatures.—Effects of low heats for retorts.—Effects of high heats and heavy pressure.—Accumulation of carbon in the retorts.—Means of removing the incrustation.—Results of carbonizing wet coal.—Fuel employed in carbonizing.—Tar used as fuel.—Luting for retort lids.—Air-tight lids.—Stoking machine.—West's charging and drawing apparatus.—Cost of labour for carbonizing 69

Progress of retorts and settings.—Retorts first employed by Murdoch.—As described by Accum.—Clegg's rotary retorts.—Arrangement of gas apparatus.—Clegg's web retort.—Iron retorts.—Setting described by Clegg, jun.—The defects of iron retorts.—Grafton's brick oven.—Clay retorts.—Their introduction.—Various dimensions and forms in use.—Theory of retort setting.—Practice of retort setting.—Various systems.—General details of retort setting.—The advantages of the coke hole considered .. 81

Hydraulic main.—The purpose of this apparatus.—Best method of taking off the gas.—Divided in some works into various distinct compartments 106

Condenser.—Its object.—Winsor's claim for the introduction of the condenser.—Malam's condenser.—Perks's condenser.—The air condenser.—Effects of cold on gas.—Wright's condenser.—Condensation of hydrocarbons.—Graham's condenser.—George Livesey's experiments with the condenser.—Pélouze and Audouin's condenser.—Dr. Bowditch's experiments.—Considerations on the effects of condensation.—The breeze box 108

Exhauster.—Object of the exhauster.—Invented by Broadmeadow.—Grafton's exhauster.—Jones's exhauster.—First employed in London.—Beale's exhauster.—Anderson's exhauster.—Körting and Cleland's exhauster.—Economy and means of working the exhauster.—Exhaust governors.—Effects of excessive exhaustion 118

CONTENTS.

Washer and scrubber.—Effects of ammonia in gas.—Wilson's washer.—Lowe's scrubber.—The value of ammonia disregarded.—Purification by means of sulphuric acid adopted, and its evil effects.—Mann and Walker's scrubber.—Dimensions of scrubbers.—Anderson's brush scrubber.—Paddon's scrubber.—Whimster's scrubber.—The method of appreciating the value of ammoniacal liquor.—The Twaddle hydrometer.—The alkalimeter.—Table of specific gravities of liquids 126

Purifiers.—Murdoch's first experiments in purifying gas.—Lime adopted by Clegg and patented by Winsor.—Malam's treble purifier.—Reuben Phillips invented purification by moistened hydrate of lime.—John Malam's improvements in purifiers.—Cockey's rotary valve for purifiers.—Walker's valve.—Dimensions of purifiers.—Means of lifting the covers of purifiers.—Körting's reviving apparatus.—Method of constructing purifiers of brickwork 135

Purification considered.—Degree of purity defined by the Board of Trade.—Wet-lime purification.—Dry-lime purification.—Its superiority.—Oxide of iron.—First patented by Laming.—Revivification discovered by F. J. Evans.—Patented by Hills 146

Various kinds of oxide.—To ascertain the value of oxide as a purifying agent.—To determine the quantity of sulphur in spent oxide.—To prepare sulphate of iron and lime.—To prepare test papers.—Gas Referees' instructions for testing the purity of gas.—Delicacy of the operation of testing gas 146

History of the gasholder.—The most primitive kind of store for gas.—The gasholder invented by Lavoisier.—The form of holder first adopted.—Wooden vats at one time used for the tanks of holders.—Clegg's rotary gasholder 158

Gasholder tanks.—The choice of site of a gasworks.—The best description of ground for erecting tanks.—General construction of tanks.—Value of Portland cement and puzzolana.—Excavation of tanks.—Concrete.—Caissons used in forming tanks 163

Gasholder tanks.—Prices of brickwork.—Specification of a brick tank.—Form of contract by Longworth.—Specification of tank by Methven.—Cost of tank.—Specification of tank by H. Brothers.—Puddle may be dispensed with in building tanks.—Specification of a tank built without puddle.—Prices of tanks.—Stone tanks, description of the method of constructing.—Composite tanks introduced by Wyatt.—Specification of tank at Redheugh applied by Douglas.—Concrete tank constructed by George Livesey without puddle.—Cast-iron tanks.—Advantages of.—How constructed.—Specification of and cost of a cast-iron tank.—Weights of cast-iron tanks.—Cost of tanks of various dimensions.—Description of holder working in cast-iron tank.—Annular tanks.—One formerly at Blackfriars Station.—Another at the Surrey Consumers' Work.—Compound tanks.—Description and price of.—One erected at Halifax.—Advantages and economy of compound tanks in certain localities 172

Gasholders.—Opinion divided as to the best method of constructing gasholders.—Necessity for the trussing of roof of holder considered.—Influence of high winds on holders.—Controlling cause of pressure in a holder.—Tables of the weight of iron employed in the formation of gasholders.—Thickness of the sheets of holders.—Rule for ascertaining the approximative weight of holders.—Table of pressures.—Specification and cost of a gasholder.—Method of erecting gasholders, and precautions to be observed.—Prices of holders.—Telescopic holders.—Derivation of name.—The necessity for first employing telescopic holders.—The economy of construction as compared with single holders for a given capacity.—Specification and drawing of telescopic holder by H. Brothers.—Specification of gasholder at Redheugh by V. Wyatt.—Quantities.—The largest gasholder in existence designed by N. F. Kirkham.—Large gasholders as adopted by the late Imperial Company described.—The price of the Imperial holder 200

Station meter.—The necessity for the.—The wet meter always used for gasworks.—The meter described.—Prices of station meters.—Parkinson's station meter the largest in existence.—The tell-tale.—Causes of variation in the measurement of the meter 220

CONTENTS.

	PAGE
The governor.—Its object.—The first invented by Clegg.—Crosley's dry governor.—Ordinary governor.—Parkinson's governor.—Serious accident caused by a governor.—Hartley's governor.—Braddock's governor.—Dimensions of governors.—Means of preventing oscillations.—Peebles's governor.—The rule for calculating the construction of the governor.—District governors.—Stevenson's and Cathels's	228
Pressure gauge and register.—The gauge described.—King's gauge.—The pressure register invented by Crosley, the table for.—Wright's pressure register	234
Photometer.—Its object.—As applied on the Continent.—Count Rumford's photometer.—Ritchie's photometer.—Bunsen's photometer introduced by Mr. King.—Requisite apparatus for photometrical experiments.—Scale and rule for making a photometer.—Wright's, Letheby's, and Evans's photometers.—Hartley's experiments in photometry.—Instructions of the Metropolitan Gas Referees on the method of testing the illuminating power of gas.—Tables for correcting the volume of gas.—Sir Humphry Davy's experiment.—Crookes's radiometer.—Lowe's jet photometer	238
Mains of various materials proposed.—The Chameroy main.—Cast-iron mains.—Turned and bored joints.—Lead joints.—Durability of mains.—Method of preserving mains.—Leakage from mains.—Means of ascertaining obstructions in mains.—Weights of mains.—Cost of laying mains.—The bladder valve.—Cost of mains per yard.—Length of mains in London.—Number of lamps supplied in London.—Dimensions of mains.—Barlow's table of the discharge of gas through mains of various diameters and lengths at different pressures	251
Gas mathematics.—How to accommodate the pressure to the demand.—To find the quantity of gas that will pass by change of pressure.—To ascertain the flow of gas by tubes of different diameters.—To ascertain the quantity delivered of different specific gravities.—Table of the square roots of the specific gravities of gas.—Table of the square roots of pressures	268
Services.—Means of preserving.—Wrought-iron services.—Lead services.—The service cleaner.—Bottle syphons.—Apparatus for drilling mains.—Dimensions of services.—Quantity of gas delivered by services of various diameters	270
Consumers' meters.—Clegg's first attempts at making.—Clegg's first patent meter.—Malam's meter.—Perfected by Clegg and Crosley.—Congreve's meter.—Clegg's pulse meter.—Malam's dry meter.—Bogardus's dry meter.—The formation of the Dry Meter Company.—The Sullivan Meter Company.—Defries' dry meter.—Richards's dry meter.—The general application of meters.—Sales of Gas Act.—Compensating meters.—Warner and Cowan's meter.—Lighting public lamps by meter.—Average meter system.—Advantages of the dry meter.—Various systems of supplying meters.—Necessity for a periodical investigation of meters.—The motive-power meter	272
Gas burners.—Argand's patent.—Murdoch's first burners.—Argand burner employed for gas.—The fishtail invented.—Experiments of Drs. Christison and Turner.—Various controlling influences on gas burners.—Effects of heavy pressure.—Atmospheric influence.—Varied light from different consumption of gas.—Bannister's experiments.—Kirkham's experiments.—Effect of small burners.—Wood's experiments.—Blackburn's experiments.—Henry Woodall's report.—Light obstructed by glasses.—The Farmer theorem.—Best material for the construction of burners.—Gas Referees' report on burners.—Experiments.—The dimensions of the Standard London Argand burners	287
Regulators.—The object of.—Dry regulators.—Former opposition to regulators.—The wet regulator.—The public lamp regulator.—Giroud's rheometer, experiments with.—The necessity for regulators	301
Gas fittings.—Early prejudice against gas.—General application at the present time.—The durability of gun-barrel.—The dangers attending the use of soft metal pipes when placed in inaccessible localities.—Table of the delivery of gas through small pipes.—Elegance of modern fittings in dwellings.—Necessity for care in the construction of hydraulic chandeliers	306

CONTENTS.

Residual products.—The value of coke.—The variable production of coke of the Metropolitan companies and quantity used as fuel.—Prices realized for coke.—Difficulties attending the sale of coke.—The applications of breeze as fuel.—As applied at the Dover Works.—Manufacture of the sulphate of ammonia.—Cost and returns from the process.—All the ammonia in the liquor not indicated by the acid test.—Applications of tar.—As manufactured in works generally.—Products from a ton of cannel. —Net returns.—Tar paving.—Attempts to make gas from tar.—Foul lime, its application to agricultural purposes 307

The composition of coal gas   316

Carburating gas.—Ibbetson's patent.—Donovan's patent for enriching hydrogen with hydrocarbon vapours. —Lowe's patent for enriching coal gas.—Stephen White's hydrocarbon process.—Mansfield combined hydrocarbon vapours with air to produce light.—How applied.—Operation of the hydrocarbon process. —The Photogenic Gas Company.—The Air Light Company.—The Eupion Gas Company.—Dr. Bowditch's system of enriching gas.—Causes of failure.—Carburated air employed in America 320

Considerations on the establishment of gasworks.—Geographical position.—Nature of contract for public lighting.—Restrictions enforced on companies.—Capacity of coal stores 324

Gasworks of various magnitudes.—Experimental apparatus.—Works for small establishments.—Factories. —Works for small towns or villages.—Rules to be observed in designing gasworks.—Works of three million feet per annum.—Operations of gasworks in Scotland.—Works of sixty million feet.—Cost of apparatus.—The Redheugh Gasworks.—The Beckton Gasworks.—Plan of.—The extent of site of.— The means of transport at.—Retorts at.—The system of purification at.—The gasholders at.—The number of persons employed at.—Views of the Beckton Works from the river and from the clock tower.—Derivation of the name of Beckton.—The ex-Imperial Company's Works at Bromley.—The capital and operations of the Metropolitan gas companies 328

Chimney shafts.—Heights of stacks.—Dwarf chimneys.—Edinburgh shaft.—The damper .. 342

Competition of gas companies, and results.—Necessity for all companies to be authorized by Parliament .. 348

Influence of barometric pressure and temperature on gas   352

Retort setting   354

The importance of good settings.—Facilities afforded for the proper setting of retorts.—Setting of retorts in a works of the first magnitude.—Settings, of one iron retort, of three retorts, of five clay retorts, of six retorts, of seven retorts, of ten retorts.—Tindall's method of supplying air to furnaces, where coal is employed as fuel 355

A PRACTICAL TREATISE

ON THE

MANUFACTURE AND DISTRIBUTION OF COAL GAS.

INTRODUCTION.

DURING several centuries a class of philosophers denominated alchemists devoted themselves, with the utmost ardour, to the purpose of attaining certain results, of which the means of transmuting the baser metals into gold was in most cases their chief, if not their only object; and, incredible as it may appear, these operations were not only regarded as possible, but actually believed, and this by men of intelligence, to have been accomplished. Moreover, we find among the writings of some of these enthusiasts instructions are given how to proceed to obtain so desirable an object, but it is almost needless to say that they are fallacious and without any basis of truth. A further object of their fruitless researches was to obtain the "philosopher's stone," by means of which, it was believed, "the greatest disease could be cured, sorrow and evil and every hurtful thing avoided, by help of which we pass from darkness to light, from a desert and wilderness to a habitation and home, and from straitness and necessities to a large and ample estate."

These were the chimeras vainly pursued during many ages by alchemists, and if we had not painful proof before our eyes that there are people who believe in the virtue of spirit-rapping and spirit-writing at the present day, we might be allowed to doubt the existence of alchemists in former times. But there are spirit-rappers now living, and beyond all doubt there were once alchemists, of whose peculiar operations it is recorded, that it "is an art without an art, that begins with lying, is continued with labour, and ends with beggary," a description equally applicable to spirit-rapping.

Nevertheless, alchemy has not been altogether useless in the development of science, for its absurd hopes and foolish attempts kept alive a spirit of inquiry regarding the metals, which really led to the discovery of the gases, and consequently to the ultimate expansion of chemical knowledge. Of the truth of this we may soon convince ourselves by a candid review of the discoveries which first proved that the gases had weight, and were material bodies, regarding which the ancients appear to have been wholly ignorant. This fact was discovered and demonstrated in the year 1630, by Jean Rey, a physician of Perigord, who noticed that both tin and lead increased in weight during alchemical calcination; and he attributed this augmentation of weight to the absorption of air, which he therefore supposed to be ponderable like other material bodies.

This discovery was followed, in the year 1665, by the publication of Hooke's 'Micrographia,' one of the most remarkable works ever published, for in it we find developed the germs of our existing theory of combustion, although for a time the solid facts of Hooke were buried beneath the purely imaginary phlogistic theory of Stahl and his disciples. "The air in which we live," says Hooke, "is the dissolvent of combustible bodies, and this solution or combustion does not begin until the bodies are heated, and then the effect of that solution causes the increase of heat which we call fire; but the solution of the combustible body is due to an inherent substance which is contained in the air, and this substance resembles that which is fixed in saltpetre, though it is not exactly the same. Part of the combustible body is converted into air by this solution and rendered volatile, part is resolved into light matters that are carried off by the air, and part remains fixed." Now, making allowance for the difference in language, we must admit that the changes which occur during ordinary combustion are as completely explained in the above description as they could be by the most learned philosopher of the present day.

Here we may observe that at the period in question, and in opposition to the theory mentioned, all things were believed to be composed of four elements, namely, earth, water, air, and fire. But these were not understood in the same sense of the terms of the present time, as by earth was understood all solid matter; by water all liquids; by air all matter in a state of vapour, or gas; and fire was applied to that we call heat or caloric, by which solids are rendered liquid, and liquids become gaseous. Air was considered to be an element of extreme subtilty, with a tendency to rise from the earth; that it entered into the formation of some substances, and that it was a principal constituent of others. But a still greater vagueness existed with regard to its characteristics and functions, for although it was known that air was in some way necessary for the support of fire and animal life, yet it seems to have been considered that aëriform substances, differing in character from the atmosphere, were not essentially distinct, but differed in consequence of some admixture.

Hence our atmosphere was recognized principally in consequence of its effects, as displayed in the production of heat and cold by the action of the air, or in the motion communicated to bodies by the wind, as in windmills, and the ripples or waves upon water; but the ideas of its materiality thus derived were indefinite, and perhaps not much unlike our own at the present day with regard to what is called the "electric fluid."

Among the accidents arising from this want of knowledge were the mysterious and sudden deaths which occurred in mines or other subterranean or enclosed places, by the accumulation of carbonic acid or other noxious gases; and as no assignable reason could be given for these direful calamities, superstition supplied the want by attributing this to the work of evil spirits, who tempted men into these fatal localities, and there treacherously caused them to perish; and by alchemists this was termed the "spirit sylvester," or wicked spirit.

Among the last of the alchemists was Paracelsus, who distinguished himself by rejecting many of the theories of his contemporaries, and to whom is due the merit of first producing gas or "air" by the action of the solution of sulphuric acid on limestone; but to which circumstance he attached no importance.

Closely following Paracelsus was the celebrated Van Helmont, of Brussels, of one of the most ancient families of Europe, who dedicated himself at first to alchemy, and subsequently to more rational philosophy, and preferred his laboratory at Velvorde to all the state and splendour of the court. His writings were produced by his son soon after the death of the father, and were published first in Latin and afterwards translated into English in 1662.

From these we learn that Van Helmont was the first to recognize the existence of other aëriform bodies differing entirely from the atmosphere, and that he produced them in various ways, as by combustion, fermentation, and the action of acids on limestone. With reference to the first, he says that charcoal as well as other bodies disengage from their combustion this "spirit sylvester," and describes an experiment where "seventy-two pounds of oak charcoal gave one pound of cinders, and the seventy-one pounds remaining served to form the spirit sylvester." He remarks, "there are bodies which contain this spirit, of which they are almost entirely composed, and is therein fixed and solidified, and are made to leave that state by fermentation, as we observe in the fermentation of wine." "This spirit, up to the present time unknown, not susceptible of being confined in vessels, nor capable of being reduced to a visible body, I call by the new name of GAS." By some authors this word is stated to be derived from "chaos," by others from "geist," a spirit, pronounced like the Scotch word "ghaist," and synonymous with that and the English word "ghost."

In the course of the numerous investigations of Van Helmont he discovered that the gases arising from fermentation, the products of combustion, and the "spirit sylvester," were identical in their nature (carbonic acid gas), to which he gave the name of "gas sylvester;" further, that it would not support life, and by it flame was extinguished. In the author's writings he distinguishes between condensable gases or vapours and incondensable or permanent elastic fluids. In his experiments on air he refers to its weight and elasticity, and when describing the air thermometer, details with much precision the effects of temperature and pressure on air. Referring to fire, he describes it as the "destructor and artificial death of things."

"Nothing," says Van Helmont, "acts so powerfully upon us as gas, as demonstrated in the 'Grotto des Chiens,' and the asphixis from charcoal." Thus the demons or wicked spirits which had been previously so much feared were in this way explained, as it was the gas sylvester which killed the workman in the mine, or the labourer in the cellar.

But most remarkable amongst his statements is that "flame is incandescent gas, or a vapour lighted," and although the assertion was correct, he failed to demonstrate its accuracy.

Lastly, in Van Helmont's works is described the following experiment, which has since engaged the attention, and been repeated times out of number by chemists. He says: "Place a lighted candle in a dish, into which pour water to the height of two inches, then cover the candle by a glass bell, when you will presently observe the air in the bell to diminish in volume and the water to rise and take its place, until the light is extinguished." The conclusion drawn by the author from this experiment was, that as a vacuum was produced, it was immediately occupied by the water; but he does not say if the flame destroys the air, nor that the latter serves to aliment the former.

It is a matter for surprise that the various facts produced by Van Helmont should have been allowed to pass unregarded by the chemists who succeeded him, and, moreover, that in the various experiments where any hydrocarbon was submitted to destructive distillation, the fact of the vapour igniting did not convey some useful lesson.

Bernoulli, of Basle (1690), adopted means to produce, collect, and retain the aëriform body, carbonic acid, by submitting limestone to the action of strongly acidulated water, but without ascertaining its nature.

We have shown that although Van Helmont understood the existence of various gases, he possessed no means of storing or confining them; this want was supplied by Boyle in the experiment hereafter mentioned, and which may be considered the starting point of pneumatic chemistry, as well as the first production and storage of hydrogen gas, the properties and importance of which were not recognized for upwards of a century afterwards. He says:

"A glass globular vessel having a long neck, of the capacity of about three ounces, was filled with equal parts of oil of vitriol and water; six small nails being placed therein. Another similar vessel, but having a wider neck, so that the first could enter it freely, was also filled in like manner. The vessel having the smaller neck was then temporarily closed, reversed, and plunged into the larger, when directly was seen to rise in the upper globe aëriform balls, which in assembling displaced the liquid. Speedily all the water in the upper vessel was expelled, and its place occupied by a body having the appearance of air," and was believed by the author to be that produced artificially by the action of the liquid on the iron.

The Hon. Robert Boyle, who was described by his contemporaries as an ornament to his nation, although, as it appears by his works, he had some leaning towards alchemy, pursued his scientific investigations with remarkable industry, as evinced in the numerous works on various subjects from his pen. He said that he "devoted himself to chemistry, not to produce medicines nor as alchemist, but as a philosopher." Among the remarkable observations of Boyle were "that iron rust was caused by a corrosive effluvia of the air, and the study of these products will one day make known the composition of air;" and further, that "there are some vital substances existing in the atmosphere, indispensable to combustion, respiration, and fermentation." Again, " it is surprising that there is something in the air which by itself is sufficient to support the flame, and once this matter consumed, the flame is immediately extinguished, and yet the remaining air does not lose any of its elasticity."

In Hofer's 'Traité sur la Chimie,' a work of the greatest reliability, as in the most important points the Latin language is copied literally from the original, the author says, "When reading this part of Boyle's works, we expect at each moment to fall upon this *vital substance* of air, as he repeatedly approaches closely to that point, and avoids it continuously, occasioning the greatest disappointment." Thus we find that Boyle was close on the track of the discovery of oxygen, which, however, remained one of Nature's secrets until discovered by Dr. Priestley.

But whilst chemists were thus endeavouring to become acquainted with these hidden laws, as if to direct their attention to the proper source, and towards the realization of the art under consideration, inflammable gas issuing from the earth, as also in mines, was then discovered.

The earliest description of the former is met with in a communication to the Royal Society in the year 1667, by Mr. Thomas Shirley, wherein he mentions that his attention was directed, about eight years previously, to what was considered to be a spring, "where the water did burn like oyle," and "did boyle and heave like water in a pot;" but, on investigation, he found this to "arise from a strong breath, as it were a wind," which ignited on the approach of a lighted candle, and "did burn bright and vigorous."

This was the first of many similar observations in England on carburetted hydrogen, a gas identical with that now so universally employed for lighting, issuing spontaneously from the earth. At one time it was considered to arise from supernatural agency, but is now well known to be occasioned by the heat of the interior of our globe being in such proximity to the coal, or other bituminous substances, as to effect a gradual and constant distillation, the resulting gas issuing from fissures in the earth. To the same operation is due the existence of burning fountains, where the gas had by accident, or perhaps by lightning, been ignited, causing continuous fires, near which religious temples were often erected for the purpose of worshipping that element.

Amongst the earliest accounts of the ignition of gas in coal mines, is a communication from Mr. Jessop to the Royal Society of London, in 1674, wherein is related the particulars of an accident which occurred to a workman named Michel, in a mine in Yorkshire. This man descended a coal-pit with a torch in his hand, and on arriving in the gallery he was suddenly surrounded by a vivid flame, which burned his clothes, face, hair, and hands. Having been withdrawn from thence, he declared that he heard no noise, though the workmen who were employed in the vicinity were terrified by a frightful explosion, accompanied by a vibration of the earth. In the case mentioned there was an accumulation of carburetted hydrogen gas, along with atmospheric air, properly called "firedamp," or fire vapour, "damp" being a Dutch word for vapour; which exploded by the approach of the flame. An accumulation of carbonic acid gas, by which a person is speedily asphyxied, is called chokedamp, or choke vapour; the terms are frequently but erroneously used indiscriminately.

In the 'Philosophical Transactions' for 1733 is detailed "An Account of a Damp Air in a Coal-pit of Sir James Lowther, sunk within Twenty Yards of the Sea." According to this record, some men, in sinking a pit, "came to a black stone bed, which, contrary to expectation, afforded very little water; but, instead thereof, a vast quantity of damp corrupted air, which bubbled through a quantity of water then spread over that part of the pit and made a great hissing noise, at which the workmen, being somewhat surprised, held a light towards it, when it immediately took fire upon the surface of the water, and did burn very fiercely." After describing the precautions taken to carry off the gas by means of a pipe, and referring to its continuous flow, the writer observes: "The said air being put into a bladder and tied down, may be carried away and kept some days, and being pressed gently through a small pipe at the flame of a candle, will take fire and burn at the end of the pipe so long as the bladder is pressed to feed the flame, and when taken away from the candle after it is so lighted, it will continue burning until there is no more air left in the bladder to supply the flame." "The air when it comes out of the top of the pipe is as cold as frosty air." "It is to be observed that this sort of vapour does not take fire except by flame; sparks do not affect it, and for that reason it is frequent to use flint and steel in places affected with this sort of damp, which will give a glimmering light that is a great assistance to workmen in difficult cases."

A great progressive step was made by Dr. Hales, as we find described in his 'Vegetable Staticks,' published in 1726, the first experiment on the destructive distillation of coal. Therein he states that when distilling 158 grains of Newcastle coal he obtained 180 cubic inches of air (gas) weighing 51 grains. The same author also discovered that iron filings, when intermixed with oil of vitriol, produced scarcely any gas, but that on the addition of water, an inflammable gas was abundantly evolved from the mixture, which was ignited by a flame. In this experiment hydrogen was again produced, but an additional fact of its inflammability was learned, and it was subsequently called inflammable air. A further improvement in pneumatic apparatus was effected by Dr. Hales, by which the distilling retort could be placed on the fire, and the gas evolved collected in a receiver by means of an hydraulic lute.

Again, in the 'Philosophical Transactions' for 1739 was published an extract from a letter written by Dr. John Clayton, supposed to have been addressed to Mr. Boyle (already mentioned, who died in 1691); and if this be correct, Dr. Clayton's observations were made about the same time as those of Mr. Shirley. Dr. Clayton relates that he saw a ditch near Wigan wherein the water "would seemingly burn like brandy," but on investigation he discovered that it arose from a "shelly coal, and the candle being put down into the hole, the air catched fire and continued burning." The Doctor subsequently made some experiments on the distillation of coal, which he effected in a retort on an open fire. Using his own words, "At first there came over only phlegm, afterwards a black oil; likewise a spirit arose which I could in no ways condense, but it forced my lute and broke my glasses. Once, when it had forced my lute, coming close thereto, in order to try to repair it, I observed that the spirit which issued out caught fire at the flame of the candle, and continued burning with violence as it issued out in a stream, which I blew out and lighted again alternately several times." He then describes how he filled several bladders with the gas, and further remarks, "I kept this spirit a considerable time, and endeavoured in several ways to condense it, but in vain; and when I had a mind to divert strangers or friends, I have frequently taken one of these bladders and pricked a hole therein with a pin, and compressing gently the bladder near the flame of a candle till it once took fire, it would then continue flaming till all the spirit was compressed out of the bladder, which was the more surprising, because no one could discern any difference in the appearance between these bladders and those which are filled with common air."

The foregoing account is of the utmost interest, describing as it does the manufacture, the means of storing gas, as also the production of light therefrom, the pressure necessary to expel it to be inflamed, the facility of lighting and extinguishing, and the fact of its being kept a considerable time without losing its inflammability; all most important lessons of instruction, which however demanded considerable time before they were practically applied.

As light and heat come equally within our province, we have to direct the reader's attention to another theory of the cause of combustion, which for a period held its position, and was accepted by all the learned men of the day in opposition to the clear, simple, and, as now proved, correct view of Hooke.

This theory was introduced in 1700, by a German named Becker, and enlarged by another of the name of Stahl, and it is difficult to say whether the temporary celebrity or the absurdity of this theory was the greatest. The principle of this was based upon the existence of a supposed substance called phlogiston, derived from the Greek, signifying flame, which was supposed to combine with metallic oxides to produce metals, and with oxidized products to form simple bodies; thus oxide of iron, when combined with phlogiston, formed metallic iron; and this metallic iron when burnt, again became oxide of iron, the phlogiston being at the same time set at liberty in the shape of flame, light, heat, and fire.

Such was the celebrated phlogistic theory of George Ernest Stahl, that prevailed nearly all over

Europe for more than half a century, and which, during the whole of that time, might have been upset with one single experiment made with weights and scales; since, by the conversion of oxides into metals and other like experiments, the addition of the supposed substance phlogiston must always have been attended by a loss of weight; or, in other words, phlogiston must be a substance that weighs less than nothing, a conclusion too absurd to require refutation.

Nevertheless, the theory of Stahl not only held its ground, but became diffused throughout the whole range of art and science, so that even the practice of medicine itself acknowledged this influence; and hence all diseases of an inflammatory or febrile nature were supposed to arise from an excess of phlogiston, and were, therefore, put under what is still termed the "antiphlogistic treatment."

We might perhaps now wonder that such men as Priestley, Scheele, Cavendish, and Watt, could resign themselves to unproved and undemonstrated assertions; but when we come to examine the phlogistic theory clearly our wonder ceases, and we feel astonished at the singularly adroit manner in which that theory had been made to fit and explain all the facts then known. Thus, as we have seen, these things which we now call metallic oxides, were supposed by Stahl to be simple substances which required only a certain amount of a substance called phlogiston to convert them into metals, and the nature of the metal thus produced was supposed to depend upon the quantity of phlogiston added. Consequently the alchemical notion of making gold was not controverted, for as gold was supposed to be a metal which contained more phlogiston than any other, the difficulty of making it was resolved into the means of causing phlogiston to unite with a calx or metallic oxide in very large quantities.

Thus taking such a calx as the oxide of zinc, and combining it with a certain quantity of phlogiston, we obtained metallic zinc; and putting this zinc into nitric acid it became a calx and united with the acid; but as no inflammable gas was given off, the phlogiston remained in the residuum or nitrate of zinc; and to prove that it did so remain, we have only to mix the residuum with any combustible matter and to apply heat, when a violent deflagration would ensue, accompanied by the evolution of light and flame, in which shape the phlogiston was supposed to escape and disappear.

When, however, the zinc was put into strong sulphuric acid there was no action, because no substance was present to combine with the phlogiston, although the acid was ready to combine with the zinc calx; if, however, in this state of things water was added, the phlogiston would produce a kind of air called phlogisticated air, which being burnt parted with the phlogiston in the shape of heat, and reproduced the water; an explanation which exactly fits the production and combustion of what we now call hydrogen gas, and to which Stahl's theory assigned the name of phlogisticated or inflammable air.

The metallic oxides being destitute of phlogiston, could not of course yield it, and therefore when Priestley discovered that the peroxide of manganese yielded a gas or air by heat, he gave it the name of dephlogisticated air, as if it were common air deprived of its phlogiston; whilst Cavendish communicated to the Royal Society his opinion that it was water deprived of its phlogiston, consequently what we now call hydrogen was then looked upon as water with a deficiency of phlogiston. But perhaps the best mode of showing the confused ideas then prevalent amongst scientific men will be to quote a letter written by the celebrated James Watt to Dr. Priestley on this very subject. "Let us now," says Watt, "consider what obviously happens in the deflagration of the inflammable and dephlogisticated air. These two kinds of air unite with violence, they become red hot, and upon cooling totally disappear. When the vessel is cooled, a quantity of water is found in it equal to the weight of the air employed. This water is then the only remaining product of the process, and water, light, and heat are the products. Are we not then authorized to conclude that water is composed of dephlogisticated air and phlogiston deprived of their latent and elementary heat; that dephlogisticated or pure air is composed of water deprived of its phlogiston and united to elementary heat and light; that the latter are contained in it in a latent state so as not to be sensible to the thermometer or the eye; and if light be only a modification of heat or a circumstance attending it, or a component part of the inflammable air, therefore dephlogisticated air is composed of water deprived of its phlogiston and united to elementary heat."

INTRODUCTION. 7

If now we substitute the words hydrogen gas for inflammable air, and oxygen gas for dephlogisticated air, we see how far Watt's ideas carried him in the direction of comprehending the exact composition of water; and it may serve to show how nearly a man may approach the truth without actually reaching it, for had not Watt been blinded by his phlogistic notions, it is almost certain that he would have unveiled the real composition of water, and thus anticipated the discovery of it by Lavoisier in the following year.

It is, however, to Lavoisier that we stand indebted for a clear view of the composition of water, as well as for the annihilation of the phlogistic theory; and although both Cavendish and Watt had clearly worked the matter out, they were unable to divest themselves of their preconceived opinions so as to grasp the result of their labours, and like Lavoisier bring forward an intelligible explanation of the phenomena which their labours had so eminently assisted to develope. However much, as Englishmen, we may therefore regret the loss to our national *amour propre;* still, in justice to the distinguished French chemist Lavoisier, we must accord to him the merit of simplifying our views regarding the nature of the primary gases, hydrogen and oxygen, as also for his other great works; and this we may do with all the more willingness when we come to reflect upon the splendid discoveries of Sir Humphry Davy, and the unrivalled essay upon the nature and composition of flame which he has left us, and which is so intimately associated with the world-renowned "safety lamp." In point of fact, then, if France can claim the honour of elucidating pneumatic chemistry, England may justly claim the merit of having rendered it subservient to the use and benefit of mankind at large.

Having explained the theory of combustion as accepted during many years, we now return to the progressive operations connected with our subject. For this purpose it may be stated that on perusing the works of the most celebrated chemists published in the middle of the last century, we find the great discoveries of Van Helmont, Boyle, Bernoulli, and others, had been neglected, if not entirely forgotten; for, excepting the mention of "damps or vapours, which infest subterraneous places, and are inflammable and uninflammable," and inflammable air (hydrogen), all other aëriform bodies were considered to be identical with our atmosphere; and these chemists, when describing the destructive distillation, as now called, of wood shavings, express the greatest astonishment "that a quantity of air five hundred times greater than the body from which it was extracted could have existed therein;" it is moreover remarked, that the shavings retained the same form and shape after the distillation that they possessed before the operation; and they further observe, since air is disengaged in such abundance from so many bodies, it must necessarily be part of their composition.

Analysis as then practised consisted simply in submitting the substance under investigation, whether oil or milk, or any similar matter, to that degree of temperature to expel its most volatile constituents, whilst with other substances more refractory, as wax, rosin, wood, &c., the heat of the retort was continued to the point of destructive distillation, for the purpose of observing the nature of the residuum in the retort, termed the *caput mortuum*, a name derived from the alchemists, whilst the volatile constituents were never regarded, except when alluding to the quantity of air given off during the process. One experimenter informs us that this air was permitted to escape for fear of breaking his retorts. Another states that when the heat arrives at a certain point, air is disengaged, and from this arises the greatest danger during the operation, which, being ignited accidentally, may have occasioned the "spirit sylvester" so much dreaded and feared in those days of superstition.

In the 'Chemical Essays' of Dr. Watson, published in 1767, mention is made of an experiment, differing from Hales's by neglecting the quantity of air produced, but giving the details of the residuals. He says:

"I took 96 ounces of coal, and putting them into an earthen retort, distilled them with a fire gradually augmented until nothing more could be obtained from them. During the distillation there was a frequent occasion to give vent to an elastic vapour, which would otherwise have burst the vessels employed in the operation, and the result was as follows.

"Weight of Newcastle pit coal, 96 ounces.

"Weights of the products .. .. $\begin{cases} \text{Liquid .. .. .. 12 ounces.} \\ \text{Residue .. .. 56 ,,} \\ \text{Loss of weight .. 28 ,,} \end{cases}$

96 ounces.

"The matter which is lost during the distillation of both wood and pit coal has been called *air*, and it certainly has one, at least, of the most distinguishing properties of air—permanent elasticity—for bladders may be inflated with it as with common air. It does not begin to be separated from wood or coal till the lighter of the two oils appears, and then it rushes out with great violence, and unless a proper vent be given to it the strongest vessels will be burst by it."

He further remarks: "The air which issued with great violence from the retort was inflammable, not only at its first exit from the distillatory vessel, but after it had been made to pass through two high bended glass tubes and three large vessels of water." He states that he collected the air in bladders, that it retained its inflammability for a long time, and burned like the air separated from some metals by solution in acids. Then follows a series of useful queries with the view of exciting others to investigate the matter, in order to obtain further and more important discoveries. .

We have seen that the term "gas," as applied by Van Helmont, was not retained by chemists; indeed, Boyle, in his 'Sceptic Chemist,' seems to ridicule the term by saying "gas or blas, or whatever they may call it;" but it appears in the 'Chemical Dictionary' of Macquer, published in 1771, as a "name given by chemists to invisible parts which escape from certain bodies. The mineral noxious vapours called damps may be also ranked among the vapours called gas."

It may be here stated that both Sir Isaac Newton and Volta expressed their opinion that flame was produced by the ignition of vapour, without, however, proving the fact; and Neuman observed that when pitch was held over flame, it emitted a vapour which took fire from a lighted paper; yet still the inflammability of gas disengaged from resin, wood, coal, and other bituminous substances, excited no attention. Up to this period, therefore, but little progress was made towards lighting by gas; and, as will be shown hereafter, the appliances for the realization of this object had yet to be invented. In due time the one was subservient to the other.

Contemporaries with Dr. Watson were the renowned chemists, Black, Priestley, Cavendish, Scheele, Lavoisier, and others, who devoted themselves to the development of the science they did so much honour to, and whose labours were productive of the most remarkable results.

The first of these great men in the order of time was Dr. Black, who produced carbonic acid gas, and on investigation found its nature and qualities to be directly the opposite of those of atmospheric air; this being the first recorded instance in which the gas was obtained in an unmixed state, and the positive knowledge of its dissimilarity to our atmosphere acquired. To this, the name "fixed air" was given, signifying air that had been fixed in the body from which it was derived; which term was subsequently applied by chemists, for several years, to all gases, thus repeating the ideas of Van Helmont.

The primitive steps in pneumatic chemistry adopted by Boyle, Hales, and others, were far from being satisfactory for the study of gases, but by Priestley's invention of the pneumatic trough-sliding trays, inverted jars, and other appliances, the necessary mechanical means for that purpose were to a limited extent obtained, and from these all the various hydraulic seals of a gasworks, whether in the hydraulic main or valve, or luting for the covers of purifiers and other apparatus, were all derived. And from this source, as will be hereafter shown, undoubtedly the great French chemist, M. Lavoisier, conceived the idea of the invention of the gasholder.

We have described how closely Boyle approached to the *vital substance* in atmospheric air, which however was discovered by Dr. Priestley on the 1st of August, 1774. The method of experimenting which he adopted, consisted in exposing a quantity of red precipitate of mercury to the action of the sun's rays concentrated by a burning lens; the red precipitate was contained in a small flask filled up with quicksilver. "I presently found," he says, "that by means of this lens air was expelled from it very readily. Having got about three or more times as much as the bulk of my materials, I admitted water to it, and found that it was not imbibed by it. But what surprised me more than I can well express, was that a candle burned in this air with a remarkable vigorous flame." And thus was first discovered oxygen gas, or as termed dephlogisticated air, and the foundation of the science of chemistry laid.

Cavendish, at the same epoch, instituted investigations into the inflammable air (hydrogen) of Hales and others, and found it to be the lightest of all matter; that it was evolved in different quantities according to the metal acted upon by the acids and water; and he also observed its other various characteristics. After some years of labour and experiments, Cavendish, in 1784, made the astonishing discovery that this gas was one of the components of water.

During this period extraordinary discoveries were made by the great chemists named. By them the compositions of our atmosphere and water were determined; the cause of combustion, with its resulting gases, was made known; and innumerable other marvellous scientific facts, foreign to our subject, were ascertained.

The discovery made by Cavendish induced Lavoisier and Meusnier to determine the composition of water by analysis. For this purpose they passed steam through a red-hot pipe, whereby they obtained a quantity of inflammable air, which, when mixed with atmospheric air or oxygen, exploded on being ignited. They further determined, from the similarity of the gas so obtained to hydrogen, that this gas only was produced, and the oxygen deposed. Reasoning from analogy that carbon possesses a considerable affinity for oxygen when at a high temperature, they afterwards introduced charcoal into the pipe, which, being heated to redness, resulted in a considerable increase of the quantity of gas, although of a somewhat different and mixed nature.

The knowledge of the explosive nature of hydrogen when intermixed with atmospheric air being acquired, endeavours were speedily made to render this a source of motive power, for which purpose several patents were obtained, and undoubtedly the scheme must have presented great prospects of success; yet during the greater portion of a century from the date of its first invention, although it has engaged the attention of some of our cleverest men, motive power obtained from explosive compounds has not reached that point to be pronounced satisfactory.

In 1781 Lord Dundonald procured a patent for "extracting or making pitch, tar, essential oils, volatile alkali, mineral acids, salts, and cinders from pit coal;" thus embracing all the products of coal except the most valuable—gas, which patent was carried into active operation during several years.

In 1786 the celebrated Lavoisier and other French chemists, in opposition to the opinion of their friends and collaborators Priestley and Cavendish, decided on adopting a general expressive chemical nomenclature, and by them the word "gas" was revived, and selected to apply to all permanent aëriform bodies. Hence its adoption into every civilized language either as *gas* or *gaz*.

But, with all these marvellous achievements, the art of gas lighting was not materially advanced, as will be well understood from the following extracts from Dr. Priestley's 'Experiments and Observations on Different Kinds of Air,' published 1790. The author, when treating on inflammable air, so far as relates to our subject, remarks: "There are different kinds of inflammable air, as has been observed by most persons who have made experiments on air. That which has been commonly observed is, that some of them burn with what may be called a 'lambent flame,' sometimes blue, sometimes yellow, and sometimes white, like the flame from wood or coal in a common fire; whereas another kind always burns with an explosion, making more or less of a report when a lighted candle is dipped into a jar filled with it. Of the latter kind is that which is extracted from metals by means of acids, &c., and of the former kind is that which is expelled from wood, coal, and other substances by heat. It is observable that when wood is heated in an earthen retort the first portion" (of inflammable air) "burns with a lambent white flame, like that from burning wood in an open fire." Hence it appears that at this period there did not exist the most vague idea of the use of inflammable air as a means of artificial light.

c

It is stated in Matthews' 'History of Gas Lighting' that, in 1784, a Mr. Diller exhibited in London and other large cities, that which he termed "philosophical fireworks," produced by the combustion of inflammable gases; we may suppose this to have been by an application of hydrogen, with probably certain mechanical means, more recently known as Chinese fireworks; since it is certain that had any public application of the use of carburetted hydrogen been adopted, it would have come within the knowledge of Dr. Priestley previous to the publication of the work referred to.

In our brief narrative we have described the first ideas respecting gases, the primitive bases of pneumatic chemistry, the earliest productions of inflammable air, the spontaneous emissions of carburetted hydrogen from the earth and their ignition, the manufacture of coal gas, together with the means of storing it, and, on a small scale, its application to lighting. All this had been achieved, but there was still wanting the mind to conceive the mode of applying coal gas to useful purposes, and the intelligence to carry it to a successful issue.

HISTORY OF GAS LIGHTING.

The honour and merit of the first application of coal gas to useful purposes are, beyond all question or doubt, due to Mr. William Murdoch, who, in the year 1792, then residing at Redruth, in Cornwall, directed his attention to the quantities and qualities of different kinds of gases obtained by distilling various mineral and vegetable substances; and who, from observations he had made on the burning of coal, was induced to try the combustible properties of gas; and in order to explain his first operations we cannot do better than give Murdoch's own words, which appeared in 'Nicholson's Journal,' vol. xxi., published in 1808:

"It is now nearly sixteen years since, in the course of experiments I was making at Redruth, in Cornwall, upon the quantities and qualities of different kinds of gases produced by the distillation from different mineral and vegetable substances, I was induced by some observations I had previously made upon the burning of coal, to try the combustible property of the gases produced from it, as well as from peat, wood, and other inflammable substances; and being struck with the great quantities of gas which they afforded, as well as with the brilliancy of the light and the facility of its production, I instituted several experiments with the view of ascertaining the cost at which it might be obtained, compared with that of equal quantities of light yielded by oils and tallow.

"My apparatus consisted of an iron retort, with tinned copper and iron tubes through which the gas was conducted to a considerable distance; and there, as well as at intermediate points, was burned through apertures of varied forms and dimensions. The experiments were made upon coal of different qualities, which I procured from distant parts of the kingdom for the purpose of ascertaining which would give the most economical results. The gas was also washed with water, and other means were employed to purify it.

"In the year 1798 I removed from Cornwall to Messrs. Boulton, Watt, and Co.'s works for the manufacturing of steam engines at the Soho Foundry, where I constructed an apparatus upon a larger scale, which during many successive nights was applied to the lighting of their principal building, and various new methods were practised of washing and purifying the gas.

"These experiments were continued with some interruption until the peace of 1802, when a public display of this light was made by me in the illumination of Mr. Boulton's manufactory at Soho on this occasion.

"Since that period, I have, under the sanction of Messrs. Boulton, Watt, and Co., extended the apparatus at Soho Foundry, so as to give light to all the principal shops where it is in regular use, to the exclusion of all other artificial light.

"At the time I commenced my experiments I was certainly unacquainted with the circumstance of the gas from coal having been observed by others to be capable of combustion; but I am since informed that the current of gas escaping from Lord Dundonald's tar ovens, had been frequently fired; and I find that Dr. Clayton, in a Paper in vol. xi. of 'Transactions of the Royal Society,' so long ago as the year 1739, gave an account of some observations and experiments made by him, which clearly manifest his knowledge of the inflammable property of the gas, which he denominates the spirit of coals; but the idea of applying it as an economical substitute for oils and tallow does not appear to have occurred to this gentleman, and I believe I may, without presuming too much, claim both the first idea of applying and the first actual application of this gas to economical purposes."

Such is the modest and clear account given by Mr. Murdoch of the origin and the progress of the first attempts at gas lighting, which statements were further confirmed by Dr. William Henry, of Manchester, and various others, also by Mr. James Watt, jun., hereafter mentioned, thus leaving no possibility of doubting the accuracy of the above narrative.

One of the earliest writers on gas lighting (Matthews), when describing "the public display of gaslight" referred to, says: "The illumination of Soho Works on this occasion was one of extraordinary splendour. The whole front of that extensive range of buildings was ornamented with a great variety of devices, that admirably displayed many of the varied forms of which gaslight is susceptible. This luminous spectacle was as novel as it was astonishing; and Birmingham poured forth its numerous population to gaze at and to admire this wonderful display of the combined effects of science and art. The writer was one among those who had the gratification of witnessing this first splendid public exhibition of gas illumination, and retains a vivid recollection of the admiration it produced."

Several French authors, ignoring the claims of Murdoch, have accorded the merit of priority in the application of coal gas for domestic purposes, to their compatriot Philipe Lebon (generally, but erroneously, written Le Bon), who, according to some, in the year 1786, and others 1798, "patented his thermolampe, a kind of stove or oven, in which he distilled wood or coal, in order to obtain the necessary heat for warming dwellings or workshops, and at the same time the gas for lighting them."

With the sincere desire to ascertain the truth, and award merit to whom it may be due, we have diligently searched the French Patent Register and can state positively that no patent was obtained by Lebon previous to 1799; therefore these assertions are founded on error. Moreover, to judge from the description given hereafter, it is certain that Lebon, when specifying this patent, had no definite system with reference to the application of gas, either for lighting or heating; therefore the mistake in awarding him the merit of priority on that account is clearly shown.

The following is a literal translation of the specification of the patent in question, accompanied by a copy of the corresponding drawing.

For a New Method of Employing Fuel (combustible) with greater Utility either for Heating or Lighting, and to obtain certain Products therefrom.

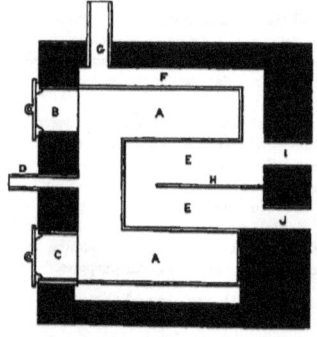

"*Description of the Apparatus.*

"Fig. 1 is a section of the apparatus.

"A A is a box (*caisse*), having two ends of pipe, B and C, hermetically closed, into and from which the fuel may be introduced and withdrawn.

"D is a third pipe for the purpose of conveying the vapours or gas which are emitted from the fuel enclosed in box A A.

"E is a furnace placed in the box A.

"F is a chimney or division from the furnace in connection with the pipe, G.

"H a division which separates the furnace into two parts, leaving a passage for the air.

"By means of the pipe, D, the gases and vapours can be conveyed and distributed, or compelled to pass any number of condensers or baths that may be thought convenient, and submitted to all the known means of purification and analysis, and to collect the various products:—in a word, this pipe allows the gas to be disposed of at will.

"Hydrogen gas in a more or less state of purity, according to the means of purification employed, is principally obtained.

"The vapours, on being exposed to cold condensers, are reduced to a liquid state, and contain acids, oils, and other substances analogous to the fuel placed in the box, A. This is converted into coke (*charbon*) and can be employed in the furnace to heat or decompose new fuel and produce the same effect.

"The opening, I, which can be closed at pleasure, is for the purpose of introducing the coal (*charbon*) on the division H; and opening J, placed in the lower part of furnace, allows a passage for air to support the combustion.

"All being thus disposed, and the box, A, containing the fuel to be decomposed, and the coal of the furnace ignited, the action of the heat causes a considerable quantity of vapours and gas to be discharged."

A few words in explanation of this specification may be necessary. The box A A, probably intended to be constructed of sheet iron, was the retort; B and C, the mouthpieces; D, a pipe for the emission of the gas, answering the purpose of the modern ascension pipe; E the furnace, with H, a plate instead of fire bars; F, a flue to the chimney G; I, furnace door; J, orifice for air, which was intended to pass over the top of the fuel on the plate, H, so as to support the combustion. It need hardly be observed that the apparatus is of the most impracticable kind, but we must remember this was the first crude idea.

A memorandum attached to the specification states that "either wood, coal, oils, roots, grease, and other combustible matters can be employed, the results being always the same."

Two years afterwards—i. e. August, 1801—Lebon obtained a certificate of additions to the patent just referred to. In his description he enters with tolerable accuracy, considering the limited knowledge of chemistry at that period, into the products of combustion.

Here, for the first time, we meet with the word "thermolampe" (a stove to give light and heat); and, as it will subsequently appear, Lebon attached but little importance to be derived from gas as a means of lighting, the main substance of the patent being to obtain motive power by the explosion of inflammable gas when intermixed with atmospheric air. With the description is given a drawing of three cylinders, with their corresponding pistons, "similar to fire" (steam) "engines," explanatory of the method of operation. In the details he proposed to light the explosive compound by the electric spark, which was the subject of an important patent fifty years afterwards.

The specification is very lengthy, and proves the patentee to have possessed great intelligence. There creeps in, however, with his calculations of profit, the erroneous assertion that the hydrogen of the gas would, during combustion, unite with the nitrogen of the atmosphere and so produce ammonia. He concludes by stating how the gas may be employed for lighting, and his description is as follows: The combustion of the inflammable gas is effected in a glass globe, supported on a tripod, and cemented in such a manner as to prevent the products of combustion escaping; there are three pipes—one of small dimensions to convey the gas into the globe, another for conveying air into the same, and a third to carry off the products of combustion. Provision is also made that the pipe for the air, as well as that for the noxious products, may both be extended to the exterior of the apartment lighted; and he adds, that taps or valves may be placed on each of these pipes, in order to regulate the supply.

Although, about this period, there are some contradictions amongst authors as to dates, all agree that in the winter of 1802 Lebon for the first time exhibited lighting by gas at a house in Paris, and from Winsor's 'Traité' we are led to conclude that this was carried into effect according to the description in his certificate of additions just mentioned. The experiment created the greatest wonder and amazement amongst the thousands who witnessed it, and naturally called forth the praise of the local journals. From them the fame of the application or discovery was rapidly diffused by the Continental press, and thus accidentally came under the notice of Winsor, a German by birth, whose proper name according to Schilling was Wintzler, who was destined to play a most prominent part in establishing the first gas company, and probably in accelerating the general use of gas for domestic purposes.

Winsor was then at Frankfort, whence he proceeded to Paris, and there witnessed several times "the wondrous effects of common smoke being made to burn with greater brilliancy and beauty than wax or oil." His first object was to become acquainted with the method of obtaining this light, but Lebon was by no means communicative, even by the "offer of 100 Louis d'or," his conditions simply being, that before he would disclose the secret of these "wondrous effects," a subscription should be raised for 200 of his stoves, amounting in value to about 10,000*l.* There being no prospect of such conditions being realized, Winsor left Paris, as he said, "with the deepest regret at having failed in the object of his journey; however, firmly resolved, whatever trouble, time, and expense it might cost, to find out the causes of such important effects and results."

Perseverance and energy were qualities that Winsor possessed in a remarkable degree, and it is possible that he may have acquired the necessary knowledge from his own researches; but whether from this or other sources, he the following winter exhibited at Brunswick, in the presence of the reigning duke, Charles William Ferdinand, and all his court, a series of experiments on the products of the distillation of wood, and the means of lighting with gas obtained therefrom.

Having thus far met with success, no time was lost by Winsor in endeavouring to turn his "invention," as he termed it, to a profitable account. In one of his pamphlets he says, "the thought of introducing the discovery for the advantage of the British realm struck me like an electric shock;" and so at the commencement of the year 1804 he appeared in London, and exhibited and lectured on "Gaslight" at the Lyceum Theatre.

There he described the various advantages to be derived from the new light; its freedom from sparks, so often the fruitful source of fires, the absence of smoke, and the intensity and steadiness of the flame. He also made experiments to prove the difficulty of the flame being extinguished by the wind, and numerous other points favourable to gas lighting. And here began the earliest troubles of his undertaking, for, speaking English indifferently, he engaged an assistant to read his lectures, and this person at times, when the audience were in the theatre, absented himself with the manuscript of the lecture in his pocket, to the great disappointment of the public; but after a few good-natured words of apology from the principal sufferer, they dispersed. These events must have been very galling to Winsor, as their relation forms a prominent feature in one of his many pamphlets.

The annexed drawing is copied from Parkes' 'Chemical Catechism,' published in 1810, and will give an idea of the most approved experimental apparatus then in use. It was, doubtless, with such simple means that Winsor's lectures were illustrated.

In the month of May in the year 1804, Winsor obtained a patent for "An improved stove, oven, or apparatus for the purpose of extracting inflammable air, oil, pitch, tar, and acids from, and reducing into coke and charcoal, all kinds of fuel, which is also applicable to various useful purposes."

Possessing this, his next step was to issue a pamphlet, replete with the most extravagant absurdities respecting the "British Imperial patent stoves, ovens, and utensils for producing sevenfold heat and beautiful light, oil, pitch, coke, and pure inflammable gas, without spark, smoke, soot, or ashes, from which all the accidents of fire arise." In this pamphlet he states that, "Since the beginning of the world mankind has lost above 80 per cent. in all combustibles, by the mere evaporation of smoke. This smoke, which often proves troublesome and dangerous to health and houses, is now discovered to contain the most valuable substances, and, if properly extracted, gathered, washed, purified, and resolved, we gain not less than five costly products, viz. oil, pitch, acid, coke, and gas." The extreme extravagance of the foregoing statement needs no comment. After showing a profit, from the application of the invention, of 64*l*. 10*s*. for an outlay of 6*l*. only, and asserting a number of chemical absurdities, he concludes by the following advertisement:—

"Such noblemen, gentlemen, public or private companies, or individuals, who, from witnessing the uncommon effects of common smoke, shall be convinced of the great utility of this discovery, and may have an inclination to embark some property in this valuable speculation, may be furnished with the sight of a written plan for a threefold grand national establishment." And here we come to the first introduction of the *National Light and Heat Company*.

Winsor, continuing his lectures, which there is every reason to believe were remunerative, began to agitate for the formation of a company, for the proper and full development of gas lighting. Extreme boldness was one of his characteristics, but his extravagant assertions as to the enormous profits to be derived, created doubt and suspicion concerning the enterprise he advocated. He was very deficient in the knowledge of chemistry, of which deficiency he was not only unconscious, but considered himself capable of discussing scientific matters with the most intelligent men of the day, at one time asserting that our atmosphere was "too strong a medicine," and at another, "that an admixture of coal gas made it more congenial to our lungs;" and that "gas could never inflame when mixed with air." These and a host of other absurdities no doubt greatly retarded his operations; but he was confident, enthusiastic, and energetic, determined to achieve a success, and his cause was good.

Following his first pamphlet above referred to, appeared another, published in 1805, from which the following extracts are taken: "I have made great improvements in my patent light-stoves, purified the gaslights from all scent, and increased their lustre. Persuaded of immense advantages and encouraged by numerous friends, I beg leave to offer to you the enclosed plan for a profitable national company; and, with the patronage of you and your friends, a national concern will soon be raised to open a mine of wealth to Britain, and add to the despair of our foes in their devices for our ruin. The 5*l.* deposit will suffice for realizing the plans in London and its environs, and all further sums wanted will prove but a small reduction from speedy profits." In another circular he seriously asserted that a deposit of 5*l.* would secure an annuity of 575*l.*

One is almost induced to suppose that this enthusiast believed at the time in all he stated, for, judging from his writings during twelve years, his continued exaggerations lead only to the opinion that serious, sober reflection did not form part of his nature; he hastily came to a conclusion, and this, once formed, was not to be shaken. His productions of later years were more reasonable, still, however, bordering on the fabulous, as when asserting that the "National Light and Heat Company was more solid than any other establishment whatever; it exceeds in benefits the water companies, by giving five valuable products from the destruction of a very great nuisance—that of smoke. These products amount to 20*l.* in every chaldron of coal." "Being founded on a patent invention, it is certainly more secure than all our flourishing insurance, and even water companies."

Having briefly alluded to Winsor's first efforts in reference to the development of gas lighting, let us now consider the difficulties attending the task he had undertaken. Firstly, he was proposing a marvellous innovation—one without any precedent, of the most extraordinary magnitude, and requiring a proportionate capital for its realization. Unlike most other great changes, in which the progress is gradual—step by step—here the change from oil and tallow to that of gas lighting was proposed to be immediate, and this without the public having been prepared by any previous practical application of the latter.

Moreover, the explosive nature of mixed gases had recently become known, and vague, yet powerful, apprehensions were entertained as to the danger likely to arise from their employment for the ordinary purposes of life, which was a mighty obstacle to the introduction of the new light. Added to this were the numerous interests of trade and commerce connected with the supply of oil and tallow, especially that of shipping, which has since been so signally benefited by the new art to which it was then supposed antagonistic. These and many other difficulties stood in the way of the innovation, which was said to be "wild, absurd, and extravagant in its conception, and fraught with the greatest danger in its execution;" and so preposterous appeared the project that "Winsor's gaseous proposals for enlightening the inhabitants of London," his "fire of wind" and "smoke lights," were popular subjects for ridicule and satire, as may be seen in some of the light literature of the day.

But such objections in no way disheartened Winsor, for, when they appeared in any newspaper, he replied, substantiating his position and views, through the same medium; and when the columns of a journal were closed against him he resorted to pamphlets, of which he produced several; and in spite of the gross absurdities abounding in his numerous communications, there is often so much good-humour and wit in them as favourably to impress the reader in his behalf.

But notwithstanding all the great disadvantages under which he laboured, and his extravagances and absurdities, yet by dint of his indomitable perseverance we find, towards the close of the year 1806, that Winsor was supported by a large and influential body of subscribers or shareholders, numbering amongst them members of the nobility, magistrates, bankers, merchants, &c. At that period he occupied premises, adapted for offices and laboratory, in Pall Mall, on the site of the present Carlton Club, and here he continued his lectures and experiments, to which a charge for admission was made to the general public, but members of Parliament were admitted gratuitously.

The first public street lighting with gas took place in Pall Mall, on the 28th of January, 1807, and was an event that naturally dispelled many of the doubts existing as to its practicability; and in the month of June following a commission of the subscribers was nominated to verify the official experiments in gas lighting. These were made on the 3rd and 4th of July, in the presence of the president, Mr. Grant, Baron Wolff, the Duke of Athol, Dr. Jenner, and others; and the results were so highly satisfactory that a meeting of the subscribers was held a few days afterwards.

On this occasion twenty-six members formed a committee of shareholders or subscribers, who, after expressing their gratification at the results of the official experiments, and considering that the enterprise presented such extraordinary advantages, alike for the nation and the people, unanimously resolved—
"That 20,000*l.* should be placed at the disposal of the committee, to enable Mr. Winsor to make his experiments on a greater scale, and, above all, to obtain their great object—a charter of incorporation of the company, which, by its resources, would render the discovery of the greatest advantage for the nation and Government."

At the same meeting other resolutions were passed, relating to the officers of the projected company, their finances, &c., and amongst these was the following: "That the committee will hold their first regular meeting on Wednesday, the 12th (July), at Mr. Winsor's house, at two o'clock precisely;" and this may be considered the first official assemblage in connection with the Chartered Gas Company.

Speedily after this we find the petition for incorporation presented to the king; and, as the document will undoubtedly possess interest, the following, translated from the 'Traité pratique de l'Éclairage par le Gaz inflammable,' by F. A. Winsor, is given:—

Extract of Petition presented to the King.

"Sire,—The committee of the subscribers of Mr. F. A. Winsor, established by deed to verify his experiments in lighting by gas and the utilization of the products arising from the distillation of coal, respectfully represent to your Majesty—

"That the members believed it their duty to make investigations on the discoveries of Mr. Winsor, and of the general advantages which may arise therefrom, conducted under the direction of a company incorporated by a charter.

"That the novelty and importance of the subject prevent them at present from entering into very minute calculations, but they ask permission to lay before your Majesty some experiments made by Mr. Winsor in the presence of some of the members of the commission (who are named, with date of experiments).

"The results of these experiments agree perfectly with those made and published three years previously by Mr. Winsor, which we now have the honour to submit to your Majesty.

	lb.	oz.
"Thirty-six pounds, or 2 pecks, of Newcastle coal produced		
3 measures (pecks) of coke or purified coal	24	2
Oily tar	3	12
Ammoniacal liquor	4	6
Inflammable gas	3	12
	36	0

	s.	d.
"The 3 pecks of coke (being a fuel which gives twice the quantity of heat of coal), but only estimating it at the same price as coal is worth	1	3
The oily tar, which is superior to all others, but putting it at the ordinary price	0	6
The ammoniacal liquor	0	6
The inflammable gas	4	2
Total	6	5
And as the 2 pecks of coal cost	0	10
There is a profit of	5	7

"And in consequence a gain of 670 per cent. is effected; from this has to be deducted the expenses of labour, of pipes, of lanterns, &c.

"That we are not yet enabled to calculate exactly the expenses, but there is not the least doubt of the immense advantages of the enterprise.

"That after such precise details, it would be easy to calculate the profits to be derived; but as this depends on the privileges obtained, and the sale of the products, we do not present them to your Majesty.

"The members of the commission are fully convinced that Mr. Winsor's gas can be employed with safety, to light streets, houses, lamps, and can be applied in general everywhere where heat and light are required. That his coke produces double the heat of coal, and that, on account of its cleanliness, is preferable to all other fuel; and on every account his coke and tar are superior to those made up to the present time.

"The members are of the firm persuasion that the general idea they now present of this discovery will remove all doubt, and that the sale of the products obtained by the operations of Mr. Winsor will, in a short time, produce two millions sterling per annum.

"That it may be permitted to the members of the commission to observe to your Majesty that a great and rich company, sustained by exclusive privileges, is the only means capable of obtaining all the possible advantages. That in the case of not establishing a company, rich, incorporated, and privileged, other persons will adopt more or less the process of Mr. Winsor, with some variations, according to their different views, which would probably result in serious and grave accidents, &c.

"For these reasons the committee pray your Majesty, in the name of the subscribers of Mr. Winsor, that your Majesty will be pleased to grant a Royal Charter of Incorporation for the exclusive privilege of the utilization or lighting by gas, for the sale of the products, and for the apparatus and operations.—Signed, J. L. GRANT."

At this period Winsor's plan was to raise a capital of one million sterling, and that the company should have the exclusive privilege of lighting by gas in all the British possessions.

The petition was shortly afterwards submitted to the king at a Privy Council, by whom it was decided—"That his Majesty could not grant the charter of incorporation requested in the petition until an Act of Parliament had first been obtained authorizing the company." This was a matter of disappointment to the shareholders. However, the groundwork of a bill was immediately prepared, but as the session of Parliament was far advanced its presentation was of necessity deferred.

The year 1809 is remarkable as the epoch at which the first application to Parliament was made for an Act to incorporate a gaslight company. On this occasion the capital proposed for the "National Light and Heat Company" was 500,000*l*., and the promoters asked for power merely to carry on their operations with an exclusive privilege over the whole of London; so moderating their former pretensions.

The application was opposed by Murdoch, aided by Watt, on the ground of right of priority in the discovery of the application of gas for lighting; and this opposition gave rise to a long and minute investigation before a committee of the House of Commons, commencing May 5, 1809.

The leading witness on behalf of the company was Mr. Accum, a chemist, who had been for some months assisting Winsor, and under whose superintendence the "official experiments" had been conducted. He deposed to the superiority of the coke, as prepared by Winsor, to all others; that its heating power, compared with coal, weight for weight, was as three to one; that there was neither smoke nor smell attending its use, even in close apartments; and that it contained no sulphur. He stated that he was acquainted with Murdoch's mode of producing gas as it was before the public, but conceived that Winsor's was greatly superior. He described the operation of distilling coal, and the quantity of gas obtained therefrom; its advantages when used for lighting; its superiority over all other artificial light; its safety, salubrity, cleanliness, and cheapness. He also maintained, with considerable exaggeration, the value of the other residual products, laying particular stress upon the assertion that all those excellent qualities were obtainable only by Winsor's process.

Many other witnesses were examined, some of whom deposed to the superiority of the asphalte derived from the tar, for japanning; others, that the tar was better than that generally met with for ship-building purposes; and one witness stated that Mr. Winsor's coke was not to be equalled for iron-smelting purposes, as with it the metal became so fluid, "that, in a manner of speaking, it would run through the eye of a needle." A chemist detailed the various compounds which could be procured from the ammoniacal liquor, according to Winsor's process; and a surveyor delivered to the committee his estimate of the cost of mains, apparatus, lanterns, columns, &c., for supplying the parish of St. James, Westminster, with gas, which amounted to 26,546l. 2s. Singular enough, according to this estimate, no means of storing gas were provided. There were to be six stations, situated about 700 or 1000 yards apart, in each of which six stoves (furnaces) were to be erected, and at these the gas was to be manufactured, and delivered therefrom as required.

The absence of provision for storing gas would almost appear to be an oversight at the time; but such was not the case; for in Accum's evidence we find that, when allusion is made to the danger of gas, he says: "If there is no reservoir the gas cannot accumulate; there is no reservoir in Mr. Winsor's plan; the pipes contain the gas without reservoir." Again, in a pamphlet published in 1806, an opponent of gas lighting is represented to ask questions, suggesting difficulties, which are answered by Winsor. One of them was the following:—*Question*: I suppose so many thousands of lights are turned off from all the houses at a certain hour of the night; what becomes of all the gas flowing in the tubes? —*Answer*: If we turn off a million of lights in London at once, the gas will return to the trying pipes in the parish furnaces, and assist in the carbonizing of coal; but it will be much better employed to flow into the street lamps, which will give ten times more light than at present, and prevent the nightly depredations in the metropolis."

On this occasion Accum underwent a severe cross-examination by Mr. (afterwards Lord) Brougham, not, as has been asserted, because the learned counsel had any doubt about the practicability of gas lighting, since it had been carried into successful operation by Murdoch, whom he represented, but simply through the witness's want of knowledge on many matters connected with the subject he so strongly supported, and the mysteries he associated therewith.

This was particularly the case when Accum, attempting to describe the means of ascertaining the illuminating power and size of the flame of a candle, said, to "obtain the exact size of the flame," he "placed a piece of window-glass between that and the wall of the apartment, and the flame was depicted on the wall;" and when pressed by Brougham to describe the means, he replied, "It is part of my business to be paid for information to those who are tormented with the desire of gaining knowledge;" and, again, on another occasion, "I beg leave to ask the chairman whether I have a right to answer the question without a fee?" But one of his greatest blunders was in asserting, and repeating the assertion, that a circle of four inches diameter possessed only twice the area of another of two inches diameter. These replies, with other observations equally ridiculous, gave Brougham an opportunity of "roughly handling" the witness, as described.

Amongst the witnesses examined on behalf of Murdoch were Mr. (afterwards Sir) Humphry Davy, who stated that the production of gas from the distillation of coal had been known for thirty or forty years. As secretary to the Royal Society he knew that Count Romford's medal had been awarded by the president and council to Mr. Murdoch for his paper on the economical application of gas; that the coke produced by Winsor was not superior to others, and that sulphur existed in the one and the other. Other witnesses deposed that there was no superiority in the tar and other products by Winsor's process, and that this was not different to that ordinarily practised in the distillation of coal.

Mr. Lee, of the firm of Phillips and Lee, proprietors of one of the most extensive cotton mills in Lancashire, stated that their premises had been lighted with gas for three or four years. He also entered into the greatest details regarding the consumption, safety, salubrity, and other advantages of gas for lighting, and his straightforward evidence was of considerable importance to Murdoch.

By the evidence given on this occasion by Mr. James Watt, jun., son of the great James Watt, we learn that he had been acquainted with Murdoch about thirty years; that the latter communicated to him, about 1794 or 1795, his ideas respecting the combustion of inflammable gas from coal, but that the first practical experiments were made in 1798, when a retort was constructed with a pipe of about thirty or forty feet in length, and to the end of it he applied burners of various dimensions, and gave light during the night to one of the buildings of the Soho Foundry.

The witness added, that he

"Was satisfied with the experiment so far as the brilliancy of the light went; but no precise experiments were made by him at that time, as to the economy; but he took it from Mr. Murdoch's general statement;" and further, "that Mr. Murdoch proposed that a patent should be taken out for this invention, which, however, was not done. I told him that I was not quite certain if it were a proper object for a patent, and I was induced to be rather nice upon the subject of patents from the circumstance of being at that time engaged in carrying on the defence of a patent, which my father had obtained for improvements on the steam engine, and which had then occupied more than four or five years, and had been attended with a great expenditure of money." "For that and other reasons, I advised him not to take out a patent. Amongst those reasons were, that it was known to me that the current of gas obtained from the distillation of coal in Lord Dundonald's tar ovens had been occasionally set fire to; also that Doctor Watson and others had burnt the gas from coal after conducting it through tubes, or had burnt it as it issued from the retort. Therefore no patent was applied for. I considered that Mr. Murdoch was the first person who had suggested the idea of an economical comparison between the light of gas and that obtained from oils and tallow; and the first person to whom the idea had occurred of applying it to purposes of public and private use; but I was not clear, after the legal difficulties I had seen raised respecting my father's patent for the steam engine, that it would form sufficient ground for maintaining a patent. However, I thought it a subject for consideration, and advised Murdoch not to prosecute the experiments for the present until the question respecting the steam engine had been decided, when we should have an opportunity of considering the matter more maturely."

"In consequence of this conversation, Murdoch acquiesced in my sentiments, and nothing was done until the end of the year 1801, at which period my brother, being at Paris, wrote me, saying, that if we intended to do anything with Murdoch's light, no time should be lost, because he had heard that a Frenchman, of the name of Lebon, was at the same period endeavouring to apply the gas obtained from the distillation of wood to similar purposes," adding "that Lebon had in view to light up a part of Paris with it."

"In consequence of this we agreed that it would be proper to resume our experiments and to bring the question of economy to a fair trial. Accordingly, these experiments were resumed, and at the period when the peace took place a public display of these lights was made in the illumination of Mr. Boulton's manufactory.

"We continued our experiments from that period, and in the course of the following year (1803) we erected an apparatus for the purpose of giving light to a part of our works; and till 1804 we continued making experiments upon the gas produced, also upon the different coals out of which gas was to be made, upon the manner of making the retorts best adapted for the purpose, upon the materials and construction of the pipes, upon the size and construction of the gasholders, upon the economy of different sorts of burners, and upon various other points which may escape me at this moment."

Mr. Watt also gave an account of the various gas apparatus that had been erected by his firm in different parts of the United Kingdom, and the progressive advancement in the means of constructing such apparatus. He further stated that he "had never foreseen any cause to impede the operations of the firm until he heard of the probable establishment of the 'Incorporated Gas and Coke Company,'" which he was there to oppose. Lastly, he referred to the absolute necessity of storing gas in gasholders;

and the great economy in fuel, and other advantages, to be derived from the continuous working of the retorts, and accumulating gas ready to be employed when desired.

Altogether, the evidence of Mr. Watt is of the most interesting nature, as it confirms the statement of his friend and collaborator, and narrates historically their efforts and gradual progress in gas lighting; and by mentioning its practical application in various places, he removed the doubts concerning its safety, whilst his suggestions on the use of gasholders were of the utmost value.

The evidence being finished, Brougham replied in a very able speech, the conclusion of which merits repeating. He said: "Of such experiments and calculations, of such impossibilities and contradictions, is composed that which the courtesies of this place oblige me to call the evidence of Mr. Accum; and it is *his* mode of trying experiments that the committee is called upon to compare with the methods of Mr. Murdoch and Mr. Watt, and it is to the statements of *that* gentleman that the committee is asked to give credit, in direct contradiction to the plain, simple, intelligent, and consistent accounts of these respectable persons." The inquiry terminated in the defeat of the projected company.

But whilst Winsor was thus engaged in directing public attention to the utility of, and endeavouring to establish a company for, the general development of gas lighting, others were actively employed in carrying the new art into practical operation; for early in 1804 Mr. Lee (already referred to) had his dwelling lighted by gas, in order to test its various advantages before deciding on its general adoption in the extensive cotton mills of his firm. The following year the greater part of those premises were thus illuminated, and a few months afterwards the whole, comprising upwards of 900 lights, was completed, under the direction of Mr. Murdoch, on behalf of the firm of Boulton and Watt.

A most interesting account of the application of gas was communicated by Mr. Murdoch to the Royal Society, in 1808, for which, as mentioned, he was awarded Count Romford's gold medal. The description was speedily copied into the various public journals, and attracted much attention, leading unquestionably to important consequences, as it proved by facts and actual operation, on an extensive scale, the practicability of what had been so often regarded as visionary and absurd.

The construction of gas apparatus became a part of the business of the Soho Works for a few years, and by that firm gas lighting was introduced into various manufactories; among which may be mentioned those of Messrs. Burleigh, of Manchester; Mr. Gott, of Leeds; Mr. Kennedy, also of Manchester; Messrs. Peter, of Kirkland, near Leven, where cannel coal was first practically employed for the manufacture of gas, and where we are informed part of the original apparatus is still in use. These and other gasworks had been erected and were in operation at the time of the parliamentary investigation referred to. But it appears that neither Boulton and Watt nor Murdoch continued long to follow this branch of industry, but devoted themselves more particularly to the manufacture of steam engines, in the construction of which, as we learn from Bourne's 'Treatise on the Steam Engine,' Murdoch added to his great reputation by the invention of the eccentric and rod, for giving motion to the slide valves of engines; an invention alike remarkable for its simplicity, efficiency, and general application. By this and his other great merits Murdoch very deservedly acquired a competent independence.

Mr. Samuel Clegg, who had been a pupil at the Soho Works, and subsequently distinguished himself in a remarkable manner by the invention and construction of apparatus used in gas lighting, applied himself, in the year 1805, solely to the manufacture of gas apparatus; and during that year the cotton mill of Henry Lodge, Esq., near Halifax, was by him lighted with gas. The following year he lighted in like manner that gentleman's dwelling-house, where the first attempt was made to purify gas by introducing lime into the tank of the holder. In 1807 he lighted the manufactory of Messrs. T. and S. Knight, of Longsight, near Manchester.

In 1809 Mr. Clegg erected a gas apparatus in a large manufactory at Coventry belonging to Mr. Harris, in which he introduced a paddle at the bottom of the tank to agitate the lime. Early the same year he communicated to the Society of Arts his plan of an apparatus for lighting manufactories with gas, for which he was presented with a silver medal. In 1812 the extensive cotton mills of Ashton

Brothers, at Ryde, near Stockport, were lighted. Here he introduced the wet-lime machine and hydraulic main; and the same year the premises of Mr. Ackerman, a publisher in the Strand, were lighted with gas by him; of which works we present an engraving, copied from the frontispiece of Accum's 'Treatise on Gas Lighting,' published 1815, which is thus described:

Gasworks constructed by S. Clegg for lighting the premises of Mr. Ackerman, 1812.

Fig. 1.—The furnace, with the retorts, A A, placed side by side. The door of the furnace is also shown.

Fig. 2 is the receiver for the tar to collect the bitumen and other products condensed after the distillation. This is a cast-iron cylinder closed at the top, with a small hole to allow the air to escape when the liquid enters the vessel.

Fig. 3 is the lime apparatus to purify the crude gas and render it suitable for burning. This contains lime mixed with water, through which the gas is caused to pass and become purified.

Fig. 4 is the gasometer for receiving and retaining the purified gas and to distribute it at will. This consists of two principal parts—viz. first, a large internal vessel, closed at the top and open at the bottom, constructed of sheet iron, and destined to receive the gas; second, an external cast-iron vessel of larger capacity than the former nearly filled with water, and in which the other vessel is suspended.

At the commencement of the operation the internal vessel, being at the bottom, is entirely filled with water; but as the process of distillation takes place in the retorts, the gaseous products evolved from the coal are conveyed by the vertical syphons, B B, to the horizontal pipe, C, with which they are connected.

The liquids collect in the condenser, C, where they remain until there is a sufficient quantity to pass by the pipe D, which is attached to the condenser, C. The ends of the pipes, B B, are immersed in the liquid contained in the condensers, whilst the gases and vapours, after surmounting the pressure, pass in the pipe E, which has a serpentine form, within the outer vessel or gasometer, as shown by dotted lines, and so passes out at bottom, E, which conveys the tar and other substances condensed in the serpentine pipe into the tar receiver (Fig. 2), whilst the permanent gas passes by the pipe, F, into the lime machine, where it is purified.

This done, the purified gas is conveyed from the purifier by the other pipe, G, to the gasholder, by means of a vertical pipe, into the gasometer, where the gas, being exposed to a large surface of water, is washed and purged of any impurities it possesses after leaving the lime machine, where it has been agitated with that substance mixed with water.

The gasometer rises in proportion to the quantity of gas accumulated, whence it passes to the burners to be supplied.

L is a barrel containing a mixture of lime and water for supplying the lime vessel by the bent pipe. A bent lead pipe supplies water to the gasometer; also to the barrel L.

O is an overflow pipe to carry off any excess of water. P is a crank for occasionally stirring the lime and water in the purifier. R is a dial set in motion by an endless cord passing over a pulley fixed on a shaft worked by one of the wheels of the gasometer. This dial is divided, and shows the number of feet contained in the gasometer. T is an iron door for hermetically closing the retort. U is a wedge by which the door is secured.

In the 'Monthly Magazine' for April, 1805, appeared a communication by Mr. Northern, of Leeds, giving an account of some experiments made by him on the production of coal gas, which he concludes by observing: "I have great hopes that some active mechanic or chemist will, in the end, hit upon a plan to produce light for large factories and other purposes at a much less expense by coal gas than is at present incurred from oil."

Mr. Pemberton, of Birmingham, made a series of experiments on the production of coal gas and its application to lighting, and exhibited gaslights in a variety of forms in front of his establishment in one of the main streets of the town; and in 1808 he erected a gas apparatus for Mr. Cook, a toy manufacturer, by whom gas was employed for soldering purposes. From that time Mr. Pemberton devoted himself to the construction of gasworks.

It was not, however, by practical mechanical men alone that gas lighting was thus slowly, yet surely advanced, for men of science also contributed their aid, and most conspicuous among these was the celebrated chemist, Dr. William Henry, who, at the latter part of 1804, delivered a course of lectures at Manchester, in which he showed the mode and facility of producing gas from coal, with the advantages of its use, as a means of obtaining artificial illumination; also the means of burning it in an Argand lamp, as had been already done by Murdoch. He subsequently analyzed the composition and investigated the properties of coal gas, with that careful regard to accuracy which characterized all his inquiries; and in his numerous experiments he produced gas from a variety of bodies, as wood, peat, different kinds of coal, oil, wax, &c., and endeavoured to arrive at the relative value of each description of gas, for the purpose of lighting.

From repeated investigations of various compounds of hydrogen and carbon, Dr. Henry ascertained the ingredients which contribute to the illuminating power of coal gas, and he advanced many suggestions, which conduced to improvements in producing gas, as well as in its purification from substances that might be offensive or injurious to health. He likewise ascertained the quantity of atmospheric air requisite for its proper combustion, with the quantity of the resulting carbonic acid, and numerous other important facts in connection with gas lighting; and to Dr. Henry is especially due the merit of first suggesting lime as a means of purification.

Returning again to the operations of Winsor, we find amongst his subscribers were gentlemen of the greatest intelligence and perseverance, who, although defeated in their first application to Parliament, were far from being discouraged. The following year, therefore (1810), they again applied to be incorporated as "The London and Westminster Gaslight and Coke Company," and although they had to contend against great opposition, and incurred considerable expense, they succeeded in obtaining an Act of Incorporation, and a declaration that his Majesty, if he saw fit, might grant them a charter within three years, constituting them a body politic, for and during a period of twenty-one years thereafter.

The bill, however, as introduced the previous year, was materially altered—conditions were imposed limiting the powers of the company to London, Westminster, Southwark, and the suburbs. It was, moreover, stipulated that, if required, the company should contract with the various parishes of London, Westminster, and Southwark, to furnish a stronger and better light, and at a cheaper rate, all expenses included, than such parishes could be supplied with oil if lighted in the usual way. The capital of the company was limited to 200,000*l*., to be raised in shares of 50*l*. each, and they were prohibited from exercising any of the powers granted by the Act until half the capital was subscribed. Other restrictions were introduced referring to the management of the company, such as the appointment of the governor, deputy-governor, directors, the value of votes, &c. And thus was the first gaslight company established "for the purpose," as was said, "of making a great experiment of a plan of such extraordinary novelty."

Incidentally, it may be here observed that the Royal Charter of Incorporation was granted to the company in April, 1812, from which period, for nearly sixty years after, it was called the Chartered Gaslight and Coke Company.

The gas company being incorporated, and the first capital available, operations were commenced, with the view of realizing the great profits so fondly hoped and expected. Hitherto, all had been merely experiment, the simple production of illuminating gas on a very limited scale, without any consideration as to cost; but now it became imperative to carry the new art into practical operation on a comparatively gigantic scale, and, above all, to make it commercially a profitable enterprise, and this under the greatest disadvantages; for, although gasholders had been recently introduced, and cast-iron water-mains were then just being adopted, beyond this almost everything had to be invented, calculated, designed, and constructed, in order to arrive at a successful issue.

The indispensable requirements for the manager—or, more properly speaking, engineer—of such an undertaking, were good mechanical knowledge and skill, some acquaintance with physics generally, an inventive mind, together with considerable energy and perseverance. No mean combination of qualities at any time, but at that period exceedingly rare; and, unfortunately for the first success of the company, the directors did not understand these qualifications to be essential for their engineer; thus their choice for that post fell on Accum, aided by Hargreaves—a chemist and a medical man. Some reason may be assigned for the appointment of the first, but it must ever be doubtful whether Hargreaves' knowledge of veins and arteries could be of any assistance in the laying of mains and services.

The engineers appointed, premises, consisting of a wharf adjoining the Thames, in Cannon Row, Westminster, were then obtained, where costly apparatus was constructed and more costly experiments entered into; without, however, attaining any definite object. Whether from caprice or necessity, the locality was found inconvenient, and was relinquished; and in an evil hour for the future interests of the shareholders of the Chartered Gas Company, the directors decided on taking premises in Peter Street, Westminster; afterwards others in Curtain Road; and lastly in Brick Lane, St. Luke's; all of which stations were situated in populous neighbourhoods, where ground for extensions had subsequently to be purchased at fabulous prices, and all were at a considerable distance from that indispensable requirement, water-side communication, for railways were not then even thought of.

The choice of premises was a most egregious and unpardonable error, by which, during the fifty-seven years the manufacturing operations were carried on at those stations, an incredible amount was expended for loading and unloading and cartage of coal, &c., from the various wharves to the works; and there is no exaggeration in estimating this needless expenditure at half a million sterling—a loss which the slightest foresight or exercise of common sense at the time would have prevented.

Meanwhile the operations of Accum and his colleague were of the most unsatisfactory nature; absorbing, by their fruitless operations, the greater portion of the first instalments of 10*l*. per share, without advancing towards the accomplishment of the wishes of the directors and shareholders of the company. Difficulties, disappointments, vexations, and anxieties, quickly succeeded each other, and such was the discredit of the undertaking, that the shares of the company would hardly realize one-fifth of their nominal value.

At this critical epoch, the commencement of 1813, the company had the good fortune to engage Mr. Samuel Clegg, already mentioned, as their engineer, who, by his experience, knowledge, ability, and energy, was eminently qualified for the position. From that time their operations assumed a practical form, and, under his able direction and superintendence, the works at the three stations were constructed, the mains supplying a limited portion of London were laid, and the other necessary details executed; and on the 31st of December of the same year, Westminster Bridge was lighted with gas.

The first mains laid by Clegg were two inches diameter, but these being speedily found too small, were replaced by others of larger dimensions. The capacity of the first gasholder at Peter Street was 14,000 cubic feet, and in 1814 the united storage of the three stations was 52,000 cubic feet.

An accidental explosion, which occurred about this period at the Peter Street station, by which Mr. Clegg was injured, seemed to menace the progress of gas lighting, as it gave opportunity for the confirmation of the prophesied dangers attending its use. This accident was described by Clegg as follows: "There was a vault near the gasometer in which the lime machine belonging to the apparatus was contained; the workmen on letting the lime water out of the vessel removed too great a quantity of it, and allowed the gas to escape at the valve where the lime water was drawn off into the vault. Where the light originated I am not prepared to state, whether from some person coming in with a candle, or by a communication from the flue of the retorts, (there was a connexion of that kind), I cannot say; the effect of it was, that it blew my hat off my head and destroyed it, and blew it all to pieces, and knocked down two nine-inch walls, and injured me very much at the time, and burnt all the skin of my face and the hair off my head, and I was laid up for a fortnight or three weeks by it. I was the only one hurt on the premises." A committee of the Royal Society was appointed to inquire into the cause of the accident, and their investigation and report were eventually conducive alike to the interests of gas companies and the public, by leading to some useful alterations and modifications of the apparatus, while they inspired confidence and dispelled fears.

The parochial authorities of St. Margaret's, Westminster, were the first to contract to have their streets lighted with gas, and in April following this was carried into operation.

By the practical application of gas to the lighting of public streets, as well as for a few private houses, popular opinion was speedily changed, as we find by the establishment of the City Gas Company in 1814, which commenced operations in a yard in Fetter Lane, but being objectionable to the inhabitants, legal proceedings were instituted against the company, which compelled them reluctantly to remove the year following to premises at Blackfriars, formerly occupied by the New River Company as a wharf for landing timber, and as offices, and workshops for boring wooden pipes used for water-mains. Cast-iron mains for this purpose were first introduced about 1810, but the laying of wooden pipes was not entirely abandoned until some time afterwards.

Clegg's great success was not, however, sufficient to counteract the difficulties created by his predecessors, as we find in June, 1814, the Chartered Company applied to Parliament for an extension of time to raise their capital, which was granted; and from this period the career of the company was one of comparative success, their business increased in an astonishing manner, demanding a considerable augmentation of their establishments, apparatus, mains, &c., for which purpose they again applied to Parliament, in 1816, for powers to extend their capital by a further sum of 200,000l.

The usual parliamentary investigation took place on this occasion, and several witnesses, manufacturers, and shopkeepers of varied positions, deposed to the economy and other advantages of gas lighting. By them its cost was estimated at from one-third to one-fourth of the light obtained from oil-lamps. The saving of the labour, the cleanliness, security from fire, and the facility of distinguishing colours, were also described as important advantages of gas lighting.

Police officers from Worship Street and other places gave evidence of the great importance of gas lighting in preventing robberies, and they mentioned instances where the perpetrators of these crimes had been apprehended by its aid; also that three-fifths of the oil street-lamps were extinguished before midnight; and when the lamplighters went round to trim their lamps in winter, the thieves used to borrow their ladders for the purpose of committing robberies, and that a nobleman's mansion had been thus plundered.

Towards the end of 1814 the South London Gas Company, afterwards called the Phœnix, was established, also a company at the east end of London; neither of which, nor the City of London Company, obtained any powers or privileges from Parliament for some years afterwards.

The necessity of supplying gas by measure suggested itself to Mr. Clegg at a very early date, and in December, 1815, he obtained a patent for a rotary retort, a governor, and a "gauge or rotative gas-meter," all of which will have our attention when referring to the apparatus under their relative headings. In this year appeared the first publication on the manufacture of gas, by Accum.

From this epoch to 1819 very considerable improvements were introduced in the construction of gas apparatus; these being principally effected by Clegg and John Malam. To the former we are indebted for the hydraulic main and hydraulic valve, the first employment of wet lime (cream of lime) purifiers, and the first idea of the gas-meter; to the latter, who was draughtsman at the Peter Street station in 1817, is due the merit of rendering the gas-meter a practical instrument, and of effecting great improvements in retort-settings, in gasholders, and in the combination of three wet-lime purifiers. At this early date the firing stage, with its coke-vault, was in existence.

Mr. Winsor endeavoured to establish gas lighting in Paris in 1816, and, to forward his views, he translated Accum's work into French, adding some original matter descriptive of his operations in London when engaged in a similar undertaking, from which we have freely extracted. But beyond lighting an arcade called the "Passage des Panoramas," he was not successful in his enterprise. Paris was not lighted for some years afterwards.

From this period Winsor was not much before the public, but to the credit of the Chartered Company, it must be stated that they, in consideration of his services, allowed him an annuity of 200*l.*, to the period of his decease in 1831, which was afterwards continued to his widow.

Early in 1817 Clegg, after having ably carried out the duties devolving upon him, retired from the service of the Chartered Gas Company, for the purpose of erecting gasworks to supply Birmingham, Bristol, Chester, Kidderminster, Worcester, and other towns.

At this time one of the most important of the early inventions in connection with the purification of coal gas was patented by Reuben Phillips, of Exeter. This consisted in causing the gas to pass through layers of hydrate of lime, commonly called dry-lime purifiers. The importance of the invention was not, however, fully recognized until several years after the expiration of the term of his patent, and then was only forced, as it were, on the attention of gas companies by the great nuisance arising from the transport of the foul cream of lime, or "blue billy," through the streets. As a remedy for this the dry-lime system was adopted. The apparatus, as proposed by Phillips, was however defective; the purifiers being without bottoms, were sealed at the lower part with water, and the pipes of communication passed from the top of one purifier to the top of another, having, of course, suitable lutes.

The City of London Gas Company applied to Parliament this year to be incorporated, with a capital of 200,000*l.*, to supply London and Westminster. Gas companies now began to be established in many of the large towns and cities of England. In some places, Birmingham amongst them, the works were commenced by private individuals. The Imperial Company of London was also founded, but like most other similar undertakings, made no application to Parliament at its outset for statutory powers.

The progress of gas lighting now began to be of a more satisfactory nature. Proper means of purification were adopted, thus removing all complaints of bad odour or insalubrity. Experience had proved all the former fears of danger to be without foundation; and the brilliancy of the light, together with the other numerous advantages it possessed, caused gas to be in great demand for street lighting. Shops, manufactories, and numbers of private dwellings began to be supplied; and such was the extraordinary extension of the business, that again, in 1819, the Chartered Company made another application to Parliament, to be enabled to increase their capital by a further sum of 200,000*l.*

David Gordon and Edward Heard (one of Winsor's assistants at the Lyceum Theatre, the delinquent already referred to) this year obtained a patent for compressing gas into strong copper or other vessels, fitted with peculiar valves, ingeniously constructed so as to regulate its emission. Into these vessels, it is stated, the gas was compressed so that 30 cubic feet occupied the space of 1 cubic foot only. Sometimes, in the operation, the pressure was continued to such a degree that the gas was resolved into a liquid state; and it is worthy of remark that it was in the liquid from one of these reservoirs that Faraday discovered benzole.

This patent led to the formation of the "London Portable Gas Company," which commenced their operations in Clerkenwell. From their works vessels of an average capacity of about 2 cubic feet, containing the compressed gas, were carried to the premises of their consumers; in some cases it was burned direct from the reservoirs, which were rendered ornamental for the purpose; in others, connection was made with the gas and the fittings. The operations of this company gave rise to other similar establishments in England, and subsequently on the Continent.

At this period Messrs. Taylor and Martineau acquired the patent of Mr. J. Taylor, obtained in 1815,

E

for producing gas from oil. An apparatus was erected at Apothecaries Hall, London, for supplying that building, and shortly afterwards Covent Garden Theatre, Whitbread's Brewery, and the Post Office then in Lombard Street, were lighted with oil gas.

John Grafton, who was formerly a pupil of S. Clegg, in 1820 patented a means of making clay retorts, which, by his description, appear to have been already in use, for he says, "The material heretofore tried for retorts has generally failed, in consequence of these retorts being made in one piece, which caused them to break to pieces shortly after the fire was applied." He proposed to make clay retorts in several pieces, jointed by a "cement not liable to be destroyed by the fire." Haddock about the same epoch patented the means of purifying gas from ammonia, by passing it through dilute acid.

Gas lighting now began to be extended to the Continent, the first general application having been made by M. Pauwells, who, in compliance with an order of the French Government, erected a small works for lighting the Luxembourg Theatre, at Paris, and the neighbouring district. It was also introduced into Germany about this period, by Blochmann, who lighted the theatre at Munich with gas, and who subsequently erected gasworks at many towns in that country.

Speedily after this two gas companies were established for lighting Paris, one founded by Pauwells, called the "Compagnie Française," the other founded by Messrs. Manby and Wilson, entitled the "Compagnie Anglaise." The works of the two companies were erected almost simultaneously, and Paris was lighted for some years by them. Afterwards, at different periods, four other small companies were established at various points in the suburbs, to which the mains of the original companies did not extend. Gradually these small works increased in importance, until eventually they became formidable opponents to the others; but the losses and inconvenience attending active opposition soon became apparent, and it was very wisely arranged to district the city, so that each company supplied only a certain locality, a proceeding that was conducive to the interests of all concerned. In 1855 all the Paris companies were united under the title of the "Compagnie Parisienne de l'Eclairage par le Gaz," with the exclusive privilege of supplying the city and suburbs for a term of fifty years, at the end of which term the works, mains, &c., will become the property of the municipality.

Following the example of the London Portable Gas Company, compressed gas was introduced into Paris; but after some years of profitless operations, the system was abandoned, and uncompressed gas substituted. For this purpose small gasholders were placed in the consumer's premises, each having a pipe on which was a tap and union joint in communication with the street. The gas was conveyed from the works by a van fitted up as a dry gasholder, and on this being connected with the pipe of the consumer's holder, the gas was expelled therein. The extraordinary expense attending these operations may be conceived when it is stated that each van contained only about 300 or 400 cubic feet of gas; and in some instances a journey of 10 miles with horse, van, and attendant was made to deliver that small quantity; and, to supply a minor theatre for the night, five or six vans of gas were necessary. It is not, therefore, surprising that the concern was not remunerative.

Another means of employing portable gas, also introduced like the former into several large Continental towns, was to have a vessel similar to Gordon's, but of much larger capacity, placed in the premises of the consumer; another vessel, containing the compressed gas, was conveyed there, when the two vessels were connected and left until the pressure therein became equalized; and of all the methods of applying portable gas, this was the most unprofitable.

In 1820, John Malam patented the first dry meter. This apparatus consisted of an outer box or case containing six bellows, similar to those ordinarily employed for domestic purposes, with their nozzles meeting in the centre. The description is not intelligible, and the whole appears the result of a hasty idea never carried out; and it is singular that a practical man like its author should have presented to the world such an unfinished instrument.

The following year, in May, an Act of Parliament was passed, authorizing the erection of an oil gasworks at Bow, for lighting Whitechapel Road and its vicinity; and about this time the most strenuous efforts were made by Taylor and Martineau, to whom reference has been already made, to introduce oil gas into cities and towns not yet supplied, as well as to places where gasworks were already established. They were so far successful that oil gasworks were erected at Edinburgh, Liverpool, Bristol, Dublin, Colchester, Taunton, Norwich, Hull, and Plymouth; but all of these companies, in the course of a few years, after sustaining considerable loss, gladly resorted to the manufacture of coal gas, and abandoned the other.

In January, 1822, Sir William Congreve, the Government inspector of gasworks, made a report on he state of those establishments in and about the metropolis, and as that report describes the magnitude of the works at the period, a portion of it is here given:

"1. The Peter Street works consist of eighteen gasometers; the average contents of each may be stated at 15,000 cubic feet.

"2. The Dorset Street works contain fewer gasometers; several of them erected in one building, of the *enormous size of* 40,000 cubic feet each gasometer.

"3. The Brick Lane works are nearly on the same scale as those of Peter Street, consisting of sixteen gasometers, some of them *working in coal tar*.

"4. The Whitechapel works contain two gasometers.

"5. The Mile End works contain six gasometers.

"6. The South London works are in two divisions, one in St. George's Fields, the other near Southwark Bridge; this latter contains two gasometers of about 15,000 cubic feet each, and ten more are to be erected.

"7. The Curtain Road works contain six gasometers."

The following year he reported that "The Imperial Gas Company were erecting at their Hackney station two gasometers of 10,000 feet each, and are about to erect four more of the same size. At their Pancras station they have marked out ground for six gasometers of 10,000 cubic feet each." In the report is also mentioned "that at the Whitechapel works *two large canvas bags*, of about 15,000 feet each, were for some time used as gasometers, and that a smith's forge was placed near them," and he hints at the direful consequences that might arise from the ignition of this "artificial volcano."

Opposition existed at a very early period between the Chartered and City Companies, and, subsequently, with the Imperial; the mains of two and sometimes three companies being placed in some of the streets of the metropolis, involving an expenditure of two or three capitals for the supply of a street or locality, with all the contingent personal expenses and loss by leakage, which contributed so largely to the want of success of the metropolitan gas companies during many succeeding years. The importance of this did not escape the notice of Sir William Congreve, who made a report to the Secretary of State on the various disadvantages arising from this competition; which resulted in an Act of Parliament limiting each company to its particular district, and imposing a fine of 20*l.* for every light supplied by either of the companies out of its defined limits.

The business of the Chartered Company continuing to increase, they made a fourth application to Parliament in 1823, for powers to extend their capital by a further sum of 300,000*l.*, making a total of 900,000*l.*; and this capital remained without alteration until the year 1867, a circumstance which conveys the impression that, during the interval, energy was not amongst the characteristics of the directors.

In 1823 the Ratcliff Company was incorporated by Act of Parliament, with a capital of 100,000*l.*; and, in the same year, John Malam patented his beautiful arrangement of dry-lime purifiers, hereafter referred to.

The introduction of oil gas into many of the cities and towns of the realm, as stated, with the fallacious report of their success, were conducive to attempt the formation of a metropolitan company for a like purpose; and early in the year 1824 application was made to Parliament for an Act of Incorporation for the "London and Westminster Oil Gas Company," empowering them to raise a capital of 500,000*l.*, to manufacture and supply oil gas in the cities of London and Westminster. According to the usual practice, a select committee was appointed to investigate the merits of the proposed bill, as also the grounds of the different petitions presented against it. The commencement of these proceedings

E 2

was in May, and such was the extensive nature of the inquiry, that they lasted throughout the whole of the session of 1824, and part of that of the year following.

Amongst the supporters of oil gas were some of the most eminent scientific men of the day—Sir Humphry Davy, Michael Faraday, Richard Phillips, editor of the 'Annals of Philosophy,' T. W. Brande, lecturer on chemistry, Sir William Congreve, the Government inspector of gasworks, &c.; who all gave evidence on this occasion more or less favourable to oil gas, with a view to demonstrate its superiority to that produced from coal.

The principal advantages were represented as follows:

"The illuminating power of oil gas, as compared with coal gas, is (according to the various witnesses) as from three or four to one; or, in other words, oil gas yields three or four times the light of coal gas.

"The light yielded by oil gas being so much greater than coal gas, the quantity required will be reduced in the same proportion; and, in consequence, the manufacturing apparatus,' mains, pipes of distribution, and capital, would in like manner be diminished.

"The operation of manufacture is more simple, cleanly, and healthy for the workmen employed.

"The gas can be procured from oil unsuitable for burning in lamps, therefore possessing little value.

"The heat evolved from oil gas, for the light obtained, is less than that from coal.

"Oil gas is entirely free from the impurities—sulphuretted hydrogen, ammonia, and carbonic acid—which exist in coal gas; and, on 'these accounts, oil gas only is suitable to be used in private houses and places of public resort.'"

Coal gas was also asserted to be more susceptible to explosion than the other; and, according to Sir William Congreve, "the most important feature in the use of oil gas is its safety, as from the very narrow limits of the explosive mixture that can be formed of this gas and atmospheric air, it is scarcely possible that an accident can arise." This gentleman evidently imagined that the explosive compound was formed by the hydrogen only of the gas intermixing with atmospheric air—an error corrected by the witnesses produced in favour of coal gas.

To demonstrate the danger of coal gas, the Peter Street accident, and the explosion of a gasholder which took place at Manchester in 1819, with other minor accidents, were brought very conspicuously before the committee; and, to show the impurities of coal gas, samples of corroded copper pipes, and metal goods that had been tarnished by its action were exhibited.

Against this formidable array of witnesses for the promoters, the coal gas companies had to contend, and the Chartered, being the oldest and most extensive, was the first that engaged the attention of the committee in reply. The petition presented by that company was admirably worded, and stated plainly the petitioners' reasons for opposing the measure proposed. It gave a concise yet comprehensive view of the rise and progress of that company, of the prejudices and difficulties they had surmounted, of the sacrifices they had made, and the hazard and expenses they had incurred in the course of their operations, from all of which the public had derived great advantages in many respects, and, from their experience, all other gas companies had profited.

The investigation was of vital importance to the metropolitan gas companies, for on the result depended the value of the large capitals invested in that industry. Truth and justice were on their side, therefore it was necessary the evidence should be clear, precise, and minute, so as to render it satisfactory to the committee, and carry conviction with it.

The scientific witnesses on behalf of the coal gas companies were Mr. Herepath, of Bristol; Mr. Dalton, Mr. J. T. Cooper, Professor Leslie, Dr. Fyffe, and Mr. Anderson, of Perth, all well-known chemists; and Mr. George Lowe, engineer of the Chartered Company.

The testimony of these gentlemen varied widely from that of the witnesses engaged on the other side. Mr. Herepath deposed that the light of oil gas, as compared with that from coal, was as $2\frac{4}{10}$ to 1; that coal gas could be purified so as to leave no smell when burning; and he suggested the means of removing the ammonia by washing with dilute sulphuric acid.

Mr. Dalton, president of the Philosophical Society of Manchester, gave evidence to the effect that coal gas could be rendered free from sulphuretted hydrogen, carbonic acid, and ammonia, by the ordinary process of purification; that oil gas was alike in nature to that obtained from coal, but contained more olefiant gas; that inasmuch as coke was used in the retorts in the manufacture of oil gas, both carbonic acid and sulphuretted hydrogen might be produced; and that the oil gas apparatus not having the means of removing these impurities, they were likely to exist in that gas. Referring to the explosive qualities of the two gases, he stated that an explosive compound was formed with one part of oil gas to eighteen or twenty of atmospheric air, or with one of coal gas to eight or ten of air; thus proving the former to be the more explosive and dangerous. The illuminating power of the gases he considered to be as two to one. The difference in the evidence of this gentleman and Mr. Herepath on this point was undoubtedly due to the nature of the coal from which they derived their gas.

With the view of giving evidence on this important occasion, Professor Leslie, Dr. Fyffe, and Mr. J. T. Cooper had made collectively some extensive experiments on the production of the two gases and their varied properties, and consequently they were enabled to speak with all confidence. By them the evidence of Mr. Dalton was confirmed in its various points, proving that coal gas could be easily deprived of its impurities; that oil gas possessed only double the illuminating power of coal gas, and that the former was more explosive and dangerous than the latter.

Various other witnesses were examined on practical points. One of them, a chemist before mentioned, Mr. Adam Anderson, deposed that in 1822, when it was determined to light the city of Perth with gas, he had been requested to obtain information in order to decide whether oil or coal gas was the more desirable to adopt. For this purpose he visited several works for the manufacture of both kinds; but the information he obtained was so conflicting and unsatisfactory that he resolved to make a small apparatus capable of producing the two descriptions of gas, and to judge for himself. With this he experimented for three months, and in the end was induced to adopt the use of coal for the purpose. He stated that, at the period of his giving evidence, coal gas was in use at Perth, and that taking the testimony of Taylor and Martineau, *oil gas was four times more costly than the other;* and he estimated the illuminating power of the two kinds to be as 100 to 80½. We may conclude that Mr. Anderson employed a good "parrot" or cannel coal to obtain gas of such a superior quality, and as all the other witnesses on both sides referred to that which was obtained from Newcastle coal, the discrepancy is explained.

But the most important witness on behalf of the coal gas companies was Mr. George Lowe, whose examination extended over several days. That gentleman was a native of Derby, and had been entrusted, about six years prior to this investigation, with erecting and putting in operation the gasworks for supplying that town; and, subsequently, when oil gas was brought before the public, he entered into investigations on purpose to arrive at the truth, with the view of adopting that which offered the greatest advantages; and by these investigations he was convinced that oil gas could not be employed as a commercial enterprise in competition with the other. These opinions he had published through the 'Philosophical Magazine,' the 'Quarterly Journal,' and other periodicals. From his researches, therefore, he was well qualified to give evidence on this important occasion.

In his evidence, Mr. Lowe, in addition to the confirmation of preceding witnesses in support of coal gas, proved that the cost of the first material (at 3s. per gallon), to produce 1000 feet of oil gas, would be upwards of 30s.; that by the apparatus then in use at the Chartered works the coal gas could be, and was, purified so as to deprive it of all the noxious qualities mentioned by the witnesses favourable to the other; and such was the surety of the operation, that if the man in charge neglected his duty for twenty-four hours together, the gas would still be delivered in a pure state. He further deposed that to supply a public lamp with oil gas, admitting its illuminating power to be double that of coal gas, it would require 2½ feet per hour; and taking an average of 4000 hours, or 10,000 cubic feet per annum (according to the price charged for oil gas, 50s. per 1000 feet), it would amount to 20*l.*; whereas gas companies only received 5*l.* or 5*l.* 5s. for this lighting, which included all expenses. Hence it was obvious that, under any circumstances, the supply to the public lamps would be ruinous to any oil gas company in the world.

THE DRY GOVERNOR INVENTED.

In illustration of the explosive powers of the respective gases, he had prepared models, so that experiments were made in the presence of the committee, which naturally produced more effect than simple assertions. The evidence of Messrs. Anderson and Lowe on the relative cost of oil and coal gas, which had been studiously avoided by their opponents, was of the most vital importance; and as the company were determined to be prepared on all points, Mr. John Evans (father of the late engineer-in-chief, now one of the directors of "The Chartered Gas Company,") being the superintendent of the Peter Street station, was called, and he deposed that, if necessary, the whole of the works belonging to them could be converted and made suitable for the manufacture of oil gas for the sum of 30,000*l*.

The result of the inquiry was, that the powerful evidence brought by the coal gas companies in the defence of their interests, completely defeated the scheme of the London and Westminster Oil Gas Company; but their victory involved them in very large expenditure.

During the period of this parliamentary inquiry some oil gas companies had adopted coal for the manufacture of gas, and witnesses gave evidence to that effect. The Edinburgh Oil Gas Company applied to Parliament in 1827 for certain powers to enable them to convert their works into coal gasworks; and in the inquiry on that occasion it was given in evidence that the average produce from a gallon of oil was 75 cubic feet, for which they only received payment for 55 feet. It was stated that their operations during a period of five years were attended with a loss approaching 60,000*l*.

At the first establishment of gas lighting old gun-barrels were used for services, subsequently pipes for the purpose were made which were very costly. In 1825 Cornelius Whitehouse obtained a patent for "certain improvements in manufacturing tubes for gas and other purposes," and consisted in the production of these pipes by machinery. For this purpose the iron was cut to the size necessary, and rudely bent to the desired form; it was then heated to the welding point, and afterwards drawn through a mandril, so that the pipe was formed and welded at one operation. By this invention wrought-iron pipes were produced at a very moderate price, thus affording great facilities for the general advancement of gas lighting. For many years these pipes were known by the name of "gun-barrel," a name derived from those they had superseded. In 1830 Whitehouse's patent was acquired by Mr. James Russell, from which period his name has been associated with the invention.

The first attempt to render hydrogen gas, obtained by the decomposition of water, available for illuminating purposes, was patented about this period by Ibbetson, but the perusal of his specification leads us to imagine that he did not understand much of the subject. Broadmeadow at this time patented an exhauster, which consisted of a gasholder working in water; his object being to reduce the loss of gas by escapes at the mouthpieces of the retorts, and elsewhere in the manufacturing apparatus.

An addition to the metropolitan companies was made in 1824, by the establishment under a deed of settlement of the Independent Company, which was founded by the late Mr. Laing, who had been formerly in the service of the Chartered Company. The Independent commenced lighting in 1826, and for a short time competition existed between them and the Chartered Company; this, however, ended in a districting arrangement. The former obtained their Act of Incorporation in 1829, with a capital of 120,000*l*.; and the district described and assigned to them in their Act is the one until very recently lighted by that company.

In 1825 Samuel Crosley invented the dry gas governor, an instrument which, in various forms and shapes, has since been the subject of almost innumerable patents. Clay retorts also began to be employed in Scotland, where they subsequently became generally adopted many years before they were used in England.

Gas lighting had made such marked and rapid progress that in 1829 it appears there were about 200 gasworks in various towns and cities of the United Kingdom. In that year the Imperial Company made a second application to Parliament for power to increase their capital 250,000*l*., and the British Company, referred to by Sir William Congreve as the Whitechapel works, were incorporated with a capital of 200,000*l*. At the commencement of their history the latter company gave promise of considerable prosperity, and great exertions were made by them to obtain the power of lighting the city of

London; and so assured were they of succeeding in this endeavour, that mains were laid from their works to the city boundary of Aldgate, and year after year they applied for permission from the civic authorities to extend their supply within the limits of the corporation domain, but without avail.

Although, at the commencement of their history, gas companies were by no means commercially successful, the greater part of them never having made profits, while a few paid dividends which had never been earned, their experience did not deter the formation of rival undertakings in the metropolis as well as in the provinces; and we find, in 1830, the Equitable Company established, and erecting works near Pimlico, whence their mains were promptly extended through a large portion of the West End of London, and as far eastward as the western boundaries of the city, enabling them to compete with the Chartered and Imperial companies in their respective districts.

In the same year Michael Donovan invented the means of rendering hydrogen gas an illuminating agent by combining the vapour of hydrocarbons therewith.

In 1830 Clegg patented his pulse meter, for measuring gas without the intervention of either water or flexible material. Subsequently Dickson and Ikin patented the means of exhausting by steam, which they termed a steam conductor, trap, or decoyer of the gas from the purifier into the gasholder.

Two additions to the gasworks of the metropolis were made about this time by the establishment of the London and South Metropolitan companies in 1833. The former was carried out with remarkable energy, laying their mains in competition with the Chartered, Imperial, City, Equitable, Phœnix, and South Metropolitan companies, and for about twenty-four years was engaged in ruinous competition, without realizing profits. The South Metropolitan Company during eight years was unprofitable, since then it has been one of the best managed works of London.

In 1834 gas was introduced into the United States, New York being then lighted by it; Rotterdam, Amsterdam, Havre, Caen, Amiens, and Nantes, were also lighted at this period. This year the "lime light" was patented by Gordon and Deville, and, when first produced, was supposed to be destined speedily to supersede gas lighting, causing some anxiety to those interested in the latter. According to the specification of the patent, a jet of hydrogen and oxygen gases combined, was ignited and caused to impinge on a small cylinder of lime, to which a slow rotary motion was given for the purpose of presenting continuously a fresh surface to the jet. By this means the lime became incandescent and produced a light of great intensity. The invention excited the greatest admiration, and was exhibited in several public places of resort, as well as private establishments; but the expense of producing the gases was a fatal objection to its adoption for general use.

By Brunton's patent of this period we are first introduced to mechanical stoking. The object of this invention was to admit the coals by means of a hopper into the retort, which was open throughout, the coal as well as the coke being pushed forward by a piston, the rod of which passed through a stuffing box in the mouthpiece lid, the coke falling into a box sealed with water at the farther end. The system was tried in a few instances, but being found impracticable was abandoned. A very similar plan was patented by Barnett the following year. Several systems of mechanical stoking are now in the course of trial; which of them will become successful time alone can decide.

In 1835 Carter invented the valve known by his name; but this, like many other simple and excellent inventions, was not appreciated for several years afterwards.

This year Chaussenot patented "an improved construction of the lamps or apparatus used for burning gas, for producing a better combustion of the gas," which consisted in the application of two chimneys to the Argand burner, the one being concentric to the other, so that the air necessary for the combustion was caused to descend between the chimneys, and thus, becoming highly heated, the illuminating power of the gas was considerably increased. Other persons have since claimed this invention, and in some instances it has been patented; but in consequence of the excessive heat causing the inner chimney to melt, it has never been generally adopted.

Further improvements in dry meters were introduced at this time by Bogardus, an ingenious American, Sullivan, Noon, Patterson, Defries, and N. Taylor, all of which will have our attention hereafter. Amongst the inventions connected with the manufacture of gas of this period was Burch's double door for retorts. The inner door, or partition, formed of fireclay or iron, was placed within the retort, just beyond the ascension pipe, the object being "in the first place to keep the mouth of the retort and the ascending pipe cool; and secondly, to confine the heat within the retort as much as is practicable without causing any improper pressure of gas therein." The advantage claimed for this arrangement was, "the effectual prevention of the choking or 'baking up' of the ascending pipes." Although this system has never been practically applied, it appears to present advantages worthy the attention of engineers, more particularly in preventing radiation to a great extent at the weakest point of the retort—the mouthpiece.

A further step towards stoking by machinery was made by Heginbotham in 1838, who proposed to pulverize the coal intended to be carbonized, and to convey it by means of a hopper into the mouth of the retort. An "apparatus composed of a central shaft passed entirely through the retort, and around this shaft was formed a worm or screw," to which a rotary motion was given, "in order to cause it to propel the coal through the retort, and keep it in constant motion, thus constituting a self-acting gas-generator, capable of feeding and discharging itself without the necessity of removing the mouthpiece of the retort." The same principle was proposed by Hompesh a few years afterwards, and since then it has been occasionally reproduced in the Patent Register.

The Commercial Gas Company was added to those of the metropolis about this period, and from its first establishment has been remarkably well managed and profitable.

The "Bude light," patented by Gurney and Rixon, in 1839, excited considerable attention. This invention consisted in supplying "a jet or stream of pure oxygen to the interior of the flame of either oil, wick, or inflammable gas lamps," by which means the light was considerably increased. In this, as in all other instances, where oxygen gas had to be resorted to, the expense attending its production prevented the adoption of the system.

Following this patent another was obtained by Cruckshanks. The peculiarity and novelty of the invention consisted in rendering platinum incandescent by the flame of hydrogen gas, and obtaining light therefrom. This was effected "by placing a cage or cone of fine platinum wire over the flame, so as to enclose it, and to touch the outer surface of the flame;" thus the platinum wire became incandescent and yielded light.

Amongst the events of this period—viz. 1840—was the invention of purifying gas from sulphuretted hydrogen, by causing it to pass through oxide of iron; for which a patent was obtained by Mr. Croll, who also proposed to revive the foul oxide by "heating it red hot in an oven, and stirring and roasting it for some hours."

The motive-power meter patented this year by Lowe is an instrument that has never been properly appreciated, as by its use all premises could be well supplied, even supposing no pressure to exist in the mains containing the gas. A part of the patent in question refers to enriching coal gas "with naphtha or other hydro-carbonaceous liquid, so as to increase the illuminating power of the gas."

In 1842 gas lighting was introduced into Spain by M. Charles Lebon, and two years afterwards Algiers was also lighted by him. About this time Messrs. Manby, sons of one of the founders of the "Compagnie Anglaise" at Paris, formed a local company for supplying Madrid with gas. Following this example, gasworks were subsequently erected in the principal towns of that country, nearly all of which were carried out with native capital.

At this period the exhauster was first applied by Grafton, to remove the pressure from the retorts.

The first employment of clay retorts in England, on an extensive scale, was at the Brick Lane station of the Chartered Company, in 1844, under the direction of Mr. Croll. These retorts were obtained from one of the most accredited manufacturers in England, but in consequence of their defective nature the great experiment resulted in complete failure, to which circumstance may be due the great prejudice

DRY METER INVENTED.

which existed for some years against their use. This year the dry meter, as now universally employed, was invented and manufactured by the writer of the present work.

Joseph Cowen, of Newcastle, at this epoch obtained a patent for improvements in the manufacture of retorts for generating gas for illumination. The object of this invention was, " in the first place, to make clay retorts which shall be better able to withstand changes of temperature, and consequently less liable to crack." " To effect this," he says in his specification, " mix with Newcastle or Stourbridge fireclay, or any other suitable clay for the purpose, sawdust, pulverized wood, charcoal, carbon obtained from the interior of the retorts, and other carbonaceous matter, in such proportions as the quality of the clay may require." Retorts made according to this patent possessed the advantage of not cracking, but in some instances they became porous, on account of the carbon which entered into their composition being destroyed, thus passing from one evil to another. There is, however, reason to believe that this patent led the way to the present mode of making retorts.

In 1845 another undertaking was founded to supply the western part of the metropolis, called the Western Gas Company. The works were constructed by Mr. Palmer, formerly engineer of the Equitable Company, and who had been for many years connected with gas lighting.

This enterprise at its commencement was by no means successful, for being in competition with the Imperial, they met with the strongest opposition from that company; while from the position of their works—adjoining a wealthy and aristocratic neighbourhood—they were subjected to continuous legal proceedings, for real or imaginary nuisances. In some instances, after permission had been obtained, and the company's' mains laid to supply a district, they had to be taken up again; and such was the precarious position of the undertaking that, in 1851, a large amount of capital was issued in 20l. shares at a discount of 50 per cent. With this fresh capital the works were reconstructed; a change of management improved their operations, their rental rapidly increased, and, in 1853, they began to pay dividends, from which time their career has been completely prosperous, and about two years ago the company amalgamated with the Chartered.

Amongst the incidents of this epoch, the patent of Du Buisson merits particular attention, on account of the intimate knowledge he possessed on the subject of which he treated. His patent was for "new and improved methods for the distillation of bituminous schists, and other bituminous substances, as well as for the purification, rectification, and preparation, necessary for the employment of the products obtained by such distillation for useful purposes." In his specification the patentee describes the process of distilling schist or other bituminous rocks, and obtaining oils therefrom; the means of purifying these oils, and dividing them into three classes of different specific gravities, one of them being suitable for burning in lamps with a double current of air. He further mentions that paraffin is obtained in large quantities, from which "excellent candles" may be made; and in describing the means of rectifying the raw oil, says, "it should be agitated with sulphuric acid, which will facilitate the deposit of tar, the supernatant oil is then to be washed with diluted caustic soda, and afterwards to be distilled."

By this specification was anticipated the most valuable patent in connection with gas lighting, the mode of operation and the products of both being alike; but Du Buisson specified bituminous schist, whilst his successor employed coal for the purpose.

In 1847 Mansfield obtained a patent for "an improvement in the manufacture and purification of spirituous substances and oils," hereafter referred to when treating on carburetting air.

On the first introduction of meters the price of gas was 15s. per 1000 feet; about the year 1832 this was reduced to 12s., and was gradually further diminished until 1849, when the price charged by all the metropolitan companies for common gas was 6s. per 1000 feet. The high price charged in the earlier periods of gas lighting undoubtedly held out inducements for the formation of rival enterprises, and with what success we have endeavoured to explain; and, as we have likewise shown, the unprofitable and depressed condition of existing companies did not prevent the formation of others. Hence, in 1848, we find another enterprise projected, in opposition to the Chartered and City companies, to supply the city of London at 4s. per 1000 feet.

F

GREAT CENTRAL COMPANY FORMED.

With the view of preventing this competition, the old companies, unfortunately for them, somewhat tardily reduced their price, first to 5s., and subsequently to 4s. per 1000 feet; but, in the meantime, their opponents had gained extraordinary advantages, having presented petitions to the civic authorities, signed by a large majority of gas consumers, praying that the projected company should be authorized to lay their mains within the city, to supply the petitioners with gas. This, together with the powerful influence the promoters of the projected company possessed, resulted in the formation of the Great Central Gas Consumers' Company.

According to the contract entered into by that company with the city authorities, all their mains were to be laid during the night time, and the paving reinstated by eight o'clock on the following morning. They were, moreover, bound to have the whole laid, and their works in a condition to supply gas, within nine months of the date of contract. The contractors for carrying out the works were Messrs. Rigby, and the whole was completed, and the company ready to supply, within the prescribed period.

When we consider the various difficulties attending the execution of the works of this company—that land had to be obtained for the site; that the first contractors employed, Messrs. Peto and Betts, abandoned their engagements; that the mains and services had to be laid through a large and crowded district in an almost incredibly short space of time, and during the hours of night—it must certainly be regarded as one of the most remarkable examples of energy, activity, and promptitude, in the annals of gas engineering.

At the commencement of their operations considerable doubt existed as to the possibility of the company proving remunerative; but as they had entered into a contract with their engineer, Mr. Croll, to be supplied with gas, delivered at the consumers' meters, at the rate of 1s. 9d. per 1000 feet, success, so far as regarded the company, under the fulfilment of such conditions, could not be doubtful. Contrary to general expectation, the undertaking was commercially successful from the commencement; but the establishment of this company affected very seriously, for some time, the value of gas property, particularly that of the metropolitan companies.

Moreover, the epoch was exceedingly fertile in the production of various schemes with the view of superseding coal gas. The most remarkable of these was the patent of Shepard for a "means of producing gases by the action of currents of electricity on water having chemical matter dissolved in it." According to a prospectus issued in Paris, preparatory to the formation of a company (La Compagnie Générale du Gaz Electrique) for working this patent, by the process it was stated hydrogen could be produced at $3\frac{1}{2}d.$ per 1000 feet, which, in order to be available for lighting, was afterwards to be enriched by hydrocarbons, as well understood. The capital proposed for the enterprise was no less than two millions sterling; and, as invariably happens under like circumstances, the scheme had its scientific supporters, amongst them some eminent French *savants*, who gave the most flattering accounts of the invention. However, on investigation, it was ascertained that in the decomposition of water and ingredients, the "chemical matter dissolved" therein being oxalic acid and sulphuric acid, one-third of the resulting gases consisted of carbonic acid, the other being hydrogen; thus forming a compound almost valueless for the purpose intended, whilst the cost of production was more than one hundred times greater than that stated in the prospectus. In consequence of this discovery, the gigantic bubble exploded.

Another cause of alarm was the proposed introduction of the electric light. This is produced by a powerful voltaic battery, having at the end of each of its electrodes a piece of charcoal in the shape of a pencil; these are placed with their points opposite to each other, and separated by only a very short distance. The battery being charged the charcoal is ignited, and a most intense light is produced, but one by no means suitable for domestic purposes. This light was exhibited in various places of amusement in the large towns of England and the Continent. The cost of the material necessary for producing the requisite electricity, however, prevented the system being generally applied.

The Electric Light and Colour Company was projected about the same time. According to the prospectus of this undertaking, the residues of the process were to be employed as pigments, and were to realize such prices that the cost of light would be merely nominal. The means of producing hydrogen by decomposing water through electro-magnetic agency was also brought before the public; but the power employed in the process was not compensated by the value of the gas produced.

About the same time Gillard was busily employed with his platinum light at Narbonne and other small towns in France, which for a time held forth great expectations. But of all these projects not one was prosperous.

The success of the Great Central Company caused another competing company, the "Surrey Gas Consumers' Association," to be added to the number of those of the metropolis.

The means of lighting and extinguishing all the public lamps of a district simultaneously by electricity, was patented by Petit in 1856, and subsequently by Baggs, Denny Lane, and others. The practicability of this is by no means questionable, the main difficulty consisting in the expense of laying the necessary wires; and undoubtedly the subject is well worthy the consideration of practical men, as by this process a saving would be effected in all the public lights of about three-quarters of an hour's lighting per diem, this being approximatively one-twelfth part of the gas supplied for that purpose; whilst the number of men now employed would be considerably diminished, as they would then be required only for cleaning the lanterns and attending to the burners.

In concluding this part of our subject, it may be of some interest to refer to the effects of competition as practised amongst gas companies during many years in the metropolis. This was of such a nature that in the largest and best streets, as well as in those of the most unpretending kind inhabited by the very poorest classes, the mains of three competing companies were laid, and frequently all of the same diameter, placed close together and often crossing each other. Thus at times it was utterly impossible, at least by the sight, to distinguish the mains of the respective companies; consequently it often occurred that the gas was supplied to a consumer by one company, whilst another received payment for the same, and this state of affairs existed for years.

The imperative orders of the directors of the opposing companies at that time were to prevent their rivals extending their list of consumers, and contracts were often made at one-third of the proper amount; whilst some of the contracts entered into with the parishes for supplying the public lamps at 5 feet per hour during upwards of 4000 hours in the year, including lighting and repairs, at the period when the gas by meter was 9s. per 1000 feet, amounted to only 2l. 12s. per annum. This being in reality about one-third of the true value.

Moreover, by the numerous changes, together with the carelessness of the companies' officials, and the secrecy observed by this ruinous rivalry, there were many consumers, as it subsequently transpired, who were not upon the books of either of the companies; several of whom were old established firms of reputed respectability, who had burned gas for lighting their premises for years without paying one shilling. The plea when discovered generally being that "nobody called for the amount, and how could we pay?" In such cases, as gas property was but indifferently protected, a compromise was effected, and thus the matter ended.

We may here observe that the history here given of the operations of the gas companies of the metropolis, is identical with that of many other companies; for during the short career of gas lighting there has been scarcely a town of importance in the United Kingdom in which a ruinous competition in gas at one period or another did not exist. Similar observations apply in like manner to the Continent and other parts of the world. In some few instances the opposing companies have been successful from their first establishment, but in others, for a series of many years, loss and continued failures have been the consequence. As examples of this, the London and Equitable companies, for upwards of twenty years, never earned a shilling of dividend, the 50l. shares of the former, as stated by a gentleman largely interested in gas companies, could be bought at one period by "cart-loads" at 2l. per share, whilst other similar undertakings have barely existed, and a few only were capable of realizing a very modest dividend.

That a termination to such a state of affairs should eventually take place was only natural. Hence we find an arrangement entered into between the companies supplying the south part of London, limiting each to a particular district, thus avoiding the extraordinary leakage in the numerous mains, the bad debts arising from competition, the excessive personal expenses attached to the system of opposition, and the absurdity of investing three or four capitals to supply some districts which were hardly sufficiently remunerative to be supplied by one company alone.

Probably impulsed by the success which attended the districting arrangement on the south, four years afterwards the same system was adopted by the companies on the north, of the Thames, and from that period their prosperity may be dated, giving results alike beneficial to the general public and the companies.

It appears that when negotiating with the other companies in order to determine their respective districts, the Chartered was somewhat short-sighted, as by this arrangement the other companies had large suburban districts, to which the metropolis was and is speedily extending, whilst the Chartered was restricted principally to the centre, where little extension of lighting could be anticipated. These errors have, however, been since amply compensated, and the high position of that company maintained by the amalgamation therewith of the Great Central, Equitable, City of London, Western, Imperial, and Independent companies, with a united capital of 8,820,500*l.*, with a gross revenue for the year 1875 of 2,213,331*l.*; and by the possession of the largest and most complete gasworks and distribution plant in the world; all of which fully justifies the abandonment of the old for the more recent name of "The" Gas Company, but only after many years can it be disassociated from the original name the "Chartered," the pioneer of gas lighting.

We have now arrived at a period when the follies and losses of ruinous competition were remedied, and the comparative state of perfection in gas lighting attained. And in applying the term comparative it is with the belief that much has yet to be achieved before the science of gas lighting can be pronounced perfect; for, undoubtedly, there are mysteries in the process of the distillation of coal, and the utilization of the residual products therefrom, yet to be unveiled—secrets relating to the purification of gas yet to be unfolded, and important facts connected with the means of consuming gas hereafter to be made known.

We have arrived also at the period when reason supplanted rivalry and obstinacy in the management of business; when the directors of a gas company considered it no act of humility or submission to co-operate with those of other companies, with the view of advancing the interests of those they represented—a state of harmony which led to the important result of the general districting of the metropolitan gas companies. From that time all have been unanimously combined to secure their common welfare, and hence has arisen the present prosperous state of the various enterprises.

As we have proceeded, reference has been made to some of the most important patents connected with the manufacture of gas, and here we subjoin a few remarks on others possessing less merit. As a source of obtaining illuminating gas, the following materials have been proposed by several patentees, at various times, some being subjects of several patents, viz. peat, bone, horn, wood, sugar, seeds, nuts, cocoa-nut shells, kernels, waste hops, waste cotton, tanners' bark, waste soap, wool, and other substances.

As examples of repeated patents being obtained for the same object, may be mentioned those for dry regulators, which, with but slight modifications of the original patent of S. Crosley, have been reproduced at least two hundred times; the means, also, of carburetting air, or "air gas," have been reproduced almost as often. Hydrogen gas, either obtained by the decomposition of water through the action of incandescent charcoal, or iron, or by voltaic or magnetic electricity, has had its numerous patentees.

Mechanical stoking, in one form or another, has also been the subject of more patents than is generally known. The screw for charging and discharging retorts, Brunton's process of hopper on mouthpiece and piston in retort, have both been patented repeatedly. Ribbed retorts, in various ways, are subject to the same observation. Respecting meters, we may remark that instruments precisely similar to the first experiment of S. Clegg, prior to his patent of 1815, have been repeatedly patented, and the various impracticable schemes which are recorded in the Patent Register, with the view of superseding the beautiful simple machine of S. Crosley, are almost incredible, and some of them by no means flattering to the mechanical intelligence of their authors.

Nor can we pass over some of the extravagances and absurdities which abound in our Patent Register. One patentee proposes "to distil old or waste railway sleepers for the purpose of obtaining the tar therefrom with which they are saturated, for the purpose of preparing other sleepers," clearly proving that different persons have different notions of economy. Another gentleman proposes to compress gas into "a strong vessel, which may be shaped like a slop-pail, capable of containing from 12 to 30 cubic feet of gas, compressed to 3 or 20 atmospheres," the vessel to be carried from room to room for the purpose of *igniting domestic fires*. Another proposes to obtain oxygen by separating it from the nitrogen of the atmosphere by means of "*fans constructed on the difference of the relative density of the two gases;*" and the same gentleman refers to "blasts of oxygen in a pure and *electrified* state," whatever that may mean. Others have proposed "to receive the smoke at its exit from furnaces, chimneys, and other vents, by an arrangement of tubes, conducting such smoke to a vacuum or vacuums, by these arrangements it being prevented from issuing into the atmosphere; also to utilize the smoke so collected for *making illuminating gas*, and for other commercial purposes."

These may be taken as a few specimens of patents which adorn our Register, by no means creditable either to the authorities or to the patentees, and which suggest considerations for the amendment of our patent laws, so as, on the one hand, to prevent the possibility of absurd schemes, which excite only the mirth or contempt of persons conversant with the subject, being dignified with the name of inventions; and, on the other, the merits of real improvements being overlaid by the claims of rival patentees, to the injury of the true discoverer and the prejudice of the public.

In concluding this part of our subject, we may briefly sum up the progress of gas lighting and its results. Prior to the application of gas for illuminating purposes, artificial light was employed merely to supply a direct necessity, and only to the extent of enabling persons to perform the most ordinary duties of life; and so meagre were the means employed at that time that, for all the higher occupations, such as reading and writing, it was necessary to approach close to the lamp or candle used for the purpose.

In order to appreciate the advantages of gas lighting, as well as to make a comparison with the artificial illumination of the past, we may observe that, at the commencement of the present century, although the streets of the metropolis were supposed to be lighted with lamps placed at similar distances apart as those of the present, yet as the light yielded by each of these was not equal to a candle, and this being often enclosed within a dirty glass globe, the state of darkness can be well imagined; and so obscure were the streets that the inhabitants on going any short distance frequently carried lanterns, whilst numbers of linkmen thronged at the entrances of theatres to accompany the gentry in their sedan chairs to their houses, at the entrance of which iron extinguishers for the linkmen to extinguish their torches were provided, some of which are still in existence. Such was the danger attending the "darkness visible" of the metropolis, that after midnight robberies were committed with every impunity, and it is recorded that the Prince of Wales and the Duke of York were waylaid on their return from a gaming house, the carriage was stopped in the centre of the West End by some armed men in masks, who robbed them of all the money they possessed.

At that time, although Argand had recently invented his burner, the only method of using oil in lamps was with the simple twisted wick dipping into the oil, yielding a minimum of light and maximum of smoke, and, in imitation of the general obscurity which prevailed, the watchmen of the period were provided with immense lanterns, of which the transparent medium to protect the flame from the wind was horn, glass for the purpose being unknown; thus the light produced was so feeble that, in order to recognize a person, the lantern was held close to his face. The means of lighting generally adopted at public meetings was the old-fashioned thick-wicked candles, demanding the frequent attention of an individual to snuff them; and the divine or lecturer, in the midst of his discourse, was compelled to resort to that operation in order to have sufficient light. Singular enough, in the first edition of Accum's treatise on gas about one-half the volume is devoted to the best means of obtaining the maximum amount of light from candles. The means of forming the wick so that, by the action of the atmosphere, the operation of snuffing is dispensed with, was not then known, but we find mentioned in that work, that when a candle is burned at an angle of 30° the object was attained, and it is suggested that candlesticks can be so made to obtain such results.

The demand for gas lighting has increased in direct proportion to the advance of knowledge and experience on the subject. The first prejudice against its adoption being overcome, the removal of all fear as to its danger and insalubrity followed; its convenience, facility of use, cleanliness and economy, and the fact of gas being under the most perfect control, added further recommendations in its favour, and, above all, its beauty, and the singular merit possessed by it, that almost every class of merchandise, when lighted thereby, is shown to the best advantage, led to the rapid introduction and extended use of gas in all places of business. In dwellings, its claims to priority as an illuminating agent cannot well be overrated, while the improvements made of late years in the various appliances for its consumption, as gaseliers, pendants, brackets, globes, and other appliances, have made gas fittings amongst the most conspicuous ornaments of a house; whilst the illumination of our public buildings and streets with gas leaves little or nothing to be desired.

The introduction of embellishments of various kinds, such as plate glass for shop-fronts and other purposes, mirrors, gilding, and similar decorations, which have now become so general, is undoubtedly due to the progress of gas lighting, for without such a mode of displaying their attractions, there would have been little use for this style of ornamentation.

In confirmation of this assertion we need only call to mind the extraordinary change which we have sometimes witnessed by the introduction of gas into some ancient Continental town, with its old-fashioned open shops, which have existed in that state from time immemorial. No sooner have gasworks been erected, and a supply of gas to the streets and houses been provided, than a rapid change in the primitive state of affairs has taken place. Closed shops, plate-glass fronts, and a general air of life and brilliancy pervade the streets in the evenings, while elaborate embellishments, mirrors, and works of art, grace the interior of the premises.

Nor must we omit to refer to the advantages of gas lighting in a moral and social point of view. As a powerful auxiliary agent in the prevention and detection of crime, its claims were asserted from its infancy, and the experience of succeeding years has justified and confirmed the first impressions. The great importance attached in this respect to gas lighting is made manifest by the demand for its adoption in countries and states where civilization is not in the most advanced stages, as in many of the towns of South America, where, although the inhabitants are poor, and the cost of gas more than double the price charged in London, the public lamps are so numerous that they are frequently placed at distances of not more than thirty yards apart, thus imposing a heavy tax upon the community for the sake of attaining so desirable an object as the prevention of crime.

Darkness has always been regarded as the symbol of ignorance and crime, light as the symbol of intelligence and virtue; and it is no exaggeration to assert that the art of gas lighting has contributed its full share, during its brief but brilliant career, towards the rapid advance of civilization and progress.

CHEMISTRY OF GAS MANUFACTURE.

By Lewis Thompson, Esq., M.R.C.S.

The prospect of success in the application of chemistry to any particular art, seems to be almost in the inverse ratio of the time in which that art has been practised; thus the arts of tanning leather, extracting gold and silver from their ores, and making porcelain, offer to the scientific chemist an extremely limited field for improvement, because long-continued practice and observations in these arts have brought about ultimately the same results which might have been obtained in a short period by scientific efforts, so that it is not in the greater perfection of the results, but in the shorter time employed in procuring them, where we find the advantage of chemistry; and viewed in this light we may truly say, that the manufacture of gas, in consequence of its modern origin, offers to a scientific chemist the largest and most remunerative prospect of improvement now open in the whole range of our manufacturing industry.

But the man who would enter upon the question of improvement in the making of gas, must be one who has a firm faith in the possibility of that improvement, and bringing the aid of chemistry to bear upon existing defects, will seek to anticipate at once those changes which time and repeated efforts are sure to effect in the end.

Before, however, we attempt improvement in gas making, it is indispensably necessary to understand, as far as science can teach us, the exact nature and composition of the raw material to be operated upon, and also the probable result of the operations to which that raw material is subjected in the production of gas from it. These conditions most unfortunately cannot be fulfilled at present in anything like a satisfactory manner, in consequence of the fact, that no method has yet been devised by which coal may be analyzed or separated into its different organic constituents; although it is absolutely certain that coal consists of a vast number of organic substances merely mixed together and blended into one mass by the prolonged action of pressure, moisture, and the heat resulting from spontaneous fermentation.

The method of analysis, called by chemists "organic analysis," is in reality a mere process of combustion, and in principle does not differ from burning coals in a common fireplace, consequently we learn from it nothing as to the different organic constituents, but only a rough general idea of the ultimate component parts, from which we are unable to form the slightest conception of the plants, shrubs, trees, or whatever it was that gave birth to coal.

Thus shut out from chemical aid, we turn to botany, and here we certainly do find information of an extremely interesting nature, though defective in this respect, that the botanical evidence goes no farther than to prove that certain plants and trees were mixed up with the coal formation, but whether the coal was formed from such plants and trees remains still to be proved. Indeed the existence of the botanical relics might lead us to the conclusion that such plants and trees did not form, or even contribute to the production of, coal, simply because they are there in a shape different to the great mass of the coal, and like the relics of fishes, and certain amphibious animals also found with them, have in no way contributed to its formation.

But although equivocal on this point, the botanical relics most indisputably prove a very important fact, for all the relics are the remains of plants and trees that can only have grown in a tropical climate, and therefore we are warranted in concluding that whatever may be the nature of the organic constituents of coal, they are of a kind analogous to those which now grow in tropical climates, and this is beautifully corroborated by the evidence deduced from the anatomical structure of the animal relics above alluded to, in proof of which the following extracts from a paper contributed by two distinguished naturalists to the 'Annals of Natural History' for July, 1874, may be regarded as sufficient in a work limited as this is, to the mere outline of geological argument.

The paper in question was the result of a singularly elaborate scientific investigation made by Dr. Embleton and Mr. Thos. Atthey, of Newcastle-upon-Tyne, in the examination of the bones and skull of a Loxomma Allmanni, found at a depth of some hundreds of feet in the Northumberland coal-field, near Newcastle. The skull, we are told, "resembles rather that of the alligator than the crocodile, but the snout is broader than that of the alligator;" the length of the skull "is $12\frac{1}{4}$ inches from the snout to the occiput, and $24\frac{1}{2}$ inches to the articular condyle." The lower jaw is nearly $14\frac{1}{2}$ inches in length from back to front, and has been broken, but one half of the teeth are entire, and upwards of twenty in number, therefore the complete jaw would have more than forty teeth, six of which being "tusks $1\frac{1}{2}$ inch in length, and upwards of half an inch in width." To conclude, we are told that the loxomma resembles the alligator and crocodile by breathing air, which clearly links together, in a most remarkable manner, the two great classes of fishes and reptiles.

Here then we see the remains of an animal, the counterpart of which is now to be obtained only in the tropical regions of Africa and South America, and this animal is found buried in the coal formation of this country, where many less perfect specimens have been discovered, so that guiding ourselves by the light thus thrown upon the climate once prevalent in the United Kingdom, we are in some measure authorized to conclude that the kind of plants and trees now growing in the tropics, formerly grew in this country, and, like the loxomma, have left their relics in the coal formation.

But what are we to imagine these relics to consist of, unless we suppose them to be the remains of plants actually now growing in the tropics, and yielding such articles as indigo, cinchona, quinine, strychnine, cochineal, and many other valuable substances of which the first alone, indigo, has been released from its coally sepulchre by the agency of the gas manufacturer? Is it too much then for the chemist to view with an eye of hope that day when many, or perhaps all, of these useful substances will be extracted from coal, and the gas which is now regarded as the chief product, set down amongst the least valuable. In good truth there is yet a mine of gold concealed in the coal formation of this country, if chemistry could only furnish the spade to dig it out; but the present system of organic analysis is a mere sham and delusion when applied to coal, and therefore disgraces chemistry.

Having taken a rapid glance at the circumstances under which the vegetable constituents of coal were produced, we have now to inquire into the probable effect of heat when applied to such constituents. These we see were tropical productions, containing no doubt the kind of nitrogenous alkaloid principles for which we are indebted to the tropical regions at the present day, and in proof of this we find that the destructive distillation of the vegetable alkaloids produces ammoniacal and other compounds almost identical with those from coal; so that even here we find an argument in favour of the supposition that Great Britain once had a tropical temperature and yielded tropical products.

In the results obtained by the destructive distillation of coal as contrasted with those derived by the same means from indigo, quinine, cinchonine, and strychnine, we find a resemblance amounting almost to identity if we make allowance for the fact that in one case we are distilling a crude impure mixture, and in the other pure organic vegetable bases. Apart from the sulphurous and bituminous impurities of coal, we find in both cases aniline, quinoline, carbolic acid, paraffin oil, tar or pitch, ammonia, water, and luminiferous gases, though the relative quantity of these substances is different with each of the matters operated on; thus indigo gives chiefly aniline, whilst cinchonine gives chiefly quinoline, and strychnine, quinine, and coal give these compounds in a more equal but greatly reduced proportion.

As, however, quinoline might be had in abundance from coal tar, the possibility of converting or, we may say, reconverting, it back into quinine, cinchonine, and strychnine ought by no means to be looked upon as a romantic vision beyond the range of chemical power, for herein probably resides the first of a series of discoveries destined to change the aspect and even the name of gas making; therefore, practically speaking, these remarks are not intended to arrest the attention of dilettanti sciolists, but to secure the careful consideration of those few persons who have duly qualified themselves for the task, and are willing to set at naught the inconveniences and dangers of an employment that, in itself, has so very little to recommend it.

If we come to look at the difficulty of obtaining quinoline, or as it was once called "leukol," from cinchonine, we might wonder that such large quantities can be obtained from coal tar; but the fact is that quinoline possesses a singular power of withstanding the effect of heat, and therefore a temperature which will destroy the whole of the aniline in the process of gas making, has little or no effect upon the quinoline; consequently, if ever this substance is turned to account commercially, we may confidently rely upon our gasworks to furnish the necessary raw material.

To give an idea of the nature of the change here desired, we may examine into the composition of the different substances before us, and then see what would be the effect of combining 20 per cent. of water with the quinoline. For this purpose we will place them all in a row to show their compositions in one hundred parts, thus:

	Quinoline.	Cinchonine.	Quinine.	Strychnine.	Quinoline with 20 per cent. of water.
Carbon	83·7	77·9	74·1	69·5	69·9
Hydrogen	5·5	7·8	7·4	6·3	6·4
Nitrogen	10·8	10·0	8·6	7·4	9·
Oxygen	..	4·3	9·9	16·8	16·7

Thus we see how closely quinoline approaches in composition to strychnine, by the addition to it of 20 per cent. of water, and it is by making such comparative estimates of the composition of organic substances that we can give ourselves a fair chance of success in any attempt we make to convert worthless things of this kind into valuable products.

Having thus illustrated the manner in which chemistry can be used in giving value to waste products, let us now inquire, as far as science will allow us, into the circumstances by which these waste products escape destruction in the manufacture of gas, for we must not conceal from ourselves the fact that all these waste products are the result of an imperfection in a process which would, if it could, convert the whole of the volatile constituents of coal into illuminating gas and ammonia.

From what has been already said it becomes unnecessary to repeat how entirely ignorant we are concerning the proximate principles of coal, and the only supposition we can make amounts to no more than this, that coal contains certain substances having a composition like that of starch or sugar, with others in which carbon and hydrogen are united to the elements of water as in oil and resin; but in addition to these there are some substances that contain nitrogen, though in what form this nitrogen is combined we are quite unable to say, and almost unable even to guess, though a correct knowledge upon this point might prove highly useful to the gas maker.

We are able, however, to conjecture that in vegetables the nitrogen they contain may form an amide or ammoniacal compound, as for instance in the alkaloids and in albumen, gluten, &c.; also that it may exist as a cyanogen compound in such things as the essence of bay leaves, bitter almonds, &c.; and lastly that it may exist in combination with sulphuretted compounds, of which we need only mention the essences of mustard and horseradish, in which there are good reasons for believing that the nitrogen is present in the form of sulphocyanic acid, and if this supposition be correct we see at once the origin of the bisulphuret of carbon in coal gas, and moreover we see that this so-called "noxious compound" is not the result of a high heat, or any fault on the part of the gas maker, but is the inevitable result of the decomposition of sulphocyanic acid even at a dull red heat.

The whole question of the bisulphuret of carbon outcry, so far as the gas-maker is concerned, resolves itself therefore into this inquiry: Were there amongst the vegetables that contributed to the coal formation any of that extensive class of plants which furnish products similar to the essence of mustard? If this is answered in the affirmative, the gas maker can no more avoid or diminish the production of bisulphuret of carbon, than the chemist can do so in the destructive distillation of the sulphocyanide of ammonium; they must both produce bisulphuret of carbon, for they both operate upon the same thing.

Quitting, as altogether hopeless, the prospect of obtaining a correct knowledge of the proximate principles of coal, we will content ourselves by looking only at its ultimate constituents, in which respect it may be said to combine together the elements of both the animal and vegetable kingdoms, with one remarkable exception, the absence of phosphorus. It is true, indeed, that the ashes of peat, bogwood, and some of the lignites contain phosphate of lime and phosphate of iron, but in the ashes of true coal the phosphates may fairly be said to be wanting, although, as we have seen, the bones and relics of animals are occasionally present in coal. Still, as a general rule, phosphorus is not one of the ultimate constituents of coal, and therefore it will not require to be taken into account amongst the products arising from the manufacture of coal gas.

We may consequently look upon coal as containing the following substances in an unknown state or states of combination: carbon, hydrogen, oxygen, nitrogen, sulphur, iron, lime, silica, alumina, and magnesia, with occasional traces of potash, soda, chlorine, and iodine. Upon such a mixture the effect of heat must necessarily vary with the force or degree of that heat, and herein resides much of the difficulty experienced by the gas maker, who may be said to sail continually between the Charybdis of "low heats" and the Scylla of "high heats," for at a dull red heat the hydrogen carries off a very large quantity of carbon, and then forms an oily compound, with the production of very little gas; whilst at a bright white heat the hydrogen in great measure quits the carbon, and expanding largely, gives rise to the production of a great quantity of gas, having a very feeble illuminative power, and being therefore worthless.

With a knowledge of these facts, it might be supposed that the only thing needed to secure perfection in gas making would be to discover the best point between these two extremes of heat, and then to maintain that temperature constantly and uniformly during the whole process. A little reflection, however, will soon convince us that this uniform temperature can never produce uniform results, because the substance operated on is not uniform in itself. Let us conceive a current of water which five men are just able to carry away as fast as it arrives, and now suppose one of the men removed, the current will be too strong for the remaining four, and still more so if four are removed and but one left; yet this will be exactly the case of a gas retort charged and worked for five hours at one uniform heat. During the first hour the quantity of volatile matter will be equal to the heat and carry it off, but from that time the heat remaining the same and the quantity of volatile matter constantly diminishing, the heat is not carried off, but accumulates and acts with increased intensity upon the reduced volatile matter, which is therefore overheated and destroyed.

To prevent this evil result the period of the charge ought to be divided amongst as many retorts as possible, so that in every bed of retorts we may have retorts in the greatest possible number of conditions as to the period of the charge. Suppose the period of charge is five hours and the number of retorts in a bed is ten, we ought to charge one of these retorts every half-hour and thus regularly absorb the heat, so as to prevent any of the almost worn-out charges from being overheated; by which means we may in some measure equalize the absorption of heat from the entire bed, provided we have been so fortunate as to heat the whole of the retorts equally; and then with this equal absorption of heat we can safely calculate upon a like equality in the quantity and nature of the volatile matters sent into the hydraulic main during the whole period of the charge; a condition of things that may reasonably be regarded as the perfection of gas production, consequently our chemical researches now pass from the retort to the condenser, for the coke is incapable of improvement.

To speak in correct terms, the process of condensation is a process of purification, and it is something like a misfortune that the gas manufacturer has been led to regard condensation and purification as two distinct operations, for in reality the purification of gas commences in the hydraulic main, where we have constantly two powerful purifying agents at work, and it is highly necessary that we should understand the effects that may be produced by these agents when properly employed.

One of them is water kept at a temperature which in practice is accidental and therefore variable, but which for many chemical reasons ought to be preserved as nearly as possible at one uniform point, and that point may very advantageously be fixed at 100° Fahr., and ought to be maintained, either by the introduction of cold water into the hydraulic main, or by passing cold water through a tube fixed in the main for that purpose.

The point or heat just indicated is not arbitrary, but depends upon several circumstances, which conspire together to render this temperature the one at which both the water and the tar (the other purifying agent) are enabled to work with the greatest advantage ; thus, at 100° Fahr., water enables carbonic acid to decompose hydrosulphate of ammonia and form a solution of sesquicarbonate of ammonia, so that for every atom of one impurity sent on to the purifier, an atom and a half of another impurity is kept back in the hydraulic main, but at 160° Fahr. the simple carbonate of ammonia alone is produced, and the half atom of impurity is sent on to the purifier, so that we thus lose the full benefit of water purification in the hydraulic main whenever the temperature exceeds 100° Fahr.; and there is then a still more important loss in respect to the purifying power of the tar, because at temperatures not exceeding 100° Fahr. tar is able to retain a large amount of naphthaline and bisulphuret of carbon, but allows these things to pass on unabsorbed at a heat of 160°; consequently the temperature of the hydraulic main is very important.

The next step in purification is brought about by a reduction of temperature, or condensation as it is called, and we might suppose that the more powerfully and rapidly this was carried on, the better the result would be; but it is not so, for rapid condensation causes the gaseous vapours to assume a vesicular condition, in which state they are floated like mist or clouds through the condenser, and passing on to the purifier add greatly to the trouble and expense of purification.

To form a correct idea upon this point we need only examine the steam arising from coffee or any other hot dark fluid by means of a powerful lens previously warmed, so as to repel the vesicules, which it will then be noticed are of different sizes, averaging about the three-thousandth of an inch in diameter ; and when we have once observed the way in which these vesicules float about in the air, we have no longer any difficulty in understanding how it happens that rapid condensation may send even tar on to the purifiers, for a vesicule of this kind is merely a little balloon filled with impure gas and having a thin skin of tar instead of silk for a covering, like the case of a common soap bubble.

If, however, the condensation is carried out slowly so as to allow the heat to be gradually withdrawn from the gas, there is no disposition to form vesicules, because the vapours thus gradually cooled take on the form of drops, and either fix themselves on the sides of the condenser or fall to the bottom of the receiver ; consequently we now understand the reason why a sudden and powerful system of cooling by the application of a good heat-conductor like water, is much less efficacious than that produced by the action of air, which being an imperfect conductor removes the heat slowly. In principle, then, we see that slow condensation is the best, and perhaps the fixing of stout wires or small rods at intervals across the tube of the condenser might assist in breaking the vesicules when formed, or preventing altogether their formation by disturbing the current in the centre of the tube.

Taking it for granted that the gas has been purified as well as it can be, by the action of warm water and tar and the proper abstraction of heat, we have now to inquire into the nature of the impurities which still remain in the gas, although one of these, the ammonia, has undergone so great a change in the opinion of gas makers, that it can scarcely be looked upon as an impurity, since it is now a source of profit. Ammonia, however, can be very easily removed by a great number of methods, and in practice the mere application of an extended surface of cold water is found sufficient to effect that object, though we must not forget that too much washing with cold water will certainly diminish the light-giving power of the gas, for the volatile hydrocarbons, on which that power greatly depends, are not altogether insoluble in water; nevertheless, as the aqueous solution of ammonia thus formed can be

G 2

rendered marketable without further trouble or expense, this method is generally practised, except in small establishments, where the ammonia is wasted, a practice very much to be condemned, because ammonia is valuable, and even if saved by passing through sawdust or ashes wetted with diluted sulphuric acid, would always pay for collection.

Excluding ammonia, therefore, from the list of gas impurities, we have really but two things left that ought to be regarded as matters for the gas manufacturer to remove, because, in the existing state of scientific knowledge, there is no authentic information to show what is the nature of the other substances that ought or ought not to be removed; consequently, when we seek to compel the gas manufacturer to remove some impure things, without being able to say what they are, we begin to legislate upon a principle of the greatest absurdity; for even if the gas when burnt gives sulphurous acid in one case, from the presence of bisulphuret of carbon, it in no way follows that sulphurous acid indicates bisulphuret of carbon in all other cases; and, indeed, the little scientific knowledge we do possess on the subject points directly the other way, and seems to prove that there are sulphuretted hydrocarbon compounds in gas of a nature totally different to the bisulphuret of carbon.

If those who made the present gas regulations ever had a well-formed idea upon the impurities of coal gas, we might suppose that idea to be one which admits the existence of two antagonistic forms of sulphur, the one being a form which can be removed from coal gas, the other form that which cannot be removed; and in point of fact such an idea is not only in accordance with the practical knowledge hitherto acquired on the subject, but is also supported by the recent scientific discoveries, which teach us that there are at least two allotropic forms of sulphur, in which the physical and chemical characters of sulphur are so altered as to constitute in effect two distinct substances; and the same thing happens with phosphorus.

It would, however, be paying a most undeserved compliment to our legislators to assume that they know anything at all about allotropy, or even about the practical working of a gas manufactory; for the regulations have evidently been formed upon a supposition that coal is a thing having a fixed and uniform composition, which must always yield in the hands of the gas manufacturer a certain amount of gas, having in it a definite quantity of sulphur that can be removed, and a definite quantity that cannot be removed; and, indeed, without a supposition of this kind, an enactment which fixes a limit for sulphur that cannot be removed is no less unjust to the public upon one view of the case, than it is unfair to the gas manufacturer on another.

If coal is not uniform in composition and does not uniformly give off a certain fixed quantity of sulphur that cannot be removed, the public may well say, "Why fix a limit at all? why not leave the subject open for improvement?" And again, if coal is variable in composition, then the quantity of this not-to-be-removed kind of sulphur must vary also, and in this case the manufacturer is to be fined, because nature will not accommodate her laws to the Acts of Parliament. The truth is that coal is an extremely variable substance, and therefore this uniform limit is extremely unjust to the gas manufacturer, and may be said to compel him to use only certain kinds of coal, to the exclusion of all others; and therefore the improvement of gas will henceforth have to depend upon an improved selection of coals, and not upon any improved mode of purifying gas; for it is quite certain and susceptible of proof, that the amount of the kind of sulphur not to be removed does not depend upon high heats or upon any condition within the power of the gas maker, but upon the peculiar composition of the coal itself, in which the sulphur is present under at least two distinct forms, the one form being in combination with iron as iron pyrites, and the other in combination with organic matters, of the nature of which we are as yet totally ignorant. But this we do know, that many kinds of coal give off at a dull red heat the form of sulphur compound which cannot be arrested or removed by any process yet discovered, except such as would totally destroy the illuminative power of the gas. And with these preliminary remarks we will now enter upon the subject of purification by different agents.

The cheapest, the most effective, and in every way the best mode of purifying coal gas, is by the plan called wet-lime purification; but this plan is no longer in use in this country: it has been abandoned in obedience to the schemes of certain sanitary gentlemen who pretend to find a connection between bad health and bad smells, although it is quite notorious that the most contagious and pestilential diseases are entirely free from any offensive smell; thus hydrophobia, itch, cholera, typhus, scarlet fever, small-pox, and even the plague itself, are completely inodorous, and emit no kind of smell whatever.

The peculiar value of the wet-lime purifier depends no less upon the water than upon the lime; for dry, or unslaked, lime has no purifying power whatever, and therefore we might be justified in asserting that the purification by the wet-lime process depends quite as much on the water as on the lime; and this is strangely corroborated by the superiority of the wet over the dry lime mode of purification.

But the chemical investigation of this circumstance explains to us very clearly how the increased power is obtained; for if we pass a stream of sulphuretted hydrogen through a quantity of hydrate of lime containing no excess of water, the compound formed is merely hydrosulphate of lime, or sulphuret of calcium, if we suppose that water is formed at the same time by the hydrogen of the sulphuretted hydrogen and the oxygen of the lime, and in either case the lime combines with but one atom of sulphuretted hydrogen.

If, however, we have a decided excess of water, the lime or sulphuret of calcium then combines with another atom of sulphuretted hydrogen to form a compound often erroneously called hydrosulphate of lime, but which we see is in reality a hydrosulphate of the sulphuret of calcium, and contains twice as much sulphur as exists in the compound formed by dry hydrate of lime or slaked lime containing no excess of water; and this is the difference between the wet and dry lime purifiers, although in the dry-lime purification as great an excess of water is employed as can be conveniently used, for the lime is moistened until it adheres together by pressure like a ball of snow; and thus far the dry lime is made to imitate the wet, but can never equal it, from the circumstance that the greatest part of the sulphuret of calcium in the dry-lime purifier cannot combine with another atom of sulphuretted hydrogen, because the water is wanting by which stability is given to the existence of the hydrosulphate of the sulphuret of calcium.

The superior power of the wet-lime purifier does not, however, stop at this point, for the solution of hydrosulphate of the sulphuret of calcium forms one of the very best agents yet known for the removal of bisulphuret of carbon, with which it unites to produce bisulpho-carburet of the sulphuret of calcium, or, as it is by some chemists called, hydro-bisulpho-carbonate of lime, a name, however, which does not express its composition, because there are three atoms of sulphur in it. We will, however, return to this subject hereafter.

Not many years ago the dry-lime mode of purifying gas was universal throughout this kingdom, and as there appears to be a great likelihood of its restoration to that exclusive position in consequence of the sulphur limit established by the Board of Trade, we have an additional reason for investigating the chemical changes that occur during its use.

The substance technically called dry lime is in reality hydrate of lime mixed with an uncertain quantity of water, though it would be much better if gas managers would resort to a habit of making the mixture uniform under all circumstances, instead of relying upon the doubtful plan of compression in the palm of the hand; we have examined different samples of lime prepared by this hand process, and have found considerable differences in them, varying in fact from five of lime to two of water up to five of lime and five of water. From what has been said regarding the wet-lime purifier, it is easy to conceive that the more water we can mix with the lime the better; but this has a limit, for if we add too much water the mixture becomes pasty and will not allow the gas to pass through it; therefore we defeat our object.

The proper course for a gas manager to follow in this matter is to examine with care any sample of mixed lime which in practice he finds suitable to his purpose, and having ascertained its composition, to make up ever afterwards a mixture containing the same amount of water. By putting 500 grains of mixed lime into a closed iron tube or gun-barrel, then heating it red hot for a quarter of an hour, and weighing the lime when cold, he can gain all the information he requires upon this point, and will thus ensure certainty in the process of purifying, not only in regard to carbonic acid and sulphuretted hydrogen, but also in respect to the limited amount of sulphur that exists in what is called "other forms," and the removal of which depends almost entirely upon the amount of hydro-sulphuretted sulphuret of calcium produced in the lime mixture during purification. Supposing the entire object, or even the principal object, in purifying coal gas amounted to the removal of sulphuretted hydrogen, we need not go beyond the use of those oxides of iron which are suitable for that purpose, but to which no specific chemical name can be given; for there are hydrated and other oxides of iron, some of which will, and some of which will not, purify gas, though why this is so we know not, unless we attribute it to allotropy, which after all is only another name for the difficulty.

But the purification of gas has now assumed such a shape that the sulphuretted hydrogen is to be looked upon as a most important purifying agent, and its removal by oxide of iron or any similar substance may be said to have been forbidden by the authorities, for such in effect is the consequence of the sulphur limit. To remove the sulphuretted hydrogen from gas by any other means than such as will produce a compound capable of removing in its turn the remaining sulphur down to the assigned limit, is only to ensure the infliction of absolutely ruinous fines, therefore it is no longer necessary to speak of oxide of iron purification, because, like the wet-lime purifier, it is a thing of a past age, at least in gasworks in the United Kingdom controlled by recent Acts of Parliament.

The changes that take place during the employment of oxide of iron are extremely simple, and on that very account we have reason to regret the prohibition of a practice at once easy and economical. Thus, the thing that can be removed from the gas by oxide of iron is sulphuretted hydrogen, and when the iron has got its dose of sulphur, mere exposure to the air dislodges the sulphur and allows the oxygen of the air to take its place so as to reproduce oxide of iron, and this for a great many times in succession. To understand how these apparently contradictory results arise, we have only to notice that the sulphur in the first instance is aided by the hydrogen, which, in fact, takes away the oxygen from the iron and so makes room for the sulphur; but in the following air process there is no hydrogen to help the sulphur, and consequently the oxygen of the air unites to the iron by superior affinity and the sulphur is set free, except a small portion of it which becomes oxidized into sulphuric acid and remains in combination with oxide of iron.

With this brief notice of oxide of iron we fall back therefore upon dry-lime purification, as the only process now left us by legal wisdom, though the difficulties of that process are not trifling, because, as we shall soon see, the most complete ignorance prevails in respect to the very part of the purification which the gas manufacturer has to carry out under penalty.

It is easy enough to comprehend that hydrate of lime applied in sufficient quantity will remove all the carbonic acid and all the sulphuretted hydrogen from any given quantity of coal gas; but when this is done, the real difficulty of the gas manufacturer only just begins, for he has now, under penalty, to remove from the gas things without a name, or, at all events, with a name which has no meaning and is entirely indefinite as to its nature. No matter for the variations in a thing which, like coal, is extremely variable; no matter for the difficulty of removing things that are indefinite in their affinities, and most certainly are not all alike in kind or composition; no matter indeed for anything whatever; the gas manufacturer, like the Israelites, must furnish his stipulated work "with or without straw," and if he does not he will be "beaten," upon a principle which has antiquity and nothing else to recommend it.

Impelled by necessity to take the high road of ignorance and make a plunge into darkness, we begin by assuming that after purifying gas as far as hydrate of lime will do so, there remains in the gas nothing but a small quantity of bisulphuret of carbon. It would not require much labour for us to prove that this assumption is as worthless as the greater part of the so-called facts upon which gas legislation has been carried out; but we are content to argue the question upon either this or any other assumption that will admit the force of scientific reasoning or truthful demonstration. That bisulphuret of carbon exists in some kinds of coal gas is an admitted fact, but that it exists in all kinds of coal gas is simply a blind assumption without any basis whatever.

If one kind and but one kind of sulphur compound, no matter what it might be called, existed in gas after the gas had been purified by lime, it would be disgraceful to fix any other limit than complete and total purity from sulphur, provided the sulphur could be removed; and if it could not be removed, there certainly ought to be no penalty inflicted under plea of stimulating improvement. But let us return to the bisulphuret of carbon.

Until very lately the compounds formed by that substance had remained where they were left by M. Zeise more than forty years ago, and this too in spite of Acts of Parliament, gas referees, examiners, testers, and the expenditure of vast sums of money. It would be worse than useless for us to repeat over again the discoveries and experiments by which M. Zeise proved that at least two distinct acidulous compounds existed which had the bisulphuret of carbon for their base: it is enough that we give the names and composition of these two substances. One of the acids was called by Zeise xanthic acid, from the yellow colour of its salts, but it is now generally known in chemical works by the expression bisulpho-carbo-vinic acid, and it may be looked at either as composed of two atoms of bisulphuret of carbon, and two atoms of alcohol; or, supposing the alcohol converted into ether and water, then the acid is composed of two atoms of bisulphuret of carbon, one atom of ether, and one atom of water, which atom of water is supposed to be displaced when the acid combines with a metallic oxide, exactly as the atom of water in oil of vitriol is displaced when the acid combines with baryta or potash. In the composition of the other bisulphuret of carbon acid alcohol takes no part, and therefore we get rid of the word "vinic" in the name; it is called hydro-sulpho-bisulpho-carbonic acid, and is supposed to consist of one atom of bisulphuret of carbon and one atom of sulphuretted hydrogen, so that when it unites with a metallic base, say lime or oxide of calcium, the hydrogen then forms water with the oxygen of the lime, leaving the calcium to combine with the sulphur and form sulphuret of calcium, which, uniting to the bisulphuret of carbon, gives rise to one atom of the bisulpho-carbo-sulphuret of calcium, or, as it is called in some chemical books, sulpho-carbonate of lime. But no matter for its name; we see that it is formed by the union of the sulphuret of calcium and the bisulphuret of carbon, and this is just what we wish to do in purifying gas by lime; for the sulphuretted hydrogen of the impure gas combines with the lime put in the purifier, and by this means part of the lime is converted into sulphuret of calcium, which acts upon any bisulphuret of carbon in the gas, and produces bisulpho-carbo-sulphuret of calcium, a substance very generally found in gas-lime refuse, whether the manager knows anything about it or not. In this, as in almost all other cases of chemical action, however, the presence of water assists in causing combination, for absolutely dry sulphuret of calcium is comparatively inert, and therefore we have again to recall to the mind of the gas manager how important it is to preserve one uniform standard of water in gas lime, and from the above description he will readily understand that the hydro-sulpho-bisulpho-carbonic acid is the only one ever found in gas-lime refuse. The other, the xanthic acid, or bisulpho-carbo-vinic acid, is never found nor ever can be produced in gas lime.

Thus far we have followed in the track of M. Zeise, and of the chemical books in which that gentleman's labours have been republished, but not repeated, otherwise it would seem impossible that matters could remain in the very unsettled and unsatisfactory form they have so long maintained. At the time when M. Zeise made his experiments, neither the word nor the idea of isomery, or of allotropy had found a place in chemistry; but we now know that two things may be identical in composition and yet totally different in quality and nature. Of this kind of isomery we have a very good example in coal gas, which contains two gases having exactly the same composition, but differing from each other in specific gravity and other qualities. These gases are olefiant gas, and the gas obtained in a liquid form by the Portable Gas Company, and which in this form was confided to Mr. Faraday for analysis, and is hence sometimes called Faraday's olefiant gas, because it has exactly the same amount of carbon and hydrogen as olefiant gas, but differs in this respect, that olefiant gas appears to contain eight volumes of hydrogen, and eight volumes of carbon condensed into four volumes, whilst the Faraday gas has sixteen volumes of each condensed into four volumes, and therefore the last gas has twice the specific gravity of the first or olefiant gas.

It is very necessary that this condition called isomery should be clearly understood, because upon it most probably depend the peculiarities displayed by the bisulphuret of carbon acids that we have just been describing; and it is more than probable, had M. Zeise been aware of this peculiarity in chemical compounds, he would have carried his investigations far beyond their present limit. Up to the discovery of any other combination, all the bisulphuret of carbon compounds have as a matter of course been looked upon as containing bisulphuret of carbon; and upon this supposition, not only the names, but the composition of the different compounds which it forms have all been fixed, although many chemists have felt themselves unable to agree with that conclusion, and some eminent chemists of the French school of theory deny altogether the existence of an acid composed as the hydro-sulpho-bisulpho-carbonic acid is said to be composed. It is, moreover, quite certain that no such acid can be made and preserved long enough to enable us to settle this dispute; for when we dissolve in water any of the so-called bisulpho-carbo-sulphurets, and add an acid for the purpose of setting free the hydro-sulpho-bisulpho-carbonic acid, a yellow powder at first appears, but this is instantly followed by a discharge of sulphuretted hydrogen, and then a red oily-looking fluid settles down, which, however, can scarcely be the acid in question, because, as we see, it has parted with its sulphuretted hydrogen.

There is, however, now a hope that we may henceforth understand the nature of the changes that occur during the combination of the sulphuret of carbon with metallic oxides, for it has been discovered that this sulphuret takes on an allotropic condition and then assumes all the qualities of a halogen body, like chlorine or cyanogen, and therefore unites with metals and metalloids, but not with metallic oxides, as will shortly be explained. To this allotropic or isomeric form of the bisulphuret of carbon the name Erythrogen has been given, in consequence of the red colour of its compounds, the name being derived from the Greek Erythros, red, just as the word cyanogen is derived from the Greek "kyanos," blue, on account of the blue colour of some of its compounds. In conformity with the halogen class to which it belongs, erythrogen produces with hydrogen an acid, the hydro-erythric acid, in which substance the peculiar effect of allotropy is most singularly displayed, for though composed of two extremely volatile bodies, hydrogen and bisulphuret of carbon, yet the acid itself cannot be volatilized by heat; but if gradually subjected to a high temperature it is decomposed into sulphuretted hydrogen, sulphur and a compound of sulphur and carbon, which cannot be decomposed even by a white heat and is almost incombustible. To obtain a clear view of the erythrogen compounds we will contrast a few of them with similar compounds of cyanogen, by which it will become easy to comprehend the nature of the bisulphuret of carbon acid that exists in gas-lime refuse, and bears at present the very complex name of hydro-sulpho-bisulpho-carbonic acid. The following table exhibits the analogy:

Erythrogen	..	Sulphur . 2 Carbon . 1	Cyanogen	..	Carbon . 2 Nitrogen . 1
Hydro-erythric acid	..	Erythrogen 1 Hydrogen . 1	Hydro-cyanic acid	..	Cyanogen . 1 Hydrogen . 1
Erythride of potassium	..	Erythrogen 1 Potassium . 1	Cyanide of potassium	..	Cyanogen . 1 Potassium . 1

And hence we pass on to what is called hydro-sulpho-cyanic acid, which leads us to a somewhat similar compound of erythrogen called hydro-sulpho-erythric acid, in which sulphuretted hydrogen takes the place of hydrogen and forms therefore the identical gas-lime acid, so that instead of hydro-sulpho-bisulpho-carbonic acid we can use the term hydro-sulpho-erythric acid, and the name of the compound contained in gas lime then becomes converted into sulpho-erythride of calcium, the hydrogen of the acid combining with the oxygen of the lime to form water, whilst the calcium unites to the erythrogen and sulphur to produce sulpho-erythride of calcium, which corresponds almost exactly to the sulpho-cyanide of calcium; consequently the hydro-sulpho-erythric acid has its analogue in the hydro-sulpho-cyanic acid, and thus the difficulty of the gas-lime acid is solved by comparison with its corresponding cyanogen acid. But the resemblance of erythrogen to cyanogen does not stop at this point, for they are both decomposed when we attempt to unite them with a metallic oxide, and it is this decomposition which has cast so much obscurity upon the nature of gas-lime refuse. If we pass cyanogen through a concentrated solution of potash, part of the cyanogen is decomposed by the oxygen of the potash, so as to form potassium, which then unites to the remainder of the cyanogen to produce cyanide of potassium, the other portion of cyanogen being resolved into nitrogen and carbonic acid, and we obtain a solution of cyanide of potassium and carbonate of potash. If, however, we use bisulphuret of carbon or erythrogen instead of cyanogen, the same decomposition occurs, and potassium and carbonic acid are formed; but instead of nitrogen we have sulphur set free, and this combining with the erythride of potassium forms sulpho-erythride of potassium, whilst we get as before carbonate of potash; but in place of cyanide we obtain sulpho-erythride of potassium, the nitrogen being evolved in the case of cyanogen and the sulphur retained in the case of erythrogen. So that a striking resemblance exists between the compounds of cyanogen and erythrogen; and bearing this in mind, we come to understand in some measure the cause of that difficulty which has so long impeded the complete purification of coal gas, because it would seem that to remove bisulphuret of carbon we must employ a metal or metallic sulphuret, and not the oxide of a metal. Having, therefore, developed the composition and chemical qualities of erythrogen, we will now describe the mode of procuring it, and give a general outline of the compounds which it forms.

To begin: we must first make an amalgam of potassium and mercury, in the proportion of one part by weight of potassium and about one hundred and fifty parts of mercury, because with a larger proportion of potassium the amalgam is not fluid; these proportions having been put into a flask with a quantity of naphtha, the whole must be heated so as to melt the potassium, when a rapid combination ensues and much heat is evolved; as soon as the mixture has become cold we pour off carefully the naphtha, and having wiped the surface of the amalgam with a dry cloth, transfer the amalgam quickly into a clean stoppered bottle, and add to it a quantity of bisulphuret of carbon equal to at least three times the amount of potassium employed; then agitate the whole together rapidly until the mixture begins to become solid, when it must be set aside for some hours to cool and allow the combination to finish. After this remove the stopper, and blow a current of air into the bottle until the surplus bisulphuret of carbon is expelled, when by the addition of a little water we can dissolve out the erythride of potassium; and having filtered the blood-red solution to free it from the mercury, we have only to evaporate the fluid at a gentle heat, when we obtain the erythride of potassium as a dark semi-crystalline mass of an extremely hygroscopic nature and very soluble in both alcohol and water. If, however, to its aqueous solution we add an acid, such as the hydrochloric, we throw down the hydro-erythric acid in the form of a red-brown precipitate, which becomes black by drying, and may then be preserved unchanged, but will dissolve in a cold solution of pure potash, and reproduce the same blood-red fluid as at first. It will also dissolve in soda and ammonia, giving solutions of a red colour, by which all the ordinary metals may be precipitated from their solutions, the precipitates having various colours.

By this mode of making erythrogen, we learn that if ever coal gas is to be completely freed from bisulphuret of carbon it must be by the use of a metal or metalloid, and therefore the light thus thrown upon the process leads us to the use of metals in a high state of division, of which the mixture formed by heating the tartrate of lead to a dull red heat in a close vessel furnishes an example. The mixture thus produced consists of carbon and lead in the most minute form of subdivision, so that mere exposure to the air causes it to take fire and burn into oxide of lead and carbonic acid; and a very similar compound may be formed by carefully heating a mixture of two parts of litharge and one of lampblack, and the disposition to take fire by contact with the air is destroyed if we pour a little naphtha upon the product after it has become cold and before we remove it from the vessel in which it was made: similar compounds may be formed from other metals, as for instance copper, and therefore a large field is open to this system of purification. The great change which the discovery of erythrogen has made in our views of the purification difficulty renders it unnecessary to examine any other method of purifying gas until time has been allowed to see what changes may follow upon our altered knowledge. If by employing any of the ordinary metals in a state of extreme subdivision we can thoroughly purify gas, from which all the carbonic acid and sulphuretted hydrogen have been previously removed, we may safely calculate upon a return to oxide of iron purification, and a cheap mode of removing carbonic acid will then constitute our principal desideratum. Upon this latter point we have witnessed some experiments made by Mr. W. R. Phillips, of the Luton gasworks, which are highly interesting.

The gas at Luton in these experiments was first freed from sulphuretted hydrogen by oxide of iron, and then passed through a mixture of two parts of carbonate of potash and one of chalk, by which the carbonic acid was completely arrested and the carbonate of potash converted into bicarbonate, a substance which again becomes carbonate by exposure to a heat not much exceeding that of boiling water, consequently the same carbonate of potash may be used over and over again, for an indefinite number of times; but the experiments upon this subject are not yet finished, nor can their value be appreciated so long as the only method left open to us is in reality, though not in appearance, the method of purification by hydrate of lime.

We now pass on to the details of gas purification suitable for works subjected to no restrictions as to sulphur in other forms than sulphuretted hydrogen.

In purifying gas we have three objects in view: first, to remove the ammonia in such a way as to cause it to carry with it or remove a portion of the carbonic acid; second, to remove the sulphuretted hydrogen by an active agent having the quality of renewal or renovation of its power, whilst at the same time it possesses the advantage of converting the sulphuretted hydrogen into sulphur, which it retains and, as it were, hoards up for future use; thirdly and lastly, the carbonic acid must be removed either by a cheap agent, or one capable of having its power renewed at a trifling cost.

The first of these objects can be very well secured by the use of hydrated sulphate of lime, which at ordinary temperatures will absorb or remove from the impure gas all the ammonia, and with the ammonia an equivalent proportion of carbonic acid, thus forming a mixture of sulphate of ammonia and carbonate of lime.

The next object may be attained by the use of one of those forms of oxide of iron which will remove the sulphuretted hydrogen from coal gas, and which is best when prepared so that it is intimately blended with the sulphate of lime employed to effect the first object, because this blending prevents the oxide of iron from running together into impervious masses during the operation of distilling from it the sulphur it has collected, as will be explained hereafter.

The mode of preparing this mixture of oxide of iron and sulphate of lime is as follows: take any given quantity of green copperas (sulphate of iron), and mix with it half its weight of recently slaked lime, adding just sufficient water to make the whole damp; then grind it into a pulp in a mortar-mill, and spread the green-coloured pulp on the ground in a dry place and leave it so for twenty-four hours, when it will have become dry and of red-yellow colour; in which state it is to be broken into a coarse powder, and is then fit for use in precisely the same way as lime is used: that is to say, it is to be spread about four inches in thickness on the sieves of the purifiers, and will then remove from the gas all the ammonia, all the sulphuretted hydrogen, and part of the carbonic acid.

We have now consequently to complete the purification by removing from the gas the remainder of the carbonic acid left in it after passing through the oxide of iron purifier, and this complete purification we may carry out by means of hydrate of lime employed in the usual way. As, however, the lime is thus converted into carbonate of lime, and thereby rendered worthless, attempts are being made to discover a renewing or renovating agent for the removal of carbonic acid; and one of the most hopeful of this kind that we have yet seen is the employment of carbonate of potash, by Mr. W. R. Phillips, already mentioned, upon which we will make a few observations, after having inquired into the numerous chemical changes that occur in the working of the oxide of iron purifier.

The action of the ammonia and carbonic acid in the gas upon the sulphate of lime is an ordinary case of double decomposition, in which the two substances mutually exchange their acids at temperatures below 100° Fahr., but at all higher temperatures this decomposition will not take place;

thus at 60° Fahr. the carbonate of ammonia becomes sulphate of ammonia, and the sulphate of lime is converted into carbonate; but at the temperature produced by the reconversion of the sesquisulphuret of iron into sesquioxide, the sulphate of ammonia is reconverted into carbonate, and sulphate of lime is produced; and the application of this fact is of considerable importance towards the economical use of the renovating process, as we shall soon see by examining that process.

As the protoxide of iron cannot exist in contact with atmospheric air, but becomes sesquioxide, it follows that whatever form of oxide we employ it must always be sesquioxide, or, as some chemists call it, peroxide. When therefore a proper form of sesquioxide is subjected to the action of the sulphuretted hydrogen in coal gas, the oxygen of the oxide and the hydrogen of the sulphuretted hydrogen unite and form water, whilst the sulphur and the iron combine together and produce sesquisulphuret of iron; so that if we have charged a purifier with the iron mixture just described and passed impure gas through it until it has become foul, the mixture may then be regarded as consisting of sulphate of ammonia, carbonate of lime, and sesquisulphuret of iron.

Now, the burning of iron wire in oxygen gas is a very common experiment which shows the large amount of heat produced by iron when it is oxidized, and the conversion of the sesquisulphuret of iron into oxide of iron by exposure to the air is exactly a similar case; and therefore heat is produced by which the sulphate of ammonia and carbonate of lime react upon each other and produce sulphate of lime fit to be again used, and carbonate of ammonia, which, being volatile, flies off, and consequently requires to be condensed in water or arrested by an acid; and upon the careful management of this part of the renovating process much of the success of the process depends. We see that when the purifier has become foul, the ammoniacal and lime salts then present require only the aid of heat to be reconverted into their original condition, and an excess of heat can produce no injury to them; but the case is very different with the sesquisulphuret of iron, in which an excess of heat may prove injurious in a variety of ways.

The perfection of this part of the process would consist in the production of just so much heat in any given time as will maintain the temperature of the mass at about 180° Fahr., because at this heat the ammoniacal sulphate is completely decomposed by the carbonate of lime, the iron is completely oxidized, and the sulphur being merely thrown out of combination is neither oxidized into an acid nor fused so as to shut up the pores of the oxide of iron, and render that substance useless for purification. But this much desired temperature is very difficult to be maintained by common working men; and, in fact, we may say that it never is maintained in practice, for we have never yet examined a sample of used oxide which did not show by means of the microscope that some of the sulphur had been fused; and it is to this fusion and sealing up, more than to the increase of the sulphur, that the oxide of iron gradually loses its powers of purification, and therefore some improvement is yet needed in the renovating process; and the gas engineer who can devise a simple form of apparatus by which air can be sent in any required quantity through the foul material, so as to maintain a perfect command of the temperature, whilst the deoxidized air charged with carbonate of ammonia is transmitted through a scrubber, to produce a saleable solution of carbonate of ammonia—such engineer, we say, will render an important service to gas purification.

Admitting the existence of such an improvement, the time would still come at which the gradual accumulation of sulphur in the material would make it payable (if we may use the word) to separate the sulphur for sale; and consequently a question arises, how is that separation to be effected? With a view to prepare ourselves for this question, and also to preserve as far as possible the purifying power of the oxide, we have taken care to mix it with a substance which will bear a red heat and prevent the particles of oxide of iron from running together into nodules, or little masses impermeable to gas, and therefore unfit for purification. We are then prepared for the question, and can allow it to be answered in a variety of ways, without much risk of destroying the purifying power of the resulting oxide; which has been made up upon the principle of the little balls of clay and platinum used by the late Professor Turner for combining oxygen with hydrogen and the various compounds of hydrogen and carbon, in which cases he discovered that a mixture of five grains of spongy platinum and one of pipeclay preserved its powers under circumstances in which the pure spongy platinum became useless; and this he sagaciously attributed to the power of preventing cohesion amongst the particles of platinum by the interposition of the clay, a condition secured in the instance of oxide of iron by the interposition of sulphate of lime, as we have explained. Fully prepared for the event, we can allow the sulphate to be burnt off at a low red heat, by which we form sulphurous acid that may be passed into a proper chamber and converted into oil of vitriol in the ordinary way, our mixture of oxide of iron and sulphate of lime still maintaining its purifying power, as was most conclusively demonstrated by the late Mr. Alfred King, at Liverpool; or we may distil off the sulphur, either by a dull red heat, or still better, by superheated steam, in which case a temperature of 400° Fahr. is quite sufficient, and the purifying power is preserved. Therefore, in point of fact, we may by proper management so contrive that no offensive matter whatever is sent out of the gasworks in consequence of purification, and much of the cost of that process will be repaid by the sulphur produced.

The only remaining impurity is the carbonic acid, which, as we have said, may be removed by slaked lime with the production of carbonate of lime, a harmless and worthless substance, but entailing in this use of lime an outlay, and so an increase in the price of gas, a circumstance very much to be regretted, because gas is in reality a necessary of life, if we mean civilized life. Not content with this constant burden of expense, Mr. W. R. Phillips has wisely called in the higher resources of chemistry to his aid, and is now determining in a practical manner how far the peculiar decomposing power of the alkaline bicarbonates can be made to substitute the use of lime. It has long been known that at ordinary temperatures the whole of the alkaline carbonates, ammonia included, have the power of absorbing carbonic acid and giving it off again at a heat very little above that of boiling water; in other words, the carbonates become bicarbonates at low temperatures, and these bicarbonates are resolved back into carbonates at temperatures exceeding 212° Fahr., and herein resides the principle of Mr. Phillips's invention; he makes carbonates into bicarbonates by purifying gas, and he cheaply reconverts the bicarbonates into carbonates for that purpose.

The idea is certainly a scientific one, and can fail only on the score of expense, though it does appear to us that with a fixed outlay to begin upon, and a constant saving in the shape of lime, it cannot be very long before the saving will equal the original fixed outlay, and from that time forward the saving will be all gain, except the mere cost of labour, for the cost of fuel must be very trifling. Mr. Phillips has chosen carbonate of potash (and no doubt for good reasons), but having chosen it, the question that arises is this: how many times must it be used before it pays for its own prime cost? Let us suppose a hundred times; then, as carbonate of potash is really indestructible, the only future expense of purification will be manipulation, and the loss created by it. Now there seems fair ground for success in this attempt, and we wish it well.

We close our remarks by directing attention to the formulæ to aid the distribution of gas in pipes, to be found under the head of "Mains," which we hope may prove useful and interesting.

COAL.

Our earth, with its present configuration of "waters" and "dry land," with the various races of animal and vegetable life now existing thereon, with all the numerous substances, as granite, slate, limestone, clay, coals, &c., entering into its composition, was, previous to the existence of the science of geology, believed to have been created at one period—at one operation—and that no material change had taken place therein since its first formation.

By the science mentioned, we are, however, taught the very opposite of this, that the "crust of the earth," as termed by geologists, has undergone various important changes at successive and remote periods, that during each of these periods it has been inhabited by distinct races of animal life, with a vegetation peculiar to each epoch, and only in a few instances has any one particular race of beings existed in two distinct ages. Moreover, the various granitic and other rocks have acquired their present positions, sometimes gradually by sedimentary deposition, sometimes rapidly as by volcanic agency, but in almost every instance, except the purely granitic, we find the remains of animal or vegetable life associated with these rocks, and having in each case distinctive characters peculiar to the mode and period of their formation; so that, in fact, the history of these vital relics, when developed and thoroughly understood, will constitute a history of what may be called the childhood of the world; and the materials of that history are now being daily brought to light in excavations of almost incredible depth, and by sections showing the sequence of deposit in the respective formations which imply the existence of an inconceivable amount of time.

We learn, also, that this crust of the earth has been subjected, by earthquakes or similar causes, to violent convulsions and dislocations, by which the valleys have been exalted and "the mountain and hill made low;" and by the same action, the depths of the ocean of one age have become the summit of a hill in another, and reefs extending hundreds of miles have been upheaved, and the neighbouring land submerged; and so recently as 1822, as we are informed by Mr. Phillips, the eminent geologist, that on the western coast of South America the level of the land and sea was altered for 1000 miles in length, and in many places the ground was permanently raised. We are also taught that, by the action of rivers, the land in certain localities is gradually carried away as silt, sand, and mud to the ocean or lakes, where they are deposited, embedding creatures of the present period, which may hereafter become records of the present era.

According to the opinion of some geologists, at the remote period of the carboniferous epoch the temperate zones possessed a climate similar to that of the tropics of the present, when vegetation was exceedingly bounteous, and forests of trees of various kinds and of great magnitude, intermixed with ferns and other cryptogamic plants, covered extensive areas and flourished in marshy ground, and a warm humid atmosphere which undoubtedly existed at that period. That these forests by the changes in the surface of the earth were afterwards sunken, and subsequently covered by silt, mud, and sand brought down by rivers; which covering afterwards constituted the soil for another forest, when this again was sunken, and like the other again covered, thus forming different strata or seams as now found. That in the course of thousands or even millions of years (for geology and astronomy teach us that time and space, at least as far as regards human comprehension, are without limits), by the combined agencies of pressure, the absence of air, by heat, moisture, and chemical action, these embedded forests were converted into coal, whilst the mud, silt, and sand forming the roof and flooring of each seam, assumed the nature of schist, shale, or stone.

Other eminent geologists believe the coal beds to have been originally a description of peat-bogs, or masses of vegetable remains accumulated around the places of their growth, and that eventually these tracts of country, from causes already mentioned, subsided below their former level and were covered as described.

Again, according to another theory, the trees were uprooted, conveyed by rivers and deposited in deltas, or lagoons, as is now taking place at the mouth of the Mississippi, and covered like the forests with the same result.

Lastly, there is another kind of vegetable accumulation to which the origin of coal may be attributed. This is the turf or peat moors, as they are called in the North of England, which occur in low ground towards the estuaries of rivers and along the margin of the sea, containing a mass of vegetable matter composed of mosses and other humid plants, roots of ling, &c., enveloping trunks of trees, sometimes prostrated in particular directions, and apparently cut by art, or decayed by nature.

A difficulty which presents itself against the acceptance of these theories, is the purity of the coal and the absence of sand or any earthy matter, throughout a volume and area of such vast extent. "This enigma," says Lyell, "however perplexing at first sight, may, I think, be solved by attending to what is now taking place in deltas. The dense growth of reeds and herbage, which accompanies the margins of forest-covered swamps in the valley and the delta of the Mississippi, is such that the fluviatile waters in passing through them are filtered and made to clear themselves entirely before they reach the areas in which vegetable matter may accumulate for centuries, and eventually forming coal if the climate be favourable. In such cases there is no intermixture of earthy matter. As a singular proof of this fact, I may mention, that wherever any part of a swamp in Louisiana is dried up during an unusually hot season, and the wood set on fire, pits are burned there many feet deep, or as far down as the fire can descend without meeting with the water, and it is then found that scarcely any residuum or earthy matter is left."

Such is the generally accepted theory of the formation of coal, and in the confirmation of which, as we are informed by the author just referred to, "Liebig and other eminent chemists have, from their researches, ascertained that when wood and vegetable matter are buried in the earth, exposed to moisture and partly or entirely excluded from the air, they slowly decompose, and become gradually converted into lignite or wood coal, and by a continuance of this decomposition, the lignite is changed into common or bituminous coal." According to other writers, by the disengagement of carburetted hydrogen from the ordinary bituminous coal, it is converted into anthracite, which may consequently be regarded as a natural coke.

In addition to this evidence of the origin of coal, there are the further proofs by the impressions of ferns and foliage being found continually in combination therewith, and sometimes the trunks and branches of trees converted into coal. Specimens of these were frequently met with in Boghead cannel, and in some cases the bark was transformed into coal differing in appearance to the rest of the trunk or branch.

It must not, however, be supposed from the foregoing that geology is in opposition to Scripture, as the very reverse of this is the case, since, according to that science, our earth has undergone six distinct changes or formations, corresponding with the "six days" of creation; if, therefore, the term "day" be accepted as a period of time the explanation is clear. Moreover, so far as researches have hitherto extended in all parts of the world, although there exist the remains of a numerous animal creation, according to the respective periods of their existence, some of which far exceed in size any of the present time, yet in the earliest of these formations there is no trace of the remains of man. Therefore, the conclusion is that only at the sixth day or period (that actually passing) man was created, so confirming in a remarkable manner that part of Holy Writ.

In most northern European languages, a very similar name as "Kul," or "Kohle," is applied to the material in question, but undoubtedly our term is derived from the Saxon "Col." In languages derived from the Latin, however, the word "carbon," with a slight alteration in the orthography, is applied to charcoal, and we believe that "col" was originally employed for that substance. Beyond all doubt the ancients were unacquainted with the mineral under consideration, and although we find the expressions "burning coals" and "fire of coals" in the Bible, these, we are of opinion, arise from a defect in the translation, as coal when translated into languages of Latin origin is rendered as "carbon" or charcoal. Moreover, in these languages to imply coal, the word "stone" is affixed to the other as "carbon de piedra," "charbon de pierre," &c., which difference distinctly conveys the impression that charcoal and not coal is meant.

The earliest mention of pit coal amongst the ancient authors is by Theophrastus, in his ' History of Stones,' in which he says, "there is a fossil substance called coal, which is broken for use; it kindles and burns like wood. It is found in Liguria and in Ellis, in the way to Olympias, over the mountain. These coals are used by the smiths"—which statement leaves no doubt that the mineral mentioned by him was our common coal.

From evidence of a satisfactory kind, it is highly probable that coal proper was known to the primeval Britons, and was used by them in metallurgical operations, but we have reason to suppose that as their workings were confined to the outcrop of the seam, their knowledge and application of this valuable material must have been very limited.

That the Romans were acquainted with the use of coal is made palpable by the discovery of its cinders amid the ruins of their iron forges or "bloomeries," and it was certainly used by them in their pottery furnaces at Condata, Warrington, where quantities of Wigan cannel, coal and cinders have been found in connection with an extensive collection of pottery now preserved in the museum of that town.

The use of coal was continued—but probably to a limited extent—during the Saxon period. In the year 852 a grant of some lands was made by the Abbey of Peterborough, under reservation of certain payments in produce to the monastery, amongst which, according to the Saxon chronicle, were twelve cart-loads of "fossil" or pit "coal." They appear also to have been known and used in the county of Durham during the twelfth century. Thus we find from the "*Golden Buke*," which contains the survey instituted by Hugh Pudsey, Bishop of that See, and completed in the year 1183, that the smiths who were employed on the territories of that immense diocese were generally required to find, dig, and carry their own coal for the forge. "Coals dug from the ground" are also mentioned in some of the ancient deeds belonging to this period.

It was not, however, until the year 1234 that coal was wrought as a commercial product. In that year King Henry the Third, in consideration of a fee-farm rent of 100*l.* per annum, granted to the burgesses of Newcastle-on-Tyne a charter to dig "coals and stones," and to convert them to their own use, and some few years afterwards it became an article of export, and was called sea coal, again making distinction between common coal, or charcoal, and coal.

The use of coal became so extensive in London about 1306, that in consequence of a representation being made to the king of the vapours arising therefrom polluting the atmosphere, a proclamation was issued forbidding its further use, and, singular enough, we have witnessed a similar prejudice to its employment in small towns on the Continent.

Various patents or, as termed, special licences were granted at the commencement of the seventeenth century, for smelting iron, making glass, soap, alum, and other classes of manufacture, with "sea coal." But more particularly bearing on our subject was the "especiall lycence" granted to John Barber and another, in 1681, for a "new way of making pitch and tarre out of pit coale, never before found out or used by any other."

Thirteen years afterwards Eele and others obtained a similar grant for "a way to extract and make great quantities of pitch, tarr, and oyle" out of a sort of stone. This, after an interval of fifty years, was followed by another grant for obtaining "an oyle from a flinty rock, for the cure of rheumatick, scorbutic, and other cases."

The earliest practical details or instructions for the distillation of tar and procuring pitch, are in the specification of the patent granted to Hawkins in 1746, which are given with such clearness and precision as to be useful to anyone indifferently acquainted with the subject, even at the present day.

A remarkably progressive step was made by Baron Van Haake in his patent for a "new invented secret art or mystery in extracting and making several compositions called mineral oil and mineral tar." In the specification, he describes "a number of iron cylinders being prepared, when a proportional quantity of such mineral or coal is placed therein, and a large quantity of common burning coal is put round them and set on fire, which causes a fluid matter to run or issue from the mineral or coal in the cylinder."

Following this was the extraordinary patent of Lord Dundonald, in which he described all the products to be obtained from coal, except the most valuable, the gas; and although eventually, when carrying this into practical operation, the vapour emitted therefrom frequently ignited, it does not appear to have suggested to anyone the means of applying it to illuminating purposes. Eleven years after the enrolment of this patent, Murdoch commenced his first experiments.

Coal may be regarded as a variable admixture of carbon in a fixed or volatile state, intermixed with hydrogen and other matters already enumerated, and the value of any particular quality will depend upon the purpose to which it is to be applied.

Thus the material that contains the largest quantity of fixed carbon (at the ordinary temperature for carbonization), is, on account of the absence of the requisite hydrocarbons, quite unsuited for the manufacture of gas. On the other hand, that quality of coal which possesses a large amount of volatile matter, in consequence of the tendency of the volatile portions to pass off as smoke before being consumed, is unsuitable for boilers in the production of steam. Therefore, the selection of the material must be made according to the purpose for which it is required, and for gas making that possessing the greatest quantity of volatile matter, with a carbonaceous residuum containing little ash, is generally the most suitable coal for that purpose.

By modern mineralogists, coal is divided into four classes—namely, anthracite or glance coal, semi-bituminous, black bituminous or caking coal, and cannel, sometimes called in the western part of England splint coal.

Anthracite possesses little interest to the gas manufacturer, consisting as it does almost entirely of fixed carbon, and, as observed, it may be regarded as a natural coke.

The semi-bituminous, found extensively in South Wales and largely used for steam purposes, contains from 80 to 95 per cent. of fixed carbon; ignites with difficulty, but when fully ignited with a good draught yields an intense heat. In appearance it is lustrous, formed of small cubes, and of a friable nature. During the process of combustion, these cubes have a tendency to become detached from the mass and fall through the bars of the furnace before being consumed, and to avoid this loss of fuel, generally, a portion of caking coal is intermixed with the other.

Bituminous, or caking coal as more properly called, as its name implies, in the act of combustion cakes together, and when in the retort after a time becomes of a black doughy consistency, intumesces, and after having yielded the larger portion of its gases, begins to agglomerate and increase in temperature, until it becomes red hot, at which point the troublesome sulphur impurities are evolved.

The rich caking coals of Northumberland and Durham contain from 58 to 70 per cent. of fixed carbon, and from 2·5 to 5 per cent. of ash. Some qualities of these are unsuitable for household purposes, on account of their tendency to cake together and obstruct the passage of the air; these, as a rule, are very friable, and contain a large quantity of "slack" or small coal, which, although not adapted for the purpose mentioned, answer admirably for the manufacture of gas and coke.

The other coal which possesses great interest for the gas manufacturer, is the Cannel or Parrot coal, with which Scotland is particularly favoured; it is also found to a limited extent in the North, but more extensively in the Midland counties, of England. It is also found in North America, Belgium, and a few other places.

This is the richest of all coals for producing gas of high illuminating power, and has a smooth compact appearance, and is duller in colour than other descriptions. Its principal characteristics are, that its fracture is flinty or conchoidal; it decrepitates when submitted to the heat of the retort, does not intumesce nor alter in form during the process of carbonization, and delivers its gas principally in a lateral direction, the coke resulting therefrom showing the laminæ whence the gas issued.

In the absence of means of analyzing any particular quality of coal, we must therefore endeavour to arrive at a knowledge of its quality from its external appearance. For this purpose we cannot do better than quote the words of Mr. James Paterson, of Warrington, who, from his intelligence and his extensive researches, is eminently well qualified to be regarded as an authority on the subject. That gentleman says, when treating on caking coals:

"There are certain indications more or less defined, which, without attaching undue importance to their presence, justify at least the presumption whether the coal is suitable for gas manufacture or otherwise.

"The first of these is *cubical structure*, a condition which exhibits in itself many and important deviations from uniformity. It occurs stratified, massive, columnar, and in ovidal concretions; nor is the difference in their lithological characters less definite and marked; the same seam not unfrequently presents two or more distinctive bands, differing materially in their composition and quality. The two former classes contain the source, whence the chief, if not the sole, supply of our gas coals is derived; these are generally divided into parallel laminæ, or layers of varied thickness, by what are termed "partings," consisting uniformly of a thin deposit of mineral charcoal in the line of stratification.

"Vertical to these parallel lines is a succession of flaws or cracks, cutting through the stratum in the direction of its bearing line or level, generally observing a north and south polarity. These divisional lines produce a tendency to break it down into parallelopipedal masses, which present in some cases, such as the best portions of the Wigan Arley Mine or the Silkstone seam in the Burnsley district, very bright clean surfaces; in others, and more generally, their lustre is covered with a white sparry concretion deposited from the filtrations of calcerous water, the presence of which is however not objectionable in the case of a coal used for the manufacture of gas. These characters are, I believe, uniformly found to a greater or less extent in soft and easily frangible coal, and I have found them to be reliable indications of a good coking, and generally of a good gas coal.

"On the contrary, when the mass is compact, hard, and foliated, splitting into longitudinal plates in the line of its stratification, it will be found that the deposit of charcoal gradually disappears with the more indurated character they assume; and the vegetable remains, which formed these deposits in the soft coal, become in this converted into the bisulphide of iron, generally amorphous, but sometimes in the crystalline form. Their cross section is splinty or angular, the vertical cracks are only partially present or altogether absent; under these circumstances, the coal, as a general rule, is better adapted for steam than for gas purposes. The coke from such coal is more compact, retains more of the structure of the coal, contains in most cases a much greater, though varying, percentage of ash under incineration, and frequently, but not uniformly, requires a stronger draught for its combustion.

"*Lustre.*—This character in coal is a good and reliable evidence of its coking qualities, but less definite in relation to its gas-producing properties. All our best gas coals possess a degree of lustre more or less bright; but there are coals having a bright lustre, which are altogether unsuited for the purpose of gas manufacture, and experience alone can identify this difference—a difference which may be said to rest upon "a shade." In order to illustrate this point we will take the case of an anthracite which is highly lustrous, and compare it with the Wigan Arley Mine, or the rosinoidal coal of the Middle Mountain Mine, when a casual observer would in all probability detect no difference in their respective lustres; but to the practised eye, the former will be found of a greyish metallic white, much resembling polished steel, whilst the latter will present a black glossy, velvety surface, more or less shiny, according to the minutely divided matter which enters into their composition.

"Under a low magnifying power, there are other characteristic differences which are brought out, which would probably be more difficult to describe than to be identified by the analyst.

"It very frequently occurs that a mass of coal presents us with a distinct reciprocating series of bright black alternating with bright shining and more highly lustrous bands; and from an analysis of these, I have found the former to be slightly the most favourable for the production and quality of gas, and the latter for the production and quality of coke.

"*Fracture and Cross Fracture.*—These are distinguishing marks of good gas coal. When gently broken they will show the fractures to be either cubical, conchoidal, semi-conchoidal, or intermediate angular, presenting a fine black pitch-like or rosinoidal surface. One or more of these characters are uniformly presented by the soft and more easily frangible class of coals.

"On the contrary, should they be found dense, breaking with rough slaty or hackley fractures, presenting a black or greyish-black appearance, or when broken down into small masses, present irregular and undeterminable forms, the grain will be found to be coarse, its composition more or less earthy, and the evidences unfavourable to a good gas coal.

"Another well marked and highly important character is the *streak* or shade of colour indicated upon the surface. When scratched with the point of a nail or a blunt steel point, the eroded surface will present either a dull brown, brown inclining to black, black inclining to brown, dull black, bright black, or lustrous appearance. It will invariably be found that coal possessing either of the two former characters yields gas of a rich quality, and the more the streak approaches to a brown colour, the greater will be the percentage of hydrocarbons under distillation. This character is more readily brought out by attrition, as the rubbing of two pieces against each other and exposing the abraded surface to the light, when the depth of the brown shade will be indicated.

"From a long experience I have uniformly found this to be the most positively reliable of all the external evidences we possess in approximating the value of gas coal or cannel. I have hitherto regarded the latter indication (lustre) as evidence of their coking properties, and have generally found them to increase in that respect as the lustre becomes heightened by the streak. It does not appear, however, that these indications are reliable. I have recently had under examination a sample from the "Middle Mountain Mine" (Upholland), in which the streak is bright, but not heightened in lustre; yet this coal partakes of higher coking properties than any other that has come under my observation. The coking properties of coal may be readily and satisfactorily indicated by its combustion on the fire; open or free burning coal, under distillation, may produce a serviceable, but not a good coke; its heating power as a fuel is generally active, but not profitable. On the contrary, when the coal swells, fuses together, intumesces, and throws out jets of flame, it yields under distillation a good quality of soft coke suitable for forge, malting, or domestic purposes. Its value may be determined by incineration and carefully weighing its ash.

"The specific gravity of coal expresses its density or weight in relation to bulk or to water, as the standard unit of comparison, and, though desirable to know its weight, this affords no evidence of its composition except by inference, and should never be resorted to when more correct information can be procured.

"The amount of sulphur in combination with coal intended for the purpose of gas manufacture is a matter of considerable importance to the gas engineer, as any excess of that impurity involves a proportionate extent of additional purifying surface and materials. This material presents itself as a bisulphide of iron in two different states: in hard coals and cannels it assumes more generally the cubical form, covering the planes of intersection with minute yellow crystals in the line of stratification; however, where woody fibre is present it is almost invariably found in the amorphous state. The latter form usually accompanies soft coal, and is most frequently met with in concretionary masses of considerable bulk and weight; as a rule the crystalline form is absent.

"The diffusion of this impurity throughout a mass of coal is so arbitrary and capricious that in the majority of cases a quantative analysis cannot be relied upon, as representing the normal amount of its presence.

"These uncertainties led me to adopt what may probably be considered a more reliable test in the analysis of coals for gas purposes, viz. to note the quantity of gas which a given weight of shell lime will purify, the results of which are given in the following list, together with the other properties of each particular coal."

The following comprehensive tables, by Mr. Paterson, of the analyses of coals of well-known mines, and extensively used for the manufacture of gas, must possess great interest to engineers, as the absolute commercial value of any particular description, alike for the quantity as well as the illuminating power of the gas, the weight and degree of purity of the coke obtained, and other important information, are therein conveyed.

COAL.

ANALYSES OF CANNELS.

TABLE of the QUANTITY of GAS, ITS ILLUMINATING POWER, WEIGHT of COKE, &c., OBTAINED FROM VARIOUS DESCRIPTIONS OF CANNEL. By JAMES PATERSON, ESQ., OF WARRINGTON.

			Purified Gas per Ton of Cannel in Cubic Feet.	Specific Gravity of Gas Air at 1°.	Illuminating Power of Gas in Sperm Candles of 120 Grains.	Value of One Cubic Foot in Grains of Sperm.	Illuminating Matter in lbs. of Sperm per Ton of Cannel.	Weight of Coke in lbs. per Ton of Coal.	Ash in Coke per cent.	One cwt. Shell Lime Purifies about
Abram Coal Co.	Wigan	Cannel of the Wigan 4-feet	14,800	0·666	29·23	701·52	1483·21	1024	5·08	17,000
Alliance Coal Co.	Ditto	Wigan Cannel	12,830	0·507	20·71	497·	911·	1370	4·88	15,000
Ditto	Ditto	Stone Cannel	10,500	—	18·24	437·76	656·64	1241	9·21	8,000
Bestwood Coal and Iron Co.	Nottingham	Bestwood Cannel	11,680	0·524	18·21	437·	729·17	1209	19·83	—
Bersham Colliery Co.	Wrexham	Dersham Cannel	12,450	0·563	—	368·40	655·23	1180	3·33	—
Blundell and Sons	Wigan	Wigan Cannel	13,150	0·589	21·65	519·60	976·10	1118	4·77	14,000
Bruncker, W., and Co.	Ditto	Ditto	12,770	0·502	22·20	533·	972·34	1402	4·49	—
Broadway Colliery Co.	Mold	Broadway Cannel	10,785	0·532	16·52	396·50	611·	1268	5·33	—
Coed Talon	Ditto	Smooth variety	10,800	—	22·67	543·	838·	1344	68·70	—
Ditto	Ditto	Curley variety	13,600	—	27·75	665·	1204·	903	7·17	16,000
Crippin, W. and F.	Wigan	Cannel of the Wigan 4-feet	17,300	0·666	30·07	721·68	1763·56	920	6·13	16,000
Ditto	Ditto	Ditto W. K.	16,460	0·646	—	564·72	1328·	922	6·75	14,000
Ditto	Ditto	Ditto T. T.	14,100	0·556	19·47	467·30	941·27	1047	3·63	—
Ditto	Ditto	Ditto T. B.	12,200	0·478	14·31	345·44	308·57	1360	6·18	14,000
Dewhurst, H., and Smethurst	Ditto	Ditto	14,750	0·603	23·19	536·	1172·	1041	4·	13,000
Ditto	Ditto	Semi-Cannel	10,700	0·478	16·82	403·08	617·	1390	7·83	—
Edge Green Coal Company	Ditto	Edge Green Cannel	10,540	—	10·63	470·	708·	1348	13·78	9,000
Fir Tree Colliery	Ditto	Cannel of the Wigan 4-feet	11,500	0·523	15·75	378·	643·	1210	4·88	14,000
Guy Brothers	Manchester	Little Heston Cannel	13,300	0·575	18·77	395·	752·40	1413	64·68	14,000
High Carr Ironstone Works	Stoke-on-Trent	Goldenhill Cannel	10,533	—	18·25	558·	077·84	1241	7·40	—
Hindley Field Colliery	Wigan	Cannel of the Wigan 4-feet	12,800	0·505	22·	536·	1077·	1033	4·44	14,000
Inco Hall Coal and Cannel Co.	Stoke-on-Trent	Wigan Cannel	12,780	—	22·35	533·	979·	1411	—	—
Lawton Colliery Co.	Manchester	Lawton Cannel	8,000	0·668	22·	533·	662·	Nil	—	—
Levers, Ellis, and Co.	Ditto	Mostyn Cannel	10,650	0·555	11·32	343·68	592·88	1336	27·	14,000
Ditto	Wigan	Cannel of the Wigan 4-feet	10,175	0·533	17·75	495·	619·22	1340	17·87	14,000
Moss Hall Coal Co.	Ditto	Wigan Cannel	10,600	0·567	22·88	549·	831·34	1300	2·95	—
Pearson and Knowles	St. Helena	St. Helena Cannel	11,870	0·552	22·78	546·72	927·	1415	4·37	14,000
Peasley Cross Colliery	Ditto	Ditto	10,430	—	23·1	—	923·	1259	14·5	—
Rose Bridge Collieries	Wigan	Wigan Cannel	11,125	—	22·63	543·	963·	1414	5·	—
Royal Colliery, Shastic Heath	St. Helena	Ditto	11,675	0·577	20·56	493·44	895·	1388	5·22	—
Scot Lane Colliery	Wigan	Ditto	11,980	—	22·80	547·	895·	1411	4·6	—
Shirland Colliery Co.	Alfreton	Shirland Cannel	10,600	0·533	22·88	549·	831·34	1300	2·95	—
Simpsons and Co.	Manchester	Cannel of the Wigan 4-feet	12,900	0·565	22·88	549·	964·29	1088	5·45	—
Sutton Heath Colliery	St. Helena	St. Helena Cannel	11,590	—	19·15	450·60	947·	1411	5·42	—
Wigan Coal and Iron Co.	Wigan	Wigan Cannel	11,580	—	22·73	546·	880·	1409	4·32	—
Worsley Colliery Co.	Manchester	Bridgewater Cannel	11,050	—	19·	456·	710·	1430	7·32	14,000

Analyses of Coal. By James Paterson, Esq.

			Purified Gas per Ton of Coal in Cubic Feet.	Specific Gravity of Gas Air = 1.	Illuminating Power of Gas from Candles of 129 Grains.	Value of One Cubic Foot in Grains of Sperm.	Illuminating Matter in lbs. of Sperm per Ton of Coal.	Weight of Coke in lbs. per Ton of Coal.	Ash in Coke per Cent.	One Cwt. Shell Lime purifies about
Abram Colliery	Wigan	Wigan 4-feet	10,250	0·509	15·62	375·	549·	1307	5·66	11,000
Ditto	Ditto	Wigan 5-feet	11,000	0·454	14·00	336·	532·8	1315	7·05	12,000
Ditto	Ditto	Wigan 6-feet	10,500	0·488	15·76	378·	567·36	1361	6·88	12,000
Albert Colliery Co.	Newbold, Chesterfield	Gas Coal	10,375	0·488	14·13	338·	501·	1402	3·33	13,000
Alliance Coal Co.	Wigan	King Coal	9,325	0·473	14·30	343·	471·	1453	5·31	—
Anderton Hall	Ditto	Arley Nuts	11,500	—	18·80	355·	583·	1485	2·72	16,000
Aston Hall Colliery	Hawarden, Chester	Gas Coal	8,820	0·459	16·46	395·	483·	1384	2·38	14,000
Ditto	Ditto	Premier Gas Coal	11,250	0·451	16·17	388·	623·43	1368	3·37	10,000
Atherton Colliery Co.	Leigh	Arley Mine	11,500	0·456	14·66	350·64	576·	1422	3·77	14,000
Baaley Brook	Wigan	Arley {Mr. Alfred King and Mr. Paterson}	9,500	0·460	16·70	401·	544·	1450	4·50	—
Birnham Colliery Co.	Wrexham	Gas Coal	10,000	0·481	14·50	348·	527·	1336	4·12	14,000
Bestwood Coal and Iron Co.	Nottingham	Ditto	9,540	0·469	15·14	363·36	495·21	1292	6·72	—
Blackleyhurst Little Delph	St. Helens	Ditto	9,725	0·400	13·81	331·	460·	1470	2·75	15,000
Blainscough Colliery Co.	Wigan	Arley Mine	11,970	—	16·39	393·36	672·65	1433	3·23	—
Brinsop Hall Colliery	Ditto	Ditto	9,980	0·466	16·45	394·80	502·30	1444	3·60	—
Bryn Hall Collieries	Ditto	Hard Coal	9,540	0·511	16·28	307·	500·	1377	5·64	14,000
Ditto	Ditto	Gas Coal Brights	10,800	0·458	16·14	387·	597·	1294	5·50	—
Ditto	Ditto	Letterod W. C.	11,200	0·444	13·03	313·32	501·31	1241	5·71	14,000
Ditto	Ditto	Arley Mine	11,250	0·448	14·55	349·20	561·21	1463	2·46	16,000
Clifton Hall	Ditto	Ditto	10,730	0·575	17·30	414·	636·44	1464	3·33	15,000
Derbyshire New Main	—	Silkstone	11,020	0·452	15·43	370·32	583·	1411	3·21	15,000
Derbyshire Silkstone Coal Co.	—	Ditto	10,775	0·456	14·71	353·	543·87	1465	6·-nearly	12,000
Ditton Coal and Iron Co.	St. Helens	Middle Mountain Mine	10,055	0·513	17·26	414·	595·	1470	1·73	15,000
Dewhurst, Hoyle, and Smethurst	Wigan	Gas Coal	11,050	—	14·40	345·56	545·50	1420	3·37	—
Foodbank Colliery	Dunfermline	Ditto	10,530	0·479	14·41	366·	559·57	1245	5·21	15,000
Garswood Hall	Wigan	Wigan 4-feet, Tops	10,980	0·469	16·78	402·60	629·	1435	5·67	15,000
Ditto	Ditto	Wigan 4-feet, Bottoms	10,620	0·566	16·	384·	593·56	1346	5·79	—
Garswood Little Delph	St. Helens	Gas Coal	9,000	0·417	14·74	354·	455·17	1419	3·	—
Guy Brothers and Co.	Manchester	Amberwood Gas Coal	9,350	0·500	17·15	411·6	548·	1372	6·66	13,000
Haddock Little Delph	St. Helens	Gas Coal	8,936	—	12·75	306·	391·	1398	3·08	15,000
Hargreaves, Richard	Liverpool	Ditto	10,270	0·512	14·72	353·30	518·34	1470	3·33	16,000

COAL. 61

Colliery	Location	Seam								
Hindley Field	Wigan	Gas Coal	10,760	0·450	15·26	365·	562·	1314	5·45	13,000
Ditto	Ditto	Wigan 4-feet	10,750	0·462	15·35	368·	565·75	1374	7·77	13,000
Hindley Green	Ditto	Arley Mine	11,200	0·452	15·11	365·	585·69	1435	2·28	17,000
Ditto	Ditto	Five feet	9,470	0·479	13·35	230·4	409·10	1315	3·21	14,000
Hindley Hall (Roger Leigh and Co.)	Ditto	King Coal	11,550	0·498	15·36	368·26	608·26	1425	2·11	16,000
Ince Hall Coal Co.	Ditto	Wigan 4-feet	10,840	0·459	16·52	396·48	614·	1382	5·17	—
Kirkless Hall Colliery	Ditto	Yard Coal	9,730	0·488	15·32	367·70	511·10	1464	4·23	—
Maurers Main Colliery	Rotherham	Gas Coal	11,000	0·457	16·62	399·	627·	1366	2·02	17,000
Moss Hall	Wigan	Ditto	10,800	0·461	16·75	402·	620·23	1444	2·32	—
New Winning (Blundell and Sons)	Ditto	Orrell 4-feet	10,975	0·469	16·96	407·	638·43	1464	4·33	14,000
Ditto	Ditto	Orrell 5-feet	11,800	0·481	13·92	394·	563·	1461	5·88	12,000
Ditto	Ditto	Wigan 4-feet	10,150	0·489	14·63	351·	509·12	1400	6·13	12,000
Ditto	Ditto	King Coal Tops	10,300	0·467	14·53	349·	513·53	1460	5·22	12,000
Ditto	Ditto	King Coal Bottoms	10,000	0·477	14·48	343·	497·	1431	5·33	10,500
Pearson and Knowles	Ditto	King Coal	10,050	0·479	14·87	356·89	512·40	1432	5·27	—
Peasley Cross	St. Helens	Little Delph	9,000	0·411	12·50	300·	386·	1425	4·75	—
Pelsall Coal and Iron Co.	—	Deep Coal	11,500	0·472	15·31	367·44	603·65	1320	2·44	15,000
Rose Bridge Colliery	Wigan	Arley Mine	12,600	0·444	15·40	370·	666·	1493	1·71	17,000
Ditto	Ditto	Yard Coal	10,300	0·444	15·22	365·30	537·51	1315	2·30	—
Shirland Colliery Co.	Alfreton	Gas Coal	11,000	—	13·46	323·	507·	1312	1·85	—
Ditto	Ditto	Tipton Gas Coal	9,800	0·500	16·22	389·	545·	1402	1·77	16,000
Silkstone and Dodworth Coal and Iron Co.	—	Old Silkstone Seam	12,240	0·459	16·66	399·84	639·15	1498	0·88	17,000
Ditto	—	Parkgate Coal	11,220	0·497	15·87	381·	610·70	1430	1·62	—
Silkstone Fall Coal Co.	Barnsley	Gas Coal	11,590	—	16·98	407·	656·43	1443	0·81	17,000
Ditto	Ditto	Best Thorncliffe	11,200	0·496	17·	408·	652·89	1431	1·80	17,000
Ditto	Ditto	Lop Bed	11,670	0·440	15·17	364·	606·84	1395	2·	17,000
Snydall Hall Colliery	Ditto	Arley Mine	10,150	0·542	17·90	434·	630·	1291	3·22	15,000
South Yorkshire Coal and Iron Co.	Ditto	Old Silkstone Seam	12,300	0·498	16·73	401·52	705·32	1456	1·24	17,000
Spring Colliery	Wigan	Yard Coal	10,450	0·477	13·31	319·	476·22	1456	3·62	16,000
Ditto	Ditto	Snuithy Coal	9,800	0·494	15·22	365·	511·39	1463	2·11	16,000
Stavely Colliery Co.	—	Gas Coal Tops	9,040	—	11·52	270·	350·	1396	4·32	13,000
Ditto	—	Gas Coal Bottoms	10,100	—	12·23	294·	424·	1340	4·50	13,000
Swan Lane Colliery	Wigan	Gas Coal	9,400	0·521	16·82	404·	542·50	1314	2·42	13,000
Tawel Vale Colliery Co.	Ormskirk	Gas Coal	10,420	0·457	15·64	375·36	558·75	1300	4·22	—
Victoria Colliery	Bairford	Ditto	10,030	0·483	13·63	327·	454·	1264	2·88	15,000
White Moss Colliery	Ormskirk	Park Mine Gas Coal	11,400	0·467	15·37	366·88	586·45	1423	2·31	—
Wigan Coal and Iron Co.	Ditto	Arley Mine	11,840	0·471	15·37	367·	620·75	1449	1·79	—
Wigan and Whiston Coal Co.	Wigan	Platt Lace Gas Coal	9,750	—	13·75	330·	459·64	1412	2·94	—
Ditto	Ditto	Whiston Arley	12,300	0·494	15·	360·	632·27	1441	1·87	17,000

The author of the preceding observes "that the accompanying tables of analyses represent, as nearly as may be arrived at by experiment, the true value of the respective coals named; but it is not to be expected that similar results will be obtained practically on a large scale, when taken over the half-year's working, however judicious and well directed the management may prove to be over the carbonizing department. But the difference in the results, whatever that be, must not be charged against the coals, but traced to its proper cause, which in all probability will be influenced in many ways, such as difference of temperature, leakage from the retorts, condensation within the retorts through the agency of deep hydraulic dips, &c. All or any of these contingencies may very materially alter the conditions under which the respective distillations have been conducted."

For the purpose of appreciating the value of these preceding tables, we must direct the reader's attention to some of the various classes of coal therein mentioned. Thus we find in the first list one class of cannel which yields 17,300 feet of 30 candle gas and 920 lbs. of coke, containing a moderate quantity of ash, which renders the coke alike suitable for sale as well as for heating the retorts. This is therefore a desirable coal, producing as it does a larger quantity of rich gas, suitable either for increasing the illuminating power of poorer gases, or being consumed alone, as also for the value of its coke.

Another description of cannel gives 13,600 feet of 27·75 candles, with practically the same weight of coke as the preceding, but which, on account of the large quantity of ash combined therewith, is useless either for heating the retorts or for sale. Hence the difference of these two kinds of coal consists not only in the increase in the quantity and quality of the gas produced by the former, but also the value of its coke, and in the absence of which fuel would have to be purchased for carbonizing. Again, there are other kinds yielding an average quantity of gas of variable qualities, with a large percentage of coke, containing a minimum of ash, which are well adapted for all localities, where that residual commands a good sale. Lastly, in this list we find a cannel yielding but 8000 feet per ton, and giving no coke whatever, which material, under certain conditions, would be literally "dear at a gift;" for if we consider the expenses of carriage, and, in some instances, the Customs duties paid, it is evident that the first cost of coal enters for only a limited portion of its expense when delivered at the works; consequently, under these circumstances, only the best quality alike for the production of gas and coke is desirable, and thus, under certain conditions, a cheap inferior coal would be valueless.

In the other table furnished by Mr. Paterson, the great discrepancies referred to do not exist; but, like the other, it serves as a reliable guide in the choice of any coal in all the points of production and illuminating power of the gas, and the weight of coke obtained per ton, together with the percentage of ash, and a knowledge of the expense of purification, all points of great value; and from these tables the practical man is enabled to value approximately any coal on the list, and on ascertaining its price from the merchant, can determine whether it answers his views or not.

By the kindness of Mr. Hislop, of Paisley, we are enabled to present the following table of analyses of various Scotch cannel coals, which, in addition to the ordinary general information therein conveyed, embraces the percentage of carbonic acid and sulphuretted hydrogen in the foul gas, as also the sulphur in pounds contained in the volatile matters as well as in the coke for each ton of coal carbonized, and consequently it may be regarded as one of the most complete analyses yet published. And we may observe that the value of the communication is considerably enhanced, in consequence of the difficulty experienced in obtaining such information from various other quarters, and the readiness with which it was contributed by Mr. Hislop.

TABLE OF RESULTS OBTAINED IN THE PRACTICAL ANALYSIS OF VARIOUS SCOTTISH CANNEL COALS.
By G. R. HISLOP, ESQ., OF PAISLEY.

		Purified Gas per Ton of Coal, Cubic Feet.	Specific Gravity of Gas Air 1000.	Illuminating Power of Gas in Sperm Candles of 120 Grains.	Value of One Cubic Foot of Gas in Grains of Sperm.	Value of Gas from Ton of Coal in lbs. of Sperm.	Carbon and Hydrogen in Foul Gas per cent.	Weight of Coke in lbs. per Ton of Coal.	Carbon in Coke per cent.	Sulphur in Volatile Matters from Ton of Coal in lbs.	Sulphur in Coke from Ton of Coals in lbs.
Lesmahagow Main Coal	Lanarkshire	12,420	688	34·25	824·24	1458·88	5·25	1050	86·80	13·66	10·75
Dykehead, Airdrie	Ditto	13,126	716	38·71	929·04	1742·08	2·75	1109	84·50	9·40	6·27
Chapleside, ditto	Ditto	13,265	672	35·86	860·00	1630·22	2·25	973	81·11	3·38	2·22
Haywood, Carnwath	Ditto	11,706	596	30·55	733·20	1226·10	3·15	1135	82·80	17·35	5·15
Wilsontown, ditto	Ditto	11,120	587	29·04	696·96	1007·17	2·80	1279	92·20	7·84	3·36
Clengh, ditto	Ditto	11,694	586	28·15	675·60	1150·80	3·00	1193	88·60	10·75	5·82
Benlar, Whitburn	Ditto	10,640	581	28·06	673·90	1024·30	5·00	1385	58·20	16·55	9·65
Crofthead, West Calder	Ditto	10,535	516	25·32	607·68	914·56	3·35	1234	91·40	5·37	2·91
Kirkwood, No. 1, Coatbridge	Ditto	11,615	593	30·04	722·10	1196·20	3·50	1002	87·40	8·06	4·92
Ditto, No. 2, ditto	Ditto	10,538	578	28·41	681·80	1026·40	2·75	1062	87·20	8·51	3·36
Tonnochside, Bellahill	Ditto	10,913	576	28·32	679·68	1039·02	3·75	1250	82·80	10·30	6·27
Brodisholm, Baillieston	Ditto	10,859	564	27·86	668·60	1037·25	4·50	1245	79·80	13·67	4·92
Bracheed, Old Monkland	Ditto	10,033	570	27·06	648·40	930·90	4·30	1267	85·80	16·53	8·33
Allanton, Hamilton	Ditto	10,561	518	22·32	535·70	808·20	4·75	1272	92·45	19·04	5·37
Shieldmuir, Motherwell	Ditto	9,520	536	23·13	555·10	751·06	4·50	1242	91·80	8·56	5·82
Muirkirk, No. 1	Ayrshire	12,160	609	32·54	781·20	1356·60	2·91	1125	95·20	7·87	3·55
Ditto, No. 2	Ditto	10,658	588	28·34	680·16	1035·60	4·50	1165	89·40	10·53	4·70
Cairntable, No. 1	Ditto	11,480	602	32·83	787·90	1232·20	3·00	1164	94·04	2·68	4·03
Lesnemark, New Cumnock	Ditto	10,602	574	29·23	701·50	1062·30	3·50	1241	89·50	22·84	9·41
Knockshinnock, ditto	Ditto	10,290	540	25·55	613·20	805·27	5·00	1281	93·80	18·76	6·49
Niddry, No. 1, Portobello	Edinburghshire	11,615	634	30·62	734·88	1219·37	4·25	1087	92·80	15·23	7·17
Ditto, No. 2. ditto	Ditto	11,650	572	27·34	656·16	1052·05	4·50	1139	85·80	15·94	7·13
Kirkness	Fifeshire	13,825	610	32·13	771·12	1522·96	3·69	1034	75·10	12·11	8·28
Old Wemyss	Ditto	13,220	612	32·52	780·48	1482·28	4·30	1036	73·50	16·35	7·61
Inchgall	Ditto	13,590	656	32·72	785·20	1518·42	3·33	1026	74·20	10·52	5·60

NOTE.—The above table exhibits the results obtained from a few of the principal cannels out of upwards of 120 kinds from different districts in Scotland. The market value of cannels is very largely affected by the quality of the coke which they afford, and a large proportion of the above number contains from 80 to 85 per cent. only of carbon, which limits their use, and in the manufacture of gas renders an admixture of such as afford a superior class of coke for firing purposes indispensable, and which has the effect of greatly enhancing the price of the latter.

We regret our inability to furnish a list of recent experiments on Newcastle coals, equally complete with those contributed by Messrs. Paterson and Hislop; but as no additional information has been obtained since the following coal analyses were published in the 'Journal of Gas Lighting' some years ago, we believe that their great utility will justify their reproduction at the present time.

ANALYSES OF GAS COALS BY LEWIS THOMPSON, ESQ.

Name of Cannel	Volatile Matter.	Coke.	Ash per Cent.		Sulphur.		Volatile Matter.	Specific Gravity of Coal.
			In Coal.	In Coke.	Coal.	Coke.		
Arniston	45·5	54·5	4·18	7·66	1·70	·95	·75	1·196
Boghead	68·4	31·6	22·8	72·15	·53	·08	·45	1·221
Capeldrae	54·5	45·5	10·5	23·07	·65	·20	·45	1·227
Kirkness	60·	40·	13·5	33·75	1·40	·58	·82	1·208
Knightwood	48·5	51·5	2·4	4·66	1·10	·61	·49	—
Lesmahagow	49·6	50·4	9·1	18·05	2·23	1·14	1·00	1·222
Lochgelly	33·5	66·5	13·1	19·7	·75	·25	·50	1·320
Old Wemyss	52·5	47·5	15·1	31·78	1·30	·60	·70	1·325
Wigan	37·	63·	3·	4·76	1·25	·60	·65	1·271
Ramsay's Newcastle	36·8	63·2	6·6	10·44	1·75	·94	·81	1·290
Band of Cannel in Leverson's Wallsend	30·8	69·2	9·35	13·51	1·00	·50	·50	1·320
Band in Pelton Main	31·5	68·5	9·4	13·72	·95	·49	·46	1·320
Band in Washington Coal	27·4	72·6	9·37	12·9	1·10	·56	·54	1·326
Staffordshire	50·	50·	2·9	5·8	1·30	·52	·78	1·220
New Brunswick	66·3	33·7	7·6	1·78	·7	None	·07	1·098

ANALYSES OF BITUMINOUS COALS.

Name	Volatile Matter.	Coke.	Ash per Cent.		Sulphur.		Volatile Matter.	Specific Gravity of Coal.
			In Coal.	In Coke.	Coal.	Coke.		
CHESHIRE—Hardcastle	31·5	68·5	5·	7·3	2·10	1·10	1·00	1·230
CUMBERLAND:								
No. 1	25·2	74·5	2·1	2·81	1·30	·70	·60	1·249
No. 2	25·6	74·4	1·4	1·88	1·10	·60	·90	1·275
No. 3	30·9	69·1	4·	5·8	1·70	·80	·90	1·200
DERBYSHIRE—Staveley	40·9	50·1	2·7	4·57	1·20	·80	·40	1·275
GLOUCESTERSHIRE:								
Coalpit Heath	30·1	69·9	5·8	8·3	4·10	2·20	1·00	1·370
Whitecroft, near Lydney	34·3	65·7	11·1	16·89	3·10	1·90	1·20	1·401
LANCASHIRE:								
Arley	33·7	66·3	3·6	5·43	1·20	·60	·60	1·270
St. Helens	37·2	62·8	1·2	1·91	1·10	·54	·56	1·285
NORTHUMBERLAND AND DURHAM:								
Blenkinsopp	38·	62·	5·1	8·22	1·20	·80	·80	1·298
Dean's Primrose	29·5	70·75	2·4	3·4	1·40	·71	·69	1·261
Garesfield (Butes)	28·3	71·7	3·2	4·46	·90	·40	·50	1·290
Ditto (Cowan's)	29·4	70·6	·95	1·34	·85	·40	·45	1·259
Giford	35·	65·	1·	1·54	1·10	·50	·60	1·200
Hastings Hartley	36·6	63·4	2·	3·15	·95	·50	·45	1·278
West Hartley	35·8	64·2	4·7	7·3	1·10	·60	·50	1·260
Leverson's Wallsend	39·4	65·1	4·9	7·52	1·30	·05	·65	1·278
South Peareth	27·8	72·2	1·8	2·5	1·20	·60	·60	1·266
Pelaw Main	30·3	69·7	2·6	3·73	1·20	·70	·50	1·271
Pelton Main	28·4	71·6	1·41	1·90	1·10	·02	·48	1·270
New Pelton	30·2	69·8	1·75	2·5	1·10	·56	·54	1·265
South Tyne	36·3	63·7	3·9	6·1	2·10	1·10	1·00	1·339
Urpeth	27·8	71·3	1·35	1·89	1·00	·60	·40	1·271
Washington	31·25	68·75	2·2	3·2	1·30	·07	·03	1·260
SOMERSET:								
Nailsea	34·9	65·1	3·	4·6	2·85	1·50	1·35	1·312
Radstock	38·25	61·75	3·5	5·66	3·10	1·80	1·30	1·275
STAFFORDSHIRE—Apeldale	40·	60·	·75	1·25	1·80	·38	·42	1·267
Apeldale	38·5	61·5	1·9	3·1	1·50	·82	·68	1·307
Heathern	42·9	57·1	1·75	3·06	1·50	·70	·80	1·230
Silverdale	34·	66·	1·95	2·95	1·30	·70	·00	1·301
Woodshutts	40·2	59·8	1·22	2·04	·90	·54	·36	1·201

ANALYSES OF BITUMINOUS COALS—continued.

Name of Cannel.	Volatile Matter.	Coke.	Ash per Cent.		Sulphur.		Volatile Matter.	Specific Gravity of Coal.
			In Coal.	In Coke.	Coal.	Coke.		
WALES (NORTH):								
Ruabon, Top Yard Seam	37·5	62·5	2·5	4·	1·40	·80	·60	1·269
Ditto, Mean Coal	41·5	38·5	1·	1·71	·85	·45	·40	1·284
Ditto, Yard Seam	34·	66·	1·4	2·12	1·10	·60	·50	1·271
Ditto, Nant. Seam	37·9	62·1	1·4	2·25	1·10	·70	·40	1·269
WALES (SOUTH):								
Rhonda	22·8	77·2	2·7	3·5	2·30	1·20	1·10	1·278
Rhonda Low Main	23·1	76·9	2·1	2·73	2·20	1·10	1·10	1·280
YORKSHIRE:								
Elsecar Low Pit	37·	63·	1·1	1·74	1·20	·63	·57	1·258
Grigleston Cliff, soft	35·6	64·4	1·	2·48	1·40	·75	·65	1·255
Meriomly	37·	63·	1·6	2·54	1·10	·60	·50	1·220
Silkstone, No. 1	34·1	65·9	2·78	4·21	1·30	·80	·50	1·260
Ditto, No. 2	38·	62·	2·55	4·1	1·10	·60	·50	1·259
Ditto, No. 3	32·5	64·8	2·8	4·3	1·45	·75	·70	1·262
Soap House Pit	35·	65·	·8	1·23	·75	·40	·35	1·258
Woodthorpe	33·1	66·9	10·5	15·7	1·20	·70	·50	1·347
France Denaen (Valenciennes)	29·9	70·1	6·	8·56	2·40	1·30	1·10	1·265

From the preceding highly interesting tables we learn several important facts, amongst these that the quantities of cannel coals are exceedingly variable, both as regards the quantity of volatile matter they relatively possess, as well as the ash in their coke. For instance, according to the investigations of Mr. Lewis Thompson, Boghead contains 68 per cent., whilst Lochgelly possesses but 33·5 per cent. of volatile matter. Again, with the coke, we find in the former 72·15 per cent., whereas the Wigan contains but 4·76 of ash.

In the analysis of Mr. Paterson we find the ash of these coals to vary from about 3 to 69 per cent., whilst the samples examined by Mr. Hislop gave from 5 to 40 per cent. of the weight of the coke in ash. With these extreme differences, therefore, in the gas-producing elements of various coals, together with the amount of useless residue combined therewith, it becomes a question of the most vital importance to make the proper selection for the intended purpose.

Hitherto the specific gravity of coal has been generally regarded as an indication of its quality as a gas or coke producing agent, but on carefully examining the analyses of Mr. Thompson we learn the fallacy of this, and that no relation whatever exists between the specific gravity of the numerous kinds of coal and their constituents, which the following extract demonstrates most forcibly:

	Specific Gravity.	Percentage of Volatile Matter.
Boghead	1·221	68·4
Mortomly	1·220	37·0
Old Wemyss	1·325	52·5
Woodthorpe	1·347	33·1

Thus we find that Boghead, having practically the same specific gravity as Mortomly, contains nearly double the quantity of volatile matter of the latter; whilst the high specific gravity of Old Wemyss and Woodthorpe, together with the great difference in the volatile matter they respectively contain, is a further proof that specific gravity is no criterion by which coal can be valued.

On examination, the same want of uniformity, although to a less degree, will be found to prevail throughout the whole of the list, and although caking coals show a greater regularity in their constituents and specific gravity, yet the latter is no indication of the former. Thus the "Whitecroft" of Gloucester, with a specific gravity of 1·401, contains precisely the same quantity of volatile matter and the same weight of coke as Silkstone No. 1, having a specific gravity of 1·260.

These conclusions are further confirmed by the investigations and observations of Mr. Paterson; and we may repeat therefore, that the specific gravity of any ordinary kind of coal is no indication of its quality, and in separating that question from the theory of gas manufacture, that art is to some extent simplified.

During many years the discrepancies existing between the specific gravity of gas as compared with its illuminating power, according to previous experimentalists, were attributed either to the imperfect instruments employed at the period, or to a want of knowledge on the subject. These observations, however, cannot be applicable to the gentlemen who have favoured us with the preceding analyses, and yet in certain instances we find in these that the specific gravity does not always correspond with the illuminating power. This, however, is not surprising, inasmuch as coal gas is composed of a variable mixture of gases, having in some instances similar specific gravities but very different powers of illumination. Thus olefiant gas and carbonic oxide, possessing in fact the same specific gravity, are both contained in coal gas, but the first is one of the most powerful agents we can employ for the production of light, whilst for that purpose the last is absolutely useless. In proof of this, 9 of hydrogen and 1 of olefiant gas have the same specific gravity as 9 of hydrogen and 1 of carbonic oxide; but the first would be 22-candle gas, and the last would give no light whatever. These facts should be sufficient to prove that the specific gravity test cannot always be relied on; but, independent of these, we have to consider the degree of moisture in the gas by which its weight is increased, and the presence of atmospheric air which acts in two directly opposite ways, as it augments the specific gravity and diminishes the illuminating power of the gas it is combined with.

In addition to the numerous kinds of coals mentioned, there are other materials, as bituminous schists or shale, which are sometimes, but rarely, met with abroad, and may be employed for enriching poor gas, or as a substitute for coal when this cannot be conveniently obtained. Bituminous schists exist in many parts of the world, but some of these are worthless for the purpose in question, whilst others are very desirable. A description of this, very closely allied to Boghead in its composition, with about 70 per cent. of volatile matter, consisting principally of hydrocarbons, has recently been discovered in Australia, and there is the probability of this being introduced into the English markets. In the neighbourhood of Bahia, in South America, a similar material has also been discovered, but hitherto, we believe, has not been worked.

The writer has frequently met with various kinds of bituminous shale, some identical in appearance with that obtained in the neighbourhood of Poole, and formerly called the "South of England Boghead." A Continental company, allured by the name, once ordered a small cargo of this, but on submitting it to carbonization it was found to contain only about 30 per cent. of volatile matter, the residue being entirely worthless. Thus the fuel required to bring the material to a high temperature in order to expel the volatile matter, together with the labour and other expenses, were not compensated by the gas obtained, so that the mineral was never employed, but thrown away, and this after paying a heavy freight and customs dues for the same. Hence it is always unwise to purchase any description of coal without being first well assured of its quality from independent sources.

There are still other substances, as lignites and pitch, which bear upon the question under consideration, and are mentioned more as a caution than recommending their use.

A class of lignite, which strongly resembles coal, and by its appearance is calculated even to deceive practical men, exists in Spain and the island of Majorca. In the former place it is worked extensively and supplied as fuel for steam purposes, for which its low price renders it somewhat suitable. This material yields about 7000 feet of gas of low illuminating power per ton, possesses a large quantity of sulphur, and water in a hygroscopic state, the residual being ashes of a cubical form. An attempt, within our knowledge, was made on one occasion to employ this for the purpose of producing gas for supplying a town; but after being purified in the ordinary manner, the odour arising from the gas during combustion was such as to be unsupportable, even in open-fronted shops, and it required some time before the pernicious effects of one day's operations were entirely dispelled, and was likely to cause considerable prejudice to the company.

Pitch of various descriptions is open to the same complaint; that from Trinidad is well remembered, on account of the trouble it occasioned by a portion of it being intermixed with the usual coal carbonized, and was the source of incessant complaints for some time, although only a small portion was employed. Hence it is always desirable to make experiments on a small scale with any new material, and to consume the resulting gas in an apartment, so that any pernicious odour may be detected, and, above all, no cargo should be purchased abroad from any passing vessel which may be in the port, without first analyzing the coal.

It must not be supposed, however, that "true" coal is always to be recognized. A remarkable difficulty of this nature occurred some years ago in Scotland, and occasioned lengthened litigation and an expenditure of many thousands of pounds, without, however, arriving at any definite result. The facts were as follows:

Great expense having been incurred in sinking a pit in the neighbourhood of Falkirk for the purpose of winning coal, without success, the proprietor, after sustaining considerable loss, entered into an arrangement with a banker of the town mentioned, whereby the latter should continue the operations, and in the event of obtaining coal, that a small royalty per ton should be paid to the former. The banker entered with energy into the business, erected powerful engines and pumping machinery, and, at the end of a few months, had the good fortune to reach a seam of coal of a quality hitherto unknown, namely, the Boghead Cannel. The desirability of this for enriching poorer gases was speedily appreciated, and consequently it found a ready sale at a high price, which was subsequently much augmented on account of its adaptability for the production of "paraffin oil." Thus, whilst the banker was realizing a large fortune yearly, the original proprietor was in the receipt of only a moderate revenue.

Under these circumstances, science was resorted to, and in due course there were several chemists of eminence arraigned as witnesses, on both sides: the one side to prove that the mineral obtained was not coal, but bituminous clay; the other to support the interests of the banker, and to prove the mineral to be coal.

That it was not coal, amongst the other arguments adduced, were the absence of ammonia and the very small quantity of carbon remaining in the residual after distillation; and to prove it to be clay, evidence was given, that when two pieces of the mineral were rubbed together, or when breathed upon, the peculiar odour arising from dry clay under such circumstances was produced, and still further to convey the impression that it was clay, a celebrated chemist deposed that he "had made a brick of it." For this purpose, however, which he did not state, he must have pounded the mineral, and then compressed it into the mould, in the same manner that ashes or similar material may be formed into a block or brick. But it is certain that the assertions were not in strict accordance with the truth; nevertheless it was considered as evidence.

The supporters of the material being coal were chemists equally eminent as their opponents; and, after a lengthened inquiry, the jury took an ordinary common-sense view of the matter, and pronounced the mineral to be coal. The question was, however, far from being settled by this decision, for it was again renewed in England, and raised on the Continent; the various decisions terminating sometimes in favour of clay, at others of coal, and eventually a compromise was effected. If this difference of opinion therefore exists, it is obvious that, in some instances, it must be difficult to decide between a true coal and peculiar lignites, or bituminous clay, or shale. It may be observed, however, that the general opinion now entertained is that Boghead is a bituminous shale.

Whilst referring to this class of witnesses, many well remember the *evidence* of an eminent chemist, in which he stated that he had examined the gas supplied to a portion of the city, and ascertained, according to the experiments made, its illuminating power to be equal to one candle only per five feet. The assertion would appear most preposterous, since with gas of that quality a town would be almost in darkness, yet he told the truth so far as regarded his experiment. But the facts afterwards transpired, that on the day the observation was made, a main had been opened in the immediate vicinity of the premises where it was conducted, when of course the gas examined was intermixed with a portion of air, and thus its illuminating power was reduced to the extent stated. But the witness literally shut his eyes to the quality of the gas in the adjoining streets, which was of 12 or 14 candles, and restricted his evidence to his experiment. And it is to be regretted that gentlemen of science, who from their position and knowledge inspire confidence, and disarm all suspicion and doubt, should resort to subterfuges like that mentioned; for although the witness in question pretended his experiment was made with the gas, this was not literally true, and as he knew the deteriorating power of air, it was the less excusable. In all such cases neither the employer nor employed can make any pretensions to a high standard of morality.

Caking coal, but particularly the slack, is deteriorated alike for the production of gas and coke by protracted storage, and when exposed to the influence of sun and rain its value is diminished in a very marked manner, the loss under these conditions being considered equal to about 12 per cent. during the first three months' storage, but when under cover its depreciation may not reach that extent even at the end of a year. Coal, after being in store a long period, yields a coke of diminished size, and less gas; the fact is remarkably striking on the arrival of a fresh cargo, after having used coal which had been stored for several months.

The vast importance of employing fresh coal is experienced at Beckton Works, where, with all the advantages of its being transported by steam vessels, and the prompt method of loading and discharging, the coals are probably carbonized in some instances within a week of their being extracted from the mine, and under these circumstances the average yield from Newcastle coal, with 7 per cent. of cannel, is no less than 10,334 feet per ton. This extraordinary production might be supposed to arise from a defective indication of the station meter; but that this is not the case is clearly proved by the limited amount of leakage, or unaccounted-for gas, which is about the average of the other metropolitan companies. Therefore, in the absence of other more forcible reasons for this high production of gas per ton, it must be attributed to the speedy carbonization after the coal is extracted from the mine, and probably to some extent to its freedom from humidity.

Bearing on this subject, we find the following remarkable statement in the 'Journal für Chemie:' "M. Penot has demonstrated the great advantage derived by carbonizing dry coal in the manufacture of gas for illumination. The quantity of olefiant gas from wet coal, as compared with that from dry coal, is as two to three. M. Schwartz has confirmed the exactitude of these experiments on an extensive scale at the gasworks at Mulhouse." Hence, if this be correct, and if, as we are informed, some managers wet their coals in order to increase the bulk of coke, the slight advantage must be acquired by a considerable diminution of the illuminating power of the gas.

Hence it follows, that when gasworks are situated abroad, the supply of coal should be, as far as practicable, regular and uniform throughout the year, according to the consumption, of course calculating against the contingency of the loss of vessels. The supposed advantage of purchasing for several months' or a year's consumption, in order to profit by some economy of freight, is often fallacious, since the difference of the cost of freight will be more than counterbalanced by the deterioration of the coal; in such cases, therefore, instead of economizing, the directly opposite result is obtained.

Caking coal, when stored in confined places, is sometimes liable to spontaneous combustion. This is particularly the case when stowed away in a humid state, and abounding in iron pyrites; for these reasons stores should be properly ventilated, and when by accident a cargo is received in a wet condition, it should be employed as speedily as possible in order to prevent its ignition or heating.

CARBONIZATION.

Gas for the purpose of illumination is usually obtained by the destructive distillation, or, as generally termed, carbonization of coal; but in certain localities where the cost of that material precludes its use, other substances, such as oil, tallow, rosin, wood, petroleum, &c., are employed. As each of these substances however, demands a different treatment, and necessarily a separate description, and as their usage is very limited, our observations will therefore be confined to coal for the object under consideration.

The great desirability of coal for our purpose consists in the quantity and quality of the gas derived therefrom; the simplicity of the operations in manufacture, and the value of the residuals, particularly the coke, which, after furnishing the necessary heat for carbonizing the remaining portion, in some instances realizes more than the cost of the material from which it is derived.

On the proper carbonization of the coal mainly depends the prosperity of a gas company, and when the operation is not carried out with all due regard to economy, the success of any undertaking must suffer accordingly; for in this, in a great measure, consists the economy of coal used in the production of gas; the quality of that gas as an illuminating agent, and its comparative purity; also the quantity and quality of coke available for sale as a residue; the expenses of labour; and, lastly, the wear and tear of retorts and settings; all points for the most serious reflection of the gas manufacturer, and on which depends the question of profit or loss to a gas company.

For the object in question, the most important consideration is the degree of heat to which the coal is submitted during the process of carbonization; for should this be extremely low, then tar will be produced in abundance, with but a small quantity of gas. And in the event of the heat being too high, then a portion of the rich hydrocarbons in the gas will be decomposed, and their carbon deposited in a solid state on the interior of the retorts, thereby reducing the illuminating power of the gas. And when the operation is too protracted, then by retaining the coke within the retorts after the useful illuminating gas is evolved, a most troublesome and obnoxious impurity is formed by the sulphur in combination with the coke; and, as our every-day experience teaches us when burning that fuel in private dwellings, this sulphur, with its pernicious qualities, is obstinately retained by the coke to the last.

The means of estimating the various degrees of high temperature by pyrometers, as at present constructed, can hardly be regarded as satisfactory, consequently this most important question in carbonization becomes, to a certain extent, a matter of conjecture. We must, therefore, content ourselves by following such authorities as Kahn and Becquerel, by adopting their description of the various degrees of heat by their colours, together with the melting points of some of the metals, as represented in the annexed table.

600° Fahr.,		faint deep red, just visible in a completely darkened room.
662°	„	Mercury boils.
810°	„	Antimony melts.
980°	„	Faint red, seen under ordinary daylight.
1290°	„	Dull red; colour of red oxide of iron.
1470°	„	Bright red; colour of red oxide of lead.
1650°	„	Cherry red.
1830° Fahr.,		Bright cherry red.
1869°	„	Brass melts.
1873°	„	Silver melts.
1996°	„	Copper melts.
2010°	„	Dull orange.
2190°	„	Bright orange.
2370°	„	White heat.
2550°	„	Bright white heat.
2786°	„	Cast iron melts.

The practical man, by his continued experience, is able to determine by the eye, with tolerable accuracy, the principal degrees of temperature referred to in the table; and any person, on observing attentively a good ordinary house fire, will be enabled to distinguish six or seven of the distinct shades or degrees of heat, ranging from faint red, or 980°, which is discernible in ordinary daylight, to dull orange, 2010°, and sometimes bright orange, 2190°, according to the temperature the various parts of the fuel have acquired, which observation we recommend, in order to understand the different degrees alluded to in the foregoing table.

With iron retorts, the colour of the various degrees of heat is easily appreciated; but with those of clay, when viewed by daylight, the lower degrees are more difficult of being distinguished, in consequence of the light colour of the material composing the retort interfering with the colour of the "heat." However, when the temperature becomes elevated, approaching towards the proper point for carbonization, this impediment no longer exists.

When coal is carbonized at a temperature corresponding with 600°, a degree of heat which chars paper but does not inflame it, and is only visible in a completely darkened room, and then merely as a faint obscure red, its volatile parts are nearly all converted into liquids, as tar, or oil, and water, with the evolution of only a small portion of gas, equal to a few hundred feet per ton of coal, the coke remaining in the retort being of a soft, loose, friable nature; retaining a quantity of volatile carbon, which cannot be expelled at that low temperature. By this process the oil, or tar, from Boghead and other cannels, coals, shales, and similar bituminous substances, is obtained, which is afterwards purified and sold as paraffin oil.

In proportion as the heat of the retorts is augmented, so is the quantity of gas from the coal increased; thus at a temperature approaching to 980° the yield from a ton of Newcastle coal is about 6000 feet. At this heat, there is no material accumulation of carbon on the interior of the retorts, and only the slightest deposit of carburet of iron on their exterior, even after fifteen months of continuous operations; moreover, iron retorts, under such circumstances, undergo but little deterioration, as at the end of the period mentioned, when taken down, they are almost in as good condition as when first they were set. These facts came under the writer's notice at new works, where the settings were so wretchedly defective that a higher degree of heat than that stated could not possibly be attained, consequently a large portion of the coal was converted into tar, which in the locality was an unmarketable commodity, whilst nearly the whole of the coke produced was required for heating the retorts, and as a natural consequence the company sustained heavy losses, instead of realizing good profits.

If the temperature of the retorts be further augmented, we obtain an increased quantity of gas, with superior coke and less tar, and when about 1470° are reached, then a quantity approaching to 8000 feet per ton of Newcastle coal is produced, and this gradually increases, with a slight diminution in quality, until arriving at the point at which copper melts, 1996°, or dull orange, 2010°, which, in our opinion, is the best heat for the production of gas and coke from caking coal, when the "make" averages about 9600 feet per ton. We are aware that some engineers prefer a higher degree of heat, by which the coke becomes harder and brighter in appearance, and the change is worked off with greater rapidity; but our impression is that these advantages are obtained at the cost of the quality of the gas.

At all temperatures above 980° a portion of the rich hydrocarbons is always decomposed, and their carbon deposited on the interior of the retorts in a solid state, which deposit is directly in proportion to the temperature of the retorts, increased according to the pressure of the gas therein. On withdrawing the pressure by an exhauster, or other means, the evil, so far as regards that, is removed; but there still remains the accumulation of carbon occasioned by the high temperature, and the higher this is the more rapidly is the carbon deposited, and the quality of the gas deteriorated in proportion. It has been supposed that this incrustation in clay retorts was effected by the material itself, or by its rough surface; this, however, is not the case, but is simply the effect of high heat and pressure, acting alike, under like conditions, on iron as well as clay retorts, and would undoubtedly occur in a retort of glass, were it capable of attaining the necessary degree of heat and retaining its form.

This accumulation of carbon in retorts is prejudicial on several accounts: firstly, presenting as it does a large quantity of non-conducting material within them, which prevents the heat penetrating to the coal undergoing carbonization, thus increasing the quantity of fuel for the furnaces; secondly, the space occupied by it, when attaining any degree of thickness, of necessity diminishes the capacity of the retort, so increasing the expenses of wear and tear and labour; thirdly, the coke, when produced in contact with such accumulation, is always smaller, and of course less valuable, than when obtained in clean retorts. Another important evil is the deterioration of the gas from which it is derived.

We have shown that, under extraordinary circumstances, the manufacture of gas can be carried on without such incrustation on the retorts; but, for profitable and economical operations, it is an inevitable evil; it therefore becomes an imperative duty to employ all proper means to reduce this to a minimum, by the exhauster or other methods hereafter indicated.

From experience we have found that if this incrustation of carbon be allowed to accumulate beyond a certain limited extent, it then increases with remarkable rapidity; this was particularly conspicuous at the period when iron retorts were used, and before the employment of the exhauster. Then with new retorts it required four or five weeks to produce any material quantity of carbon within them; but this once formed, it speedily augmented in a most extraordinary manner, so that often, after retorts had been in use three or four months, they were literally half filled with carbon. And to this deposit was mainly due the rapid destruction of iron retorts.

Under the conditions of high heat and high pressure this accumulation takes place, and the richer the coal the more rapid is the deposit: it follows therefore that it should be removed before attaining any degree of importance, and which is readily done with atmospheric air. For this purpose, when the incrustation acquires a thickness of about three-quarters of an inch, which generally occurs, under ordinary conditions of heat and pressure, in about four or five weeks, the retorts are left without charging for about twelve hours, the average time necessary for cleansing them, the damper being partially closed, and the heat kept up as usual. The doors are then placed, of course unluted, leaving a space of about half an inch between the door and mouthpiece, when a current of air passes in at the bottom of the retort, the oxygen of which combines with and dissolves the incrustation, and a current of the gases produced passes off at the upper part of the mouthpiece; and frequently particles of carbon in an incandescent state are emitted. By these means the retort is cleared out, and rendered as free as the first day it was fixed. During the operation, occasionally the caps of the ascension pipes are taken off, so that the current of heated air, in passing through, cleanses them. And when the retorts are of good first-rate quality, there is not the slightest fear of either crack or fissure throughout their length, even at the end of two or three years' operations, supposing the cleansing to be periodically effected as described and the scurfing bar avoided.

Some engineers may object to the supposed loss of a few hours' inaction of the retorts by the periodical cleaning; but this loss is completely imaginary, for by the system recommended they are kept constantly practically clean, thus avoiding the serious disadvantages enumerated arising from an accumulation of carbon, and a positive and most important gain is effected.

Another method of removing this carbonaceous incrustation is, according to Edge's patent, by the application of high-pressure steam, admitted by a tap, and conducted by a cast-iron pipe to that part of the retort where the greatest accumulation exists. The steam rushes into the retort, and carries with it a portion of air, and by their combined action the carbon is speedily dissipated. Our experience does not extend to this means of scurfing retorts, but we are assured by engineers who have had the process in action for some years, that it leaves nothing to be desired, as by it a coating of carbon an inch thick can be removed easily in three hours. In practice, however, it is applied for about an hour once in a month to each retort, this being sufficient to keep them perfectly clean, and is done at a nominal cost, with very little trouble, and we believe the process requires only to be more generally known to be thoroughly appreciated.

In many gasworks the carbon is permitted to accumulate on the interior of the retort until attaining a thickness of from 1 to 2 inches, which in ordinary working occurs in about three or four months, when, with a chisel-bar, it is broken and detached from the retort; but although this is practised in many extensive establishments, it is decidedly erroneous, injurious, and unprofitable; and, as we shall hereafter show, first-class clay retorts, set in the most economical and advantageous manner, are not able to resist the action of the chisel-bar, but are often rendered utterly useless by its employment.

From the preceding observations it is obvious that all unnecessary pressure should be removed from the interior of the retort during the process of carbonization, which is usually accomplished by means of the exhauster; but in works of very limited capacity, where the cost of this apparatus and the continued attention it demands are important considerations, other methods must be adopted. This may be accomplished by having the gasholders, either in their first construction or by the application of counterbalance weights, so light as not to give a greater pressure than 2 inches; which effected, and the dip in the main not exceeding half an inch, the purifiers and other apparatus on the works being of ample size, the pressure on the retorts will not exceed 3 or 4 inches. This attained, and excessive heats being avoided, the deposit of carbon under these circumstances will be inconsiderable. We assume, in this case, that the mains are of sufficient capacity, so that a pressure of $1\frac{1}{2}$ inch at the works is capable of giving a good and proper supply to the town, and under ordinary conditions that pressure should not be exceeded.

The exhauster is usually fixed after the condenser, and is provided with a pressure gauge, and in some cases an exhaust register, in communication with the interior of the hydraulic main, which serves to indicate the amount of "exhaust" therein. The term exhaust is applied in contradistinction to pressure; the word vacuum (sometimes employed) is incorrect, and partial vacuum is too lengthy and not sufficiently comprehensive.

In modern works the seal of the dip-pipe seldom exceeds an inch, in which case the gauge of the exhauster may be worked at "level gauges," thus relieving all the pressure from the retort, except the inch seal referred to. In other cases the dip may be of several inches, when the exhauster will have to be worked in a corresponding manner; thus, if the dip be 6 inches, then, to have only 1-inch pressure on the retort, the exhauster would have to be worked to indicate 5 inches exhaust on the gauge; and only by the accurate knowledge of the depth of the seal can the action of the apparatus in question be assured.

Caking coals intended for gas making should always be well protected from the rain, otherwise they are deteriorated considerably, alike for the purposes of gas and coke; the first being diminished in quantity and quality, and the latter in size; and when charged in a wet state the steam produced carries off a considerable quantity of heat, so that an increase in the production of tar and diminution of the gas is the consequence. According to the late Mr. Bannister, whose experience and intelligence are well remembered, the carbonization of wet coal is one of the causes of the production of naphthaline. It has been asserted that in the distillation of wet coal a large quantity of hydrogen is produced, the volume of gas being consequently increased and its illuminating power diminished. This opinion, however, we believe to be founded on error, as it is evident that at the commencement of a charge the heat of the retort is not sufficiently high to decompose the steam arising from the wet coal.

Cannel is not materially influenced by exposure, and possesses the advantage that it absorbs but little humidity, and readily dries when placed before the benches, and in a short time is suitable for charging; slack coal, on the contrary, retains the moisture for a considerable time.

When caking coal is undergoing the process of carbonization, supposing the charge to be of six hours, the gas is emitted during the first hour without the form of the various pieces being materially changed. The coal afterwards commences to melt, and subsequently to intumesce or boil, ejecting the gas at every bubble, thus occasioning the spongy nature of the coke. During this period it remains a black, doughy mass, which state it retains until nearly all the gas is expelled. At the commencement of the sixth hour little gas is evolved; the mass is still black and begins to agglomerate or crystallize (if we may use the term), when it increases in temperature until it becomes red hot, and the coke is more or less brittle according to the temperature it acquires.

This slow action is due to the fact that coal is an extraordinary non-conductor of heat; for such is its nature in this respect that it would be possible to heat the exterior of a mass to the fusing point of

FUEL FOR CARBONIZING.

iron, whilst the interior of the coal would not be hot enough to cook an egg, and if a comparatively small fragment of it be exposed to a bright red heat, its interior, for some time, will not be materially increased in temperature. This is a quality possessed by coal which renders it so desirable as fuel, as by this property the combustion and heat are gradually produced. In the operation of carbonizing the conducting power of coal is but slowly increased in consequence of the volatile gas passing off at the comparatively low temperature of about 160° Fahr., carrying with it the heat communicated to the coal; and only towards the end of the charge, when the useful volatile constituents are expelled and the coke increases in temperature, does the remaining gas attain a much higher degree of heat. This action is very similar to water dropping on condensing pipes, when the steam generated carries off the heat from the passing gas.

There are a few cannels, amongst them the Wigan and Ramsay's Newcastle, which yield a coke identical in appearance to that of caking coal; although somewhat smaller than the latter, it is well suited for heating the retorts, and is a good marketable article.

The coke derived from the generality of Scotch and most other cannels has the form of the coal prior to its carbonization, and is compact and flaky; some descriptions contain a large quantity of ash, which renders them quite useless as fuel, whilst there are others having 10 or 12 per cent. of ash, and are suitable for heating the retorts; but of these a large quantity is required for the purpose, as compared with coke derived from caking coal.

By the comprehensive tables of analyses, under the head of "Coal," we learn the quantity of ash existing in combination with the various kinds of coke; but engineers differ in opinion as to the value of this when containing a large percentage of ash. Thus, according to Mr. Whimster, of Perth, coke possessing 20 per cent. of ash is commercially worthless; other engineers are of opinion that when having a larger percentage it is a marketable commodity, and according to Mr. Hislop, when containing 50 per cent. of ash, it is used as fuel by the poorest class, to whom it is given gratuitously.

In many of the works in Scotland, as well as the North of England, where cannel is exclusively used for producing gas, the retorts are heated by an inferior description called "canal coal," similar to ordinary caking coal. For this purpose the furnaces are required to be constructed somewhat dissimilar to those in which coke is employed, as, when heating by coal, the flame is a most important agent, and a proper supply of air is necessary, in order to prevent the volatile carbon passing off as smoke.

One of the principal mediums of profit to a gas company consists in the returns derived from the sale of coke; it therefore becomes the duty of the manager to exercise all care in order that the carbonization shall be effected with due regard to economy of fuel, and thus augment as much as possible the revenue from that source. To attain this object good settings of retorts are indispensable, seconded by careful management.

The average production of coke from caking coal is about 68 per cent. by weight; thus a ton of coal yields approximatively 13½ cwt. of coke, which is equal to from 33 to 40 bushels, according to its state of solidity. The average usually adopted is 36 bushels per ton of coal, but the latter quantity is often obtained; it follows, therefore, that a correct estimate of the coke used for fuel, without having a correct knowledge of the volume yielded, cannot be made.

The fuel for heating the retorts or carbonizing is invariably estimated at the proportion it bears to the whole of the coke produced. In gasworks of limited capacity, supplying villages or very small towns, all this is required for heating the retorts; and in some cases it is insufficient for the purpose, necessitating the purchase of other fuel. But as gasworks increase in magnitude, their position in this respect becomes more profitable; so that an establishment having the settings properly constructed, and economically managed, with a production of two millions and a half feet per annum, should dispose of 25 per cent. of the coke derived from carbonization; and the sales of works making twenty millions ought to be 50 per cent., and in another of one hundred millions, these should not be less than 70 per cent., of the coke produced.

It would be supposed that in works of the first magnitude, in consequence of the extent of their operations and the concentration of the heat, the carbonization, so far as regards the fuel required, would be effected more economically than in smaller works—say those producing a hundred million feet per annum and upwards. This, however, is not always the case, for the operations of medium-sized works, in this respect, are frequently more favourable and profitable than those of larger capacity, which we believe to be principally dependent on the kind of settings of retorts. The true test of good working and economy in fuel is the money test—viz. the amount received for coke per ton of coal carbonized. Necessarily, the returns from this source must always vary according to local circumstances, but a great deal depends on the management.

In small works the coke is drawn from the retorts at the end of a charge into iron barrows, and is wheeled away to the yard or spreading floor. In larger establishments a line of railway usually extends throughout the length of the retort houses, close to the benches, and on this railway run iron waggons of a capacity suited to receive the coke produced in the respective settings. The largest of works have invariably the coke-hole hereafter mentioned.

An essential in carbonization is the means of obtaining the desired degree of heat for the retorts. For this object a draught to furnaces is indispensable, and whenever the flame issues from the sight-holes of a setting, either the damper is too much closed, or the lateral flue may be obstructed. Should the whole of the beds be so affected, the main flue may be stopped or too small, or the stack or shaft not of sufficient capacity. It is essential also to the production of good heat that the firebars of furnaces should always be clear, so as to allow a free passage of air to the fuel. When ordinary good coke is used, the slag or clinker in the furnaces requires to be removed or clinkered about once in every twelve hours; besides this, the bars are cleared underneath at intervals by a suitable rake.

For the purpose of economizing fuel in the furnaces, the proper control of dampers is indispensable; and if this be neglected, strict economy cannot be attained. The damper should be so adjusted as to give the required draught and no more, for with any excess of this a waste of fuel is the consequence. To attain perfection in this respect the damper should be under the control of the workman in charge, so that whenever desirable, whether for stopping operations on Sundays or otherwise, it can be brought immediately and effectively into operation.

At the first establishment of gas lighting, it was an accepted axiom that, whenever the price of coal was high, it would be compensated by a corresponding value being obtained for the residual coke. This, however, as experience proves, does not always hold good, as the value of that article is influenced by innumerable circumstances; nevertheless, in all cases, wherever caking coal alone, or with a small admixture of cannel, is used for producing gas, except in very small works, the revenue from the sale of the residual coke, after heating the retorts, should always contribute largely towards the cost of the coal carbonized.

Different kinds of coal, whether cannel or caking, under like conditions of heat, capacity of charge, and size of retort, deliver their gas with variable facility. Cannel, as a rule, is easier distilled than caking coal, the former giving off its gas in about one-fourth less time than the latter.

The time required for any ordinary charge of coal, no matter of what description, is controlled by the heat of the retorts. A given quantity, say of Newcastle, when submitted to the action of a temperature somewhat above cherry red, and requiring six hours for carbonization, would be effectually distilled in five hours at dull orange, and in four hours when the heat is slightly above bright orange; and undoubtedly this rapid production at high heats is one of the causes of the system being favourably regarded by engineers.

Excessive charges in the retorts are conducive to low production of gas. At the closing of the door of the retort the gas should inflame; when this is not the case, and the products issue as a black smoke, the retorts are either too low in temperature, or the charge is too heavy.

Of the various kinds of caking coal the most stubborn in yielding its gas is that which is composed nearly all of slack, like some of the Newcastle coals; but these possess the advantage of yielding a superior coke, more particularly on account of its freedom from iron; consequently less slag is produced, and less labour is necessary in clearing and clinkering the furnaces. In addition to the excess of clinker produced, the presence of iron in the coke detracts very much from its appearance for sale, particularly after it has been exposed to the rain for any time.

When tar does not find a ready market it can be advantageously employed for heating retorts, thus replacing coke entirely for that purpose. In some cases, and which is recommended, the setting is built expressly, similar to those of five retorts in a bench, Plate 5, without the arch over the furnace, of which other drawings are given. This being necessary on account of the tar acting as a flux on the brickwork, but more particularly on the arch; some engineers employ the tar only in old settings; but in both cases, the method of arranging and construction is alike.

The best method of employing this fuel is represented in the annexed figures, the one being in elevation, the other in section.

The elevation, Fig. 4, indicates the position of the furnace door when coke is used, and if we suppose the furnace required to be changed in order to burn tar, this would be bricked up as shown, but

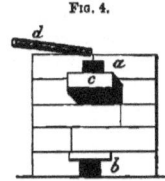
Fig. 4.

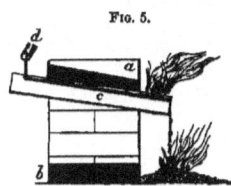
Fig. 5.

leaving the orifice a for conveying the tar, water, and air, into the furnace, and the orifice b at the bottom for the admission of air to support the combustion of that portion of the tar which falls from the gutter, both openings being about $2\frac{1}{4}$ inches square.

A channel of 2-inch angle iron, or what is better, a piece of fire-slab c, having a gutter made therein, of sufficient length to project about 3 inches within the bench, and a corresponding distance outside, is placed at an angle as represented in section, Fig. 5. The firebars being removed, the ash-pan is filled to the level of the bottom of the orifice a, with foul lime or other rubbish, on the top of which is placed a layer of breeze.

The tar for fuel should be taken direct from the hydraulic main, so that it always remains fluid, and prevents the repeated interruptions, either by clogging, or by the presence of any dirt causing obstruction, which frequently occurs when it is pumped, or carried from one vessel to another; moreover, by these means a saving of labour is effected.

From the lower part of the hydraulic main, about one inch from the bottom (in order to avoid the thick mass which sometimes forms there), an inch service-pipe is carried along the front wall and close to the top of the beds of retorts, leaving T-pieces by the side of the buckstaves wherever the supply is desirable; and from each of these T's descends a $\frac{1}{2}$-inch or $\frac{3}{4}$-inch pipe, provided with a lantern-cock on the end, this being situated about 5 feet from the ground—a height the most convenient for a man to be able to see whether the supply is regular.

About 4 inches below this lantern-cock is a $\frac{3}{4}$-inch iron pipe d, bent and cut to the proper length, so as to deliver the liquid into the gutter of slab c, a funnel of about 3 inches in diameter being placed at the top of this pipe to receive the supply of tar and water. A similar pipe is placed along the retort benches for conveying a small supply of water, having its outlets by the side of those of tar; the descending pipes, corresponding in length with the others, have each a small tap with a short piece of $\frac{1}{4}$-inch lead pipe, which can be bent so as to direct the water into the funnel.

The tar is adjusted by the lantern-cock to a stream of about one-twelfth of an inch diameter; the water is also regulated by its tap to be supplied by drops, at the rate of about 80 or 100 drops per minute. Both tar and water pass into the pipe d by the funnel, and flow to the gutter in slab c, where (supposing the furnace to be in action), by the heat contained therein, the water is converted into steam, and then into hydrogen; and the compound of tar, hydrogen, and oxygen ignites, producing a considerable flame opposite the orifice *a*, as shown in sketch, whilst that portion of tar left unconsumed falls below opposite orifice *b*, where the volatile part inflames, and the remainder assumes a solid state, and is decomposed by the current of air passing in at *b*, which air becomes highly heated.

Thus, by the flame above and the air below, the retorts are heated with the same facility as with coke; all that is required is to clean out the gutter by passing a rod therethrough when it becomes obstructed—which is not of frequent occurrence—once every three or four hours or so, and to take care that the supply of tar and water is uniform. This, when the tar is supplied as suggested, needs little attention; and the ashes drawn from *b* do not exceed half a pound during the twenty-four hours.

It is almost needless to observe that when the setting is built expressly for tar, the space occupied by the ash-pan is filled up with brickwork. When first lighting, a few pieces of wood and incandescent coke are placed in the orifice *b*, on which the tar drops. Supposing the pipes for conducting the tar and water to be placed, the alteration for converting a coke into a tar furnace is quickly made. The door of the furnace being taken off its hinges, the bottom of opening serves for the lower part of orifice *b*, whilst the upper part is the top of orifice *a*. A handy workman would do the whole without having occasion to delay a single charge.

The water, although not employed by some engineers, is of the greatest importance to the successful application of tar, as by its decomposition it becomes part of the fuel, thus aiding the combustion, besides keeping the gutters free from the accumulation of carbon, and when employing the tar, as described, no smoke is produced.

By obtaining the tar direct from the hydraulic main as proposed, no possible difficulty can arise, unless the demand for it is greater than the supply; when, instead of tar flowing into the furnaces, water will take its place, and, in consequence of the absence of fuel, the retorts will necessarily cool. Beyond this nothing detrimental can occur. But as in all gasworks a portion of the tar is required for sale, due care will therefore be observed to avoid such errors by having only a limited number of furnaces supplied.

There is some nicety about the size of the orifices *a* and *b*. If they are too large, an excess of air passes, preventing the fuel having its full effect; should they be too small, particularly the upper one, then there is a tendency to produce smoke, but a little practice will soon lead to success. The damper should also be nearly closed.

About thirty years ago tar was very generally used for carbonizing the coal, its sale being then exceedingly limited—so much so as frequently to form a source of anxiety how to get rid of it—and coke being a saleable product the application of tar as a fuel was adopted as a matter of economy. Its general application for this purpose, whether in the front or back of the bench, or both, was in the manner we have indicated. About that period, however, Mr. Paterson, of Warrington, who was then manager of the Berwick-on-Tweed Gas Company, had in use a method of blowing in the stream of tar with a jet of steam, the latter being generated in a boiler set over the main flue, from which the nozzle of the steam jet communicated with the nozzle for the tar, and thus produced a shower of fire throughout the retort bench. The method was continued for several years, and as the best cannel was used the yield of gas was very great, but the intensity of the temperature, approaching to that of a white heat, was so destructive to the retort settings as to cause it to be abandoned.

Lastly, with tar fuel there is infinitely less labour, the coal is carbonized with the same efficiency as with coke, and, as regards comparative cost, a hundredweight of tar is equal to $1\frac{1}{2}$ to 2 hundredweight of coke.

In the present advanced age of discovery, however, the products found to be eliminated from coal-tar have so far influenced its value as to render it much more profitable to dispose of the tar in the market, for which there is always a ready sale, than to use it as a fuel.

In former times, when wet-lime purifiers were in use, the lime obtained therefrom made an excellent luting for the mouthpieces, but with dry-lime purification, when properly employed, the spent material is only available after considerable exposure to the atmosphere, and then requires to be well saturated with water, a process which involves much inconvenience, and, after all, produces only an indifferent luting, and in the absence of a suitable loam in the neighbourhood it is often better to employ fresh lime for the purpose.

Among other important recent improvements is the method of dispensing with the luting for mouth-pieces, which can be employed with advantage wherever the exhauster is in use. This is the invention of Mr. Morton, and consists of a lid jointed by a hinge to the mouthpiece, and so adjusted that when closed and secured it is perfectly air-tight, without the aid of packing or luting.

Whatever may be the form of the retort, whether oval or D shape, the lid, and of course the edge of the mouthpiece, must be circular, this being the form best suited to withstand the action of the heat. The mouthpiece is therefore constructed with the flange at one end to suit the bolt-holes of the retorts, the other end being circular, and usually the socket of the ascension pipe is cast therewith. Thus, for an oval retort 21" × 14", the front of the mouthpiece will be 17 inches diameter.

Fig. 6 represents the door open, and when closed, as shown in Fig. 7, it is secured by turning the handle B, which, acting on an eccentric, presses with sufficient force to prevent any loss from leakage, at least with the pressure ordinarily existing in retorts.

Fig. 6.   Fig. 7.

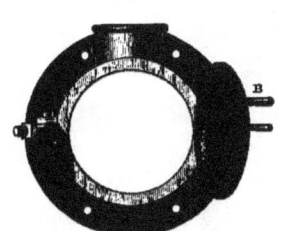

Among the advantages of these lids are, that no luting is required, thus dispensing with its cost and the labour in preparing it, besides trimming and cleaning the retort lids; avoiding the evils of accidental breakage of the luting and loss of gas; whilst the most arduous part of the stoker's duty, the putting on and taking off the lids, is dispensed with; add to this the quickness and facility of the operation, all render the air-tight lid highly desirable. We are informed that their cost varies according to their dimensions, and that the price of a mouthpiece complete for a retort 20" × 14" with 17-inch opening, is about 3l. 5s. each, and beyond that there is no other expense of either royalty or licence.

It must be observed, however, that these lids require care in their treatment, for some of them have been injured by the stoker using the shovel, from the force of habit, on the edge of the mouthpiece, when naturally they became injured. However, with moderate caution they are all that can be desired.

Within the last few years several mechanical contrivances have been patented for the purpose of dispensing with the seal or dip of the hydraulic main, and although some degree of ingenuity has been displayed in the invention of these appliances, yet, for reasons herein stated, their utility must be regarded as questionable.

The pressure existing in the retorts arises from the weight of the holder or holders containing the gas produced; the resistance of the purifying material; the washer, if employed; the station meter; and the friction in the pipes of communication between the various apparatus; which we may imagine to represent a pressure of say 10 inches. Now, if we suppose the dip not to exceed one inch, then on the removal of this by the means proposed, there will still be the heavy pressure of 9 inches on the retorts, compared with which the lesser pressure would not be material. Hence we are of opinion that no advantage can be obtained by the proposed contrivances; if, however, dip-pipes were made as formerly, with a seal of six or seven inches, then the "anti-dip" would be desirable, but it is evidently much easier to make a short seal than to attach the apparatus in question.

Bearing on the subject in question are the various machines invented for the purpose of superseding or aiding manual labour in stoking, and it is a matter for surprise and regret that the system adopted for charging and drawing retorts at the very commencement of gas lighting is still retained, and whilst most other mechanical industries have applied machinery as a substitute for manual labour, gasworks, in this respect, have made but little progress.

Of the machines for the purpose in question, the most prominent before the public are those invented by Mr. Foulis, of Glasgow, and by Mr. West, of Maidstone.

The peculiarity of the machine patented by Mr. Foulis consists in the propelling and withdrawing action necessary for the operations being obtained by hydraulic power.

The drawing machine is represented in the annexed Fig. 8, and consists of an iron frame, F, running on rails extending the whole length of the retort house. A is a cylinder capable of being raised

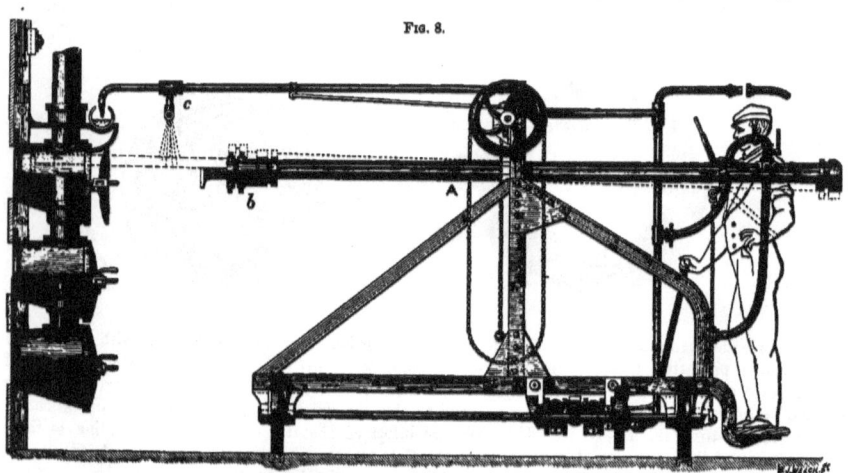

FIG. 8.

ELEVATION OF FOULIS'S HYDRAULIC DRAWING MACHINE.

or lowered with facility to any desired elevation, within which is a piston properly adjusted, and to this is attached the rod of the drawing rake, which rod passes through a gland or stuffing box, b. The water constituting the motive power is contained in a tank at a suitable height, and conveyed to the interior of the cylinder by flexible tubing, there being a suitable valve, which permits of the free entrance and emission of the water to and from the two sides of the piston as desired. Thus, the apparatus being directly opposite the retort to be discharged, on the attendant opening the valve, water is admitted at the back of the piston, by which means the rake is caused to enter the retort, and as the cylinder A rocks on a swivel at the centre, the end of the rake in entering passes over the top of the coke until reaching the desired point, when it is caused to descend, and secures the coke; the passage of the water into the cylinder is then reversed, and the rake is withdrawn with the coke. A stream of water flowing from the pipe c serves to keep the rake cool, and to quench the coke as it issues from the retort.

The charging apparatus is also actuated by hydraulic power; but in consequence of our application for a drawing having been disregarded, we are unable to describe it.

These machines are, we believe, in operation at the gasworks of Glasgow and Manchester, and preparations are being made for their application in the retort houses of the Beckton Works, where every opportunity will be given to ascertain their merits, and their success is much to be desired.

Although Mr. West's machine does not come strictly under the denomination of a mechanical stoker, it possesses the various advantages of diminishing manual labour to a very considerable extent; it avoids much of the fatigue and exposure to heat which accompany the ordinary system of stoking, besides being much quicker in operation, and not attended with the loss of coal which inevitably occurs when charging with the scoop.

For this apparatus only slack coal is used, but in the absence of this, the coal is crushed by a simple mechanical contrivance actuated by steam power, at the same time it is lifted by an endless band and scoops into a fixed hopper capable of containing about six tons, which is placed at such an elevation in the retort house as to fill the movable hopper hereafter mentioned.

The charging machine consists of an iron frame, represented in Fig. 9, running on rails in the retort house, and is provided with a hopper, H, which is filled with coal by passing it beneath the fixed hopper already mentioned, and opening a valve.

Fig. 9.

WEST'S CHARGING APPARATUS.

An iron waggon running on wheels, called the charger, C, represented in one position as immediately under the hopper, is open at the top, and, by turning the handle A, it is filled with coal. The bottom of the charger is constructed with five or six ports, having corresponding sliding plates which are closed or opened through the intervention of the rod E. The charger with its platform B being adjusted to the required level, the rod E is attached, and the waggon with its contents pushed into the retort; then by a half turn of the handle the ports are opened, and on withdrawing the charger, as represented in the engraving, the coals are deposited in an even, regular layer. This accomplished, the waggon is again filled, ready to charge another retort.

For facility of illustration, the charger is shown in two positions—namely, in the act of receiving the coal, and in discharging its contents into the retort.

Fig. 10 represents the drawing apparatus, and consists of a light iron frame running on the rails.

The rake-rods are supported at the points P, each opposite its retort, the hoe D being nearly the width of the retort is loose on the rod R, and is inserted as represented in the upper retort, and when at the required distance, by turning the handle I, the fourth part of a circle, the piece S bears against the top of the retort (as shown in the lower retort), and presses the hoe down, when the stoker draws the coke. For this purpose, passing the rake twice into the retort is all that is requisite.

We saw the apparatus in operation, and for speed, economy of labour, facility of working, and on all other accounts, it appears perfectly satisfactory. And a further recommendation to its employment is that the retort must be kept free from carbon, in order that the apparatus may work with facility, thus demanding proper attention to an important point in the economy of carbonization, which is too often neglected.

In addition to these great benefits arising from this machine, the inventor claims the further advantage of increased production of gas per ton of coal, which he attributes to the thin and regular layer of that material when in the retort. This statement, although appearing doubtful, is, however, confirmed by the accounts of the company.

FIG. 10.

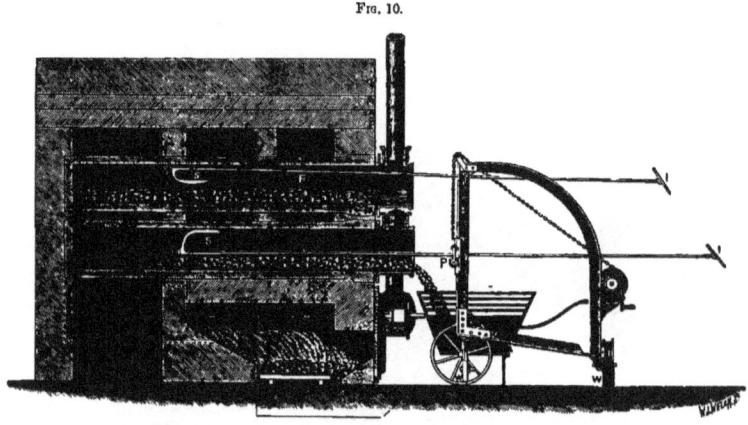

WEST'S DRAWING APPARATUS.

The cost of carbonizing per ton of coal, including attending to fires, will vary according to the price of labour, and on an average this may be estimated at from 2s. 9d. to 3s. per ton carbonized; but according to the returns of the Maidstone Gas Company, where West's apparatus is employed, this is reduced below 2s. per ton.

Our concluding observations on this part of our subject relate to matters of detail. In conducting a gasworks everything that is, or may be construed into, a nuisance should be avoided. Amongst these is the smoke which inevitably occurs on the drawing of the retorts, and this may be considerably diminished by raking out the tar which collects in the mouthpiece at each charge. At the same time, to prevent obstruction, a bent auger should be passed up the ascension pipes. The coal, when charged into the retorts, should be rudely levelled, and that portion of the coal not immediately in contact with the heated part of the retort should be pushed back to that point. All these operations, simple in themselves, are nevertheless essential to economical working; they occupy but little time, and when once systematically adopted the work proceeds as smoothly as can be desired.

PROGRESS OF RETORTS AND SETTINGS.

RETORTS.

THE history of any particular branch of science or art, describing the various errors committed in its progress towards comparative or positive perfection—if, indeed, this is ever attained—must tend to instruct and guide those immediately interested therein, and with this view we purpose to introduce the progressive steps of some of the most important subjects connected with gas lighting under their different heads.

In describing the first operations in the manufacture of gas, and the retorts employed therein, we cannot do better than refer to the writings of Creighton, the intimate friend of Murdoch, and who during several years assisted him in his numerous and varied experiments on gas lighting. These two gentlemen, together with Clegg, were engaged contemporaneously at the Soho Works, where, as before stated, gas lighting was first practically applied.

According to the author mentioned, " the retort first employed by Murdoch was of cast iron, and of a cylindrical form, as shown at Fig. 11 ; a is the retort with its cover, b the furnace, c the chimney, and d

FIG. 11.   FIG. 12.

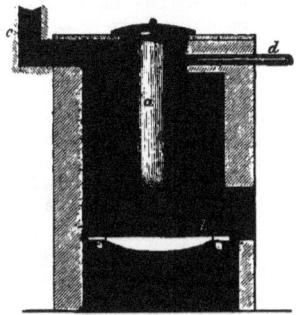
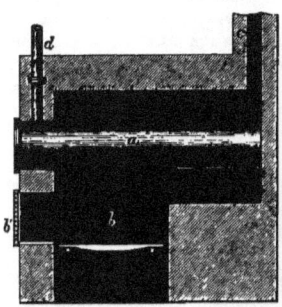

the pipe for conveying away the products of distillation. The retort, being about two-thirds filled with coal, was submitted to the action of the fire, and on its acquiring a red heat the decomposition of the coal commenced, the tar, oil, and gaseous products escaping through the pipe d, and the charcoal or coke remaining behind in the vessel." This, as well as the other retorts employed at the time, was made so as to contain about 15 lb. of coal, and was no doubt simply for the purpose of experiment.

" It is obvious," says Creighton, " that this form of retort was inconvenient, as regards the removal of the coke at the termination of each process; and to remedy such inconvenience a different construction was adopted, which we saw as early as the year 1802. Of this a representation is given in Fig. 12," one being a longitudinal section, the other a cross section within the front wall ; " a is the retort, consisting of a cylindrical vessel placed horizontally, provided with a door or cover, to charge and discharge it of its contents. The flue was so constructed that the flame surrounded the retort and afterwards made its escape from the chimney; retorts on this construction, from 1 to 20 inches in diameter, and from 3 to 7 feet in length, were found to answer tolerably well, and could be charged and discharged with facility." The various letters of reference apply to the same parts indicated in Fig. 11.

Here we may observe that it frequently happens, in our endeavours to make improvements, that our plans for that purpose have directly an opposite tendency. This was particularly the case with Murdoch at this period and for some time afterwards, for instead of advancing in the method of constructing and setting retorts, the result was directly the reverse.

Among several other kinds of retorts employed by him was one of great dimensions, resembling Fig. 11 in form, of nearly four feet diameter and a like depth, and capable of containing about 15 cwt. of coal. " In order to facilitate the discharge of the coke from this, an iron cage, formed something like a grappler, was let down into the retort previous to its being charged with coal, and when the process of distillation was completed, the grappler carrying the mass of coke with it was lifted out by means of a small crane. Another cage being then introduced, a fresh charge of coal was then thrown in, and the process of distillation continued without interruption." Thus mechanical stoking originated with the earliest operations of gas manufacture.

Our author says, " It is sufficiently obvious, from the construction of this retort, that upon being charged with fresh coal and already of a white heat, the process of carbonization will proceed most rapidly at first, but a crust of coke being speedily formed next to the heated metal, and this continually increasing in thickness, prevents the free transmission of the heat, and the decomposition of the coal is retarded more and more as the crust increases in thickness." To this should be added, that the illuminating power of the gas when so produced was much deteriorated in consequence of a portion of its rich hydrocarbons being deposited in its passage through the crust of incandescent coke.

Various other systems of retorts were tried by Murdoch, one of which resembled a decanter in form, the coal being admitted at the top and the coke withdrawn from an opening at the bottom. Another, consisting of a narrow cylinder placed at an angle of about 45°, having two mouthpieces. A third was oval, in cross section, with the ascension pipe placed within the oven, and in consequence it was subjected to all the heat of the furnace. Eventually the method of setting represented in Fig. 13 was adopted, which will bear comparison with some of those of the present day.

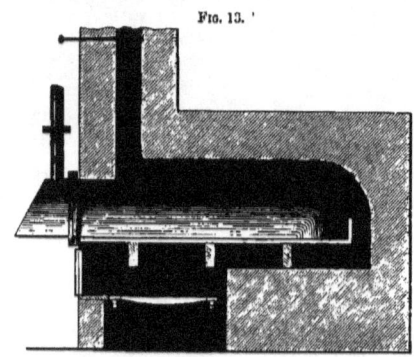

Fig. 13.

At that early period Murdoch discovered the evils arising from carbonizing at too low a temperature, and that under these circumstances tar was produced in abundance, with but a small production of gas; he also ascertained the yield from different kinds of coal, the consumption of the burners, the means of comparing the illuminating power of the gas, and much other valuable information connected with gas lighting, all previous to the year 1810.

With this degree of excellence achieved by Murdoch at that early period, it is surprising that so much difficulty should have been subsequently experienced in the progress of gas lighting. But it must be remembered that he then ceased to apply himself thereto, in order to devote his attention to the construction of steam engines, then becoming a very important branch of industry at the Soho Works. And we may infer, that after having established the new art, meeting with continued opposition, and incurring heavy expenses without deriving a corresponding benefit therefrom, he left the science he had created without any grateful recollections. Thus the results of his experiments and observations may be considered to have been comparatively lost.

There is reason to believe that the knowledge of the defect just alluded to gave the idea to Mniben to carbonize the coal in thin stratum. This he proposed to accomplish by having a large retort or oven, into which he placed sheet-iron trays, each charged with a layer of coals of about two inches thick, and when carbonized the coal and trays were withdrawn. This was the subject of a patent in 1810.

At the period of the publication of the first three editions of Accum's 'Treatise on Gas Lighting,' 1815 and 1816, little advancement seems to have been made in the subject under consideration, as we find the same and the only drawing of retorts in all these, of which a copy is annexed. The retorts A A are described as about 7 feet long, 12 inches diameter in front, and 10 inches at back. Here we are introduced for the first time to the "coke-hole," which, afterwards becoming more extensive, received the name of "vault." The ash-pan is provided with a door E, the only means adopted during many years for controlling the supply of air to the furnace; and, according to Clegg's instructions to workmen, "the bridge or row of bricks of the flue, e, of the retorts should never be hotter than a bright red heat, which may be regulated by the door of the ash-pan." The wall in front of the bench was for the purpose of protecting the men from the heat, the two circular orifices, C C, were to enable them to charge and draw the retorts, whilst the orifice H conveyed the coke into the "hole" beneath. In the adjoining vault is the tar tank provided with a gauge to indicate the quantity contained therein.

FIG. 14.

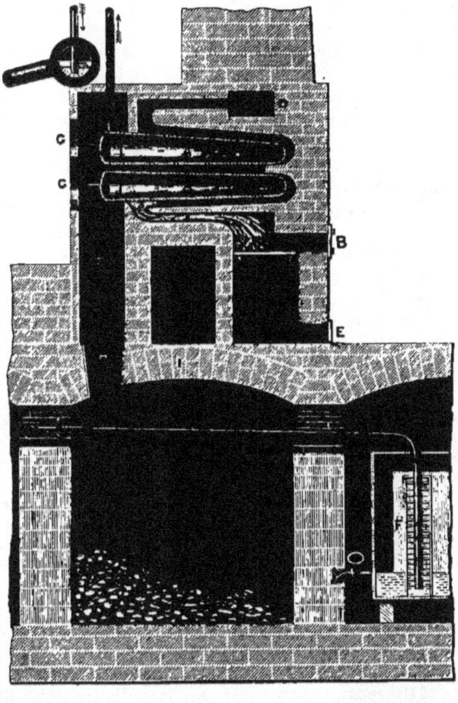

The drawing is very remarkable, on account of having the charging stage and coke-hole, the method of construction being identical with that observed in all our largest establishments of the present day.

In 1817, John Perks, of the Dorset Street Works, patented the means of arranging together a number of retorts, made of iron or other metal. The patentee proposed to place thirteen of these in one oven, and to have a very considerable length of flues passing zigzag amongst them, the charge of 1¼ bushel, or 126 lb. of coal per retort, to be carbonized in six hours, and the whole to be heated by one fire; certainly a bold attempt, especially at a time when engineers were divided in opinion concerning the comparative merits of heating five retorts by three fires, or two retorts with one fire.

At first sight this mode of setting would appear to present some advantage, and in arranging it the patentee might have had in view the circuitous flues of the steam boiler. But in both instances, whether with the boiler or gas furnace, any excess of flue must be prejudicial, particularly in the latter, where, by its introduction, a large quantity of non-conducting material of necessity exists, and the passage of the heat is impeded; consequently the heat is absorbed at these parts in the immediate vicinity of the furnace, where the retorts attain the proper temperature; whilst those situated at a distance therefrom are at a degree of heat quite unsuitable for carbonization. This evil exists, in some settings of retorts, oven at the present day, and wherever there is a superabundance of flues, as the term is, "wire-drawing" the heat, the proper temperature and strict economy of fuel can never be obtained. Beyond a fair trial, and some experience, nothing more was achieved by Perks's settings.

Our subject, at the period in question, began to be retrogressive, for although the "flue plan," with two or more retorts, was acknowledged to be a success, yet the firebricks used for flue and shield of the retorts were regarded as impediments to the passage of the heat, forgetful, as it appears, of the purpose they served. Rackhouse proposed to avoid this by submitting the retorts to the direct action of the fire, so going back to the earliest of Murdoch's experiments; which arrangement was denominated the "oven plan." The account of this given by Peckstone, when writing two years afterwards, is by no means flattering.

He says, " By adopting the oven plan we shall hardly find there are any advantages gained ; but contrariwise, the retorts, instead of lasting longer, were two-thirds of them burnt out in less than two months, owing to the lower ones being placed immediately over the fire, from the action of which nothing was left to guard them. Their form precluded every hope of carbonization being accomplished in less time than by others of a cylindrical shape; and as to the percentage at which carbonization was being carried on, what could be expected, when, after the retorts had been in action about six weeks, it was necessary to keep up twelve fires to enable the stokers to work as many retorts?

" Various have been the alterations made by different workmen in the fire work to the retorts since the oven plan was adopted, but hitherto most have failed in remedying the very serious evils which must arise from their rapid destruction. Heretofore a retort of the same shape and dimensions, when constantly used night and day, according to the flue plan, lasted from eight to ten months, but on the oven plan the retort seldom remains in a working state more than as many weeks."

Favourable results appear to have attended the system of carbonizing proposed by Maiben, as we find an adaptation of it patented by Clegg in conjunction with his meter in 1815, which he termed the rotary retort, and is described as follows :

" This retort or distilling chamber is flat and circular, made of metal plates, of about 12 feet in diameter, and placed horizontally. It is fitted up with a perpendicular shaft, to which are attached radiating arms, all being enclosed within an oven of a capacity just sufficient to allow the arms, boxes, and contents to revolve freely. The fire or furnace is placed at one side of the circumference of the oven, so that one part of the retort is heated more than another.

" The coals are placed in shallow boxes, which glide on to the radial arms. A slow motion being given to the shaft, the different boxes and their contents are gradually turned round from the cool side of the retort, where there is suitable provision for taking out the boxes containing the coke, and replacing others filled with fresh coal, ready to be carried round to the heated side where the gas is generated."

Fig. 15 represents this retort in operation. The oven (being shown in section) contains the perpendicular shaft revolving in its bearings, and to which shaft are attached the radiating arms, which carry the "shallow boxes," each containing a layer of coal of about two inches thick. The furnace is situated at the part directly opposite to the mouthpiece *a*, and is so constructed as to heat about one-third of the area of the retort, or boxes, that part immediately under the action of the fire being protected by brickwork, in order to avoid the effects of excessive heat. The ascension pipe, placed outside the brickwork, was provided with a clearing-out cap, by which we learn, as might have been expected from the low heat necessarily attained, that the formation of pitch was one of the annoyances of the system. The chain was for the purpose of opening and closing the hydraulic valve when about to charge or draw the retorts. The rod B was attached to a lever and counterbalance, to facilitate the raising and lowering of the door of retort. The gas passed, as produced, into an hydraulic box, answering the purpose of the present dip-pipe.

A charge being carbonized, the coke was allowed to remain in the retort, and, in cooling, tended to heat the coal about to be operated upon.

With all the complications, inconvenience, and loss attached to this system of retorts, it appears that they were used at the Royal Mint and several works for the supply of large towns, such as Birmingham, Bristol, Chester, and other places, where, in some cases, the arms were of cast iron. But in due time the errors of the construction and process became evident, and they were abandoned.

About this time retorts were made either square, round, oval, D-shape or kidney, or ear-shape, and we can imagine that each form had its supporters. In a setting described by Peckstone, there are six square retorts set in an oven side by side on the same level, and touching each other, and heated by a fire adjoining the outer retort. From the furnace the current of caloric first passed under and then over the retorts, consequently the heat could only penetrate at these points.

This was about the most impracticable setting that has ever been devised; for in addition to the limited surface of each retort exposed for the conduction of the heat, there was the other evil, that the furnace was placed at a considerable distance from the nearest retort, as we may suppose, with the view to its preservation.

FIG. 15.

SECTION OF CLEGG'S ROTARY RETORT.

The advantages of a thin stratum of coal being recognized, John Malam introduced wide oval retorts; these he proposed to place five in a bench to be heated by three fires, a setting which subsequently became very generally adopted.

At this epoch Grafton patented the process of employing iron retorts lined or cased with fireclay, a system which, on account of the expansive nature of the iron as compared with clay, was thoroughly impracticable; moreover the latter, during the process of baking, would contract considerably. Thus the two materials would of necessity be detached, and the iron left exposed to the heat.

The following engraving (Fig. 16) is copied from Accum's Treatise published in 1820, which conveys some idea of the magnitude of a gasworks as then constructed, and is presented for the purpose of comparison with establishments of the present day, also to show Clegg's rotary retorts in action.

"The central building exhibits the retort house; the roof is furnished with a projecting louvre to let out the smoke. The gable ends and one of the sides are of brickwork, the other side is open, and supported by iron columns. The building to the right of the retort house is the purifying house, and contains the lime machine. The trapdoor, marked a, indicates the cistern or receiver for the waste lime. Adjoining the purifying apparatus is the meter house, the front wall of which is removed to show the position of the station meter, the axis of which drives the agitating shaft of the lime machine, and for that purpose the axis of the meter and the shaft of the lime machine are connected by a strap as shown. The building to the right, at a distance, is the dwelling of the manager, and the small house to the left is a smith's shop. The retort house contains three of Clegg's rotary retorts, B B B; each of the three columns projecting above the roof being surmounted by an hydraulic valve, the chains descending into the retort house, and passing over the pulleys at the top of the columns, being for the purpose of opening these valves, whenever desirable, for charging and drawing the retorts." The gasholder represented at the left is Clegg's "collapsing holder," described by its author as follows:

"This improved gasholder is made of thin metal plates or other suitable material; it has two sides and two ends meeting together at the top in a ridge, like the roof of a house. The sides and ends are united together by hinges, and the joints are covered with some flexible material which will retain the gas, but allow the sides to fold together in the same manner that a portfolio folds or closes. Therefore this gasholder, when empty, will be folded up and flat, and the sides and ends closed together; and when full, will be opened in the form of a roof of a building," as represented in the engraving. " The bottom edges of the sheets forming the sides and ends are

Fig. 16.

AN ECONOMICAL ARRANGEMENT OF AN APPARATUS FOR LIGHTING A TOWN OR LARGE DISTRICT.

immersed in the water, in order to retain the gas within the holder. By opening out or closing the sides and ends of the gasholder, its capacity is enlarged or diminished, and this variation is obtained without a deep tank to immerse the whole of the sides of the gasholder, as required in those as ordinarily constructed." The apparatus was supported by two columns within the vessel, one of which is seen projecting above the ridge.

This was one of the contrivances adopted by Clegg in order to avoid the construction of deep tanks for gasholders; and although Accum spoke of it in terms of the highest commendation, yet, on account of the uncertainty of the durability of the flexible material, the immense quantity of material and area of ground for a holder of extensive capacity, and the variation of the pressure given by such a vessel when subjected to the influence of a strong wind, it can only be regarded as a toy, and thoroughly impracticable.

The fixed idea entertained by Clegg, of the great advantages to be derived from distilling coal in a thin stratum—which, as explained, he had already submitted to practice, but failed in accomplishing, through the defective construction of his apparatus—induced him to make another attempt to achieve success in that direction, by the invention of the web retort, of which the following is a longitudinal section, copied from his son's 'Treatise on Coal Gas.'

We are unacquainted with the period of the invention, nor does it appear to have been in operation at any gasworks. It is presented as possessing great ingenuity and considerable interest at the present time, when the charging and drawing retorts by mechanical means is engaging so much attention.

The name "web," or "woven," was derived from the construction of that part of the machine which conveyed the coal into and the coke out of the retort, when undergoing carbonization, and is represented by the letters C C. This consisted of a number of flat bars of about ¾ inch by ¼ inch, placed edgeways, parallel to each other, with bolts connecting together two sets of links, thus forming a continuous flat chain of about two feet wide and of sufficient length to pass over the two hexagonal "drums," B and B.

The retort R R, properly speaking, was embraced between the flanges, a and a, and was made of ½-inch boiler plate, protected on the under side by fire-slabs. The hopper, E, and the rest of the framework were constructed of a similar material. The two doors, D and D, requisite for renewing the web or other repairs, we may assume, were of much stouter material, and bolted to the general frame.

Fig. 17.

CLEGG'S WEB RETORT.

The two revolving hexagonal drums, B and B, gearing with each other, were driven by water or steam power, and over these passed the endless web C C, and, on motion being given to the drums, the web was caused to travel continuously in the direction of the arrows, at such speed that the coal placed thereon would pass from one end of the retort to the other in about fifteen minutes.

The coal destined for carbonization was broken into small pieces "not larger than coffee berries," of which the hopper, when full, contained sufficient for twenty-four hours' operation; when it was closed at the top by a cover sealed with a water-lute, similar to that employed in dry-lime purifiers, but which is not shown in the original drawing. At the lower part of the hopper was a star-shaped wheel F, its length being equal to the breadth of the web. This wheel was in gear with, and received its motion from, the drums B B, so that the coals were by this means delivered from the hopper on to the web, gradually and continuously, and in passing through the retort the coal was carbonized, the gas issuing by the pipe A, and the coke falling from the web into a well provided with an air-tight door, or sealed by means of water.

The furnace is supposed to have been in the front, and the heat therefrom passed through the nostrils N N, under the retort, and then over it, to the main flue, as represented in Fig. 18 in section.

According to Clegg, "The main advantages attendant upon this form of retort are, that it requires less space; the stokers (so called at present) might be spared that name; the heat would not be felt more than in a boiler house, and the retort house might be kept perfectly clean, wholesome, and free from suffocating vapour." He adds, "The system is one of great economy, and by far the most scientific process yet adopted for making coal gas. It requires no attendance except that of keeping up the furnace, and charging the hopper once in twenty-four hours. No gas is lost, no tar made. The coke produced is increased in quantity about 25 per cent., but its quality is not so well fitted for general uses, although far superior for culinary purposes." The estimated make was about 12,000 cubic feet of gas, specific gravity 490 (about 19-candle gas), from a ton of Newcastle coal.

That the system displayed considerable ingenuity must be admitted; but, unfortunately, amidst the number of excellent practical inventions produced by Clegg, there were others of an opposite character, and the web and rotary retorts may be classed amongst them.

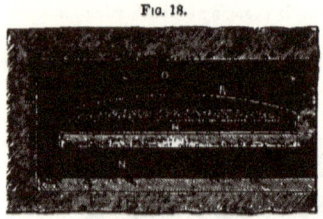

Fig. 18.

One difficulty attending the application of this system was identical with that in the horizontal rotary retort; in both cases the heat from the furnace, instead of being confined and concentrated within reasonable limits, was of necessity dissipated throughout the whole space occupied by the apparatus; so that, to obtain the temperature requisite for carbonization, a large amount of fuel must have been expended. Again, the chains forming the web would be speedily so swollen or distorted by the red heat, as to render their action quite impossible, and the stratum of coal being regulated to a thickness of three-eighths of an inch only, as proposed, would render the resulting coke of little more value than breeze. Lastly, in operation, the coke from caking coal would have a tendency to adhere to the web, and not fall as shown. In short, the web retort was one of those schemes which look well at first sight on paper, but which, when submitted to practice, is found to be useless.

No explanation can be given of the stated high production of gas, both in quantity and quality, beyond the observation, that an infatuated, sanguine mind is often innocently led to the most erroneous conclusions; which must have been the case with Clegg, for he was utterly incapable of wilfully making an erroneous or unfair statement.

Between the periods of the publication of 'Gas Lights,' by Creighton, and Clegg's 'Treatise on Coal Gas,' so far as we have been able to learn, only two works on the subject were presented to the public—viz. the 'Compendium of Gas Lighting,' by Matthews, published in 1827; and the 'Traité sur le Gaz,' by Merle, produced ten years afterwards. The second edition of Peckstone was substantially a copy of the first, and is therefore not taken into consideration.

On consulting the two authors mentioned, we find that no attempts at improvement in retort setting were made during this time; indeed, in Matthews' 'Compendium,' the most prominent setting is represented as having seven retorts in a bed, heated by four fires; and in the plan of the Imperial Continental Company's works, by Merle, the beds of four retorts each, are heated by two fires; neither of which systems could be regarded as progressive, and we are led to the conclusion that, at the epoch in question, considerable apathy existed on this, as well as on many other points, amongst gas engineers.

Iron Retorts.

The publication of the 'Treatise on Coal Gas,' by Samuel Clegg, jun., in 1841, presented a new era in the art of gas lighting. The descriptions of the various apparatus, with the excellent drawings, the vast amount of information given therein, together with the detail entered into, had never been previously attempted; and its production very deservedly excited the warmest admiration and praise of all connected with gas lighting.

Most conspicuous amongst the drawings was a setting of five retorts, heated by one fire, as represented in the following engravings, which are copied from the work in question. By whom this was designed we are not told, but we may assume that the elder Clegg contributed towards the degree of comparative excellence therein displayed.

Fig. 19 is a longitudinal section, with a part of one bed represented as broken away. Fig. 20 is a cross section just within the front wall.

As will be observed, the retorts are single, set back to back, separated by a 14-inch wall, there being also a space of about 4 inches between the end of the retort and the wall, to allow of the expansion of the former when heated. An iron retort of the length described, on arriving at the temperature necessary for carbonization, would elongate about 2 inches, and if from peculiar circumstances it were heated, and afterwards "let down," or cooled repeatedly, it would be permanently increased in length to the extent stated. This peculiar property of metals was one of the causes of the failure of Clegg's inferential dry meter hereafter referred to.

In Fig. 20 is represented an arch over the furnace which is continued throughout the length of the oven as seen in longitudinal section, having four nostrils M M M M on each side, in order to permit the passage of the current of heated air or caloric. There are also four transverse walls for supporting the lower retorts indicated by the dotted lines and the letters P P P P, one of which is shown in elevation, extending nearly to the level of the top of arch over the furnace; on these transverse walls are placed

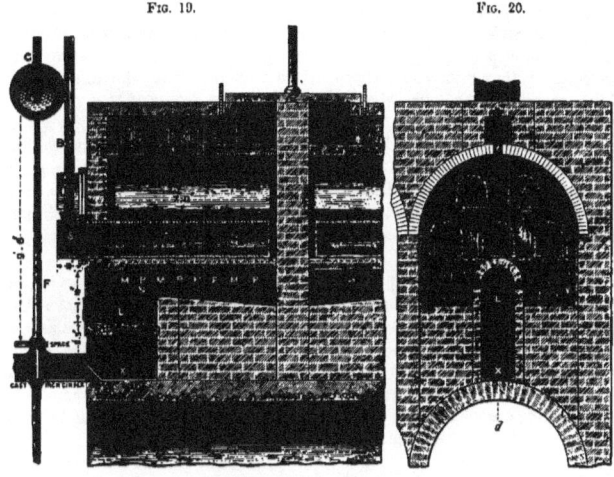

FIG. 19.   FIG. 20.

CLEGG'S SETTING OF IRON RETORTS.

fire-slabs of 2½ or 3 inches thick, covering the area level with the top of furnace arch, but leaving a space at the two sides of about 2 or 3 inches in width, extending the length of the retort, through which the current of caloric passes as indicated by the arrows.

This was the "bench" of the retorts; the term, however, is now synonymous with "bed," "setting," and "stack"; the last term, however, is generally applied to a block of settings. The lower retorts being placed on the bench, the two outer ones were protected from the action of the fire by the fireclay shield tiles, otherwise their duration would have been very limited. The piers were then built ready to receive the upper retorts.

The heat as generated in the furnace L is conducted directly through the arch, slabs, and tiles, and passes as a current through the orifices M M, thence up by the sides of the shield tiles on to the lateral flue Q, where for the first time we find the damper Z practically in operation; about thirty years after Murdoch had used it in his experiments, as already shown. And it is remarkable that in all the drawings in works on gas we have met with up to the period of the publication of Clegg's work, including that by Peckstone of 1841, the air for supplying the furnaces was always regulated by the door of the ash-pan.

Assuming the privilege of pointing out defects, we may remark that the greatest error in the arrangement consists in the nostrils M M being too much contracted. This is highly objectionable, for instead of these being of the limited capacity shown, they should be as large as possible, in order that the heat as generated may pass from the furnace to the interior of oven, with perfect freedom; the strictest care being observed to control the dampers and thus prevent any undue loss of heat. By contracted nostrils—a defect which frequently exists in settings—although the furnace may be at a white heat, it is sometimes impossible for the retorts to attain the temperature essential for carbonization.

And here we observe the marked difference between the means of regulating the air to the furnace by a door in the ash-pan, and checking its exit to the chimney by the damper. In the first instance a direct communication is established between the interior of the oven and the chimney, while in the latter method the damper becomes the means of separating the one from the other. Where the draught of a setting is suitable, the passage at that point is very limited.

Cast iron for retorts was adopted on account of the special advantages that metal presented over all other materials as regarded price and general practicability, but years of experience demonstrated the numerous inconveniences that retorts so constructed were subjected to, the most prominent of these being their liability to rapid destruction when submitted to that degree of heat most suitable alike for the economical production of gas and coke, which evil was further increased through ignorance of many matters connected therewith, and notably the cause and effects of the carbon incrusted within them.

Another defect existing in these was, that often when retorts were taken down in good condition, on cooling they would crack right across, so being rendered useless. This contingency arose from a want of knowledge on the part of gas engineers, for as the interior of the retort had a thick coating of carbon, and as this did not contract to the same extent as the iron, the latter was consequently broken by the former.

Among the other evils attending the use of iron retorts was the accumulation of the carburet of iron on their exterior; and this being a non-conductor of heat, it was requisite to remove it periodically by passing a rake through the sight-boxes, made sufficiently large for the purpose. In this operation often a portion of the carburet would fall into some inaccessible place and obstruct the passage of the heat, and frequently from this cause were the retorts rendered useless in consequence of the low temperature they had acquired being unsuitable for carbonizing. Again at times, through inferior metal, when working at a moderate heat, the retorts would assume various distorted forms, perhaps sinking between the supporting piers, or flattening at the top, or bulging at the sides, and thus displacing a shield tile, when their days were numbered, and one of a bench being destroyed, very speedily the rest would follow.

The price of iron retorts was also a very serious consideration, for although costing perhaps only 6l. or 7l. per ton, yet their durability was very limited, seldom exceeding eight or nine months, and in many cases this was restricted to one-half those periods. Thus the expense of retorts, which averaged about 6d. or 7d. per 1000 feet of gas produced, was a most important item in a gasworks, whereas, with clay retorts, one-sixth part of that may be estimated as sufficient.

Hence, as was eventually experienced, in the varied questions of production of gas, the quality of coke, the economy of fuel for carbonizing, the cost of wear and tear, and labour in charging and drawing, in all these points, great and positive advantages were realized by the use of the clay retorts. And although their general adoption was very gradual and slow, they are at length recognized at their proper value, and to the great benefits derived from them is mainly due the marvellous extension of gas lighting; for without their adoption the cost of gas would have been augmented at least one shilling per 1000 feet, and this difference of price, undoubtedly influences to a great extent the present large and increasing consumption.

It must be, however, admitted that many of the defects of the iron retorts arose from our ignorance of the various circumstances connected with their use, for had we been acquainted with the advantage of the exhauster, and applied it with the same care as at present, and had the method of setting retorts been better understood, the change from iron to clay retorts would not have been so desirable; nevertheless, that change has been productive of the best results.

There are works, however, supplying manufactories, mansions, &c., where the demand for gas is irregular, and the manufacture is discontinued for weeks or months together; in such cases, when in operation at intermittent periods, according to the general opinion, iron retorts are desirable. When employing these one great consideration is to keep them free from carbon, so as to prevent them breaking, as already explained. Iron retorts demand considerable care in setting, but as the temperature of these is not required to be so high as with those of clay, the fuel in the furnace need not be heated to the same degree.

CLAY RETORTS.

The knowledge of the difficulties associated with the use of iron retorts was not acquired until the advantages of those of clay began to be appreciated, hence it cannot be surprising if engineers were tardy in recognizing the value of the latter, but this once *un fait accompli* the progress of clay retorts was rapid, and with what excellent results is now well understood.

To John Grafton is indirectly due the merit of introducing clay retorts, with all the various advantages derived therefrom; although by the specification of his patent of 1820 it is evident their manufacture had been previously attempted and failed, in consequence of their tendency to crack. To remedy this evil he proposed to construct them of several pieces, put together with fireclay, in theory identical with the most approved manner of making clay retorts at the present day; that is, by an admixture of a large quantity of granulated burnt clay with a small portion of plastic fireclay.

Retorts according to this patent were erected at the Cambridge Gasworks in June, 1824, and the following year were put in action, at which time they also came into use at the Wolverhampton Gasworks, and very shortly afterwards were tried at Halifax, and the Blackfriars Works, London.

The first retorts or ovens constructed by Grafton were square, capable of containing about 4 cwt. or 5 cwt. of coal at each charge; subsequently they were altered to the form represented in the following figure, and increased in size so as to be capable of receiving a charge of 10 cwt. or 12 cwt. of coal, which was carbonized in four or five hours. They were built entirely of bricks of about seven or eight different forms and sizes, grooved together, set with fireclay, and afterwards covered with a coating of cement, of which pounded glass formed an important component.

According to Merle, the durability of the first brick ovens constructed by Grafton extended beyond six years' active service, yet he met with the greatest opposition from gas companies wherever he proposed them; and in some instances after they had been constructed and in operation, such was the prejudice against their use, that the workmen wilfully and maliciously destroyed them. But by dint of energy and perseverance they were gradually employed on the Continent, where we have seen them in many gasworks of considerable importance, as early as the year 1846.

The author referred to says: "At first the greatest inconvenience attending the use of these retorts was the accumulation of carbon therein; and, in order to remove this, they were put out of action, and allowed to cool, when a workman entered, and with a hammer and chisel cleared out the incrustation, which, in some instances, assumed a thickness of three or four inches." From Merle we also learn that Grafton subsequently discovered that by leaving the retort open for some time, it was cleared of the carbon, through the influence of atmospheric air. He then arranged a "simple means of getting rid of it, which succeeded perfectly. This consisted in leaving an opening at the end of the retort, which was covered by a cast-iron cap, and every week this was removed so as to permit the air to pass therethrough, when, in an hour, all the carbon was cleared out." Following closely on these observations were Grafton's experiments to ascertain the cause of this deposit, which resulted in the discovery that it was mainly produced by the gas being subjected to heavy pressure when in the retort, and of course influenced by high temperature. The knowledge of this fact led to his adoption of the exhauster, the importance of which is now so universally appreciated.

Fig. 21.

GRAFTON'S BRICK OVEN.

The writer's experience of Grafton's ovens does not justify the expression of a favourable opinion of them, the only advantages they possessed were their durability, averaging about four years, and the excellent large coke they produced; but, on the other hand, they were very costly in construction, and their consumption of fuel was enormous, being equal to about 70 per cent. of the coke produced. This, however, in justice, should be stated was mainly due to the bad construction of the furnaces, which were extravagantly large, and to the absence of dampers. Therefore, although those ovens had prepared the way for the use of clay retorts, by which they were gradually superseded, they were eventually abandoned, mainly on account of the large quantity of fuel they demanded.

Other engineers and manufacturers have since produced retorts constructed similarly to Grafton's, with the exception of the lower part, which is composed of fire-tiles, and the form slightly varied; and in some establishments, in Scotland and elsewhere, they are exclusively worked at the present day. Brick retorts are employed exclusively by the South Metropolitan Gas Company in London, and are used at many works in the depth of winter, when required only for a few days, and are then thrown out of action; on being again employed they are found to be practically free from escapes, but with the present excellence attained by some manufacturers of clay retorts, we are of opinion that, for such purposes, those of brick are quite unnecessary.

Clay retorts, and of cylindrical form, were first made and used by Mr. Herst, of Leeds, about 1826. These were about 12 or 13 inches in diameter, made in lengths of 4 feet, which were jointed together in the oven, and the ends placed on to them—a system still adopted in some works; but more frequently they are made in lengths of 8 feet or 9 feet and under, their ends being generally formed with them. About 1827, clay retorts were introduced into Scotland at the works of Kirkcaldy, Dunfermline, and Stirling, and, meeting with success, they were subsequently adopted by the Edinburgh and Leith Gas Company. At this period a manufactory for these articles was established in that country, which is still in existence.

We are disposed to believe that the success attending the first retorts manufactured in Scotland arose from some peculiarity in the clay used, so that, in the act of heating, the retorts made therefrom neither expanded nor contracted, and, in consequence, they were not liable to break nor crack when exposed to the action of the fire.

Yet, notwithstanding the gradual and increasing use of clay retorts in Scotland, gas companies in England generally were directly opposed to them, and a manufacturer of these articles, after several fruitless journeys to London for the purpose of introducing them to the gas companies, succeeded only in placing a few of them at a small suburban works then in existence. Discouraged by the opposition and the rebuffs he met with from the engineers and managers of the day, he ceased to press a business which was yet destined to be productive of such vast advantages to gas companies throughout the world.

The failure of clay retorts at Brick Lane station, in 1843, already mentioned, was a serious impediment to their adoption in London; and it was not until five or six years afterwards that the late

INTRODUCTION OF CLAY RETORTS.

Mr. Livesey made a trial of them at the works he so ably directed, and the success of the experiment contributed considerably towards their subsequent general introduction.

The publication of Mr. Livesey's success brought forth an animated correspondence in the columns of the 'Journal of Gas Lighting,' between two gentlemen of eminence, the one maintaining that clay retorts could not be generally adopted for various reasons, namely, on account of their porosity and liability to crack when in action and exposed to the necessary degree of heat for carbonization, consequently entailing a considerable loss of gas; that being a bad conductor of heat, a large quantity of fuel would be requisite for carbonizing, and, singular enough, the opponent of clay retorts recommended that manufacturers "could not devote too much care to make them of the closest texture possible." Experience has, however, taught us that if thus made, retorts must of necessity crack, and only when they are cellular or porous are they capable of withstanding the action of the heat without cracking. It must, however, be explained that the cells or pores are formed principally by the union of the granulated with the plastic fireclay.

The supporter of fireclay retorts, on the occasion referred to, met the arguments of his opponent by a statement of facts, proving incontestably, as practice has since demonstrated, that all the assertions of his opponent were erroneous; and from that period the use of clay retorts in England may be dated, and such has been their extensive application, that at the present day we believe there is not a gas-works having a production of 3,000,000 per annum for supplying towns or villages where iron retorts are used, and we may state that wherever this is the case it is prejudicial to the interests of the parties concerned.

The process of manufacturing clay retorts consists in intermixing a quantity of granulated burnt clay with a comparatively small quantity of plastic clay, from which compound the retort is formed by hand in one operation, for if it be discontinued even for a short time, the clay does not thoroughly unite at that point. Thus it sometimes happens that after the retort is delivered at the works, by an accidental shock it is separated into two parts, the fracture clearly indicating the cause, as each part has a smooth surface, instead of the cellular appearance on the breakage of the material; and we believe that the greater the quantity of granulated burnt clay in the composition of the retort, the less liable this is to break or crack, when exposed to the action of the heat.

Although the practical application and manufacture of clay retorts had their origin in Scotland, yet there is no other place, if we may judge from the numerous kinds there adopted, where such diversity of opinion concerning them exists; for in that country, at several gasworks, brick retorts of varied shapes and dimensions are exclusively used, but in the greater majority they are of clay, sometimes cylindrical, as small as 14 inches in diameter, retained from the early years of their adoption, and increasing in size at other establishments to 20 inches diameter. The square form, rounded at the corners, also the oval, are preferred by some engineers; whilst the D shape, from the "squat" 27 by 13 inches, to other varying dimensions, the smallest being 15 by 13 inches, is very extensively used by different gas companies. It is somewhat surprising, when we remember the proverbially practical character of our northern friends, that greater unanimity in this respect does not exist, if only to effect a saving of labour, the more so as, from the nature of the coal carbonized, they are not influenced by considerations concerning the size of the coke produced.

The retorts generally used in England are either cylindrical, seldom less than 16 inches in diameter, the D shape, or oval, the former usually being 18 inches wide, and the latter about 21 inches by 14 inches. On the Continent the last mentioned, with few exceptions, are almost universally employed; and from experience they are unquestionably the most desirable alike for large and small works, possessing, as they do, a large surface, almost flat, capable of receiving heavy charges, and offering great facilities for drawing the coke.

It may be observed, however, that it is remarkable that in so simple an operation as the carbonization of coal, such a diversity of opinion should exist concerning the form, but more particularly the dimensions, of retorts, as we find in the illustrated catalogue of a well-known manufacturer of these, that there are no fewer than forty-three different sizes of D retorts, of which the greater part are about 16 inches wide. That some difference in the height of those employed for cannel and those used for caking coal may be desirable, is admitted; but beyond this neither theory nor practice can warrant the extraordinary want of uniformity in the dimensions of the articles in question; and further we may observe that if engineers would only imitate the example set in the regularity of the various gauges in iron pipe, brass tube, screws used in gas fittings, by employing two or three forms of retorts only, instead of the large number mentioned, the manufacturers of these would be enabled not only to supply at a cheaper rate, but with greater promptitude.

The most desirable kind of retort for the production of coke, having reference to its sale as a residuum, must be decided by circumstances; for in localities where the demand exists for large coke, as in foundries or railways, retorts of increased capacity should be adopted, having a width of 24 or 27 inches, and for this purpose the D shape, rounded at the corners, or the oval, may be beneficially used. In cylindrical retorts especially, when of small diameter, the coke at the edges is thinner and of course smaller, which is a disadvantage.

In the earliest years of gas lighting the form of retort was supposed to influence materially the yield of gas from a given quantity of coal, and this opinion is still retained by some engineers. But with the knowledge of the varied kinds in operation, and the intelligence generally displayed in the management of gasworks, we can affirm that the form, whether cylindrical or otherwise, has no appreciable influence in this respect, the yield of gas being dependent on the temperature to which the coal is submitted when undergoing carbonization. We see, indeed, in the best managed establishments, which differ only in the form of retorts employed, the same results, practically, from the same kind of coal are obtained; whilst at the Beckton and Bromley Works, although retorts of the same size and shape are employed at both places, yet by the returns of the two companies to whom they formerly belonged, a vast difference is observed in the yield of gas per ton obtained at these works respectively. It may be concluded, therefore, that the shape of a retort considered by itself, has no influence upon the production of gas.

In the form, but more particularly in the size of retorts, labour becomes an important consideration, and in this respect those of small diameter and limited length must always involve more work in charging and discharging, for a given quantity of coal carbonized, than those of larger capacity; inasmuch as all the extra labour with the larger retort consists only in the extra quantities of coal charged and the coke drawn from the retort. With small retorts, therefore, the labour of charging and discharging must always be greater than in the larger.

The distinction between brick ovens and retorts consists in the former being considerably larger than the latter, and set in its own arch as represented. Brick retorts, on the contrary, are set in three, five, or seven retorts in a bench, according to the judgment of various engineers. They are both built *in situ*, suitable centres being provided for the purpose.

The advantages claimed by the advocates of brick ovens and retorts are, that on these cooling, the contraction takes place in all the numerous joints of the brickwork, and thus the interstices are so infinitesimally small as to be filled with the fresh charge of coal. This cannot be contradicted; but, on the other hand, the increased thickness of the material must of necessity demand a greater quantity of fuel than is requisite for a clay retort, the conditions of setting being alike. But in whatever manner fireclay may be applied in the construction of the distillatory apparatus of a gasworks, whether ovens, or brick or clay retorts, it is always infinitely superior to iron for the purpose. Therefore our observations concerning retorts will for the future be intended to apply to those of clay, unless otherwise expressed.

With regard to durability, brick retorts, built with care, are more lasting than those of clay; but any economy obtained on this score is always more than counterbalanced by the cost of construction, and the great quantity of fuel they require for carbonizing. Moreover, after two years or two and a half years' service, the fireplace of the furnaces becomes enlarged by the wear and tear of continuous working, so as to require a large percentage of fuel to heat the retorts. Durability is, therefore, not always an economy.

And here we may observe that of all the various apparatus in a gasworks, the retorts are by far of the greatest importance, and, regardless of cost, only those of the best quality should be employed, for any economy obtained at the expense of excellence in these articles must be highly detrimental to the interests of a company. The importance of this will be understood when we remember that an ordinary double retort will, during its lifetime, carbonize from 700 to 900 tons of coal. The price of the retort used for that purpose may vary from 7l. to 10l., a difference so insignificant when compared with the value of the coal, and the probable loss arising from using inferior retorts extending over a period of two or three years, as to create surprise that price should be any consideration, and that quality alone should not be sought.

During a series of years we used clay retorts manufactured by a certain firm, whose name, for obvious reasons, we withhold, which for quality nothing could excel. These have been in active service for from two years to two years and a half, and on taking them down, with the exception of cracks at the bolt-holes, they were as sound as on the first day they came on to the works, but, of course, worn at the bottom by the action of the scoop and rake; and on some rare occasions they have been withdrawn after a like period of use, without even the cracks at bolt-holes, and have been reset a second time, with new ones, with all success. The retorts referred to were single, oval-shaped, 8 feet long, 1 foot 9 inches by 1 foot 2 inches wide, and $2\frac{1}{4}$ inches thick, made in one piece, with the end complete; however, it must be also stated that they were set without transverse walls, each retort being isolated, as represented in the drawing, page 89, but without the brick arch and shield tiles.

These assertions may surprise many of our readers who are in the habit of enclosing their retorts by parallel walls hereafter mentioned, by which they are undoubtedly broken, either in consequence of the expansion, or the settlement of the brickwork, and are afterwards taken out piecemeal. In the cases referred to, these walls were dispensed with, and in taking down a setting, the ascension pipes and mouth-pieces removed, and the front wall cleared away, the retorts were taken out, as stated, generally cracked at their bolt-holes, and sometimes, although rarely, entire; but in all instances, from the bolt-holes throughout their length, there was neither crack nor fissure.

We are aware that the degree of excellence mentioned in the construction of clay retorts is achieved by comparatively few manufacturers; but as the means of attaining it are well known, depending mainly on the proper admixture of materials and careful manipulation, thus rendering the retort cellular and not of close texture, in order that when heated it shall neither expand nor contract, there is nothing to prevent every manufacturer of these articles producing them of the quality described, and gas companies should encourage such production.

Our conclusions on this subject may therefore be summed up in the following terms:

That for the economical production of gas its quantity or quality is not affected by the form of the retort.

That, for economy in fuel, clay retorts present great advantages over those of brick, and that both descriptions should be set with as little non-conducting material as possible.

That, for durability, the brick retort is superior to the clay, but that no economy can be obtained from its use; on the contrary, whatever is gained on this head is lost in the extra fuel required for carbonizing.

That the labour in charging and drawing retorts is always relatively greater when those of small dimensions are employed, and diminished as they increase in size. There is, however, every reason to believe that this, the most trying of all manual labour, will, in the course of a few years, be effected by machinery, at least in large and medium-sized works; and when we observe the numerous applications of mechanical power to various useful purposes, it will not be very creditable to the intelligence of gas engineers, if the manual labour now necessary for charging and drawing retorts is not superseded by motive power; and, as experience has always demonstrated, whenever machinery supersedes labour, it is to the advancement of civilization and to the profit of those immediately interested.

RETORT SETTINGS.

Theory of Retort Settings.

Combustion, so far as regards our subject, arises from the combination of the oxygen of the atmosphere with the carbon employed as fuel; and the more vivid this combination, the greater is the heat generated.

For the purpose of applying this heat with all due economy, and preventing any unnecessary loss thereof, we have to take into consideration four distinct points, namely:

1. The most advantageous method of obtaining the caloric requisite for carbonization.
2. The best means of conducting this to the retorts, and consequently to the coal undergoing distillation.
3. The means of preventing the heat passing off uselessly from the ovens to the atmosphere;

And, lastly, how to avoid any unnecessary loss of caloric by radiation, either from the foundations or the exterior of the benches.

With regard to the first consideration, that is, the most advantageous method of obtaining the caloric, the principal requisite is a proper draught or current. For this object the passage into the furnace at the ash-pans should be clear, the flues of ample dimensions, and the stack or chimneys of suitable height and internal area. The next consideration is the kind of fuel to be employed, whether this be coal (which, however, is rarely used for the purpose), or coke from cannel coal, or that obtained from caking coal.

When employing coal as fuel, the area of the furnace will be required of sufficient capacity for the admission of a proper supply of atmospheric air, to intermix with the volatile constituents of the fuel; by this means flame is produced, and the smoke which would otherwise exist is avoided, thus preventing the loss of heat and other inconveniences. When coke from cannel coal is used for the purpose in question, then, in consequence of the amount of ash intermixed therewith, and consequently a less vivid combustion, a comparatively large quantity is required, demanding a furnace of corresponding dimensions, in order that the intensity of heat acquired by the coke from caking coal may be compensated by the quantity in a less intense degree derived from cannel coke. On account of the uncertain proportion of volatile matter existing in the various coals used in the furnaces, and the indefinite proportion of ash in the various kinds of cannel coke, no definite rule can be applied in the construction of furnaces for these materials.

This, however, is very different with coke derived from caking coal, which invariably contains considerably less ash than the other, consequently possesses a greater quantity of carbon, and is more uniform in its composition. Moreover, this carbon being fixed, the supply of atmospheric air is required of greater regularity, and in less quantity than when consuming coal. For this kind of fuel, therefore, a furnace of small area and great depth becomes necessary, the principal advantage of its depth consisting in the fact that the air in its passage through the fuel acquires that high degree of temperature so indispensable for the indirect current hereafter referred to, and is transmitted to and absorbed in the various divisions of the oven, besides which another consideration is that the deep furnace requires less frequent attention in firing. The steam from the ash-pans is likewise decomposed when passing through the incandescent fuel, is resolved into its two gases, oxygen and hydrogen, and becomes also a source of heat.

THEORY OF RETORT SETTINGS.

The flame produced by the decomposition of water may be often observed in the operation of clinkering or clearing out the furnace, by the incandescent coke dropping into the water beneath, when steam is rapidly evolved, which, passing through the fuel, is converted into inflammable gas, producing an immense flame. This sometimes issues at the door of the furnace, from which the workman is obliged to retreat.

The fuel in the furnaces (when this is coke) should be maintained at a bright heat approaching whiteness, for as the caloric when generated is dispersed throughout the oven, and a portion of it is continually being carried off by the gas issuing from the coal, the temperature in the retorts will necessarily be considerably lower than the fuel. A large quantity of fuel may be employed in combustion, without intensity of heat being secured, which is of no practical use. A multitude of ordinary house fires would be insufficient to heat a single setting of retorts, inasmuch as the requisite high temperature for the proper distillation of the coal could not be thus attained.

Referring to the next point, the best means of conducting the heat, it must be first observed that the caloric as generated is transmitted in two distinct ways. Firstly, it is conducted in straight direct lines, radiating in every direction from the point of combustion, which, for future reference, we may term *direct* conduction. According to the other effect, the caloric is conveyed by the current or draught passing through the oven and flues in a circuitous direction; this we will call *indirect* conduction.

Undoubtedly the laws relating to the conduction of heat, whether for the production of steam or for the decomposition of coal, are identically alike in both cases, the heat being conducted or transmitted with the greatest facility where there exists the least obstruction, whether this arises from the quantity of the intervening material or its nature. These principles are well understood and taken advantage of by mechanical engineers in the construction of steam boilers, surface condensers, and such like apparatus, as also in the prevention of loss of heat by radiation, to which we will presently refer.

Prior to the general adoption of clay retorts, one of the most forcible arguments against their use, which was advanced with perfect reason, was founded on the nature of the material entering into their composition, being, as it is, among the chief non-conductors of heat. This admitted, it is evident that an excess of non-conducting material, either in the formation of the retorts or in the parts immediately adjoining them, as in the arch over the furnace, or thick walls abutting against the retorts, must be prejudicial to economical working.

Some engineers pretend, however, when once the arch is heated, that no further obstruction exists to the passage of the caloric. That this is erroneous is made evident by the arch, slabs, and shield tiles which were formerly placed to protect iron retorts from the direct action of the heat, which object by this means was effectually accomplished; if, therefore, this non-conducting material obstructed the passage of the heat in one case, it must have the same effect in the other.

From these remarks it must be observed, therefore, that any impediment or obstruction existing between the point of combustion and the retorts, except so far as may be essential to protect them from the most vivid action of the heat, must entail an unnecessary expenditure of fuel. The importance of this will be understood by observing the interior of a setting having an arch over the furnace, with the nostrils for the indirect passage of the caloric, when the difference of temperature beneath and above the arch is very marked, whilst the current of caloric rushing through the nostrils clearly indicates the serious obstruction presented to direct conduction, and the action of the heat at these points is often sufficient to fuse the retorts.

Consequently it follows that the two methods of conducting the heat, directly and indirectly, should be carefully studied.

We may therefore conclude that, in order to obtain economy in fuel for carbonizing, the existence of any unnecessary thickness of material should be avoided, and that, for the free conduction of the caloric to the interior of retorts, their surfaces, so far as practicable, should be exposed to the direct action thereof. If these views are correct, it is clear that brick retorts, having a thickness so much greater than those of clay, or any superfluous brickwork within the setting, and more particularly as

o

walls abutting on the retorts, is highly prejudicial to successful operation; we have the strongest conviction that if the cause of the extraordinary consumption of fuel for carbonizing at some gasworks were inquired into, it would be found to arise from these defects.

Moreover, we venture to assert that if the arch were removed no material variation of temperature would exist between the upper part of the furnace and the interior of the oven, and that the "cutting" heat which burns away the retort is the result of contracted passages, so causing the current of caloric to act with increased energy on the material. But it may be argued that the lower retort will need protection; this, however, we shall be enabled to show by settings actually in operation is not necessary when an open free communication exists between the furnace and oven; and in the event of the heat being excessive, this is easily avoided by making the furnaces smaller and thus economize fuel.

The same observations apply equally to the retort, the object of which is simply to separate its contents from the surrounding atmosphere; and if it were made, as has been proposed, of thin platinum, then, as only the smallest possible obstruction would be presented to the conduction of the heat to the coal undergoing carbonization, perfection in this respect would be attained. But there are difficulties of expense, and other considerations, which render this impracticable.

Following out this theory, we imagine that fireclay retorts ought to be made as thin as possible, consistent with security from breakage during transport and when in operation, and that any unnecessary thickness of material therein, or any obstruction to the free conduction of the heat as generated to their interior, either by slabs or walls abutting against the retorts, must demand extra fuel for carbonization, and be prejudicial to economical operations. In practice, however, these considerations are very often disregarded, for there are gasworks where brick retorts, having nearly double the thickness of those of clay, are exclusively adopted. Again, there are other works where transverse walls, often 9 inches thick, are placed within the oven at intervals of less than a foot, abutting against and surrounding the retorts, so that little more than one-half of their surface is exposed to the direct action of the heat.

The next question is the means of preventing the heat passing uselessly away from the furnace to the atmosphere. For this purpose the greatest care should be taken to adjust the damper to the required orifice. The damper, as is well known, is a fireclay tile, working on a flat slab, having the orifice cut therein, both of which should be nicely fitted and placed at the opening communicating with the main flue; under ordinary circumstances, when in action with a good draught, there is often little more than one square inch for each mouthpiece; hence, for six single retorts, an opening of six or eight square inches will be sufficient, but they are made considerably larger than this. In adjusting a damper it should be first closed to that point where the flame issues from the sight-holes, when, on gradually opening it until the flame ceases to issue, the proper degree of orifice will be then obtained. This is the point of separation between the interior of the oven and the external atmosphere, and should the opening therein be much exceeded beyond the actual requirements, the fuel will pass off uselessly, without any indication of the loss except in the dividends of the shareholders of the company. An instance of the waste arising from this cause, which happened at a Continental gasworks, once came under our notice, where the retorts were unprovided with dampers, consequently twice the quantity of fuel actually necessary was employed for heating the retorts. In this case, although the works were of moderate capacity, no less a quantity than four tons of coke passed uselessly away in a gaseous state daily from the chimney, a circumstance which clearly establishes the importance of this part of a retort setting, which is often sadly neglected.

The last point for our consideration is the prevention of loss of heat by radiation, either from the exterior or the foundations of the settings. For attaining this object the retort house should not be too open, the walls forming the backs of the settings when single, and those at the ends, should always be of good thickness, not less than 2 feet 6 inches, including the firebrick facings, and the thicker the better. The arches in like manner should be protected by a thickness of at least 18 inches of common brickwork on the top of the crown, and levelled throughout the length of the stack of settings. The lateral and main flues should also be enclosed at their tops and sides to a thickness of 14 inches, and the front wall of settings should never be less than that; and with attention to this point, stopped ascension pipes would be of rare occurrence. Sight-boxes, when used with clay retorts, are simply needed to observe their temperature, and any excess of these should be avoided. Two of them for a bed of five or seven retorts are all that are required, and if of $2\frac{1}{2}$ inches or 3 inches diameter, will be sufficient, their plugs being filled with fireclay. Lastly, all beds, when out of action, whether old or new, when in the vicinity of other retorts in operation, should be bricked up, the retorts having a 9-inch wall within their mouths, the furnace door and ash-pan closed in like manner, and the dampers of all settings not in use securely closed.

With these general remarks on the theory we proceed to the practice of setting retorts, reserving the description of the various working drawings connected with the subject, to be found at the end of the volume, in the appendix. And in presenting these drawings we may observe that without exception no particular theory or system is adopted, but settings such as are in use by various well-known engineers throughout the kingdom only are given, and by whom the drawings have been kindly furnished for this work.

Practice of Retort Settings.

The setting, or benches, of retorts are of two kinds, one called the "single," a term alike applying to the retort as to the setting. This consists of a certain number of retorts, each, as generally used, varying in length from 7 to 9 feet, and placed within an oven, the end of which is closed by a wall of considerable thickness, in order to prevent undue radiation of heat at that point. In small works these settings are placed side by side, extending the length of the retort house. In larger establishments they are set back to back, thus forming a continuous arch, divided by a wall in the centre, as represented in the drawing of Clegg's settings in longitudinal section (Figs. 19 and 20). These are usually employed in all medium-sized works, and are called "double benches."

Prior to 1849, single settings were exclusively used; but about that period, amongst other innovations introduced into the manufacture of gas, was the omission of the division wall, so that two single arches were converted into a "through" arch, and the retorts, instead of being, as formerly, 8 or 9 feet, were made 18 or 20 feet long, with two mouthpieces, and called "through" retorts, placed in their corresponding arches or ovens. Thus the retorts and settings are either "single" or "through," or double, as described.

The advantages derived from the introduction of the latter system are, the gain of the space previously occupied by the division wall, while, by the absence of that non-conducting material, the two furnaces assist each other and promote economy of fuel. Further, in this method the deposit of carbon on the retorts is removed in some degree by the passage of atmospheric air therethrough. However, these settings cannot be adopted in works of limited capacity, for as the two ends are required to be worked simultaneously, two gangs of men are necessary for the purposes of charging and drawing, as well as for barrowing or spreading the coke, so that less than eight men cannot work them.

Some idea of the various modes of setting retorts, and conducting the heat around them, has already been given when describing the earliest operations in this branch of gas manufacture. We now propose to describe the various methods adopted by different engineers in constructing retort settings, and applying the caloric for the purpose of carbonizing, and with the view of explaining this in the simplest manner, a few ordinary diagrams will, in the first place, be sufficient for the purpose.

In the accompanying Fig. 22, we find the direct conduction of the heat represented by straight darts, to which the slabs G are placed as obstructions to its passage, which formerly, when iron retorts were employed, were of the greatest desirability. The indirect conduction or current of heat is represented by the curved arrows; thus the heated air with its gases pass from the furnace, under the slabs, up by the sides of the retorts, so off by three or more orifices placed equidistant in the crown of the arch, to the small flue, which leads to the main flue.

In this method of setting, in consequence of the current of caloric passing directly through the oven, the full benefit to be derived from it is not obtained.

This we have assumed to be Clegg's system, and is still retained by some engineers; but is not to be recommended, on account of the extra fuel required for carbonization.

In the annexed Fig. 23, we have unquestionably the elements of the most perfect methods of setting retorts, in which no obstacle is placed to the direct conduction of the heat, whilst the current of caloric is employed to the best advantage. It possesses the further recommendation of cheapness of construc-

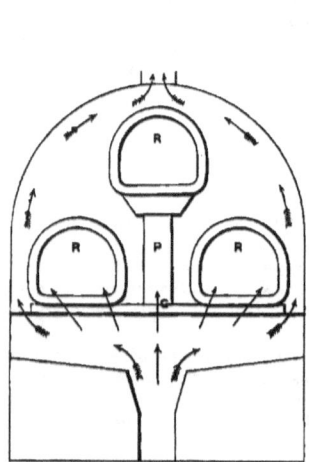

Fig. 22.

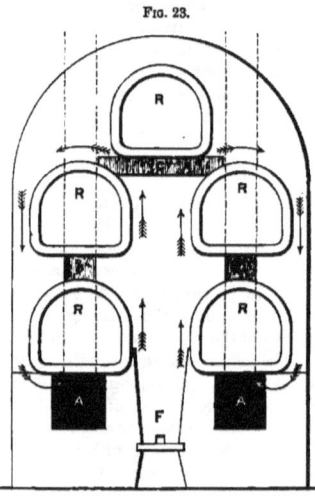

Fig. 23.

tion, and, as we shall have occasion to show, modifications of these settings are employed by well-known engineers, with all the simplicity indicated, without transverse walls, or any other brickwork beyond the quarries b b and the slab C.

As represented, F is the furnace, R R R the retorts, b b quarries which support the upper retorts, extending throughout the arch in order to prevent the current of heat passing at those points. C is a slab extending from the front wall to within a foot of the back, where an orifice of that length is left. A A are two flues under the lower retorts, communicating with their corresponding vertical flues, indicated by the dotted lines, on which, at the entrance to the main flue, are their respective dampers. The action is as follows: The heat is conducted directly without obstruction, whilst the current passing to the end of the slab C, issues through the orifice at that point, descends on both sides, as marked by the arrows, and advances to the front of the bench, where it enters the flues A A, whence it passes beneath the retorts to the vertical flues, and so to the main flue.

It must be observed, however, that the question of dispensing with the transverse walls must always be governed by the quality of the retorts, and it would be unwise in the extreme to dispense with these supports, unless the retorts are suitable. Hence in most works they are supported by transverse walls, which in all cases should be as few in number and as thin as possible consistent with security.

The simplicity of this arrangement is unquestionable, presenting as it does every facility for the caloric to pass to the retorts, and consequently to the coal undergoing carbonization. Some engineers may urge that the heat will be excessive, which objection is met by the recommendation to diminish the capacity of the furnace or fireplace, at the same time to economize fuel. Others may urge that because the principle is not adopted in some of our very largest establishments in the metropolis, consequently it possesses no advantage; but as we have had occasion to demonstrate the percentage of fuel used for carbonizing is considerably greater at these than at many of our provincial works, the objection, therefore, is favourable to the employment of the principle in question.

Fig. 24 represents a principle or system of setting retorts which is applied in the majority of large establishments. According to this method, the direct conduction of the heat appears to be a secondary consideration, inasmuch as thick arches are built directly over the furnaces, extending the whole length of the setting, while the current of caloric is caused to pass through nostrils of defined and often very limited capacity, by which means a considerable difference exists between the temperature within the furnace and that within the oven above.

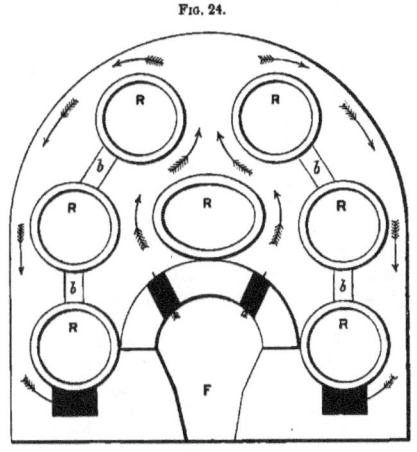

Fig. 24.

As shown in this diagram, there are four partitions, marked $b\ b\ b\ b$, continued along the whole length of the retorts, so that the caloric, in its course, can only pass through the space between the two upper retorts, where it descends, and, as in the foregoing figure, advances to the front of the setting, where it enters the flue A A beneath the retorts. In this setting it will be observed that a thick arch is placed, which must act, according to our opinion, as an obstruction to the direct conduction of heat, whilst the current can only pass through the nostrils, as indicated. In the majority of works the retorts are supported by transverse walls, in which there are openings for the current.

Another method, of which examples will be given, is to construct transverse walls within the oven, thus dividing it into two or more compartments. The heat passing from the furnace, and through the different chambers, descends to the flue at the lower part of the setting, and so passes off to the main flue.

Thus there are three distinct different principles applied in the ordinary setting of retorts—the first, in which they are heated by conduction from the furnace, and the current of caloric passing over the retorts direct to flue; according to the second principle, the caloric is conducted without any intervening body direct to the retorts, due advantage being taken of the current or indirect conduction; whilst in the third, a serious obstruction is placed to the passage alike to both means of conducting the heat.

Having expressed opinions in opposition to the practice of many engineers, we must observe that only after careful deliberation and the certitude of the accuracy of our statements has this been done. Our principal objection is to the use of the arch over the furnace or other superfluous brickwork within the setting, and that these must increase the fuel account there can be no question; besides, as we shall have occasion to show, there are engineers who with advantage dispense entirely with these as well as transverse walls. We are aware that for the purpose good retorts, that will stand the action of the heat without cracking, are indispensable, and that these exist is undeniable; we therefore urge that a more uniform and definite method in setting retorts is most desirable, for in this as in all other matters there can be only one correct system.

We now pass on to general details in the construction of retort settings, which observations are intended more especially for works of limited capacity.

General Details of Retort Settings.

The foundations of retort benches require not only to be solid in order to prevent settlement, but also to be constructed so as to prevent any undue loss of heat by transmission at these points. Their footings should extend somewhat beyond the actual position of the benches, a system observed in building ordinary walls and piers; where the soil is clay or of a hard nature, a depth of 2 feet or 2 feet 6 inches from the intended flooring of the retort house will be sufficient for the purpose, the foundation being executed with common bricks. Should the site be sandy or gravelly, a somewhat greater depth will be necessary, and if in a humid locality additional precautions are then indispensable. In these cases the foundations should be made of a depth of 5 or 6 feet, and puddled at the bottom and sides. Some engineers place inverted arches in their foundations, but as these offer no advantage over solid brickwork, they are not to be recommended.

The foundations being built within 6 inches of the level of the intended flooring of the retort house, the footings of the piers of the ovens, as well as the back wall, when single retorts are used, and the end walls, are then built, and on these are commenced the piers of the arches, which in very small works having settings of three retorts and under are only 9 inches thick, but in larger settings they are carried up 14 inches thick, with firebricks to the height of the spring of the arches. The end walls and the back walls should be also carried up to the same height, all the parts within the ovens being built with firebricks to the thickness of 9 inches, and the rest with common bricks.

For the interior of the furnace in immediate contact with the fuel the very best bricks that can be procured should be employed, for this is unquestionably the weakest part in a retort setting; and if bricks could be produced capable of resisting the heat and the fusible action of the fuel, they would be invaluable. We believe, for this purpose, the best at present known are those manufactured in Stourbridge, and the Welsh Dinas bricks. But a superior kind as regards resisting heat are made at "Nonsuch," near Ewell, and are named after the locality where they are produced; but, unfortunately, these bricks are very soft, and, in consequence, easily injured by the workmen when "clinkering." For the courses, however, beyond the action of the clinkering bar, as well as the arches, or when applied as slabs for protecting the retorts, the "Nonsuch" bricks can be strongly recommended.

For that part of the furnace in immediate contact with the fuel we have employed lumps of 6 inches thick, 10 or 12 inches long, and 9 inches wide, which have answered well, on account of the absence of joints where the ordinary bricks speedily yield, and are thus gradually destroyed.

In small settings of three retorts and under, the arches of the oven are constructed $4\frac{1}{2}$ inches thick, care being taken to fill in between them with firebricks to that height, so that not less than 14 inches of that material will exist between the arches, which rule should be observed in all settings. We knew an instance in which a 14-inch wall, forming the pier of two arches, having only firebricks at the outsides, and filled in with common bricks, the latter melted when in operation, and the arches of the settings gave way. Where the fire exists only on the one side of an arch or wall $4\frac{1}{2}$ inches thick, common clay will be baked on the other, but will be far from approaching fusibility unless in close proximity to the furnace; at this point there should always exist a thickness of 18 inches or 2 feet of firebrick material.

In settings of five or more retorts, the arches are built 9 inches thick, in two courses, and the lateral flues, as well as the main flue, when placed on the beds, are generally commenced on the crown of the arch. The lateral flue, leading directly into the main flue, is provided with a proper fire-tile damper

DETAILS OF RETORT SETTINGS.

made for the purpose, to which the brickwork should be nicely adjusted, serving as it does as a valve to control the passage of the air. The damper should either be placed level, with the extent of opening clearly indicated, or if in a vertical position, it should be suspended by a chain passing over a pulley with a counterbalance weight, or other means adopted to control its action, it being essential to know exactly the extent it is open.

Coke, as fuel, is employed to the best advantage when subjected to the direct action of the passing air. The formation, therefore, of any ledge or place within the furnace, where the coke could be placed beyond that influence, should be avoided. It is by no means uncommon for such a ledge to be built at the back of the furnace; fuel placed in such a position does not produce the best effect.

In small works, in order to repair the interior of the furnace, when this becomes much enlarged by the action of the fuel, it is advisable to build the furnace-door frame and adjoining wall in such a manner that they can be removed without interfering with the rest of the bench, so that when required, on the retorts being supported, the door-frame can be taken away, and an opening made in the wall to enable the workman to make the necessary repairs.

The furnace-door frame should always be extended to the floor, wherein it is built; its hinge or the frame should be placed at a slight angle; thus the door will remain closed without the intervention of the troublesome and useless latch.

Should the hinges and door not be made inclined, as indicated, the desired object will be attained by setting the frame at a slight angle; for which purpose its base is brought out a short distance, about 3 inches from the wall, care being taken to fill in the space between the framing and wall in order that air may not enter at that point. The simplest means of securing the door-frame is by bolts about a foot long, with nuts, the bolts being elbowed at an angle, close to the screw, and their claws or end built in a part of the brickwork beyond the destructive action of the heat.

The mouthpieces of retorts should be provided with wrought-iron lids, which are much lighter than those of cast iron, and thus relieve the men from a part of their hardest work. These lids are about a quarter of an inch thick, and to give strength to them they are hollowed or dished, having a simple lug on each side, so that they can be *dropped* into their place. These lugs should be long enough to enable the workman to take hold of them easily with his gloved hand, or while employing any other means to protect himself from the heat.

The socket of the ascension pipe should be cast with the mouthpiece, and of such diameter as to leave plenty of room for the joint. It has been supposed that by placing a short neck between the two the choking of the ascension pipe would be prevented, to which opinion, however, we do not conform. We believe the best method is to cast the socket directly on to the mouthpiece, leaving the necessary projection inside to prevent the cement dropping through.

Respecting the ascension pipes, beyond the difference in their first cost, nothing can be lost by making them large enough. In all new works they should be of 6 inches internal diameter, tapering to 4 inches, the usual size of the bridge and dip-pipes, and this we have always found large enough. When cementing the ascension pipe into the socket of the mouthpiece, a small portion of fireclay should be intermixed with the cement. The joint will in no way be deteriorated thereby, and, when desirable, it can be taken asunder with greater facility than if made with cement alone. In some instances we have observed, through the excess of sal ammoniac intermixed with the iron borings, that sockets have been split in such a manner as to render them useless, a circumstance occasioned by the expansion of the cement in setting. In all pipe-joints for gas we recommend, therefore, the admixture of a small portion of fireclay with the cement, which will prevent this casualty.

The bridge, or H, or N, pipes, are constructed of various forms, and so long as proper provision is made for clearing them when necessary, their shape is of no consequence. They are always attached to the ascension pipes by means of flanges and bolts and nuts, the joint being either iron cemented, or made with hempen washers or gummets dipped in a mixture of red and white lead, the other end generally having a spigot, which is connected to the dip-pipe with a lead joint.

DETAILS OF RETORT SETTINGS.

The dip-pipe is fixed to the hydraulic main with a flange and bolts, and hempen or cardboard washers. The flange is so situated on the pipe that, when in position, its lower part will dip about 1 inch into the liquid in the hydraulic main, whilst the upper part is 2 feet or so from the flange, and is provided with a socket for receiving the bridge-pipe. The shorter the dip the better; and if the hydraulic main could be secured always at a perfect level, then half an inch would be sufficient.

The stacks or beds of retorts are generally supported throughout their length and width by iron vertical buckstaves, held together with tie-rods. The former, by preference, are of wrought iron, having suitable holes for the rods, but usually they are of cast iron, 8 feet or 10 feet long, 1 inch or $1\frac{1}{4}$ inch thick, and 6 inches or 8 inches wide, and are set edgeways, one of these being placed in front of the centre of each pier of the beds. Thus, in "through" retorts, they are opposite each other, a tie-rod, having a screw at both ends, passing from the one to the other, and by means of which the buckstaves are bound together.

When laying firebricks (as indeed all others, when very superior workmanship is desired), they should be well wetted, the fireclay mixed almost in a semi-fluid state, used sparingly, and the bricks rubbed into their seats, so as at the same time to fill the end joints. All the intervening spaces should be well filled in, and the joints of the courses made as thin as possible.

The furnace bars and bearers are generally of 2 inches square iron, but in some works round bars are used of the same diameter. The bearers are built into the brickwork in such a manner as to leave a space of 5 or 6 inches at each end of the bars, thus allowing the air to pass freely at those points, otherwise the bars are speedily burnt away.

The best cement for attaching the mouthpieces to clay retorts is made by mixing one-third, by measure, of ordinary iron cement, without sulphur, but with excess of sal ammoniac, with two-thirds of fireclay, well mixed with water to the consistency of dough.

In fixing the mouthpieces the face of the retort is chipped, so as to present a rough surface; the bolts are then placed and embedded in the cement. The surface of the retort is then well wetted, and plastered with the same cement. The mouthpiece, being perfectly clean, is also wetted and plastered in like manner, and when placed in position is screwed up gently; afterwards, as it dries, from time to time, the nuts are tightened. The joints of mouthpieces attached in this manner seldom leak.

Ordinary firebricks are 9 inches long, $4\frac{1}{2}$ inches broad, and $2\frac{1}{2}$ inches thick, their average weight per 1000 being about $3\frac{1}{2}$ tons. The half or split bricks vary from $\frac{3}{4}$ to $1\frac{1}{2}$ inch thick, and are used for levelling the courses. Soap bricks are made $9 \times 2\frac{1}{2} \times 2$ inches, for the purpose of breaking joint when laying "headers." Arch bricks are made of the same length as the others to suit any curvature, and when ordering them it is necessary to give the thickness of the two sides.

Manufacturers of fireclay goods now offer the greatest facilities to engineers, supplying all the various pieces, as blocks, tiles, piers, dampers, required in setting retorts. Fire-tiles and slabs for guards or flue covers are kept in stock of all dimensions at most establishments, up to 20 inches square, and 3 inches thick.

The ash-pan should be of wrought iron; those of cast metal often crack, allow the water to leak, and cause considerable trouble. By some engineers these are built of brick, which answers every purpose.

In gasworks of the largest magnitude the benches of retorts are constructed on arches of equal width and depth to the ovens. This system, as we have shown, originated in the earliest years of gas lighting, and then as now these arches are sometimes used as stores for the coke until carted or stacked. The charging floor is about 6 feet 6 inches from the ground, the space beneath being occupied by the coke-hole or vault, represented in Plates 24 and 26; the coke as drawn from the retorts falls through openings into the vault beneath, where the spreader extinguishes it ready for removal. At the extensive works at Beckton an alteration has been made in the construction of this part of the works by dispensing with the arches beneath the benches, which arises from the magnitude of the operations, and the circumstance that the coke is carried away in waggons immediately it is drawn.

Engineers are divided in opinion respecting the utility of the charging floor and coke-vault. By some it is contended that the coke in falling from the height is broken and necessarily diminished in value, and that the advantages of the charging floor are not equivalent to the increased capital employed in the construction. Against these assertions it must be stated that the coke-hole dispenses with considerable labour, and that of the most painful nature, as it saves the men from exposure to the intense heat when discharging the incandescent coke, together with the labour of barrowing it sometimes a very considerable distance, when it has to be tilted and quenched. In addition to this, when the charging floor is on a level with the ground, the stokers are often compelled to discontinue their work in order to allow the barrows of red-hot coke to pass, whereas, with the other system, the carts or trucks are loaded at the coke-hole, thus saving immense labour and fatigue. These advantages, together with the increased comfort to the men by the comparative moderate temperature of the retort house when provided with a coke-vault, have all to be well weighed;—whether they compensate for the extra outlay of capital must remain a matter of opinion; but as regards the coke it is questionable if one system has any superiority over the other, as the breakage in tilting and spreading is probably equal to that of the coke falling from the retort.

When, however, we consider works of the first magnitude, as at Beckton, the furnace stage and coke-vault are indispensable; and only by the means there adopted, with the whole arrangement of piers, viaducts, rails, waterside communication, and other appliances hereafter described, could the operations of that gigantic establishment be carried out.

In the new retort house of the London Gas Company's works, the whole of the benches throughout are built upon arches corresponding in number and width with the beds, all communicating with both sides of the retort house; in addition, there is also a longitudinal arch in the centre of the stack, which extends with one interruption the length of the building. The tops of the arches adjoining the furnaces are only moderately warm; thus the whole of the coke-hole is well ventilated, and an agreeable temperature maintained, to the great comfort of the workmen employed. The saving effected by this arrangement in the construction must have been very considerable, and so far as our knowledge extends this is the only retort house built in this manner, which is worthy of imitation.

THE HYDRAULIC MAIN.

This was one of the earliest inventions of Samuel Clegg in connection with gas lighting, and, like all pneumatic appliances, is remarkable for its simplicity and efficiency; and although various attempts have been made from time to time to replace this by other means, nothing in that direction has ever been accomplished.

The principal purpose of the hydraulic main is to serve as a self-acting valve and shut-off of the gas, when opening the mouthpieces of the retorts in the act of charging and discharging; or to permit of the passage of the gas with the smallest possible resistance during the process of carbonizing. Its other purpose is to receive the first portions of the condensable vapours, of which from one-third to one-half is deposited in the hydraulic main, and, as will be found in the chapter on the "Chemistry of Gas Manufacture," Mr. Lewis Thompson considers that this part of the apparatus of a gasworks should be rendered an important means of purification.

The hydraulic main was formerly made of cast iron, but the liability of breakage at the flanges (by no means an uncommon occurrence) caused them of late years to be constructed of wrought iron, which, if slightly more costly, is amply recompensed by the increased safety, and the advantage of the radiation through the diminished thickness of metal.

Although the sectional form of this vessel has been frequently a theme for discussion, we are of opinion that this does not in any manner affect the efficiency of the apparatus, so long as there is abundant passage for the gas, sufficient area to contain the liquid necessary to give the desired seal to the dip-pipes, whenever the exhauster is out of action, and sufficient depth to meet the contingency of the accumulation of dust hereafter referred to.

Formerly the gas was taken off from the top of the hydraulic main, thus separating it from the mass of tar immediately it was produced, by which means it was supposed the purification was advanced, but, according to modern practice, for the reasons mentioned under the head of "Condenser," the gas and tar are allowed to flow away together from the hydraulic, and are thus kept in continuous contact during the process of condensation. In works of the first magnitude a main for conveying the gas runs parallel with the hydraulic, as represented in Plate 1, connections between the two, with corresponding valves, existing at intervals, and thus the gas and tar, as produced, flow from the hydraulic into the other main. In small works the same object is effected by taking the gas off from the end of the hydraulic, whence it passes to the condenser, which is sometimes formed by a series of pipes placed along the wall of the retort house.

As usually constructed each length of the hydraulic main is provided with a partition, the top of which is about level with the surface of the fluid therein; by this means, on supposing one joint in the length of the main to give, the leakage of tar therefrom would only affect the length of main in immediate contact therewith. At the Beckton Works, in the new retort houses, the hydraulic mains are of wrought iron and square in section, detached in sets of two. Thus by closing the valve of any two beds they are at once disconnected.

An objection sometimes raised against the use of the apparatus in question is the formation of pitch in its interior at the bottom; this, however, whenever occurring, can only happen when the main is exposed to the direct heat of the settings, as we have seen them sometimes placed directly on the brickwork, when naturally the temperature of the liquid will be augmented, and the deposit of thick tar encouraged. It should be borne in mind, however, that the liquid within the hydraulic, when beyond the influence of the heat of the retorts, does not exceed 180°, a temperature certainly sufficiently high to drive off all the lighter hydrocarbons, but to produce pitch the tar has to be submitted to its boiling point, which is above 500° Fahr.

We are disposed to believe that this accumulation met with at the lower part of the hydraulic occurs from another cause, that is, the dust which is always floating in abundance within the retort house. This, entering by the dip-pipes, when out of action, and detached from their corresponding ascension pipes, and mixing with the tar, gives it the appearance of pitch, and if allowed to accumulate, would eventually clog the dip-pipe, hence a precaution to be observed is to close all dip-pipes when not in use and detached, in order to prevent this dust entering, and producing the evil mentioned.

In our experience we have never met with a main clogged with pitch or thick tar, whenever it was placed at a suitable height from the beds, say 15 or 18 inches clear space between them. Under these conditions, even after a period of thirteen years' continued operations, there existed simply a layer of about $1\frac{1}{2}$ inch thick, which, on examination, was found to consist of the coal dust alluded to, intermixed with tar.

Under the head of "Carbonization" we have referred to the recent appliances to dispense temporarily with the seal of the dip-pipe during the process of carbonization.

CONDENSER.

The gas, on issuing from the retort, is intermixed with tarry and other vapours, which give it the appearance of smoke, more or less dense in proportion to the quantity of tarry matter in combination therewith. These vapours, when condensed, produce principally tar and ammoniacal liquor, the quantity of the former being controlled by the varied circumstances of the temperature of retort, the quantity of charge, the degree of moisture in the coal carbonized, and other circumstances.

The ammonia is produced by the union of its two elements, nitrogen and hydrogen, and by the presence of another element (oxygen) water is formed, and this, combining with the ammonia, produces the ammoniacal liquor, which of course is only in the state of vapour previous to condensation. A portion of the water exists in combination with the coal in a hygroscopic state before being submitted to the process of carbonization, whilst as stated from one-third to one-half of the liquids arising therefrom as well as that portion of the tar are deposited in the hydraulic main.

The greater portion of the ammonia passes off with the crude gas, and is one of its most important impurities, as well as one of the sources of profit to a gas company. There are, however, other vapours intermixed with and composing the gas, and are indeed its richest constituents, which demand the greatest care in order to prevent their condensation, and, consequently, the deterioration of the illuminating power of the gas.

The necessity of condensing the objectionable vapours was understood at the earliest experiments of gas lighting, and probably this knowledge was acquired from the inconvenience experienced by obstructed pipes, occasioned by the condensed liquid therein. In the preface of Winsor's 'Traité' the author lays particular claim to having been the first to introduce a vessel for the purpose of condensing and retaining the liquids derived therefrom, and in his patent of 1804 he refers to them as suitable " for making alum and for other chemical purposes."

The first practical application of the apparatus in question on an extensive scale was by placing a serpentine pipe within the tank of the gasholder, through which the gas passed, and as its vapours were condensed the liquids derived therefrom were conveyed to a receiver provided for that purpose, which system was subsequently improved upon, by enclosing all the pipes within a separate cast-iron tank filled with water.

Various other methods were afterwards tried; in one of these the gas, as it issued from the hydraulic main, was conveyed into a vessel, "and, by a contrivance somewhat similar to the shower-bath, it was washed and condensed before making its exit to the purifiers." The plan adopted about the same period by John Malam, consisted in having a cast-iron case, in which were a series of divisions or trays, each about 3 inches deep. These were placed the one above the other, at distances of about 6 inches apart from centre to centre, a space being left at each end alternately for the passage of the gas and water. The water was supplied from above, and on overflowing the top tray, passed to the second, and so on to the last. The gas entered at the bottom of the vessel, and proceeding in a zigzag direction over the whole of the series of trays, escaped at the top, and in its transit was cooled to the temperature of the water, and by its contact with the latter was deprived of a portion of its ammonia.

In 1817 Perks patented a vessel very similar in form to the ordinary air condenser; but the whole of the pipes were enclosed within a tank filled with water. The only merit attached to this invention being that it probably gave rise to the introduction of the system which has since been so universally applied. In these methods of condensing, the object was attained by bringing the gas in contact with water, in order, as we may suppose, to conduct the operation as speedily as possible, and in one instance, the washing and condensation were effected together, but by such processes great errors were committed.

With the air condenser, the caloric, in combination with the gas, is conducted by radiation to the surrounding atmosphere without the intervention of water, which apparatus in its simplest form is represented in the accompanying engravings, Figs. 25 and 26, the former being the elevation partly in section, the latter is the end view.

This is composed of a cast-iron chest, made in two pieces, the top and bottom, having end plates bolted to its flanges. On the top part are cast the sockets for the stand-pipes, and, within, as shown at the part in section, there are division plates, D D, between each pair of stand-pipes, which extend

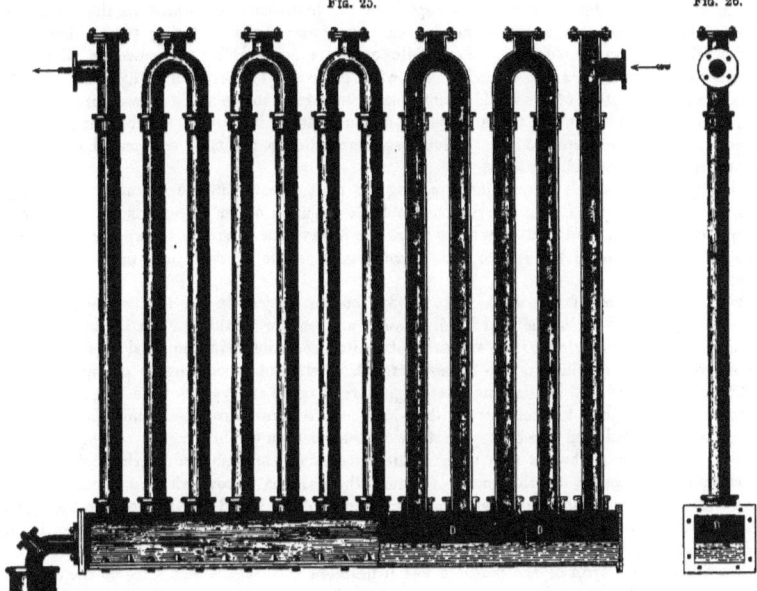

FIG. 25.   FIG. 26.

AIR CONDENSER.

nearly to the bottom of the vessel, leaving a space for the liquids arising from condensation to pass freely. The inlet pipe is in communication with the hydraulic main, the outlet being connected to the exhauster. The other stand-pipes are connected in pairs by the semicircular bends surmounting them, each bend being provided with a flange for the purpose of clearing out any obstruction that may occur. At the outlet end is bolted a bend having a clearing-out cap, which dips into the hydraulic seal, this being for the purpose of permitting the gas to escape in the event of any accidental stoppage, thus avoiding further harm.

The chest is charged to the height indicated with water, which, on commencing operations, is speedily displaced by the tar and ammoniacal liquor, and eventually the latter pass off as produced by the bend into the hydraulic seal, and thence to the tar-well. The arrows indicate the passage of the gas, which is forced to traverse all the pipes in succession, and as the heat is radiated, the gas becomes gradually cooled, or as generally termed, condensed, thus depositing the vapours in combination therewith. The simple means of radiation were, however, for many years considered ineffectual, therefore small jets of water were allowed to impinge upon the pipes, which was conducive to the more rapid action.

In large establishments, two or more rows of condensing pipes are employed, of diameter and length corresponding with the magnitude of the operations, the accepted rule for their dimensions being to provide an area of 6 superficial feet of pipe for a maximum production of 1000 feet of gas per diem.

This, however, is dependent entirely upon locality, for it is evident that the same rule cannot be applicable alike in cold and warm climates; as in the former, at times, heat would have to be conveyed to the condensers, in order to prevent the deposit of the hydrocarbons; whilst in the latter, for the purpose of bringing the gas down to the desired temperature before entering the scrubber, suitable means would have to be adopted. For while excessive condensation possesses its disadvantages, insufficient condensation has also its evils; and when gas enters the scrubber at a higher heat than is necessary, the temperature of the liquid therein being increased, its absorbent power for ammonia is in consequence reduced, and when tar is increased in temperature, its powers for absorbing some of the sulphur compounds are impaired. To arrest these impurities in the most economical manner, proper condensation is therefore indispensable.

Although the apparatus in question is among the most essential in the manufacture of gas, until within the last few years no other has received so little attention either as regards its functions or for the purpose of improvement, and only a few years ago it was considered that the apparatus which could reduce the temperature of the gas to the lowest degree in the shortest time was the most advantageous.

A curious instance of the want of knowledge on the subject in question was displayed by an engineer of eminence, who, in the year 1844, obtained a patent for "improvements in purifying coal gas, and increasing its illuminating power, and preventing its circulation being impeded by frost." This he proposed to effect by causing the gas to pass through a series of condensing or precipitating vessels, called "freezing cylinders," within which were agitators, to cause the gas to come in contact with the cold surface, and deposit its *moisture* in the shape of "icy particles, resembling hoarfrost." The patentee evidently believed these "icy particles" to arise from the freezing of the aqueous vapours combined with the gas, whereas they were nothing else than the rich hydrocarbons deposited in a solid state, and, of course, by this mode of action, the gas was impoverished in a most remarkable manner. However, by these means, at least one of his objects would be attained, for the gas being deprived of its carbon vapours, there would be no fear of the circulation of the marsh gas, or hydrogen, being "impeded by frost."

But this palpable error of the patentee was little more extravagant than some of the contrivances of his successors, who invariably attempted to reduce the temperature of the gas to that of the surrounding atmosphere, regardless whether this was in the height of summer or in the depth of winter. Possibly there may be some readers at the present day who imitate these operations, and thus, by excessive condensation, deprive the gas of its richest qualities; and it is also probable that, in some cases, in order to make up for this error, and to bring the illuminating power of their gas to the desired standard, they resort to the use of a considerable percentage of cannel coal. The folly of such operations can hardly be too much condemned.

To render the effects of cold on gas still more palpable, we must refer to the experiments of Mr. Vogel, instituted by him with the view of ascertaining the loss of illuminating power arising from the reduced temperature of gas. In these experiments the gas was passed through a U-shaped tube, immersed in one case in snow, and in the other in a freezing mixture, and was burned as close as possible to the tube, to avoid any elevation of temperature before it reached the burner. The light was compared with that given by gas at the temperature of the air, which at the time was 65° Fahr. This light being taken as 100, that given by the same gas at reduced temperatures in two series of experiments was as follows:

	I.	II.
At 65° Fahr.	100	100
32° Fahr.	76	85
4° Fahr.	33	40

After the gas had passed for some time through the freezing mixture, the walls of the tubes were covered with a crust of ice, caused by the freezing of the hydrocarbons. These figures show that the influence of extreme cold on the illuminating power of gas is very considerable, as also the necessity of preventing as much as possible it being exposed to any low degree of temperature.

Fig. 27 represents the elevation of a condenser adapted by Alexander Wright. Fig. 28 is a plan of the same, shown partly in section; and Fig. 29 is a section through the two pipes A and B.

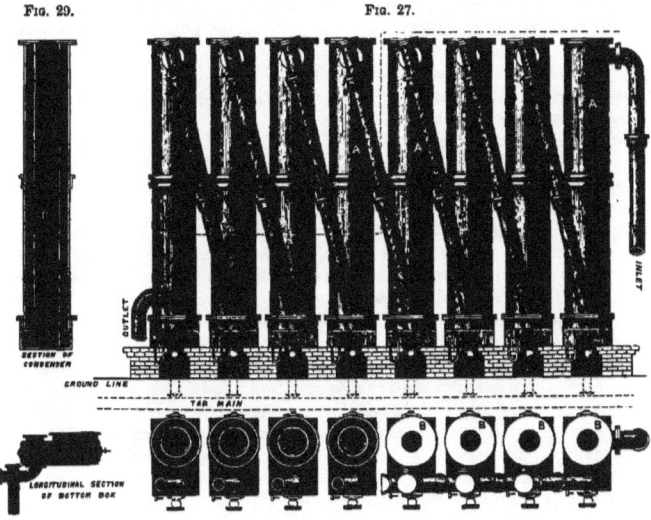

Fig. 28.
WRIGHT'S CONDENSER.

This apparatus differs from that already mentioned principally in its form, and is composed of eight large pipes or cylinders A A, each 18 feet high and 3 feet in diameter, and seven pipes of 12 inches diameter, numbered from 1 to 7, each of which connects the base of one cylinder with the top of that next to it. Within each of the pipes A A is placed another pipe, B, seen in section, and in plan, which is open throughout to the atmosphere; thus, when in action, the air entering thereby, in absorbing the heat, becomes rarefied, and, on ascending, causes a current or draught, which is considered the great advantage with this kind of construction. The gas enters at the top of pipe A, and descending, enters the bottom box, of which a longitudinal section is given, and then ascends that numbered 1, when it enters the second pipe, A, and so through all the apparatus to the outlet; each pipe A being provided with its separate box for the deposit of the condensed liquids, which pass off direct to the tar main.

CONDENSATION OF HYDROCARBONS.

It should be observed, however, that the idea of the condenser just described was obtained from one erected by Mr. Kirkham at the Pancras station of the Imperial Gas Company, in which the outer condensing pipes were 36 inches in diameter, and were connected together at the top and bottom, so that the gas, in its passage, ascended and descended alternately. In Wright's vessel, however, it always descended the larger, and ascended the smaller pipes, by which method, it was believed, the gas would be more readily condensed. In Kirkham's condenser, a door was provided at the bottom of each condensing pipe, which could be closed in cold weather, and thus, in some degree, the effective power of the apparatus was under control. This form was simplified by Warner, who divided the annular space between the large and concentric pipes, vertically into two equal parts, leaving an opening at the top for the passage of the gas. By this means the upper connections employed by Kirkham were dispensed with, and the gas followed the same course as in Wright's apparatus.

Such was our limited knowledge at the period of the introduction of the apparatus mentioned, that the fact of it being capable of reducing the temperature of the gas below that of the surrounding atmosphere regardless of the season of the year, was considered the height of perfection, and a special merit claimed for it by its author. Since then our views on the subject have become more extended, and at the present day, instead of the condenser being merely required to condense certain vapours, it has become one of the most important parts of a gasworks, and on the proper action of which depends to a considerable extent the illuminating power of the gas; it has, moreover, become a most important auxiliary in the process of purification.

Having briefly described the various kinds of condensers which at one time were very generally and are still employed by different engineers, we will now offer a few observations on the nature of coal gas, so far as regards the question under consideration, and thus lead to the present and most approved mode of condensation.

A simple gas is understood to be a permanently elastic fluid, under all the ordinary conditions of temperature and pressure, but in this respect coal gas differs very materially, being, as it is, a compound of true gases with variable quantities of the vapours of hydrocarbons. These latter constitute the principal illuminating ingredients of coal gas, and any diminution of them, either by condensation or otherwise, is detrimental to the gas as an illuminating agent.

The hydrocarbons—these rich constituents of coal gas—in addition to being affected as mentioned by extreme cold, are susceptible of being condensed and reduced into a liquid state at ordinary temperatures of the atmosphere, and, when exposed to a temperature below 50°, this condensation becomes very considerable, the action increasing with decreased temperature as already demonstrated, and once thus condensed, the resulting liquid is rapidly absorbed and permanently retained by the tar with which it is in contact, and the gas impoverished accordingly. The evil effects of this condensation are made apparent on taking down a gasholder after it has been in operation for some years, when its tank will be found to contain a large quantity of oily matter arising solely from the cause mentioned.

To illustrate the condensation of the hydrocarbons by the influence of cold, let us take an ordinary glass retort, such as is used in the chemical laboratory, and connect the supply pipe with its beak, and the outlet with the stopper part. If the retort be then placed in a basin or dish, and surrounded with ice, and the gas be allowed to pass therethrough, in a short time some of its hydrocarbons will be deposited in minute crystals on the surface of the retort, the gas being impoverished in direct proportion to the amount of the deposit. By this means, with protracted operations, the carbon vapours may, in a great measure, be extracted from the gas, leaving little more than the hydrogen and marsh gas, which, as illuminating agents, are practically useless. This experiment is more striking when applied to "air gas," or a combination of hydrogen with the vapours of petroleum, in which case the latter may be entirely separated.

This deposit of carbon vapours at low temperatures, and its injurious influence on the gas, being acknowledged, it is evident that the temperature of gas should never be permitted to attain a low degree; and we are, moreover, taught by these experiments that, in all cold climates—as in some parts of Russia, Norway, &c.—the gasholders should be enclosed within suitable buildings, and the mains laid at a greater depth than in more genial climes, in order to prevent the deterioration of the gas by the abstraction of its richest ingredients.

At the ordinary temperatures of the atmosphere, tar has a strong affinity for bisulphide of carbon, sulphuretted hydrogen, and carbonic acid; hence it is desirable to employ it for the purpose of removing these impurities. For this object the gas should be kept in contact with the tar as long as convenient, by having the condensing pipes of considerable length, and employing means to keep them at a moderate temperature; for if this be too high, the purifying power of the tar is diminished, and if it be too low, the hydrocarbons are deposited and absorbed by the tar as stated.

Another reason for observing moderate temperature in the condensation is that in proportion to the degree of the temperature, so is the strength of the ammoniacal liquor. Thus, in hot weather, under like conditions of quality and quantity of coal used, and quantity of gas manufactured, the strength of the ammoniacal liquor will be considerably lower than in very cold weather, a variation of 2° Twaddle being occasioned by the change of temperature.

It had been known for many years that the friction of the gas in passing through pipes, or that produced by the exhauster, assisted materially in the deposit of the tarry matters; but no practical advantage had ever been taken of the circumstance until within the last few years. Since then, by the action of friction, the tarry vapours have been eliminated from the gas, and in this consists one of the most important elements of modern purification of gas.

Lastly, it is well known that a sudden change of the temperature of the gas is conducive to the formation of naphthaline.

Thus, from these observations, we find the indispensable requirements for a condenser are:

That it shall be of such length as to permit of the gradual cooling of the gas.

That it shall be of comparatively small sectional area so as to produce that degree of friction to break all the vesicles of which the tarry vapours are composed, and thus eliminate the tar at this point, before the gas passes to the scrubber.

That a portion of the tar shall be retained in the condenser to act as a purifying agent to eliminate the impurities for which it has an affinity.

Lastly, that the apparatus for the various reasons stated shall be so placed that its temperature shall neither be too high nor too low, and, above all, that it should be protected from the influences of the sun, wind, rain, and frost.

For all these purposes, therefore, the form best adapted is the horizontal condenser, and whether this consists of a series of pipes placed around the retort house, either inside or outside, or whether the apparatus is similar to that about to be described, the conditions of temperature and dimensions being alike, the effect is the same.

The annexed Fig. 30 is copied from a photograph of a horizontal condenser, a description of which appeared in the 'Engineer,' in January, 1867, erected at one of the stations of the Staleybridge Gas Company, from the plans of Mr. D. A. Graham, by whom several similar apparatus of varied dimensions have been constructed at many other works.

This condenser is 65 feet long, and consists of 650 feet of 16-inch pipe, which, with the inlet and outlet pipes, the combined condensing surface, is equal to 4000 superficial feet.

The pipes are supported by a light iron frame, and placed in two tiers of five lengths of pipe, each with a slight incline equal to the diameter of the pipe in the whole length of the apparatus; the bends at the end being provided with caps for the purpose of cleaning out, should this be desirable. The crude gas enters at the top at A, and by the friction occasioned in its passage, the tar is eliminated and deposited in the pipes, when in consequence of the affinity the tar possesses for carbonic acid and the sulphur compounds these are absorbed and retained by it. Moreover, by the gradual cooling, the formation of naphthaline is avoided and the deposition of the ammoniacal liquor secured.

From the experience of Mr. Graham, who is well qualified to give an opinion on the subject, he estimates that by this apparatus he removes about 16 lb. of sulphuretted hydrogen and 5 lb. of carbonic acid for every ton of coal carbonized; the average strength of the ammoniacal liquor derived therefrom being about $6\frac{1}{2}°$ Twaddle, which of course will greatly depend on the conditions of the surrounding atmosphere.

Fig. 30.

GRAHAM'S HORIZONTAL CONDENSER.

Horizontal condensers of considerable length are now recognized as the most effective for the purposes of reducing the temperature of the gas and absorbing its impurities, but they are applied in various ways by different engineers; in many cases, as represented, as detached apparatus, and in other instances the pipes are laid around or along the retort house, either inside or outside, but always with the objects described in view.

At the Beckton Works, each condenser consists of 2600 feet of 12-inch pipe for a maximum make of 2,500,000 feet of gas, or about 3 feet surface per 1000 feet per diem. In this apparatus the engineer has evidently taken advantage of the great speed of the gas in its passage in order to acquire the friction necessary to deposit the tar, at the same time by the great length the effectual gradual cooling of the gas and the condensation of the objectionable vapours are attained. At the South Metropolitan Gas Company there are three of these condensers, the largest, constructed of pipes of 24 inches in diameter, the second of 18 inches, the third of 12 inches, and, according to Mr. Livesey's opinion, the most effective is the 12 inches, which he considers sufficient for about 500,000 feet per diem. The pipes of the apparatus at the works mentioned are, however, only about one-fifth the length of those at Beckton; thus the condensing surface in both cases is about alike.

We are indebted to Mr. Livesey for an account of some interesting experiments made by him, on condensers placed under cover, but open at the sides. According to these it appears that the temperature of the gas was reduced between the inlet and outlet as much as 30°, by a strong wind blowing on the pipes; this being about 10° or 15° lower than when no wind exists. Again, even in spring, when the sun shines, the temperature of the gas is 10° or 15° higher than when the air is clouded—circumstances which demonstrate the impossibility of controlling the temperature of the gas by the air condenser only when thus arranged.

To meet this difficulty at the South Metropolitan Works, the gas, after passing the air condenser, enters the water condensers, one of which is formed by 300 feet run of 18-inch pipes, the other of 960 feet of 9-inch pipes, placed in suitable tanks filled with water, which is so controlled as to reduce the temperature of the gas to about 50°. At Beckton a water condenser is also provided for a purpose identical with that described.

This influence of the wind and low atmospheric temperatures on air condensers has also engaged the attention of Mr. Graham, who proposes to cover them with boarding having ventilators at the upper part which can be opened and shut to suit any change of temperature, which system he has very recently carried into operation at the Middleton Gasworks; by these means the apparatus will be duly sheltered and protected from the changes of weather, and can be controlled to suit the variable make of gas whether this be large or small.

Amongst the most novel and interesting inventions for the purpose of removing the objectionable vapours of gas is the method patented by Messrs. Pélouze and Audouin—both names honourably associated with the science of gas lighting, which differs entirely from those hitherto mentioned.

Fig. 31.

PÉLOUZE AND AUDOUIN'S CONDENSER.

The mode of condensing, invented by these gentlemen, is founded upon the principle that the liquefaction of the globules held in suspension by the gas is effected either by contact of the particles with solid surfaces, or with each other; and the object is to obtain, by the aid of a very simple apparatus, occupying but a small space, the complete removal of the liquid particles carried along by the gas and vapours.

The action of the apparatus is as follows: The gas to be purified is made to flow through a series of holes of small diameter, so forming jets, which strike against a surface situated directly opposite. In the passage of the gas through the holes, the liquid molecules are brought into close contact with each other, and the operation is completed by the contact with the solid surface, upon which the tarry matter is deposited.

The apparatus is represented in Fig. 31, part shown as broken away in order to explain more clearly its action. A A A is the outer case, to which are attached the inlet and outlet pipes, I and O. Concentric with the case is fixed an annular tank B B, in the middle of which is left a free passage for the gas. This annular tank is charged by preference with dead oil obtained from tar, and in it is suspended the bell or holder C, counterbalanced as represented.

The side of the bell C for a portion of its height is formed of three or more concentric cylinders which are separated by a space of about three-eighths of an inch between them. Each cylinder is pierced, as represented at D, with a number of rows of holes each about one-twentieth of an inch in diameter, which are so placed that the gas in passing through the orifices of one of the cylinders impinges against the plate of the next; by this means all the tarry vesicules are broken up in a manner so complete, that when the apparatus is dispensed with, on a piece of writing paper being exposed to the influence of the gas issuing from a small jet, in the course of two or three seconds it is rendered almost black by the deposit of the tar; whereas, when the condenser is in operation under like conditions the paper is not soiled in the slightest degree. The tar is deposited on the side of the bell, whence it passes off by a suitable opening to the pipe p.

A pressure gauge, e, is placed for the purpose of appreciating when the holes in the bell become fouled, in which case simple means are provided for cleaning. The upper part of the vessel is a regulator to adjust the apparatus according to the various quantities of gas produced.

As regards the efficiency of the apparatus for eliminating the tar, it appears that there can be no question, whilst the general adoption it has received on the Continent is a further recommendation to its employment. To this must be added the considerations of its compactness and price, as a condenser suitable for 1000 metres, or 35,000 feet of gas per diem, occupies little more than a cubic yard of space; the price of such an apparatus being about 40l. Moreover, as they increase in magnitude, their cost according to their capacity is materially diminished.

But with these various advantages the apparatus possesses, it becomes a question whether the fears of some of our engineers, who imagine the hydrocarbon vapours to be deposited thereby, have any grounds for foundation. This, under certain conditions, appears to be decided in the affirmative in a striking manner by Dr. Bowditch, in his 'Analysis and Use of Coal Gas,' published 1867, wherein he states: "In the winter of 1853-4, I was anxious to procure a quantity of the condensable hydrocarbons which are contained in coal gas after it leaves the condenser, and before it enters the purifiers. To accomplish this, a rectangular tin-plate vessel about 14 inches long, 4 inches high, and the same width, was constructed and fitted with a series of diaphragms of wire gauze, having 1600 meshes to the square inch." Of these wire-gauze partitions, ten were placed in the longitudinal box in such a manner that the gas would pass through them. He says: "Having selected a night when it froze hard, the box was exposed to the air and a stream of crude gas from the purifying house of a gasworks was passed through it. In the morning a considerable quantity of condensed hydrocarbons was found floating on the surface of the water, which sealed the bottom of the box, and required nothing but separation from the water to be ready for examination. I mention this experiment *apropos* of sulphuretted hydrocarbons, which, I know, many chemists have failed to condense from gas, whatever amount of cold they have employed. The *only* way, I believe, to condense the hydrocarbons of gas—at all events a way which never fails and costs but little trouble—is to pass them through *sufficiently fine apertures* properly cooled. Gas which will deposit nothing in a half-inch tube, 4 feet long, immersed in a mixture of snow and salt, will part with its hydrocarbons readily if the tube contains diaphragms of wire gauze, such as I have mentioned. It is worth recording here that the first time I used this apparatus I separated two cakes of paraffin, each larger than a five-shilling piece and about the same thickness. The gas from which the paraffin was separated was made from common coal in *iron retorts*."

By this statement, it is obvious, that when submitted to a cold temperature the hydrocarbons are deposited by the mechanical action in question, but whether the same occurs at higher temperatures remains doubtful. Moreover, according to reports we have seen of several French engineers who have had Pélouze and Audouin's condenser in use for a lengthened period, the illuminating power of the gas is in no way affected by its use. With these opposite opinions, we leave the question to be decided by more general experience.

More recently Messrs. Pélouze and Audouin have patented the means of eliminating the ammonia by injecting water, in the state of fine mist or spray, into a vessel through which the gas passes as hereafter explained.

On account of the great irregularity in the production of gas, together with the variable influences of most climates, perfection in condensation is very difficult to attain. However, according to our present knowledge, whether the condensing pipes be placed in a continuous line against the wall of the retort house, or other buildings, as adopted by some engineers; or whether they be placed in a similar position, but in lengths parallel to each other, or as horizontal condensers; or whether they are arranged as a series of stand-pipes, with their communicating bends, as formerly very generally practised, certain precautions are necessary in order to ensure successful operation. For this object, the whole of the apparatus should be protected from the influence of the wind, rain, intense cold, and the sun's rays; for, by the action of the first three, the condensation might be so rapid as to remove a considerable portion of the illuminating constituents of the gas, and to which cause may probably be assigned the deposition of naphthaline, which is really the solidification of hydrocarbon vapours; and by the influence of high temperature in certain localities, that degree of radiation of heat, so essential for the removal of the objectionable vapours, cannot be obtained; when under these circumstances it will be necessary to cool the condensing pipes by causing jets of water to flow upon them.

The condenser in some establishments is one of the principal ornaments of the works, and it may be objected to make any alteration therein, but for effective working this should not be regarded. Moreover, the thermometer should be employed in every gasworks, attached to the outlet of the condenser, so that the temperature of the gas at that point could always be observed, and should it acquire that degree of cold which would be detrimental, then by the application of steam to the exterior of one or more of the pipes, or at one part of the pipe (there being a jacket arranged for that purpose), the necessary temperature to retain the hydrocarbons in suspension will be obtained, and, this accomplished, the form or method of applying the apparatus will not be material.

It may be argued that supposing the carbon vapours to be prevented from condensing at this point, and the weather to be severe, they would afterwards be deposited. For these apprehensions there are some grounds, but under no ordinary conditions can this be so prejudicial as during the process of manufacture, for if the gas be stored, it will be only that portion in immediate contact with the holder which would be subjected to the influence of the atmosphere; besides, as the gas is always entering at a comparatively high temperature, the condensing action will be in a measure counteracted, whilst the temperature of the mains at the depth they are laid is not influenced to any practical extent by atmospheric changes.

In the earliest years of gas lighting it was attempted to purify gas by passing it through a thick stratum of breeze, placed either in suitable vessels called "breeze-boxes," or in the ordinary purifiers. Like many other similar experiments of the time, this was probably made without any definite object, or obtaining any positive result; but the process was very serviceable, as it eliminated a large portion of the tar, which, however, was not appreciated, and in consequence of the nuisance connected with the removal of the residual, and the odour arising therefrom, the system was discontinued.

At the present time, however, this long-neglected appliance appears requisite, for although the modern mode of condensation leaves little to be desired, yet, as the scrubbers show in the course of time, there is still a quantity of tar deposited within them, and impedes their action in various ways, as by stopping the pores of the coke, or causing obstructions to the passage of the water, and directing its course in one particular channel. All this we believe would be remedied by placing a breeze-box after the condenser, for the purpose of catching any tar that may be in combination with the gas before it enters the scrubber, when the action of that important apparatus would be rendered more effective, the liability to obstruction or clogging, and the labour of renewing the material, be considerably reduced; whilst the residual derived therefrom could be employed in the manufacture of compressed fuel, pavement, and other purposes, thus contributing towards, or perhaps defraying, the expenses attending the process, but whether this would produce the effect of the *fine apertures* referred to by Dr. Bowditch practice alone can determine.

THE EXHAUSTER.

This vessel, as its name implies, is for the purpose of exhausting or withdrawing the gas from the retorts. The earliest mention thereof is in the patent obtained by Broadmeadow in 1824, wherein he describes the means "of exhausting or drawing the gas, either directly or indirectly, from the retort, oven, or other apparatus, by means of an air-exhausting apparatus, which may be either in the form of that which is usually called a pair of bellows, or an air-pump, or any other convenient form, which is placed between the retort, oven, or other gas generator, so preventing the waste usually incurred by the escape of gas."

Having improved on this method, Broadmeadow the following year acquired another patent for "an apparatus for exhausting, condensing, or propelling air, smoke, or other aëriform products." In his specification he says, "This invention consists in causing, by means of a suitable power, two gasometers, suspended one at each end of a beam or lever, to ascend and descend alternately in suitable tanks containing water. The interior of each gasometer is provided with an inlet and outlet pipe, the mouths of which are above the level of the water in the tanks. The inlet pipes are provided with valves, which open inwardly, and the outlet pipes with valves which open in the reverse direction. When the gasometers are set in motion the gas is alternately drawn into them until they become filled, and is then expelled therefrom." The patentee also describes the means of enclosing each gasholder, to the centre of which is attached a rod working through a stuffing box, so that it operates both in ascending and descending.

Although no drawings accompany the specification, the description is perfectly intelligible; but as the object was simply to reduce the loss of gas through leakage in the retorts, when those of iron were exclusively employed, the economy to be derived not being adequate to the expenses of the machine, it was never practically applied.

We have already referred to the inconvenience experienced by Grafton through the accumulation of carbon deposited in his ovens, and the means he adopted to remove it. But not satisfied with this, and suspecting it to arise from the pressure within the retorts, he entered into a series of experiments for the purpose of being assured on this point. In these experiments he submitted the gas to excessive pressure during the process of carbonization, which resulted in a remarkable increase of incrustation of carbon within the oven, this being upwards of 10 cwt. derived from the carbonization of 67 tons of coal. He then adopted the opposite means, by withdrawing all the pressure, except about half-an-inch seal on the dip-pipes of the hydraulic main, when, with the same description of coal employed as in the former experiment, he discovered that after four months' continuous working scarcely any deposit appeared.

The general impression is that, on Grafton becoming acquainted with these important facts, he obtained a patent for the exhauster as an instrument for preventing the accumulation of the troublesome incrustation. This, however, is erroneous, for although the experiments alluded to were made about 1840, and the patent relating to the exhauster was obtained the following year, yet in the specification of this there is no reference to the removal of the carbon; and it appears that Grafton's only object was to cause the tarry vapours given off during the first hours of a charge, to pass into a vessel which he termed a "decomposer," by which means he expected to decompose these vapours, and obtain a larger yield of gas.

The decomposer was an oven, having circuitous divisions charged with coke, and kept at a good heat by a furnace beneath. The gas was conveyed from the oven where the coal was carbonized into the decomposer "by means of an exhausting vessel," but he says "the exhausting apparatus itself forms no part of my invention, but only the application of any suitable exhausting apparatus for the purpose aforesaid." The claim of the specification is "for the application of an exhausting apparatus to the decomposer, to facilitate or compel the passage of the gas through the said decomposer."

It is a singular circumstance that although Grafton discovered the cause of the deposit of carbon on the retorts, yet in practice, according to his patent, precisely the same effect would be produced by the deposit of the hydrocarbons in the decomposer, and consequently deteriorating the gas. Thus, although he ascertained the means of removing the carbon in one instance, he deposited it in the other.

As the patent of Grafton possesses some interest, we give a copy of the drawings of the exhauster, of which Fig. 32 is a longitudinal and Fig. 33 a cross section.

T T is a cast-iron tank, in which are three inlet pipes, $m\ m\ m$, and a like number of outlet pipes, $n\ n\ n$, fixed within the tank. The inlet pipes are connected with the main pipe I, each of them having on its top a clack-valve, o, opening outwards, thus preventing the gas passing otherwise than in the direction marked by the arrow. The outlet pipes have similar valves, p, opening outwards, with a clearing-out box, b. The tank is filled with water or tar to the height indicated, and within it are suspended three

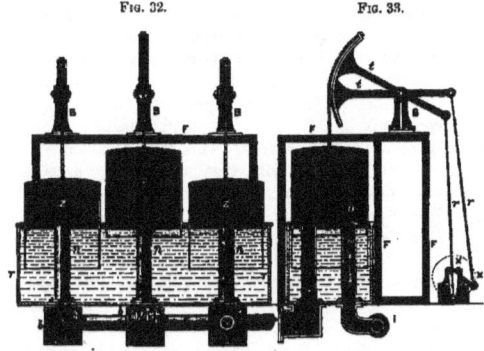

FIG. 32.   FIG. 33.

GRAFTON'S EXHAUSTER.

gasholders, $z\ z\ z$, the chains of suspension being attached to and passing over their respective curve levers, F F F is a frame for supporting the brackets, B B B, on which freely work the levers, $t\ t$, and to these are connected the rods, $r\ r$, of which, for the sake of simplicity, only two are shown; x is the shaft which extends the length of the apparatus, which shaft is provided with three cranks, placed at equal angles to each other, these cranks being opposite to the centre of their corresponding gasholders, and connected with their rods, $r\ r$. On a rotary motion being given to the shaft the three holders are alternately raised and lowered by the action of the cranks, thus receiving and expelling or exhausting the gas.

The reader will recognize a remarkable similarity between this and the apparatus employed some years afterwards at the Pancras station of the Imperial Company. The shaft of Grafton's exhauster with its cranks has only to be placed on the top of the frame, so as to work directly with the gasholders, and they become identical. Although it is evident that Grafton, at the time of obtaining the patent referred to, had no intention of applying the exhauster as a means of preventing the incrustation of carbon within the retorts, he very speedily understood its importance; for so early as 1843, exhausters were erected by him and were in operation at several Continental works.

The value of exhausters became about this period to be better understood, and a modification of John Malam's meter wheel, hereafter mentioned, was then adopted. For this purpose an oblong cast-iron case divided into two compartments was made, and in each of these was placed its respective wheel or drum, furnished with a hollow shaft having orifices in communication with the divisions of the wheel, which shaft protruded through the case into a box in connection with the inlet pipe. On the other end of the shaft was a toothed wheel, which received its motion from the engine. Thus by the revolution of the wheel, the gas was caused to pass through the water into the hollow shaft, and through the respective compartments alternately. After passing the first wheel it was conveyed to the second, and hence off to the outlet.

The apparatus, as described, was generally used combined as a washer and exhauster; but such was the want of intelligence at that period, that sometimes there was neither pressure gauge to indicate the amount of pressure or exhaust, nor the means of controlling or regulating the action of the machine; consequently, the gases of the furnaces, or atmospheric air, would often be intermixed with the gas, and at other times the apparatus would be in operation without producing any beneficial effect. Under such conditions it was not surprising if the operations of gas companies were not remunerative.

The progress of the exhauster, like the introduction of clay retorts, was very slow in England, which is the more remarkable as both the one and the other were practically applied at some of the larger works in France, years before they were used in London; and it was not until about the year 1847 that the instrument in question was employed for the first time at the Brick Lane station of the Chartered Company, under the superintendence of Mr. F. J. Evans. We believe the first exhauster there erected was that represented by the accompanying figure and commonly called the "two eights," so named on account of the form of its most prominent parts. This apparatus consists of an oval-shaped cast-iron case, provided with its inlet and outlet passages. Within this are two solid pieces resembling the figure 8 (which we will call propellers), their length being equal to the depth of the case, each having an axle working in suitable bearings on the front and back of the machine.

Fig. 34.

Jones's Exhauster.

The axles of the propellers protrude through the back of the case, and on them are keyed two toothed wheels of equal diameter and number of teeth; by this means their position in relation to each other is constantly maintained. Motion being communicated, the propellers rotate in the direction of the arrows, and in revolving, their ends are always in contact with the periphery of the case, whilst at the centre they are so adjusted that no material quantity of gas passes at that point. Thus by the revolving of the propellers the gas is exhausted.

F.G. 35.

Beale's Exhauster.

This machine is difficult to explain properly without several drawings or a model. It is very beautiful in conception, and was largely patronized by the engineers of many works during several years, until superseded by the apparatus now to be described.

The exhauster which next merits description is Beale's, shown in the annexed engraving, and consists of a cylindrical cast-iron case, with its inlet and outlet pipes attached. Within the case is another cylinder, the axis of which is eccentric to the case, its shaft protruding through a stuffing box at the back, and on which is placed the driving pulley. The inner cylinder is provided with two sliding plates, or pistons, which divide the vessel into two parts; and, as the cylinder revolves, these pistons slide in and out, so that their ends are kept constantly in contact with the periphery of the case. The gas being received on one side of the pistons and expelled from the other, is by this means exhausted. It is stated that these exhausters, when well constructed, give from 70 to 80 per cent. of effective power, and they are certainly the instruments most universally applied for the purpose.

Gradually, the utility of the exhauster becoming recognized, it underwent several changes in form and mode of action. In 1850 we saw for the first time the piston and cylinder, with clack-valves, applied for the purpose at the Newcastle Works. A few months afterwards we witnessed in operation, at the Independent Works, London, a double-cylinder exhauster, acting with slide-valves, similar to two steam engines placed side by side, the pistons receiving their motion from cranks on a shaft situated on the top of the frame; and from our experience during several years with this kind of apparatus we are convinced that all possible requirements are thereby fulfilled.

The piston and cylinder have been applied in various ways for the object in question: sometimes with a single cylinder without a governor, and running at a very high speed, in order to avoid the effects of the oscillation; sometimes with slide; at others with clack or seat valves, according to the judgment of the constructor.

Other engineers have adopted the one cylinder, in combination with an equalizing gasholder, counterbalanced to the desired pressure; thus any momentary excess of gas is drawn into the holder, which is again withdrawn therefrom by the next stroke of the exhauster. In some cases the holder driving the machine is placed so as to control the throttle valve of the steam engine.

The accompanying engraving, Fig. 36, represents a piston and cylinder exhauster as arranged by

FIG. 36.

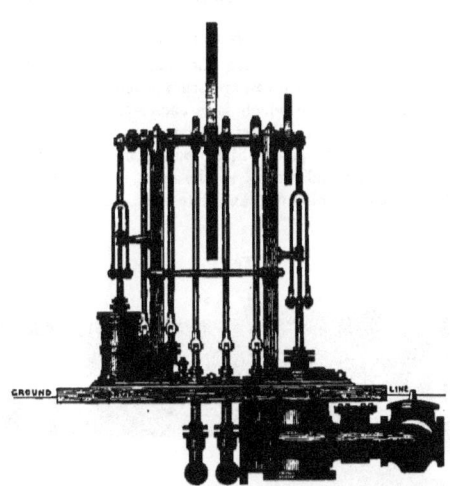

Mr. George Anderson, in which the engine and exhauster are both attached to the same bed-plate, and actuated by the same shaft; and on the latter are the necessary eccentrics for working the valves of the two machines, as well as for driving the tar, ammoniacal liquor, and feed, pumps. The whole is arranged in a remarkably simple and complete manner, which prevents the uncertainty of action, and saves the wear and tear of bands or straps, and the friction and noise of toothed wheels, or other kind of gearing. The combination altogether, embracing several different objects effected by very simple means, is a recommendation to this class of apparatus.

CLELAND'S EXHAUSTER.

The capacity of the exhauster cylinder, as compared with that of the engine, is as about six to one; the valves are the ordinary slides, and the very few parts of the apparatus ensure a more reliable action than can be attained with the generality of these machines. Mr. Anderson claims, by the use of this exhauster, considerable economy in the fuel employed, and cites certain instances in corroboration of his statement.

We next direct the reader's attention to an apparatus for the object in question, but differing from all others alike in construction and mode of action—namely, the steam-exhauster invented and patented by Mr. Cleland, of Liverpool, and since materially improved by Messrs. Körting Brothers—which operates without the aid of any intermediate mechanical appliances, the exhaustion being effected simply by the admission of a jet of steam into a peculiar arrangement of pipes enclosed within the main delivering the gas.

For the purpose of explaining this apparatus we may observe that it is well known that a current or stream of air, when moving through the atmosphere in a state of calm, produces another current or stream around it. Thus, if we blow into a still atmosphere, the current of our breath produces another current in the same direction. The action is rendered more demonstrative by placing a small tube within and concentric to another of greater diameter and length, both their ends being open, when, by blowing into the smaller tube, a current of air is caused to pass through the larger tube. And to approach still closer to the invention under consideration, if a jet of steam be allowed to pass into the smaller pipe, a current of air or gas would flow through the larger, which is precisely the principle of Mr. Cleland's exhauster, and, as it possesses considerable ingenuity and novelty, it merits particular attention.

The action of this exhauster in its simplest form is illustrated by the annexed diagram. Within

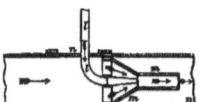

the main n, conveying the gas, is placed a funnel-shaped vessel, m m. The pipe l, terminating with a small orifice, or jet, depending on the size of the pipe in which it is enclosed, is fixed as represented, and when it is required to exhaust the gas, a current of high-pressure steam is caused to enter the pipe, which, issuing from the jet, passes in the direction of the horizontal arrows, at the same time drawing in or exhausting the gas, which flows as indicated by the arrows placed at angles to each other. The operation is so simple as to need no further explanation.

Fig. 37.

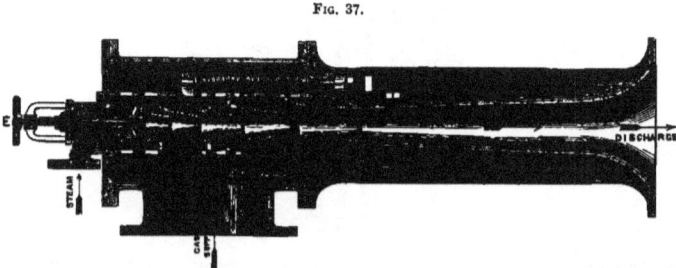

Körting and Cleland's Exhauster.

Fig. 37 represents a section of the apparatus as perfected by Mr. Ernest Körting, of Hanover, and as generally constructed. The gas, on entering by the inlet, is carried by the velocity of the steam through the several nozzles, thence in the direction of the arrows to the discharge, the water arising from the condensation of the steam being sometimes employed for the scrubber. By means of the spindle E the supply of steam is controlled in proportion to the make of gas, and a sliding sleeve, capable of being adjusted by the screw and nut on the exterior, serves to close a portion of the passage into the exhauster during the summer months, when the make of gas is diminished, thus regulating the instrument, so as to effect an economy of steam, and necessarily fuel.

These apparatus are generally constructed to employ steam of a pressure of about 45 lb., which is found to be the most effective and economical, requiring about six gallons of water per 10,000 feet, or say one ton of coal, against a back pressure of 12 inches.

The remarkable simplicity of the arrangement, the absence of noise during the action, the small cost of the apparatus, the limited space it occupies, are all strong recommendations for its use; but more important than these is the efficiency with which, as we are informed, it accomplishes its work; whilst the rapid introduction of these exhausters into gasworks of every magnitude must be regarded as a proof of the excellence of the system, which is further confirmed by the opinions of two metropolitan engineers of eminence, who have these machines in use at the works under their charge.

The same description of apparatus is employed for reviving the oxide of iron or other material in the purifiers without removal.

Various other methods of constructing exhausters have been adopted, some differing so slightly from those already described as to require no mention; others again, as the blower, similar to that used in the blast furnace, have been employed with doubtful success; and there are others, possessing more or less merit, which we have not space to describe. We now enter on the methods of working the apparatus under consideration, and the results.

As already stated, the deposit of carbon within the retorts, particularly at very high heats, is increased according to the degree of pressure, resulting in the deposition of the richest constituents of the gas therein, consequently its quality is materially diminished, and other inconveniences are experienced. To obviate these evils, and to ensure successful operation, the exhauster, under some conditions, is indispensable. With this machine all pressure upon the retorts can be entirely removed, but how far this is consistent and necessary will be shown. In illustration of the importance of the exhauster, let us suppose a works having, say, 20 inches of pressure (water being always understood) on the retorts, by no means uncommon where telescopic gasholders and small connecting mains exist. Under these circumstances, with high heats, the deposit would be very rapid; but if the pressure be reduced to, say, 1 inch, then it will be comparatively insignificant, whilst the yield of gas will be increased by the reduced pressure about 12 per cent., and its quality will be at least equal, if not superior, to that obtained without the exhauster.

This, we may be permitted to observe, is not an assertion based on theory or on mere statements or reports, but is the actual result of the experience of the writer, who, on one occasion, by an accidental breakage, was deprived of the exhauster during a period of two months whilst awaiting its repair, the pressure on the works being as described, when it was then ascertained that the difference in the yield of gas, with and without the instrument, was to the extent stated. Moreover, it may be remarked that, when working with the full pressure on the retorts, no extraordinary leakage was observed.

On the other hand, when the pressure on the retorts is limited to about 4 inches—as often happens in small well-constructed works where there are no telescopic gasholders and the apparatus is of ample capacity—then the utility of the exhauster becomes doubtful, and in our opinion the slight advantage derived therefrom under these conditions is not equivalent to the expenses of fuel, labour, and wear and tear attending its use, and the interest on the capital invested.

Hence, with heavy pressure on the retorts, the exhauster is of the first necessity, as it reduces materially the quantity of carbon deposited therein, and by the absence of this non-conducting material an economy in the fuel for carbonizing is effected; it reduces the loss by leakage arising from unsound joints, imperfectly luted mouthpieces, and defective retorts, and a larger yield of gas is consequently obtained from the coal. The greater the pressure, the more important become the functions of the exhauster, but when the former is limited as stated, then the utility of the latter is questionable.

The exhauster is generally placed between the condenser and scrubber, and in combination therewith there should be one or more pressure gauges in direct communication with the hydraulic main. In some large establishments, for the more perfect surveillance of the operations of this apparatus, an exhaust register is connected therewith. This instrument is similar to the pressure register, and to it is attached a paper each day, on which is recorded the amount of exhaust during every period of the twenty-four hours, and any accidental increase or decrease of this, if only momentary, is faithfully indicated.

The degree of pressure on the hydraulic main, however, does not indicate that on the retorts, which will depend on the seal of the dip-pipes. For if we suppose the combined pressure to be, say, 24 inches, and the dip-pipes to be 4 inches within the liquid, then on removing 20 inches of this pressure, the gauge communicating with the hydraulic main would indicate zero, or "level gauges," but there would still exist a pressure of 4 inches on the retorts; and by further exhausting, so that the gauge would indicate 3 inches exhaust, there would remain an inch pressure on the retorts. This is a point where considerable liability to error exists, as many engineers only take into account the pressure in the hydraulic main, which, as shown, is deceptive. The better method is to adjust the dip-pipes to the seal desired, say 1 inch, and work the exhauster to zero.

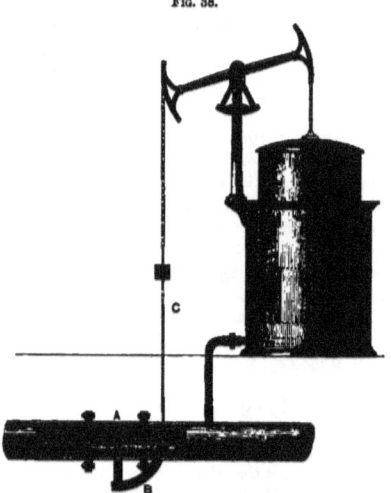

Fig. 38.

The question of the necessity for the anti-dip-pipes has been already referred to.

Mercurial pressure gauges, although sometimes useful, should never be applied in connection with the exhauster, as they are not sufficiently delicate for the purpose; water gauges, or the exhaust register, are alone reliable.

The irregularities in the production of gas, together with the uncertainty of the action of the steam engine, occasioned by the increase or decrease of pressure in the boiler supplying, render an automatic means of controlling the operations of the exhauster indispensable.

For this purpose in some cases the throttle valve of the steam engine is connected with a regulator. By these simple means the speed of the exhauster is made to correspond with the quantity of gas produced. But the method most generally adopted is to place a by-pass pipe in connection with the two sides of the exhauster, and in the centre of this by-pass is attached a throttle valve, connected to and operated by an ordinary gasholder.

This arrangement, as manufactured by A. Wright and Co., Westminster, is represented partly in section in Fig. 38. The throttle valve A works on a spindle, which, protruding through a stuffing box, is fixed to the quadrant B, and this is connected to the governor by means of the rod C, and chain which carries the counterbalance weights requisite to give the desired pressure. Thus, so long as the supply of gas is equal to the speed of the exhauster, the holder remains elevated as shown, but whenever the exhaust becomes excessive, the holder is depressed and the valve opened, when the effect is that an equilibrium is established at both sides of the exhauster, and the same gas is acted upon a second time.

In another exhaust governor patented by Mr. Hunt, the rod represented by the letter C is placed within and attached to the interior of the holder, whence it passes through an inch pipe direct into the main. There are various other apparatus for the purpose, which our space does not permit us to mention.

EXCESSIVE EXHAUSTION.

There are few subjects connected with gas lighting on which there is such a difficulty of forming an opinion as the exhauster, for, on reflecting on the various kinds described, we should suppose that the gasholder and tank, having the least friction, would be the best suited for the purpose; but after years of experience this was superseded. The friction of Malam's wheel, at any slight increase of velocity, was highly objectionable, and it is not surprising that its use was discontinued. The piston and cylinder being so universally employed in the construction of the steam engine, would suggest this as the best form for the exhauster, but this is not confirmed by practice. To judge from theory, we should be led to suppose that the friction attending Beale's exhausters would be such as to prohibit their use, but experience has demonstrated the contrary; and the fact of this apparatus being so extensively employed alike in the very largest establishments, as in those of limited capacity, is the strongest of all arguments in its favour; and in one instance within our knowledge, these have been in operation for twenty years without requiring any repair whatever, beyond being properly cleaned every summer.

One great evil to be avoided when using the apparatus in question, is the excessive exhaustion, for whenever this occurs atmospheric air or the gases from the furnaces are inhaled; under these circumstances the gas produced will be deteriorated in a marked manner, as only 1 per cent. of air reduces its illuminating power no less than 6 per cent., hence a slight pressure should always remain on the retorts, and the lighter the better, so long as the dip-pipes are properly sealed. And for the purpose of demonstrating the baneful influence of excessive exhaustion, we present the following results of experiments made by Messrs. Audouin and Bérard, with the view of ascertaining the loss of the illuminating power of gas by an admixture of atmospheric air therewith:

MIXTURE OF AIR AND GAS.

Quantity of air per cent.	Light yielded, pure gas being estimated at 100.	Light lost by the addition of air per cent.	Destructive power of each percentage of air.
1	94	6	6·
2	89	11	5·5
3	82	18	6·
4	74	26	6·5
5	67	33	6·6
6	56	44	7·33
7	47	53	7·57
8	42	58	7·25
9	36	64	7·11
10	33	67	6·7
15	20	80	5·33
20	7	93	4·65
30	2	98	3·26
40	1	99	2·47
45	0	100	—

From the preceding table we find that by an admixture of from 6 to 7 per cent. of air, the illuminating power of the gas is diminished one-half, and, as stated, that with the addition of only 1 per cent. of air, it is most materially deteriorated. Hence the necessity for the engineer or manager to be assured that the whole of the dip-pipes of the hydraulic main are properly submerged at the maximum degree of exhaust, so that, on the retorts being open, air cannot enter thereby, and that this exhaust should never be exceeded.

The perfection of these apparatus consists in the accomplishment of their duty with the smallest amount of fuel, particularly in localities where coal is expensive, when fuel becomes a very important question, and the choice of the most economical exhauster a matter for serious reflection.

THE WASHER AND SCRUBBER.

The gas, on issuing from the condenser, if properly arranged, will have deposited, so far as practicable, the whole of its tar and a portion of the other impurities, but there still remains a quantity of ammonia, carbonic acid, sulphuretted hydrogen, and other sulphur compounds, which it is essential to eliminate before supplying it to the public. To effect this, different processes are employed, to which we will now refer.

Ammonia, when existing with gas, is principally injurious on account of its destructive action on the metals—copper and brass—composing the fittings and parts of meters. In the earliest years of gas lighting the pipes for conveying the gas into the interior of dwellings were generally of copper, and by the ammonia combining with the metal a highly explosive compound was formed, which caused an obstruction in the pipes, and the friction necessary to remove this was often sufficient to cause ignition and serious accidents in consequence occurred, which were amongst the first inconveniences experienced from this impurity. In small quantities ammonia is by no means injurious to health, but often has rather an opposite tendency; and Mr. Lewis Thompson, with a sanitary view, recommended that a minute quantity of this should always be retained, in order to combine with the sulphur compounds, so as to produce the sulphate of ammonia, and thus counteract the pernicious effects of the sulphur. There is reason to believe that since the period when these suggestions were made, improvements have been introduced in the purification of gas which render such precautions no longer necessary.

Although, at the first attempts to introduce gas lighting, the greatest importance was attached to the value of the ammonia derived from coal in the manufacture of gas, and comprehensive evidence was adduced before a Committee of the House of Commons, demonstrating its numerous applications and uses; and although in the petition of the National Light and Heat Company to the King, the revenue to be derived from this residual was represented to be equivalent to about 36s. per chaldron of coal carbonized, yet for many years its value was practically disregarded. This, however, appears to have arisen from the want of a market for its sale, as Peckstone, writing in 1820, when speaking of the ammoniacal liquor, says, "This product has been manufactured into muriate of ammonia and some species of salts, but there has not yet been found a sufficient demand for the quantity so manufactured."

The ammonia for several years was not only unprofitable, but was an obnoxious, troublesome drug, causing the greatest difficulty in its disposal. In many cases it was allowed to flow into the sewer, river, or sea, according to the position of the works, thus often giving rise to serious complaints, and contributing greatly to retard the progress of gas lighting. Since that period science has taught us the vast importance of this residual, alike for agricultural as for numerous industrial purposes; and the superior intelligence now engaged in the direction of gas companies has turned the same to good account, and although very far from reaching the fabulous value given to it by the pioneers of gas lighting, ammonia is no longer a drug, but a residual of considerable value. In every gasworks, therefore, of any importance, whether at home or abroad, it must be regarded as one of the sources of profit, either by manufacturing its compounds on the premises, which is strongly recommended, as the process and plant for some of these are extremely simple, and the products very portable, and for which instructions will hereafter be given, or by disposing of the ammonia either in its liquid state, or in combination with sawdust saturated with sulphuric acid, to manufacturing chemists, who are established for that class of industry.

The first patent indirectly bearing on the subject in question is that of Wilson, obtained in 1817. He states: "These relate first to the purification of coal gas from sulphuretted hydrogen by means of ammoniacal gas, and may consist in causing a stream of ammoniacal gas to mix with the coal gas in its passage to the gasometer. The mixed gases may be caused to pass through a perforated plate placed under the surface of the water contained in a suitable vessel." Hence the idea of the washer was understood in those early days, and the system then proposed as a purifying agent has since engaged the attention of some very clever men, but hitherto without any positive beneficial result.

Water at ordinary temperatures possesses the property of absorbing upwards of 700 times its own volume of ammoniacal gas; at 32° Fahr. it absorbs 780 times its bulk, but in proportion as the temperature increases its affinity is impaired, and totally ceases at 130° Fahr. Hence the necessity of keeping all the apparatus as washers or scrubbers moderately cool, in order that the water shall possess the requisite power of absorption.

This knowledge of the affinity possessed by water for ammonia has been taken advantage of for the purpose of removing that impurity from gas and rendering it a source of profit. The most simple means for this object is the washer, similar in principle to that proposed by Wilson, represented in Fig. 39, which consists of a cylindrical or square vessel closed at the top and bottom, having a perforated plate or dashboard therein; through which the inlet pipe *a* slightly protrudes, the outlet pipe *b* being affixed to the cover. The level of the water is always maintained a short distance above the dashboard, there being a gauge *c* attached for observing this. Thus the gas in its passage has to force its way through the water, in which operation it is separated into innumerable globules corresponding in size with the orifices in the dashboard.

FIG. 39.

SECTION OF WASHER.

The smaller these orifices, therefore, the more complete will be the purification of the gas, in consequence of its numerous particles being brought into immediate contact with the liquid, and the greater will be the quantity of ammonia absorbed by the water.

Defects attending the use of the washer, as formerly applied, consisted in the great depth of the seal or dip of the inlet into the liquid, which was sometimes as much as 6 or 9 inches, also the large spaces left in the dashplate, these in some cases being 2 or 3 feet in area. Thus, in addition to the unnecessary pressure placed on the retorts, the globules of gas, when passing through the liquid, were of great magnitude, and as only their exterior was influenced by the water, the purification was very imperfect. It is desirable, therefore, when constructing these vessels, that the dip should be limited to two or three inches, when the exhauster is employed, but less than this in its absence; and that the holes of the dashboard should not exceed a quarter of an inch in diameter, placed as closely together and as numerous as possible, without materially weakening this part of the apparatus.

In the washer lately invented by Mr. George Livesey, the orifices are each about the tenth of an inch in diameter, of which there is an immense number; hence by these means the most minute particles of the gas are subjected to the influence of the water, the result being that 25 gallons of 10-ounce liquor are obtained from every ton of coal carbonized.

A system of washer has been invented by Mr. Cathels, wherein the gas is caused to pass through long narrow channels, situated just beneath the surface of the water; consequently fresh surfaces of each globule of the gas in its transit are brought into contact with the water, by which the ammonia is eliminated. That gentleman employs three of these vessels, placed at different elevations, in order that the liquid of the second can flow by gravitation to the first, and that of the third to the second washer. The foul gas enters the lowest, and passes off at the highest vessel, which contains comparatively pure water. Hence the main object of all washers of modern construction is to bring the gas in the most minute particles into communication with the water, so that the ammonia may be absorbed.

On the influence of washing gas, chemists are divided in opinion, some of whom maintain that by the action of the water a portion of the volatile hydrocarbons is eliminated from the gas, and of course its illuminating power deteriorated. Others assert that weak ammoniacal liquor has no prejudicial effect; with this, therefore, gas may be washed without injuring it. There are, again, chemists who believe that water, supposing this to be of moderate temperature, whether clean or otherwise, has no influence on the illuminating power of the gas. This opinion is corroborated by observing the water after having been in a wet meter for a number of years, which, in some instances, gives no greater indication of deposit of hydrocarbons than is found in a dry meter placed under similar circumstances; for under certain conditions, often unaccountable, dry meters after lengthened service contain a large portion of condensed hydrocarbons. However, excessive washing should be avoided.

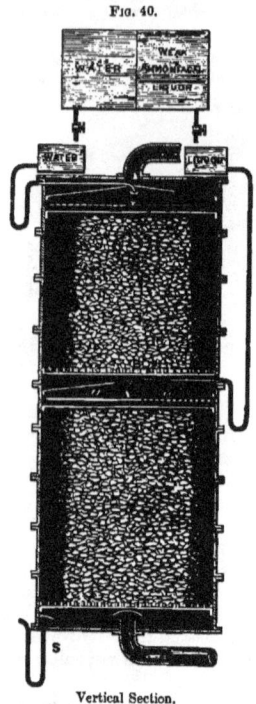

FIG. 40.

Vertical Section.
LOWE'S SCRUBBER.

One important advantage possessed by the scrubber is that the liquor can be retained therein until acquiring the requisite strength either for sale or manufacture, but in order to eliminate the whole of the ammonia two or more of these vessels will be required, the gas first entering that most impregnated as described.

Washers of various kinds were generally employed until 1846, when George Lowe obtained a patent for several objects, among these, "for revolving perforated pipes for supplying water or purifying liquor to the surface of the coke contained in suitable vessels;" the object of the patentee being to present a large wet surface for the absorption of the ammonia.

Fig. 40 represents a sectional view of the apparatus, as described by Lowe, and consists of a cylindrical vessel closed at the top and bottom, having the inlet for the gas below, and its outlet at the top. Within a short distance from the bottom are placed a number of bars, supported on the outer case by a corresponding flange, and in the centre of these is fixed a shield to prevent any coke or dust falling into the inlet pipe beneath. On the lower part of the cylinder being charged with coke, as shown, the cross bar is then fixed, and to this is attached the revolving perforated arms $a\ a$, having at their centre a funnel which receives the liquor from the tank above marked "liquor." The column of water within the funnel, acting similar to Barker's mill, produces the necessary rotary motion to spread the water equally over the coke.

The upper part of the cylinder is in like manner provided with the bars for supporting the coke with which it is charged, similar to the lower division. There are also the revolving perforated arms $b\ b$, which receive their motion from the tank indicated "water," the liquor from the two sources passing off by the syphon S. The action of the vessel is as follows:

The foul gas entering at the bottom passes through the whole of the coke, and this being saturated by the liquid issuing from the perforated arms, a portion of the ammonia is absorbed by the passing liquor, according to the varied circumstances of the height of the scrubber, the temperature of the liquor, the area of wet surface exposed to the action of the gas, and the quantity of this undergoing the process of purification. This was the first introduction of the scrubber, which has now become so universally adopted.

Although the value of the ammonia was not generally appreciated for some years, the utility of the scrubber became recognized, but merely for the purpose of removing the impurity in question, and in the majority of works without any attempt to derive any profit therefrom; and, very unfortunately for the shareholders, this system is still carried out at many works, by which a serious loss to the companies interested must always occur.

With the view of more effectually removing the ammonia from the gas, about thirty years ago some of the metropolitan companies, as well as a few of those of the provinces, adopted a solution of sulphuric acid as a means of purification, applied in a similar manner to the water in the washer, before described; and, so far, the method was successful. But a serious difficulty attended this, for as sulphuric acid possesses a powerful affinity for the hydrocarbons, these were deposited, and the illuminating power of the purified gas was materially diminished; but after some years of operations this defect was ascertained and the system abandoned. To this method of purifying gas by acids may, however, be probably attributed the care subsequently bestowed on the utilization of the products in question, for as large sums were realized from the sulphate of ammonia, obtained by that system of purification, the value of the impurity became recognized, and gradually other means were adopted to secure the profit. A further impetus was given by the establishment of chemical manufactories, first for the manipulation of the tar, and subsequently for the gas liquor.

These circumstances, added to the discovery of immense quantities of guano in Peru, and its general application to agricultural purposes, together with the knowledge that the same fertilizing agent could be derived from the refuse of gasworks, all tended to encourage the utilization of the residual in question. Nevertheless the progress of this was by no means rapid.

In works of small and medium capacity the scrubbers are of varied dimensions. In some instances, for works of 10,000,000 feet per annum, they are not more than 12 feet high and 3 feet 6 inches in diameter, in which case often the ammoniacal liquor is allowed to flow uselessly away. Where larger vessels are provided, as in works of medium capacity, the liquor is pumped two or three times into the scrubber before it acquires the degree of strength necessary to become a commercial commodity.

As usually employed, the scrubber is substantially similar to that described, and consists of a cast-iron vessel, either cylindrical or square, seldom exceeding 30 feet in height; the defect of this limited height under certain conditions will be hereafter explained. This is divided by horizontal partitions of cast iron into several separate compartments, each of which is provided with its door in the side of the outer case for placing and removing the material necessary for the proper distribution of the water.

The separate divisions are charged with coke, bricks, drain pipes, tiles, thin boards or other substances, presenting a large surface to the action of the liquid and gas, according to the judgment of the engineer; but, unquestionably, coke is the material most generally employed for the purpose. When in action, water is allowed to enter in a limited stream at the top of the scrubber, and is distributed over the area by some mechanical contrivances, as that described by Lowe, or tumbling vessels, rose jets, or other means, and in descending, percolates through the coke, thus presenting a large area of wet surface, whilst the gas in ascending is broken up, as it were, continually offering fresh particles to the absorptive action of the water, and by these means the ammonia is gradually eliminated.

To Mr. Mann, formerly engineer of the City of London Gas Company, we are mainly indebted for the present state of perfection attained in scrubbers, by which the foul gas enters at the bottom of the vessel, and is emitted from its summit without any trace of ammonia, the whole of this being removed at one operation; and in addition to this advantage, the ammoniacal liquor as it issues can be obtained, concentrated to 20 or 25 ounce liquor, according to the supply of water. In order to secure these important results, that gentleman ascertained that the height of the scrubber was the first consideration, and that its horizontal area should be controlled by the magnitude of the operations, and that the liquor should be distributed with scrupulous regularity throughout the whole area and depth of the apparatus.

Hence we find that these apparatus, as manufactured by Messrs. Walker, of London, for works of large magnitude, are upwards of 60 feet in height, and of a diameter of 18 feet, representing a cast-iron tower surmounted by a penthouse, which contains the gearing, the reservoir, and the means of controlling the supply of water, as shown in Plate 10, which represents two of a set of five scrubbers existing at the London Company's works. The action is identical with that of the other vessels mentioned, but the advantages of the height will be understood when it is stated that at about one-fourth of that, the liquor acquires a strength of 16 ounces, at half the height it becomes 20 ounces, and when issuing from the top of the scrubber the gas is entirely free from ammonia, which result, as we are informed, is guaranteed by the firm mentioned.

From this we learn that at the height of about 15 feet two-thirds of the ammonia contained in the gas are removed, and at about 30 feet high four-fifths will be absorbed, the remaining portion of the height being requisite to remove the last traces.

These apparatus are employed at several metropolitan and other works in England and on the Continent, and are of the utmost importance alike for the freedom of the gas from ammonia when subjected to their influence, as well as the concentrated state of the issuing liquid, which is admirably suited for manufacturing upon the premises where produced; to which may be added that the power required to give motion to the distributors and pumping the water is merely nominal, and as the whole of the ammonia is eliminated, it must be admitted that these scrubbers fulfil all possible requirements.

Here may be observed that the average strength required for the ammoniacal liquor when supplied to manufacturing chemists is 10-ounce; of this, as already stated by Mr. Livesey's scrubber, 25 gallons are obtained from every ton of coal, and as this realizes on the average about 7s. per butt, the return from this source under these conditions is about 1s. 8d. per ton of coal carbonized. With Mann's scrubber, however, we find the liquor concentrated to 25-ounce strength, consequently each gallon of this possesses twice and a half the quantity of ammonia contained in the other, and is therefore much more valuable for manufacturing purposes. The yield of ammonia from different coals is variable, hence with some kinds, the results mentioned may be exceeded, with others they will not be equalled, but in all cases this impurity should be regarded as an important source of profit.

It must not, however, be supposed that in medium size or small works the ammonia can only be eliminated by having the scrubbers of the great height mentioned, for in these, as in all the other apparatus of gasworks, their dimensions will be controlled according to the quantity of gas passing. By the best authority we are informed that each of the large vessels mentioned is capable of effectively removing the ammonia from a maximum production of one million cubic feet of gas per diem; thus we learn that its capacity is equal to 15·5 cubic feet per 1000 feet, which furnishes us with the basis for calculating the dimensions of a scrubber of any desired magnitude. For this purpose we will assume the diameter to be one-fourth the height, then taking as example a works of a maximum production of 50,000 feet per diem, or about 10,000,000 per annum, the capacity of the scrubber would be 775 cubic feet, or 6 feet 3 inches in diameter, by 25 feet high; or, for a make of 100,000 feet, the dimensions would be 8 feet by 32 feet; and again, a scrubber for 250,000 feet daily maximum production would be required 10 feet 6 inches in diameter, and 44 feet high, as represented in Plate 9. It is understood that only the dimensions of the effective part of the vessels are given, the penthouse not being included in the measurement.

All these dimensions are considerably larger than generally adopted in practice for works of the respective capacities enumerated, and we are led to believe that the frequent complaints of the inefficiency of the apparatus under consideration have arisen mainly from their limited dimensions. Next in importance to their capacity is the necessity of employing means to distribute the liquor regularly over the whole area, to avoid as much as possible the tar entering into the scrubber, and to place the porous or other material in such a manner as to prevent the water running in channels instead of being dispersed over the whole area of the apparatus. One advantage of coke for the object is the extensive area of its pores for the action of the water; but, on the other hand, these are susceptible of being closed or obstructed by the tar. Some engineers fill the scrubber with their boards, drain pipes, shavings, or any other material which presents a large surface.

The liquor derived from the condenser will always be found too weak for sale to the manufacturers, consequently it has to be passed through the scrubber or washer in order to attain the desired strength. According to Mr. Graham, the average produced by his condenser is $6\tfrac{1}{2}°$ Twaddle, but has fallen as low as 5° during the hottest weather, and has reached as high as 8° during the coldest days of winter; with ordinary vertical condensers, the strength of the liquor will be very considerably lower than this.

Among the most modern appliances for eliminating the ammonia from the gas are the two inventions of Mr. George Anderson and Mr. Paddon.

The first of these, called by Mr. Anderson the "brush scrubber," is represented in section Fig. 41, and consists of a square tank in a vertical position divided into four compartments, 1, 2, 3, and 4, each of which contains a drum, c, revolving with its axle, which protrudes through a stuffing box at the back. To the axles are attached toothed wheels in such a manner that the whole of the drums are geared together; thus on motion being given to one, the others are set in action, revolving as indicated by the arrows. Around these drums are attached pieces of cane or whalebone, D D D, radiating as in an ordinary rotary brush, so presenting a large surface for action. Thus these brushes in revolving dip into the liquid which washes off the ammonia, and on issuing therefrom they present a large wet surface for absorbing it.

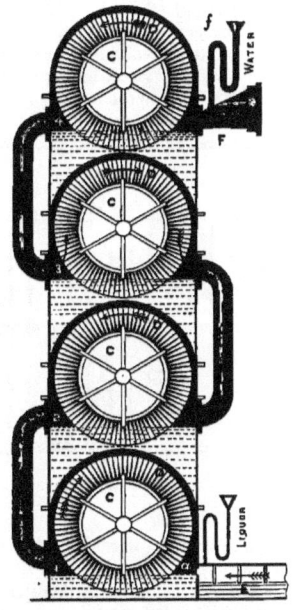

Fig. 41.

Vertical Section.
ANDERSON'S BRUSH SCRUBBER.

On water being admitted into the funnel f, it passes downwards, supplying the four chambers successively, and when flowing from E the vessel is charged. This accomplished and motion being given to the drums, the foul gas enters chamber 1, and so in succession as marked by the arrows, and passing off at chamber 4, by the pipe F. It is obvious that with this apparatus the liquid can be retained within the vessel until acquiring any desired strength, which is one of its advantages, and, as regards its efficiency, we are informed that with those in action the whole of the ammonia is eliminated at the third compartment.

The apparatus employed by Mr. Paddon for the purpose consists of a cast-iron tank of about 20 feet long and 10 feet wide, square at the bottom and semi-cylindrical above, as shown in Fig. 42, which represents a longitudinal section, and Fig. 43 a cross section.

In this tank is adjusted a shaft, $a\ a$, extending the whole length of the tank, revolving in suitable bearings, $b\ b$, having two intermediate supports, S S. On this shaft are fixed eighteen perforated discs, each of the diameter of about 9 feet 9 inches, revolving freely within the tank. Attached to the shaft is a toothed wheel c, working in a pinion e, the spindle of which passes through a stuffing box, and to this spindle is fixed the driving wheel d, which receives its motion from the engine; and on the tank being about half filled with water it is ready for action. The gas enters by the inlet F, and in passing through the apparatus, on coming in contact with the great extent of wet surface of the revolving discs, the ammonia is absorbed thereby, which, as the discs revolve, is washed off and retained by the water in the vessel.

There are three of these vessels in operation at the Gasworks at Brighton, for a maximum daily make of 1,500,000 feet; the first, in which the crude gas enters, is placed somewhat lower than the second, and this again lower than the third. Thus as the liquid in either of the vessels becomes saturated to the desired extent it is allowed to flow by gravitation.

That this system of scrubber possesses some advantages there can be no question, principally because the same amount of absorbing surface for each revolution must always exist; which observation applies to Mr. Anderson's arrangement. Both of these apparatus also possess the advantage of the scrubber, that is, of retaining the liquid until it acquires the desired strength.

Among the other methods proposed to eliminate the ammonia, is the steam scrubber of Mr. Cleland; and although there is no novelty in the application, it is mentioned on account of the system being recently adopted at some works. Theoretically we should imagine the application of steam for the purpose to be ineffective, for the reason, as already stated, that water when at a high temperature ceases to absorb the impurity under consideration, and only when its temperature is reduced to about 50° or 60° is its action really effective. It follows, therefore, that so far as regards the absorption of ammonia, steam answers no purpose, but whether it assists in eliminating the tar is beyond our knowledge, and as we were not fortunate enough to receive a reply to our applications for information on the subject, we are not in the position to furnish further particulars of the process.

Lastly, we have the invention of Messrs. Pélouze and Audouin, by which the gas intended for purification passes into a closed cylindrical or square tower. In this there are a series of small jets, and from each of these issues a fine spray of water or mist of about 30 inches in diameter and from 7 to 10 feet

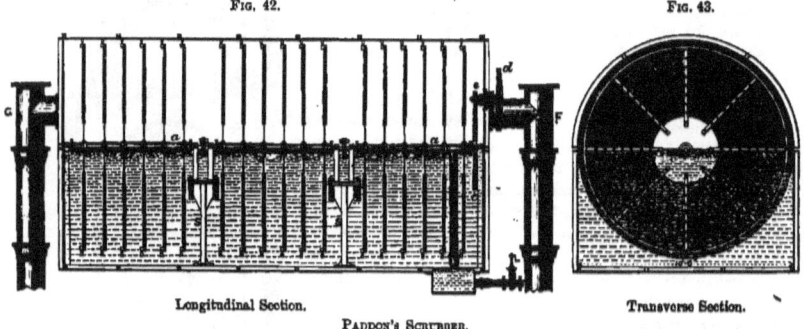

Fig. 42. Fig. 43.

Longitudinal Section.   Transverse Section.

PADDON'S SCRUBBER.

high, which, coming in contact and mechanically held in suspension with the gas, is carried therewith and afterwards condensed.

The water is converted into spray, by the action of a small quantity of the gas produced being compressed; a jet of this meeting the limited supply of water at the point of issue produces the mist necessary for the purpose of combining with the ammonia. This system possesses the advantages, that the gas and water meet in the most minute particles, that the latter can be supplied at any desired temperature, and can be employed until acquiring the necessary strength.

We have shown, therefore, that there are several means of removing the ammonia—namely, the various kinds of washers, Anderson's brush scrubber, Paddon's apparatus, and that of Pélouze and Audouin, by all of which the liquor can be retained until reaching the desired degree either for manufacture or sale. By Mann's scrubber the liquid is obtained in the most concentrated state at one operation, and we have the impression that if our suggestions as to the dimensions of the ordinary scrubber be observed, the apparatus will be found efficient in works of every magnitude.

Mr. Whimster has very successfully combined an exhauster with the washer, of which, as we are informed, there are several in use at gasworks throughout Scotland. This apparatus is somewhat similar to Grafton's exhauster, but unquestionably much superior for the purpose of a washer, on account of the large surface exposed to the action of the water, by which the ammonia is absorbed. In construction it is a wheel, divided into twelve chambers by a corresponding number of curved partitions radiating from the centre, the inlets of which are on the periphery of the wheel, and the outlets at the centre beneath the water line. The wheel is supported by a cylinder concentric to and attached to the shaft, and enclosed in a cast-iron case. The shaft revolves in suitable bearings, and carries a cog-wheel of larger diameter, which gears with a pinion, receiving its motion from its axle, in connection with the engine. By this instrument the pressure can be controlled with precision and uniformity, whilst by the extensive surface exposed to the action of the water, the greatest portion of the ammonia is eliminated.

Having thus attempted the description of the most prominent apparatus before the public for removing the ammonia, we have now to direct the reader's attention to the means of appreciating the value of the various solutions, whether arising from the condenser, washer, or scrubber.

The value of the ammoniacal liquor is dependent on the quantity of ammonia intermixed therewith, which is generally determined either by ascertaining its specific gravity with the Twaddle hydrometer, or by saturation with sulphuric acid.

The Twaddle hydrometer is usually supplied with the glass jar necessary for observation, both of which can be purchased for a few shillings. When operating, the jar is nearly filled with the liquid under investigation, when the hydrometer is gently placed therein, and when it becomes steady, note the degree on the scale level with the surface of the liquid, which being doubled, expresses the strength in ounces; for instance, if the degree be 4, this will represent 8 "ounce liquor," signifying that eight ounces of strong sulphuric acid will be required to neutralize one gallon, and this, by reference to the annexed table, will be found equal to 1020 specific gravity, one foot of which weighs 10·2 lbs.

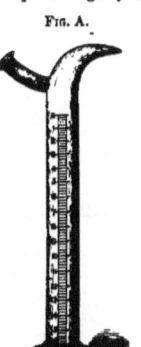

Fig. A.

For the other method an alkalimeter will be required, which is a glass vessel divided by a corresponding scale into sixteen equal parts, as represented in the accompanying engraving, Fig. A. The test used for the operation is called "10 per cent. acid," and is prepared in the following manner.* One pound of strong sulphuric acid (specific gravity 1845) is mixed with distilled water till the solution measures exactly one gallon, the specific gravity of which will be 1063·5. Then first fill the alkalimeter to the line marked o with the ammoniacal liquor to be tested, and pour it into a porcelain shallow basin; wash out the instrument with clean water, and pour the washings into the basin. Now fill the alkalimeter to the point marked o with the test acid, which then pour gently into the basin, and at the same time carefully stirring up the mixture with a glass rod; when the effervescence ceases, place a blue litmus test paper in the liquid, then add very cautiously more acid by small portions at a time, stirring briskly after each addition until the test paper assumes a neutral grey or slightly red tint, when the neutralization is effected. Then read off from the alkalimeter the quantity of test acid used, if this should be, say six parts or divisions, then the liquor will be 6-ounce liquor accordingly. It may be necessary to use several test papers, as the sulphuretted hydrogen evolved affects the colour. Further, to prevent the acid running down the side of the alkalimeter, the delivery pipe should be slightly greased before pouring out the acid.

This, although the usual method adopted, is very far from being accurate, as we shall have occasion to show when treating on the residual products.

The following table will be found serviceable for many purposes, particularly when using the

---

* All the requisite apparatus and chemicals for this and other experiments are to be obtained of A. Wright and Co., Millbank Street, Westminster.

Twaddle hydrometer, also for ascertaining the quantity of ammonia in any given solution, as well as appreciating its "ounce strength":

TABLE SHOWING THE DEGREE ACCORDING TO TWADDLE, THE SPECIFIC GRAVITY, AND WEIGHT PER CUBIC FOOT PER GALLON, THE WEIGHT PER GALLON, WITH THE OUNCE STRENGTH OF AMMONIACAL LIQUOR.

Degrees Twaddle.	Sp. Gr. and Weight per Cubic Foot in Ounces Avoirdupois.	Weight per Gallon. Lbs.	Ounce Strength.	Degrees Twaddle.	Sp. Gr. and Weight per Cubic Foot in Ounces Avoirdupois.	Weight per Gallon. Lbs.	Ounce Strength.
Water.				Water.			
0	1000	10·	0	12½	1062·5	10·625	25
½	1002·5	10·025	1	13	1065	10·65	26
1	1005	10·05	2	13½	1067·5	10·675	27
1½	1007·5	10·075	3	14	1070	10·7	28
2	1010	10·1	4	14½	1072·5	10·725	29
2½	1012·5	10·125	5	15	1075	10·75	30
3	1015	10·15	6	15½	1077·5	10·775	31
3½	1017·5	10·175	7	16	1080	10·80	32
4	1020	10·2	8	16½	1082·5	10·825	33
4½	1022·5	10·225	9	17	1085	10·85	34
5	1025	10·25	10	17½	1087·5	10·875	35
5½	1027·5	10·275	11	18	1090	10·9	36
6	1030	10·3	12	18½	1092·5	10·925	37
6½	1032·5	10·325	13	19	1095	10·95	38
7	1035	10·35	14	19½	1097·5	10·975	39
7½	1037·5	10·375	15	20	1100	11·	40
8	1040	10·4	16	20½	1102·5	11·025	41
8½	1042·5	10·425	17	21	1105	11·05	42
9	1045	10·45	18	21½	1107·5	11·075	43
9½	1047·5	10·475	19	22	1110	11·10	44
10	1050	10·5	20	22½	1112·5	11·125	45
10½	1052·5	10·525	21	23	1115	11·15	46
11	1055	10·55	22	23½	1117·5	11·175	47
11½	1057·5	10·575	23	24	1120	11·20	48
12	1060	10·6	24	25	1125	11·25	50

In concluding this part of our subject, we cannot impress too forcibly the necessity of profiting by the residual in question, for, by permitting it to flow uselessly away, the loss is as direct as permitting its money value to run to the sea, and as this may be estimated to be equal to about at from 1s. to 2s. per ton of coal carbonized, the loss to a company can be readily understood; and, as hereafter shown, when this residual is manufactured on the works where it is produced, it becomes a still more important source of revenue.

PURIFIERS.

ALTHOUGH, as already observed, the purification of gas commences at the hydraulic main, is continued with the condenser, and further advanced with the scrubber, we have adhered to the old manner of referring to the vessels which eliminate the sulphur compounds, carbonic acid and sulphuretted hydrogen, as the purifiers, and as we have moreover retained the order of description of previous writers on gas, it will therefore not be surprising if some degree of irregularity in our progress is apparent.

From the very earliest experiments of Murdoch, the necessity of purifying coal gas was made evident. This, we may suppose, was suggested in the first instance by the smoky appearance of the flame as the gas was consumed in the primitive burners, and by the nauseous odour arising from it when escaping from the pipes without being burnt. But it does not appear that any apprehensions existed at the time of the impurities being insalubrious, nor was this amongst the objections raised by the opponents of gas lighting; however, since then scientific men have at times magnified the effects of these to a marvellous extent, but without tendering sufficient proof of the correctness of their assertions.

According to Creighton, the first attempt at purification consisted in washing the gas with water. Matthews states that Murdoch, during his experiments, succeeded in removing all the disagreeable odour, but that when thus "purified," the gas gave very little light. Since that period other clever minds have been occupied in endeavours to attain a similar object—namely, to remove the odour from coal gas. But never was labour and intelligence more misapplied; for, supposing the object to be accomplished without in any way prejudicing its quality, then the danger attending the use of gas would be such as to realize all the most extravagant predictions of the first antagonists to this description of artificial light. The peculiarly disagreeable odour of coal gas, when unconsumed, has been the greatest safeguard in its use, and deprived of this, gas could not possibly have been generally applied, on account of the numerous accidents to which its employment would inevitably have led.

Clegg first adopted lime as a purifying agent, in the apparatus for lighting the dwelling of Henry Lodge, Esq., at Halifax, in 1806. For this purpose the lime was introduced into the gasholder tank, where, of course, it settled at the bottom, and simply exercised its influence on the water contained therein; and as this would absorb only the most insignificant portion of the impurities of the gas, as thus applied, the lime was altogether useless. Three years afterwards, when constructing the apparatus for supplying Mr. Harris's establishment at Coventry, Clegg introduced a paddle at the bottom of the tank, to agitate the lime; and the same year he employed a separate machine as a purifier, which was erected at the works of the Stoneyhurst College, Lancashire.

The first patent in connection with this subject was that of Edward Heard, already referred to, which was for "certain means of obtaining inflammable gas from pit coal, in such a state that it may be burned without producing any offensive smell," and is described as follows: "The object of the invention is to withdraw the sulphur from gas obtained from coal, and for this purpose the patentee stratifies lime with the coals in the retort, stove, or other close vessel, in which they are placed for operation, or suffers the gas, when produced, to pass over lime previously laid in an iron or other tube, or any other shaped vessel adapted for the purpose, and exposed to heat." He continues, "The fixed alkalies or alkaline earths, as carbonate of lime, barytes, &c., when deprived of their carbonic acid, may be substituted for lime, also such metals or other oxides as iron, manganese, zinc, copper, lead, &c."

Here we have mentioned for the first time the use of oxide of iron as a purifier, which material has since then become so universally employed for the purpose. However, it appears that Heard followed no principles in defining the different purifying agents, but seems to have chosen them at hazard.

In 1808 Winsor patented the means of "refining inflammable air or gas, so as to deprive it of all disagreeable odour during combustion, and rendering the gas itself salutary for human respiration when properly diluted with atmospheric air." Attached to the specification is a memorandum by which the patentee "surrendered and yielded up to His Most Excellent Majesty George the Third, the recited letters patent." The whole is so absurd as to lead to the impression that this was a *ruse* employed to bring gas into more favourable notice.

The following year Winsor, however, acquired another patent, in which he repeated the principal points of the former, but added, that he proposed to pass the gas "through coolers, and then through vessels filled with lime water or cream of lime, which will withdraw the sulphur and other offensive particles." Hence we learn that this patent was obtained the same year that Clegg practically adopted the lime machine; whether the circumstance was accidental or otherwise it is difficult to say.

Fig. 44.

MALAM'S TREBLE PURIFIER.

The earliest illustrations of gasworks represent the lime machine as a cylindrical vessel, about 3 feet in diameter, placed in the centre of the yard, having a crank handle for the purpose of agitating the lime. Subsequently, as we have shown, it was driven by the station meter, and necessarily the pressure on the retorts was much increased, with all its attendant drawbacks—points which were only understood after the experience of many years.

The advantages of the cream of lime for purifying gas being recognized, John Malam applied himself to the construction of an apparatus, represented in the engraving, suitable for carrying on the process with all economy and efficiency. The sketch is copied from Peckston's 'Practical Treatise,' published in 1819, where it is described as "a vertical section of a treble purifier, placed on a foundation of brickwork. A cast-iron cylindrical vessel, enclosed at the top and bottom, is embraced between the letters A, B, C, and D, which is divided into three compartments. E, F, and G are the three interior chambers, bolted to the top of the respective vessels by the flanges shown on the Plate. It will be observed that the bottoms of these vessels branch out with a wide flange, by which means the gas is acted upon by a greater portion of the purifying mixture than if the side of it fell in a perpendicular line. H H is the axis on which the agitator I I is fixed." Two of these are not shown, as they are omitted on the original. "K K K are the cylindrical vessels into which the purifying mixture is admitted by the bends L L L. M M M are the pipes which convey the gas into the interior chambers, whence it passes, after purification, by the pipe N, to the gasholder. O is the feed pipe for conveying the purifying material from the vessel where it is prepared into the apparatus; P P P are openings for supplying the respective vessels by the slide valves Q Q Q. In the drawing, the height to which the purifying mixture rises is shown in the two lower vessels, but the upper one is not charged. The agitator is driven by hand."

"When required for action, the cream of lime is admitted through the orifices P, into the pipes L, hence to the three vessels to the height indicated. The gas enters the lower interior vessel G G, and forcing its way through the liquid, passes under the dash-plate, and by means of the bent pipe M, enters the vessel F, where the action is repeated, and the gas escapes into E, and so off by the pipe N, to the gasholder. When the lime in the machine becomes foul, that in the lowest compartment is allowed to flow away, and is replaced by that in the second vessel, which is then filled by the contents of the third, the latter vessel being supplied with fresh material."

The principle here described is still retained in the construction of wet-lime purifiers, although they are rarely employed, but instead of being combined in one vessel, they consist of three distinct apparatus in combination with each other, as hereafter shown.

A great progressive step in purification was introduced by Reuben Phillips, of Exeter, who, in 1817, obtained a patent "for a new and improved method of purifying gas for the purpose of illumination." He says:

"I take any quantity of well-burnt lime, and pour water upon it until it falls into powder. I then mix it with a further quantity of water, in order to bring it into such a state that the particles of lime may adhere slightly to each other, but not to such a degree as to prevent the free passage of the gas between them. This mixture must be placed six inches deep more or less, on movable perforated shelves, in a vessel, the top of which is guarded by a water joint, and underneath is a pipe to allow the passage of the gas that way, so that the gas may pass from the bottom of the vessel to the top through the perforated shelves and lime mixture, or from the top to the bottom, as may be found most convenient, the purification being effected by the gas being caused to pass through the layers of lime mixture: but where the quantity of gas to be purified is very large, I arrange a set of these vessels, consisting of five or nine, or more, according to the size of the gasworks, each vessel containing one or more shelves. These vessels are placed in any way which convenience may require, but I prefer a circular arrangement. The vessels, being without bottoms, stand in a cistern of water or other fluid about six inches deep, so that the gas cannot pass that way."

The drawings accompanying the specification represent an arrangement of eight purifiers, similar to those in use at the present day, in combination with the valves, four of these being in operation, whilst the others can be renewed with fresh lime.

This patent is specially referred to on account of the great merit it possesses, as it indicates in the clearest manner possible the method of employing the hydrate of lime in a condition well moistened as a purifying agent; in a part not quoted it gives the first idea of deodorizing the foul material, and suggests for the first time the method of constructing the water-lutes for the covers of purifiers. The great error committed by the patentee was in forming the bottoms of the vessels by a tank of water, which was regarded by Peckston as exceedingly dangerous, and prejudicial to this method of purifying.

It may be here observed that notwithstanding the clearness of description in Phillips's specification, there are few processes in gas manufacture which have been so imperfectly applied as dry-lime purification. The term is certainly a misnomer, and no doubt has frequently led to its improper application, for the writer has often witnessed it employed literally as *dry* lime, and sometimes carefully sifted prior to being placed in the purifiers. Lime, when thus employed, is almost useless, and after being in contact with the gas the usual time is simply slightly discoloured; but when thoroughly moistened, and in such a state that when pressed in the hand it possesses a doughy consistency, it is used to the best advantage.

To the genius of John Malam we are indebted for the perfection of Phillips's apparatus, and the production of the beautiful and simple arrangement of purifiers which has now become almost universally employed, alike for the oxide of iron, lime, sulphate of iron, and other solid compounds used as purifying agents.

His arrangement, as generally applied, consists of four purifiers in connection with each other, and with a central valve by which they are controlled, in such a manner that three of them being in operation, the fourth is shut off, thus affording every facility for discharging the foul material, and recharging the apparatus for purifying. The combination is of the most simple and perfect kind, and since its invention the only important alteration made therein has been the substitution of the rotary slide, or dry valve, instead of the hydraulic valve.

For the sake of simplicity of explanation, we propose to describe Malam's invention as generally applied in practice, that is, by dispensing with one of the pipes mentioned in his specification, and attaching the outlet to the upper part of the cylinder forming the valve.

Fig. 45 represents two out of a set of four purifiers in connection with the central valve, that indicated B in elevation, the other, C, being in section. The centre valve is also shown in section, and consists of a closed cylindrical vessel E, supported by nine vertical T pipes, open at their lower ends and

Fig. 45.

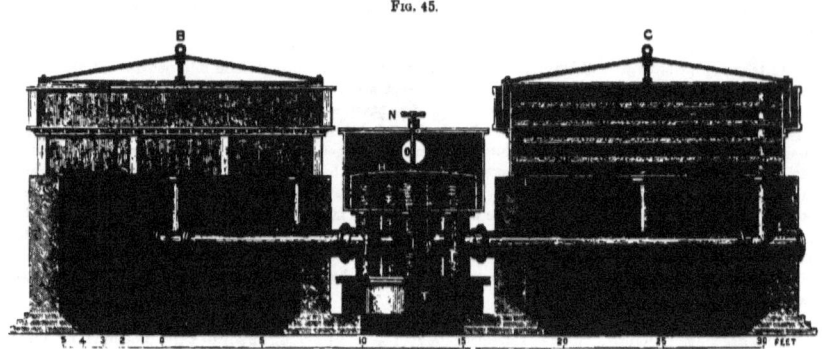

MALAM'S ARRANGEMENT OF PURIFIERS.

standing in the tank T, which is filled with liquor, and receives any condensation, thus serving as a syphon for all the connections. The positions of these vertical pipes are represented in Fig. 46, eight of these (those numbered) being in communication with their respective purifiers, and form the inlets and outlets thereof. The centre pipe I is the inlet to the valve.

Within the cylinder E is a holder H, of somewhat less than half the height of the former, attached to the centre rod which passes through a stuffing box of the cover, and on the top of the rod is a handle N, for the purpose of lifting and lowering and changing the position of the holder when desired. At a distance of about three inches from the bottom, the holder is divided into six irregular formed and distinct chambers, as represented in plan in the following engraving; and in order to avoid confusion, these chambers are not shown in sectional elevation. Thus the gasholder H can be raised to such a height that all the vertical divisions will clear the pipes, and be still sealed at its circumference by the three inches space below.

Fig. 46 is a plan of the same purifiers with central valve, parts of those marked A and D being removed to economize space.

The purifier C has its cover removed in order to show the internal construction. As represented, the bottom is formed of twelve iron plates with flanges of about 2¼ inches wide and secured by bolts and nuts six inches apart, strengthened by brackets placed between them, the bolt holes being cast in the flanges midway between the brackets. Around the edge is a corresponding flange, to which the vertical plates forming the sides are bolted, which plates are put together with lap plates, the whole of the joints being made with iron cement. Half of the first tier of grids is represented as placed in its position.

When in action the gas to be purified enters by the pipe I, into the centre of the valve; from there the gas is conveyed by the intervention of the gasholder (the top of which is supposed to be removed in order to explain the operation), into the pipe No. 1 to the purifier D, when after passing through the lime or oxide, it returns as indicated by the arrow by pipe 2. This being connected by the compartment with that marked 3, the gas passes to purifier A, and returning by the outlet, which is in the same com-

Fig. 40.

partment as the pipe 5, it then enters purifier B, returning by the pipe No. 6, in which division there is an orifice in the top of the holder, and through this the gas issues to the orifice O in the case E, and hence to the store or gasholder. By this means both the inlet and the outlet of the purifier C are shut off from the gas, affording the opportunity of changing the material ready for use whenever that purifier may be requisite.

When in the course of time the material in the purifier D becomes ineffective, this is thrown out of action, and that marked C being charged with oxide, or lime, and its cover secured, is then brought into use, which is accomplished by lifting the holder, so that the vertical divisions clear the pipes, and turning it in the direction of the outer arrows one-fourth of a circle, and allowing it again to descend to its place, which effected, A will then become the first purifier, C the last, and D rendered inoperative.

By these simple and effective means the various purifiers are changed as desired, always leaving one out of action, for the purpose of removing the foul, and replacing it by fresh purifying material, ready for use.

It might be supposed that an apparatus so perfect in principle as that described, which experience has clearly proved, would have been readily adopted by gas companies; but this was not the case, partly arising from the circumstance that for several years after the invention, wet-lime purification was alone adopted; for by the evidence of George Lowe on the Oil Gas Bill, in 1825, only wet lime, or the cream of lime, was employed for purification. Indeed, the dry-lime process was really forced on the companies, at least those of the metropolis, about 1841, by the difficulty experienced in transporting the foul lime, or "blue billy," on account of the odour arising therefrom.

The system of water-lutes on the tops of the covers, and movable pipes, as shown in Phillips's drawings, were retained at many works for several years after the expiration of Malam's patent, and indeed were represented in three editions of a standard work on gas lighting, clearly demonstrating the apathy of the engineers of that epoch. However, in the course of time Malam's combination was introduced, when the invention was accorded to the engineer of a large town in the Midland counties. We believe that neither Phillips nor Malam derived any advantage from their inventions connected with purifiers —a lot which falls too often to inventors, who achieve much for the public welfare, without advancing their own interests or deriving any profit from their abilities. Amongst these may be named Ami Argand, the inventor of the burner bearing his name, who was unjustly deprived of the fruits of his intelligence and ingenuity, and, beyond the fact of being recognized as the inventor, never realized any benefit therefrom.

Fig. 47.

Cockey's Valve.

Malam's hydraulic valve then became very extensively applied by gas companies at home and abroad, but in the course of years the difficulty attending its use, that of permitting a portion of foul gas to pass during the operation of changing the position of the purifiers, together with the foul liquor therein being always in contact with the purified gas, suggested the means of replacing the hydraulic by the slide valve, and it is somewhat surprising, with the various kinds of rotary valves which had been applied at different times to the dry meter, that the hydraulic was not superseded by the rotary valve at an earlier period.

For this improvement we are indebted to Messrs. Cockey, of Frome, who, in 1857, patented the slide or rotary valve now to be mentioned, which is applied to combinations of two, three, or more purifiers, but it will be sufficient to illustrate it in conjunction with Malam's arrangement of four purifiers; this done, its applicability to combinations of two or more vessels will be understood.

Fig. 47 is an elevation of Cockey's rotary valve for four purifiers. To the lower part, A, are attached the inlet and outlet flanges, and within are the necessary pipes to carry off any condensation that may arise in the various divisions of the seat of the valve, answering the purpose of a syphon to all the connections.

B is the seat of the valve, having, as represented in the annexed figure, eight divisions corresponding with the respective sockets for connecting with the inlets and outlets of the four purifiers. C is the cover of the valve accurately adjusted to its seat so as to render it perfectly gas-tight. This cover has a series of vertical divisions similar to those of Malam's hydraulic valve already described, with this difference, however, that the Messrs. Cockey have followed Malam's original idea of placing a tenth pipe, so that the gas is received, conveyed to the purifiers, and delivered *under* the cover, which is open to the atmosphere, and provided on its periphery with teeth, for the purpose of being actuated by a pinion, but when of small dimensions it may be turned by a handle.

The following figure (Fig. 48) is the valve seat with the eight divisions referred to, numbered from

1 to 8, its bars being shown in shade. The cover is represented with the top removed in order to show the action, and is distinguished by being perfectly light. The letters A, B, C, and D refer to the purifiers in connection with the respective sockets. The central passage, I, is the inlet, the concentric space, O, the outlet. The action is precisely the same as the hydraulic valve already described as indicated by the arrows, with the only difference that the outlet is formed by the concentric ring, thus the gas is conveyed from division numbered 6 thereto, and so off to the holder; and, like the other, by turning the cover the fourth part of a circle, the action of the purifiers is changed.

Fig. 48.

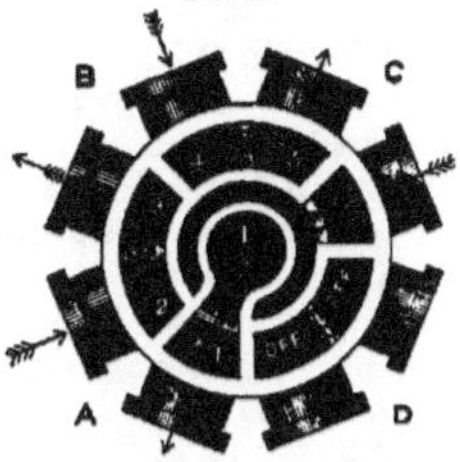

On first placing these valves, their surfaces should be well smeared with tallow, when, according to our experience, they will remain perfectly sound, and will neither get bound nor set, even after many years' active service.

In the central valve as manufactured by Messrs. Walker, of Donnington, the cover is enclosed with a cast-iron case, to which is attached the outlet, and the divisions in the seat are reduced to nine instead of ten, by this means considerably reducing its area, and being identical in principle with the hydraulic valve described. Moreover, according to the method of construction adopted by that firm, the gratings of the valve seat are always covered by corresponding surface bars, which are attached to and form part of the cover; thus any deposit upon the gratings is avoided.

Some manufacturers have departed from the original method of constructing these valves, by making them with a conical seat, somewhat resembling an ordinary tap. We are unable to understand that any advantage is gained by the change; but, on the contrary, in our opinion, the conical valve possesses several drawbacks, for in it there are three different surfaces to adjust and make sound, instead of two, as in the other kind. The cover of the conical valve is more weighty, and by its form becomes to an extent wedged into its seat. To counteract this, and in order that the cover may be turned, a screw is required to relieve the weight; and this, in the hands of unskilful persons, is productive of considerable mischief, for, by the relieving screw, the valve may be rendered unsound, and even useless, which defect cannot occur with the flat-surface valve. Besides, in all kinds of mechanical construction, the first consideration should be simplicity; the flat surface is the simplest, and in our opinion far superior to the conical-seated valve.

By Malam's combination, alike in the hydraulic as in the dry valve, there exists this peculiarity, that by turning the cover only the eighth part of a circle, instead of a fourth as indicated, either two or three purifiers can be thrown out of action, according to its position. This method, although sometimes applied, is capable of leading to error and confusion, particularly when in the hands of ordinary labourers, who of necessity are sometimes employed; and undoubtedly the best plan is to arrange the apparatus so that three purifiers shall be always in action. By adopting this course, whenever the make of gas becomes diminished, either a diminished quantity of purifying material may be used at each charge, or the purifiers changed less frequently, and for the strict observance of economy, as well as for the proper purification of the gas, when the lime, oxide, or other material is supposed to be sufficiently saturated, the test paper should be applied from time to time; and on this becoming slightly fouled at the fifth division of the purifiers (we imagine three of these to be in use, each of them subdivided and provided with test cocks), then the purifiers should be changed. In practice, however, such rigid economy is never observed.

The test taps are usually placed on the covers of the purifiers, and are generally one-eighth inch cast iron, as brass corrodes very rapidly; but when the purifiers are not divided in the centre, test taps may be placed midway in the side of the purifier, so as to ascertain the purity of the gas before passing through the last apparatus.

In order to determine the magnitude of purifiers, the first consideration is the maximum daily production of the works for which they are intended, and as gasworks invariably increase in importance from year to year, it is always advisable to make the apparatus of ample magnitude, thus avoiding any future alterations in them for at least a protracted period. Although by observing this an addition to the capital is made, yet on the other hand the labour is diminished, and there is the further advantage that gas when in a quiescent state is in the best condition for depositing its impurities. This being directly opposite to the action of the condenser, where speed and friction are essential to break up the tarry vesicules and deposit the tar.

The maximum daily make being ascertained, the general rule in small and medium works is to allow a purifying area of one square yard for each thousand feet of gas (which we believe should be increased to ten feet), or, as Mr. Newbigging states in his 'Handbook for Gas Engineers,' ten square yards of grid per ton of coal carbonized in the twenty-four hours. This with four purifiers, three of which are always in action (the system most universally adopted), gives 7½ square feet of active surface per 1000 feet maximum production. Hence under these circumstances, and supposing the greatest make to be 100,000 feet per diem, the area of the grids will be 1000 feet superficial, or 250 feet surface, to each purifier, and if these have four tiers of grids would give a surface of 62½ feet to each grid, and purifiers 8 feet square internally would be required for the production in question. According to the same rule, for a make of 220,000, the purifiers having four tiers of grids, the area of each would be 144 feet superficial, or 12 feet square. Thus as a bushel of quicklime is generally considered the average for the purification of 10,000 feet of gas, and as this when reduced to hydrate and moistened measures about three cubic feet, then for purifying 100,000 feet of gas 30 cubic feet of hydrate will be required, and supposing this to be placed in layers of three inches thick, would give in round numbers 125 feet, therefore to charge one of the vessels 60 cubic feet of hydrate would be necessary, corresponding with the surface of the grids and the quantity of gas purified by the 8-feet purifier. Under these conditions, an apparatus of the dimensions indicated, will only require to be changed by renewing the lime in one of the vessels, and altering the position of the valve or valves every second day.

From this basis the capacities of purifiers can be determined according to the periods of changing, but it is always advisable to have them of ample capacity and thus avoid unnecessary labour and the admixture of atmospheric air with the gas, which small purifiers and frequent changing always occasion. The depth of the purifiers, for the reason stated, should also be determined; this we find in some modern purifiers to be equal to about 10 cubic feet per 1000 feet make; hence assuming this to be correct, the depth of the purifier is ascertained by allowing one foot in depth for each tier of grids, into which calculation the space beneath the bottom and above the top grid in the cover is included, which gives ample space for any description of oxide and allows the gas to be in a quiescent state as desired.

In placing the purifying material in several layers, the object is to avoid compression and consequently obstruction of the passage of the gas. These remarks apply to all kinds, but more particularly to what is termed dry lime, for when this is well moistened to the desired degree and laid in a thick stratum, the lower part may readily be so compressed as to impede the passage and increase the back pressure. There are also some descriptions of oxide that are very compact in their nature, containing only from 20 to 30 per cent. of oxide, the rest being earthy matter, which are also subject to be influenced by compression.

Pressure gauges should always be placed in a conspicuous position in connection with each of the purifiers, in order that any obstruction from the material may be detected. Any excess of pressure, from whatever cause it may arise, causes a corresponding amount of labour to be thrown on the exhauster, and is equivalent to a waste of the quantity of fuel necessary to counteract that pressure. But in the absence of the exhauster the evil is still more serious, on account of the loss by leakage and the increased deposit of carbon within the retorts.

In modern works, the largest purifiers seldom exceed 25 feet in width, their increased size generally being gained in their length. The reason for this is obvious, as in the limited width less labour is incurred in carrying the purifying material to and from the apparatus. An error frequently committed in small works is in making the purifiers circular; these, although giving less trouble in fitting up, are dearer in construction, and occasion considerable trouble in choosing the various irregular-shaped grids which should be avoided.

CONNECTIONS FOR PURIFIERS.

Of late years an important improvement has been introduced into this part of the gas apparatus, by dispensing with the iron grids formerly employed, which were always breaking, and substituting the wooden grid. These consist of a number of battens, each of about three feet in length, three-eighths of an inch thick and two inches wide, placed edgeways and parallel to each other, and separated by pieces of wood or washers of about three-eighths of an inch thick, placed at three points, one at each end and the other in the centre. At these points pass ⅜-inch bolts, their length being equal to the width of the grid, with their corresponding nuts which secure the whole together, the two outer battens being somewhat stouter and of harder wood than the others. Wooden grids are usually three feet by one foot six inches, a size and weight that a man can handle with facility, and are strongly recommended on account of their durability, and can be made anywhere at a most moderate cost.

The hydraulic seals of purifiers should always be of good depth, never less than 18 inches; indeed, these should be the deepest of all the seals on the works, in order that in the event of any obstruction the gas cannot escape at these points. But with a proper quantity of self-acting syphons on a works, a stoppage can only arise from naphthaline, which gives timely notice of its accumulation on the pressure gauges.

No gasworks, even of the most limited capacity, can be properly conducted with less than two purifiers, each of which in very small works should be subdivided and connected in such a manner that the gas in passing ascends through the purifying material in the first compartment, and descends through that in the other; one purifier being in action whilst the other is being charged ready for use.

It is a general rule, which within certain limits may be accepted as correct, that the connections of the purifiers should be of such dimensions, that their diameter in inches is equal to the square root of the area of one of the purifiers in feet; thus those for the first purifier referred to should be 8 inches, and for the second 12 inches in diameter. This rule, however, is only applicable to connections varying from 3 inches to 12 inches in diameter, in which the smaller kinds are augmented in consequence of the increased friction and their liability to stoppage. On the other hand, as the delivery of gas is greatly increased in proportion as the diameters of pipes are augmented, so for any increase in the size of connections above 12 inches, a proportionate reduction in their dimensions may be allowed.

In explanation of this, let us take a pipe 12 inches in diameter, of such length and pressure that it will deliver, say, 15,000 feet of gas per hour. If we take another pipe of 18 inches diameter, and adopting the 12-inch for basis, according to their respective areas, the 18-inch pipe, under the same conditions of length and pressure, should deliver somewhat less than 34,000 feet per hour; whereas in practice it delivers about 42,000 feet. Again, a 24-inch pipe, under like conditions, should only supply 60,000 feet, whereas it really delivers 84,000 feet per hour.

Hence we find, by the augmented diameters of connections, and their respective sectional areas, that the 18-inch delivers 25 per cent., and the 24-inch delivers 40 per cent. more gas than the 12-inch; therefore, under such varying conditions, the rule mentioned is no longer applicable, and the inverse of this would have to be observed. In which case the maximum quantity of gas required to be passed per hour being known, then by reference to the table, and instructions in the chapter on mains, the diameter of the pipe will be ascertained, and the area and depth calculated as indicated, will give the dimensions of the purifiers; but in this calculation due allowance will have to be made for the various bends in the connections. The dimensions of the connecting pipes being ascertained, the area and cubical contents of the purifiers are calculated in the manner already described.

Various contrivances are adopted for lifting the covers of purifiers, the most simple of these for very small works being to attach a beam from wall to wall directly over the centre of purifiers, supposing them to be in a straight line with each other. On the beam are two rollers affixed to an iron frame, representing in form the letter U, and on the lower part of this frame is a boss in which works freely a long screw, having at its end a hook or other suitable means of connecting to the cover of the purifier. Attached to the screw is a small hand-wheel, by turning which the cover can be raised and carried out of the way by means of the rollers on the beam. In other works a swing crane is placed in the centre and commands the four purifiers, the covers of which are lifted when desired and swung round. Sometimes cranes are affixed to a wall, each of these controlling two purifiers.

The means employed by Mr. Brothers for lifting the covers are represented in Figs. 1 and 2, Plate 17; the former representing the traveller in end, and the latter in side elevation. The traveller consists of two wrought-iron standards and two lattice girders, supported by their corresponding wheels running on rails, thus forming a frame capable of being conveyed to any part in the line of rails, which is affected by the hand-wheel shown in the end elevation. This method of raising the covers by the screw, wheel, and pinion, permits a great weight being lifted with comparatively limited power, whilst the strain being close to the standards, admits of proportionably slight girders.

Another method is represented in Fig. 4, where a chain is actuated by a wheel and pinion, the chain passing over two pulleys and under a sheave attached to the cover.

Messrs. Cockey employ a frame composed of standards connected together by a bridge, in the centre of which is a screw actuated by a bevel wheel. A rail is affixed to each side of the purifiers, and on these run the standards and bridge, the covers of the purifiers being raised by turning a winch similar to the preceding.

In works of large magnitude double girders are often placed directly over the purifiers, extending the whole length of the purifying house; in other cases the girders only embrace two purifiers, as represented in Fig. 5, in the Plate already referred to. On the girders is a crane actuated by two or more men, according to the magnitude of the apparatus.

We believe the first application of the hydraulic rams for the purpose in question is due to Mr. Reid, the engineer of the Edinburgh Gasworks, but we are unacquainted with the manner it is adapted. An arrangement of the hydraulic ram, as constructed by Messrs. Robert Dempster and Sons, is shown in Fig. 3, and as may be observed, in the centre of the girders is fixed the hydraulic ram, to the piston of which is secured the cross head carrying two friction wheels, and over these pass the chains—one end of each of these being attached to the girder and the other ends to the cover. A small pump, situated at the left end of the girder, and in connection with the ram, is worked by the lever beneath the connecting rod. When required for action, communication is made between the pump and the water supply, when, by means of the lever, the pump raises the ram piston and lifts the cover; the power in this case required to lift a given weight is governed by the proportions between the areas of the piston of the pump and that of the ram. This system possesses the advantage of requiring only one hydraulic ram for any number of purifiers situated on the same line of rails.

The hydraulic ram is applied direct to some of the circular purifiers at Beckton; the merit of this adoption is due to Mr. Wyatt. Each of the covers of these purifiers has a framing resembling the trussing of a gasholder, and in the centre is a plate which reposes on the top of the ram also situated at that point. When the cover is required to be raised, water is admitted into the ram, and on attaining a certain height the cover is supported by the guide standards. On the purifying material being changed, the ram raises the cover so as to clear the standards, when both descend to their positions.

A similar system has been adopted by Mr. Holman, and although very simple and effective, its employment does not appear very general. Perhaps one difficulty consists in the expense, as a separate ram is required for each purifier, and in addition to this the covers have to be strengthened in such a manner that the centre plate shall carry all the strain, whereas in large covers as ordinarily made there are two or more points of suspension.

Among the recent inventions in connection with gas apparatus is the system, patented by Körting, of reviving the oxide in the purifier without removing the cover. This has been repeatedly attempted by employing an ordinary fan, or with steam, as proposed by Mr. Fish; but so far as we are aware it has met with but indifferent success. According to the process in question, the air is injected by the action of a jet of steam, or blower, identical in principle with the steam exhauster already mentioned; but for its employment a slight alteration in the ordinary purifier, by the addition of an extra inlet and outlet connection with their corresponding valves, becomes necessary. The blower is attached to the extra inlet, and connected to the steam pipe, and provided with a regulating tap, with air passages or inlets controlled by suitable screws. The noxious vapours pass by the outlet arranged for that purpose.

When it is desirable to revive the material in the purifier, on this being put out of action and its cover secured, the valve in connection with the extra outlet is opened, and the air passages of the blower are adjusted to the extent required. This effected, the steam is admitted, carrying with it a current of air, which, passing through the purifier, carries with it the noxious vapours from the foul material. During the process the oxide becomes heated in consequence of the chemical change in its nature, and, on the action ceasing, it returns to its normal temperature. In using the apparatus and employing new material, a proportionately small quantity of air is admitted, when the steam in combination therewith prevents the oxide overheating.

The advantages of the process are, that the oxide can be revived three or four times without removal or fresh material being added to the purifier, but if continued beyond that the oxide has a tendency to cake together and increase the back pressure. Thus by this means a considerable saving of the most disagreeable part of the labour is effected. Less space for spreading the oxide as used in the ordinary way is necessary. But the most important advantage is that the noxious vapours from the purifiers can be conveyed into a special chimney shaft, as practised by Mr. Woodall, the engineer of the Phœnix Gas Company, London, by which means the odour arising from the purifiers is dissipated in the air above, thus avoiding complaints that might otherwise arise.

In reply to our inquiries, we are informed by engineers by whom it is employed, that the Körting reviving blower has answered all their expectations.

Purifiers for dry lime or oxide are now frequently made of 50 or 60 feet in length, and, for reasons stated, seldom more than 25 feet in width. In works abroad the bottom plates are sometimes dispensed with, except two plates to which the connecting pipes are attached, the whole of the foundation being formed of brickwork or masonry, which is perfectly levelled and rendered with Portland cement. The side plates are provided with flanges in the usual way, and when placed these flanges are imbedded in cement, and the tank completed in the manner ordinarily practised. The joint of the iron and cement is then further secured by a course of bricks laid in cement, and the whole has a coating of tar. In localities where the cost of the transport of iron is expensive, a considerable saving may often be effected by constructing the whole of the lower part of the purifiers of brickwork, when only the luting box and cover need be of iron, and indeed there is no reason why this construction should not be adopted wherever economy is a consideration; the only argument that can be possibly adduced against it is the diffusive action of gas, but which however, in our opinion, if the purifiers are only built with the most moderate care, cannot be detrimental. An error sometimes committed in erecting purifiers is to sacrifice utility to appearance, by placing the vertical flanges of the lutes inside, consequently the space is often so restricted as to occasion considerable trouble in getting the covers into their positions.

PURIFICATION.

By the combined action of the hydraulic main, the condenser, and scrubber or washer, the whole of the tar and ammoniacal liquor, together with a portion of the carbonic acid and sulphur compounds, are, or at least should be, eliminated from the gas; but there still remains the larger portion of the two last-mentioned impurities to be removed before it is in the condition to be delivered to the consumer.

We may observe concerning the purity of gas, that it can be considered and pronounced commercially pure, and in the ordinary sense of the term, free from all noxious elements, when, after a lengthened exposure to the usual tests for carbonic acid, ammonia, and sulphuretted hydrogen, no indication of these impurities exists. Gas of this degree of purity is capable of being consumed in dwellings with ordinary ventilation, without being in the slightest degree offensive or injurious to the inhabitants, nor during its combustion will its products be prejudicial to metals, paper-hangings, drapery, or other articles of utility or ornamentation, with which they may come in contact, to any greater degree or extent than would arise from lamps or candles when burned in the apartments and producing a light equivalent to that of the gas. This, to all intents and purposes, may therefore be termed pure gas; and when burned in that state, according to the extensive researches and experiments of Mr. Lewis Thompson, for a given amount of light obtained from the respective materials, gas is even less injurious in a sanitary point of view than either oil, tallow, or wax.

But this state of excellence, for reasons hereafter stated, does not satisfy all the requirements of the British Legislature, who a few years ago arrived at the conclusion that gas cannot be pure so long as sulphur in any form in excess of 20 grains per 100 feet exists therein. Accordingly, in all recent Acts of Parliament obtained by gas companies, this stipulation as to purity is always inserted, and what is very remarkable is the fact, that at the period when these stringent regulations were first decreed, the means of complying therewith were unknown. Moreover, whilst occasioning considerable trouble and expense to gas companies in order to obtain such results, no possible advantage has arisen from these restrictions, inasmuch as their most strenuous supporters could not distinguish, in an ordinary apartment, the difference in the effect produced by the combustion of gas obtained by the ordinary process of manufacture, and containing 30 or 40 grains of sulphur, other than sulphuretted hydrogen; and that degree of purity demanded by the Board of Trade.

Under these circumstances, together with the excessive expense which in some localities would have to be incurred in removing the small portion of sulphur which may be in excess of the maximum (as now adopted at some establishments by compulsion), we will limit our observations in the first place to the means of depriving gas of the impurities mentioned, according to the ordinary methods in practice, so that it may be rendered commercially pure as described.

Wet-Lime Purification.

As already stated in the "Chemistry of Gas Manufacture," the most perfect system of purifying gas is by means of wet lime, or, more properly speaking, by the solution called the "cream of lime," and although the odour arising from the foul material after being employed is sufficiently offensive to prevent the adoption of this mode of purification in populous neighbourhoods, there are, however, other localities where the isolated position of the works permits of its application.

For this purpose three purifiers are generally employed, similar in construction and action to those represented in Fig. 44, but instead of being placed above each other, they are detached, as shown in the annexed engraving, which represents the positions of the vessels, and they are so situated that, when

DRY LIME PURIFICATION.

desired, the liquid in the highest can flow to the second, and from there to the third, and for the sake of explanation they are numbered 1, 2, 3, and 4.

Like that already described, the vessels are provided with agitators, actuated by the bevel wheels on the top of the axles in connection with suitable gearing which is driven by a steam engine. Thus, by the mechanical action of the agitators or stirrers, the lime is always maintained in a state of solution, and as the gas is caused to pass through very contracted spaces, it is broken up into small particles, and these coming in contact with the lime, the impurities are absorbed.

The vessel No. 4, in which the lime is dissolved, is open at the top, and provided with an agitator worked by gearing and engine similar to the purifiers, but of course is only in operation during the process of preparing the solution. This being charged with water to the desired extent, the lime, previously slaked, is then gradually thrown into the water, when by means of the stirrer it is broken up and dissolved, ready for use. In the course of operation when the contents of No. 3 are no longer effective as a

FIG. 49.

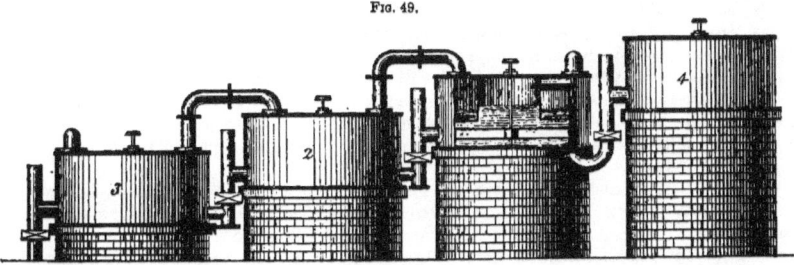

purifying agent, they are allowed to flow, by a properly sealed pipe, into a pit or other place especially arranged for the purpose; when the foul lime of No. 2 is allowed to flow by gravitation into the vessel 3, and that of No. 1 into vessel 2; which effected, the solution of lime is then supplied to the desired height in vessel No. 1. Thus the foul gas first meets with the foulest lime and passes off at the upper vessel, where the lime is in a state of comparative purity.

The quantity of lime necessary for this process is usually estimated at the half of that of the dry-lime system, that is, a bushel of good quicklime on the average is considered sufficient for the purification of 20,000 feet of gas, about 50 gallons of water being intermixed with the slaked lime to reduce it to the proper state of solution.

This solution of a bushel of lime measures about $8\frac{1}{2}$ cubic feet, which gives the datum of $4\frac{1}{4}$ feet required in each of the vessels to purify 10,000 feet of gas, therefore if we suppose them to be charged once per diem (one purifier being charged with fresh lime) for a works producing 100,000 feet, then three purifiers, each 6·5 feet diameter, containing 45 cubic feet, or about 18 inches depth of solution, would be necessary. For another works, producing 500,000 feet daily under like conditions, each purifier would be required of 12 feet in diameter, 4 feet deep, and containing 225 cubic feet of cream of lime; but it may be desirable to change these vessels twice a day, or if the locality permits even oftener, when of course the dimensions of the apparatus are diminished accordingly; but it should be remembered that large apparatus, in consequence of the slow passage of the gas therethrough, facilitates the purification.

But although the purification by wet lime is more efficacious and economical than the other, there exists the difficulty of disposing of the foul material, technically called "blue billy." Of this a portion may be employed as luting for the mouthpieces of retorts where that is used, but there still remains a large quantity which yields a most offensive odour, and in consequence is extremely difficult to dispose of, but wherever this can be effected, and the works are distant from any population, the wet lime process may be adopted with advantage.

Dry Lime.

The next method of purification is by means of dry lime, as generally and very erroneously termed, since, as stated, the dry hydrate has little affinity either for sulphuretted hydrogen or carbonic acid, and is therefore practically useless. To render the hydrate available it is required to be well moistened shortly before being used, to that degree that, when pressed in the hand, it has the consistency of dough or snow.

The moistened hydrate of lime is placed on the sieves or grids of the purifiers to a thickness of three or four inches, the whole being well and completely covered and of uniform thickness. Some purifiers are subdivided, thus the gas has to ascend and descend through the lime, passing as it were through six distinct purifiers, although but three are employed.

It is generally accepted as a rule that a bushel of quick lime, which, when slaked and rendered into hydrate, increases to about $2\frac{1}{2}$ times its original volume, is sufficient by the process in question to purify the gas derived from a ton of coal. This, however, must greatly depend on the quality of the lime, for there are some descriptions which contain a large percentage of argil; and although admirably adapted for the manufacture of hydraulic cement, they are ill suited for the purification of gas.

Another consideration is, that some coals contain considerably more sulphur compounds than others, therefore no general rule can be defined; but when properly treated, and the lime of average quality, the quantities stated alike for wet as dry lime, may be accepted as a good approximation to that required for the purification of gas.

In the ordinary method of purifying, the foul gas is admitted first into the purifier the most impregnated with impurity, thence it passes to the second, and so to the third, where the lime is comparatively clean. When purifiers are of ample magnitude, the foul lime if properly moistened remains an active agent in the purification to the last, therefore it is only when the gas begins to indicate signs of impurity at the fifth division, supposing six divisions of purifiers to exist, that they should be changed, and by the employment of apparatus of ample capacity only can the lime be employed to the best advantage.

According to Mr. Lewis Thompson, "when lime in a purifier has become foul, it still contains in the interior of its little nodules or lumps a certain quantity of pure lime to which the foul gas has not yet penetrated, so that, if we throw the purifier out of action and keep it closed, this internal lime will gradually act upon the hydrosulphate of the sulphuret of calcium which is on the outside of the nodules, and the whole will thus be converted into simple sulphuret of calcium, that has a power of combining with additional sulphuretted hydrogen to form hydrosulphate of the sulphuret of calcium; but to ensure this the lime must contain a quantity of water, without which the hydrosulphate of the sulphate of calcium cannot exist. Hence it is that 'blue billy' stinks so horribly as it dries."

The superiority of the dry-lime process consists in the facts that by it the pressure in the retorts is materially decreased, which is a great consideration in small works where the exhauster is not employed, and the foul residual creates no particular nuisance when moderately distant from habitations, and is readily disposed of for agricultural purposes; whilst by this process, as with the other, all the various impurities of the gas can be eliminated.

Oxide of Iron.

The odour arising from the dry-lime process from works established in populous neighbourhoods, as those of the Metropolis some years ago, rendered it desirable to obtain other means of purification for the purpose of avoiding the numerous complaints occasioned thereby.

Oxide of iron had been proposed by Heard as a purifying agent as far back as 1806. In 1840 Mr. Croll patented the means of freeing gas from sulphuretted hydrogen, by causing it to pass through the black oxide of manganese, or the oxides of iron or zinc moistened with water, which materials, after being saturated with the impurity, when submitted to a red heat in an oven, were to be revived and rendered again fit for use, but the expense of the process precluded its application.

To Mr. Laming is unquestionably due the merit of giving the first publicity to the process of reviving the oxide of iron by exposure to the atmosphere, in order that the same material may be used and again revived repeatedly, for which purpose he obtained a patent in France in 1849. A few months afterwards, Mr. F. J. Evans, of the Chartered Company, in making experiments with the oxide as taken from the purifiers, accidentally discovered the same principle of revivification of the material when exposed to the action of the atmosphere, which he carried practically into operation, and to that gentleman is generally accorded the merit of first appreciating the value of the discovery and the economy to be derived from its application.

In the month of November following, Mr. Hills obtained a patent for various objects, amongst them for re-oxidizing the oxide of iron by submitting it to the action of air after it had ceased to absorb the sulphuretted hydrogen, in order that the same material could be used repeatedly to purify the gas. This patent gave rise to considerable litigation, but in all cases the questions were decided in favour of the patentee. From the first introduction of the process, the oxide of iron purification met with favour and became generally adopted, owing to the absence of the offensive odours attending other processes, and its greater economy.

There are different descriptions of oxides, but those most extensively used are the natural in combination with earthy and fibrous material, which are sold at various prices corresponding we may assume with their respective qualities. Others again are intermixed with sawdust, and in the selection of these, the manager should be guided by the experience of others, as it is somewhat difficult to arrive at a fair conclusion on the value of this material, unless by experiment, as hereafter indicated.

Some firms arrange to supply fresh oxide gratuitously to large establishments, on the condition of having the privilege to cart away the foul material. Other gas companies have an arrangement for the supply of oxide either by weight, or measure, or by the quantity of gas purified, or the quantity of coal carbonized; however, in this, as in other matters, efficiency of operation should be the first, and price the secondary consideration. The quantity of sulphur absorbed by the several descriptions of oxides is variable; some will retain only about 35 per cent. of their weight, and on being gently roasted the sulphur is expelled, when the oxide becomes again a purifying agent, but of less activity than before. According to a circular of a chemical company, established principally for the supply of oxide and manufacture of the spent oxide, their material absorbs twice its weight in sulphur. The author has employed the material commercially known as oxide of iron, intermixed with sawdust, which answered every expectation on the grounds of efficiency, economy, and freedom from nuisance, but all oxides of iron are not effective for the purification of gas, therefore, although this material may in some instances be employed with advantage, it cannot always be relied on.

In explanation of this, to quote Mr. Lewis Thompson, he says, "I send you three samples of sesquioxide of iron, one of which is absolutely pure, another is from this district, and the other is auriferous, but not one of which will purify gas. Those who try to explain this peculiar condition generally ascribe it to a change in the position of the atoms composing the substance, but this change is only based on supposition, for we have no proof that the substance is altered in any other way than in regard to its affinities. Thus phosphorus shines in the dark and will take fire if it is exposed to the air, but this same phosphorus will neither shine nor take fire in the air if it has been heated in a close vessel to a temperature of 300° Fahr. The best way is to admit that we know nothing about it, just as we know nothing about the reason why we live or die. We can understand no reason why one sample of oxide of iron should purify and another not purify gas, but such is the case, and hence arises the necessity for examining each sample. It has been pretended that the hydrated and precipitated oxide was the only useful kind, but this is wrong in every way, for there are anhydrous oxides of iron that will purify, and hydrated and precipitated oxides that will not purify gas."

Oxide is always slightly moistened and charged in much thicker layers than lime, but to what degree must depend on its nature and weight, together with the size of the purifiers. The thicker the layer the better, so long as the pressure is not inconveniently increased. When taken from the purifiers, the spent oxide has a dull black purple appearance, and is then laid a thickness of eight or ten inches on the ground for the purpose of revivification, being occasionally turned over and broken up from day to day, in order to expose all its particles to the action of the atmosphere, and when it resumes its original colour it is suitable to be again employed. Fresh oxide has a tendency to ignite when first used, for this reason only a portion of new material should be intermixed with the old as occasion may require.

As it is desirable to know the capacity of any particular sample for absorbing sulphuretted hydrogen, as well as to ascertain the quantity of sulphur in the spent oxide, the methods of ascertaining these points are here given.

To ascertain the value of any particular Oxide as a Purifying Agent for absorbing Sulphuretted Hydrogen.

There is but one way in which the value of oxide of iron for the absorption of sulphuretted hydrogen can be estimated, and it is equally applicable to lime and all other purifying agents; it is as follows, and requires the use of two bent tubes, as shown below in the annexed diagram, with other apparatus, as weights, scales, &c.

The tube A is charged with pieces of chloride of calcium, and adjusted with corks so as to fit into a Wolf's bottle, *d*, as shown, also to receive the end of a bent pipe that is fixed in a large U-shaped tube B. This U tube is to have one of its legs and the bent part of it filled with 200 grains of the oxide in question, which has been previously well mixed with an equal weight of sawdust; the other leg to have a small piece of cotton wool put into it, and upon this sufficient chloride of calcium to fill the leg, which is then to be closed with a cork, through which a small pipe passes, as shown in diagram.

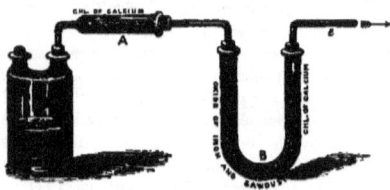

This U tube, when charged, is to be carefully counterpoised in a balance before it is fixed to the tube A; after which acid is to be poured very gently into the Wolf's bottle, upon sulphuret of iron, or any other sulphuret that will cause a gradual production of sulphuretted hydrogen, which is to be continued until the sulphuretted hydrogen passes freely out of the small tube *e*, fixed to the tube B. After this evolution of sulphuretted hydrogen has continued for about ten minutes, remove the tube B and weigh it with its contents, so as to ascertain how much sulphuretted hydrogen has been absorbed. This gives the value of the oxide as a gas-purifying agent.

To ascertain the quantity of free Sulphur in spent Oxide.

This is acquired by dissolving the sulphur combined with the oxide with bisulphide of carbon and then evaporating the liquid, when the crystals of sulphur give the weight of that contained in the oxide. The operation, as described by Mr. Wood of Hastings, is shown in the annexed figure, 50, which represents a pipette A, the flask B supported on a stand, the connecting pipe C, the vertical condenser D, and the bottle E. The flask is very thin and light, must be perfectly clean, and is carefully weighed without the cork with a delicate balance, and to this is nicely adjusted by a tight-fitting cork the pipette A and pipe C. Within the pipette is placed a small piece of tow which acts as a filter, and on this is placed 100 grains of the oxide under examination in a perfectly dry state; all being connected, as represented, it is ready for operation. Then pour a small quantity of bisulphide of carbon into the pipette, which will filter through the spent oxide and drop into the flask below, carrying with it the sulphur contained in the oxide; when by applying a small jet of gas beneath, the spirit is quickly evaporated and carried along the pipe C to the condenser, and received in the bottle E, leaving the sulphur behind in the flask B. The bisulphide is added very slowly, and as the operation proceeds on dropping from the pipette it becomes paler, and when almost colourless the sulphur may be considered to be worked out of the oxide; the spirit then being evaporated, the flask is detached and again weighed with its contents, and whatever excess of weight it may have gained by the presence of the sulphur, is the percentage of that contained in the oxide. This system, however, may be attended with some degree of risk, which is avoided by the method adopted by Mr. Marshall, who first dissolves the sulphur, for which purpose the pipette A only is attached to the flask. This effected, the flask is placed in a porcelain dish containing water, when the pipette is removed, and the flask connected to the condenser and bottle by the pipe C. The water is then heated by a gas-burner, the bisulphide evaporates, and is condensed and collected in water contained in the bottle E.

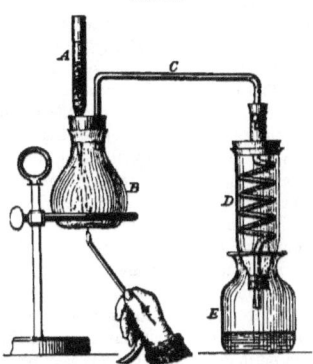

Fig. 50.

In the "Chemistry of Gas Manufacture" the method of purifying gas by means of the sulphate of iron and lime has already been described. We have only to add that, for the sake of economy, the sulphate may be of the kind known as "black copperas," which is quite as good as the green, only as it comes from the bottom of the crystallizer it is rendered black by the presence of a little soot and dirt. The black is about half the price of green copperas.

A system adopted by some managers is to prepare the sulphate of iron on their premises and intermix this with lime and sawdust, for which object about 40 bushels of sawdust of hard wood, as elm or beech, is well intermixed with 10 bushels of air-slaked or ground lime, and the whole left in a heap for twenty-four hours, when any small nodules of lime which have escaped the slaking will be broken up. For the purpose of preparing the sulphate of iron a wooden tank of suitable capacity, lined with lead, is provided, and in this is placed a quantity of old iron scrap of any kind, in the proportion of about $\frac{1}{4}$ cwt. of iron to 90 gallons of water and 10 gallons of sulphuric acid, but avoiding the admixture of any other metal. One half of the water is first poured into the tank, and to this is added a corresponding part of the acid, and in the course of a few hours the rest of the water and acid are added, and the whole left for three or four days, until all disengagement of gas ceases, when a strong solution of the sulphate of iron will be formed, with crystals of the same deposited at the bottom of the tank.

The lime and sawdust are then intermixed as when mixing mortar, a portion of the solution is gradually added, and the whole thoroughly mixed to the consistency of ordinary mortar, when it is allowed to dry and oxidize under cover. After a few days add in a similar manner the remainder of the sulphate of iron and keep under shelter as before, and on the completion of the second oxidation it will be ready for use. Earthenware pipkins and wooden pails, pitched inside and outside, are employed for the liquid. A gentleman who has employed this prepared oxide for some years speaks highly of its efficiency and economy, and like the natural oxide, this is capable of being revived, by exposure to the atmosphere, from twenty-five to thirty times, when it will have absorbed about 50 per cent. of its weight in sulphur.

Here we may observe that different engineers employ different methods of purification, but in the majority of works lime is used for the purpose, either as the cream of lime or the hydrate of lime properly moistened, commonly called "dry lime," by either of which processes the whole of the impurities can be removed so that the gas is rendered commercially pure. In other works the use of lime is unknown, the purification being effected by the condenser, scrubber, and oxide of iron; and in a few cases the carbonic acid is eliminated, so far as practicable, by means of the ammoniacal liquor in the scrubber or washer, and sometimes a percentage of cannel is employed to compensate for the deterioration of the gas by the presence of a small quantity of carbonic acid. Again, in other establishments, the sulphate of iron and lime are exclusively employed.

Then, as regards the ammonia; instead of eliminating this in the washer or scrubber, the gas is caused to pass through an ordinary purifier charged with breeze or sawdust saturated with dilute sulphuric acid, but used in a dry state, a process to be recommended in small works, by which the residual may be preserved and rendered a marketable commodity, but in works of 20,000,000 feet per annum and upwards the scrubber or washer, or both combined, should eliminate the ammonia.

These and other processes are employed, each of which has its supporters, who invariably maintain that the system adopted by them, respectively, is the best and cheapest; and it must be admitted that, under certain conditions, they may be in the right, as the expense of lime in some few localities will preclude its use as a purifying agent, when oxide of lime will be resorted to. In other places lime may be of that price, and the works so situated, as to permit of the gas being purified under the best conditions, whilst in some localities the odour arising from this system would be an impediment to its use. The sulphate of iron and lime has its supporters, principally, we suppose, on account of the economy of the process and the little odour arising therefrom. Hence it is obvious that the means of purifying gas must, in many cases, be governed by circumstances; nevertheless, in our opinion, more unanimity in this respect should exist among engineers.

We venture to suggest that the process of purification should be commenced in the hydraulic main, as proposed in the "Chemistry of Gas Manufacture;" that the tar should be kept in contact with the gas at a moderate temperature during the process of condensation, and that this should be effected gradually, after which the gas should be caused to pass through a thin stratum of breeze, to retain any tar and prevent it passing to the scrubber; that the scrubber or washers or other similar apparatus should be of ample capacity, but employing as little water as possible to eliminate the ammonia, and by having large purifiers, so that the gas may be in a comparatively quiescent state, so essential for depositing its impurities; under these conditions whatever material may be employed, whether lime, or oxide, or the sulphate of iron and lime, or a combination of these, will depend on the cost of the respective materials in the neighbourhood of the works; it being understood that oxide is suitable only for eliminating the sulphuretted hydrogen; and when using this, lime is essential to the removal of the carbonic acid; whilst lime will take out the whole of the impurities.

The tests usually applied for ascertaining the purity of gas are the following: namely, for ammonia, turmeric, or reddened litmus paper, and hydrochloric acid; for sulphuretted hydrogen, acetate of lead paper; and for carbonic acid, lime water.

To prepare Turmeric Paper.

Turmeric paper is prepared by pouring three ounces of spirit of wine upon half an ounce of turmeric powder in a clean stoppered bottle; shake it daily, and at the end of four days let it repose, so that the brown powder will settle at the bottom, and leave the liquor above clear. Then place the liquor in a clean plate, into which dip some pieces of white filtering paper, and when they are well soaked, hang them on a line to dry. This done, the paper is cut into strips of the desired size, generally about $2\frac{1}{2}$ inches long and $\frac{1}{2}$ inch wide, which are then put in a wide-mouthed bottle, well corked and kept in a drawer away from the light. This, like the litmus and acetate of lead tests, should be slightly moistened before being used. Turmeric paper, when exposed to a jet of gas containing ammonia, is changed from its yellow colour to a reddish brown. Ammonia is also detected by dipping a glass rod into hydrochloric acid, and holding it in a stream of the gas, when, if ammonia be present, a white cloud of sal ammoniac will be produced.

To prepare Blue Litmus Paper.

In a stoppered bottle, as before, put half an ounce of pulverized litmus, and to this add three ounces of cold water, which should be well shaken, and left some time to dissolve. It is then filtered, and the paper prepared as stated in the preceding, observing the same care in preserving the slips, which are also applied as tests for determining the presence of acids.

PURIFICATION.

To prepare Red Litmus Paper.

Take a quantity of the blue solution mentioned, and add cautiously, drop by drop, a small quantity of very dilute sulphuric acid, until a piece of filtering paper, when dipped therein, acquires a distinctly red tinge when dry. Pieces of white filtering paper are then soaked in the solution, placed in a clean plate, after which they are hung up to dry, and cut in slips similar to the turmeric paper. These slips must be kept well protected both from air and light.

To prepare Acetate of Lead Paper.

In a bottle containing two ounces of distilled water place a quarter of an ounce of crystallized acetate of lead, and when thoroughly dissolved saturate pieces of white bibulous paper as described, and when dry cut in slips, and bottle as already stated. On this test being exposed to gas containing sulphuretted hydrogen, it becomes darkened and eventually brown; if after an exposure of three minutes the paper does not become discoloured, the gas may be considered free from this impurity. When applying either of these tests, the paper is slightly damped, and held within about a quarter of an inch from a fish-tail burner from which the gas is issuing in full stream.

To prepare test for Carbonic Acid.

Into a bottle containing a quart of water put four ounces of quicklime, allow it time to dissolve, shake it occasionally, afterwards let the lime settle, and, when perfectly clear, transfer the liquor to another bottle, which should be kept securely corked. For testing, a piece of $\frac{1}{4}$-inch lead pipe or indiarubber tube may be attached to a small tap, when the gas under examination is allowed to blow through a quantity of the lime water placed in a wine-glass or other suitable vessel. Should the water remain clear after a trial of two minutes, the gas may be pronounced free from carbonic acid; but in the presence of this, the lime water becomes cloudy or milky, according to the quantity of the impurity.

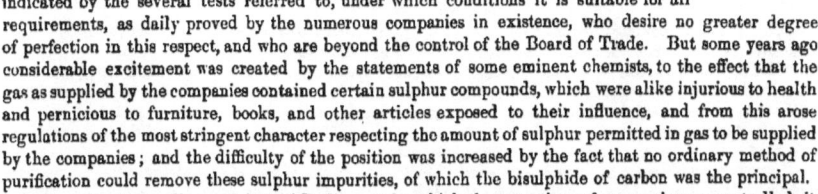

In many gasworks the constant test shown in the annexed figure is employed: for that purpose two pieces of test paper, acetate of lead and turmeric paper, are suspended within the vessel, their ends dipping into two cups containing water, by which means the papers are always kept moistened. A burner, consuming about two feet per hour, is attached, and the gas in its passage to the burner issues from the two jets, and impinges on the two slips of paper; the case being of glass, allows the test papers to be observed. When gas is well purified by the ordinary process, sometimes these will remain in the apparatus twenty-four hours without becoming soiled.

We commenced by referring to gas as generally recognized to be commercially pure—that is, free from ammonia, sulphuretted hydrogen, and carbonic acid, as indicated by the several tests referred to, under which conditions it is suitable for all requirements, as daily proved by the numerous companies in existence, who desire no greater degree of perfection in this respect, and who are beyond the control of the Board of Trade. But some years ago considerable excitement was created by the statements of some eminent chemists, to the effect that the gas as supplied by the companies contained certain sulphur compounds, which were alike injurious to health and pernicious to furniture, books, and other articles exposed to their influence, and from this arose regulations of the most stringent character respecting the amount of sulphur permitted in gas to be supplied by the companies; and the difficulty of the position was increased by the fact that no ordinary method of purification could remove these sulphur impurities, of which the bisulphide of carbon was the principal.

Accordingly, in all recent Acts of Parliament by which the operations of companies are controlled, it is enacted that the gas shall not contain sulphur in any form exceeding 20 grains per 100 cubic feet of gas, and more recently gas referees were appointed, by whom the degree of purity of the gas to be supplied is from time to time decided.

These restrictions were for a time the source of considerable anxiety to the companies interested, subjected as they were to severe and ruinous penalties, if enforced, for any excess of sulphur beyond that stipulated; and although it was well known that the sulphide of calcium possessed the power of absorbing the objectionable compounds, yet in practice it could not be accomplished with certain uniformity. Under these circumstances the ablest of chemists were consulted, without, however, attaining the desired object, namely, the removal of the sulphur in other forms than sulphuretted hydrogen to the required extent, regularly and certainly, so as to purify the gas within the prescribed limits.

For the means of solving this difficulty we are indebted to Mr. R. H. Patterson, formerly one of the Metropolitan Gas Referees, who made the important discovery that in order to eliminate the compounds of sulphur other than sulphuretted hydrogen, by the aid of the sulphide of calcium, the gas should be first freed from carbonic acid, and that with this existing in the gas the action of the sulphide of calcium was uncertain and variable. This fact ascertained, and duly carried into operation, all difficulty in eliminating these sulphur impurities was at an end, for instead of having an indefinite quantity of sulphur, often as high as 40 or 50 grains per 100 feet, it is asserted that, by Mr. Patterson's process, they can be reduced within the prescribed limits, and indeed down to 10 or 11 grains of sulphur per 100 feet of gas. For this purpose the method of purifying in some establishments is reversed, when, instead of the foul gas first entering the foulest purifier, as formerly, by the process in question it is caused to pass into vessels called carbonaters, containing clean hydrate of lime, by which the carbonic acid is absorbed, and from thence the gas passes into other purifiers, containing the sulphide of calcium, and subsequently to the oxide of iron purifiers to eliminate the remaining sulphuretted hydrogen.

This process, although effective, is very costly, and, as hereafter shown, the operation of gas testing of the present day, as ordered by the Gas Referees, is one of great nicety of manipulation, and demands some skill, but whether *le jeu vaut la chandelle* public opinion must decide.

The following is a copy of the instructions of the Metropolitan Gas Referees to the Gas Examiners, as to the means of testing the purity of the gas supplied by those London companies which are under the control of the Board of Trade:—

"TIMES AND MODE OF TESTING FOR PURITY.

"I.—SULPHURETTED HYDROGEN.

"All the gas that is consumed in testing for illuminating power shall be passed through an apparatus in which slips of bibulous paper, moistened with a solution of acetate of lead, are to be suspended, and exposed to the direct action of the gas continuously during the whole time that the testings for illuminating power are being made. If, after the testings for illuminating power, any discoloration of the test paper is found to have taken place, this is to be held conclusive as to the presence of sulphuretted hydrogen in the gas. Fresh test slips are to be placed in the apparatus every day.

"In the event of any impurity being discovered, one of the test slips shall be placed in a stoppered bottle and kept in a dark place at the testing station, the remaining slips shall be forwarded with the report to the Referees.

"II.—AMMONIA.

"In this test the gas must be passed through the testing apparatus before entering the meter, so as to prevent the possibility of any ammonia being taken out by the water contained in the meter.

"The apparatus consists of a glass cylinder, in which are placed glass beads, through which the gas is made to pass at the rate of about half a cubic foot per hour. When ten feet of gas have passed, the supply is cut off by a self-acting movement attached to the meter. A set of burettes, properly graduated, is also provided.

"The maximum amount of ammonia allowed is 2·5 grains per 100 cubic feet of gas; and the testing shall be made so as to show the exact amount of ammonia in the gas.

"Two test solutions are to be used—one consisting of dilute sulphuric acid of such strength that 25 measures (septems) will neutralize 1 grain of ammonia; the other a weak solution of ammonia, 100 measures (septems) of which contain 1 grain of ammonia, NH_3.

"The correctness of the result to be obtained depends upon the fulfilment of two conditions:

"1. The preparation of test solutions having the proper strength;

"2. The accurate performance of the operation of testing.

"To prepare the test solutions, the following processes may be used by the gas examiner:

"Measure a gallon of distilled water into a clean earthenware jar, or other suitable vessel. Add to this 94 septems of pure concentrated sulphuric acid, and mix thoroughly. Take exactly 50 septems of the liquid and precipitate it with barium chloride in the manner prescribed for the sulphur test. The weight of barium sulphate which 50 septems of the test acid should yield is 13·8 grains. The weight obtained with the dilute acid prepared as above will be somewhat greater, unless the sulphuric acid used had a specific gravity below 1·84.

"Add now to the diluted acid a measured quantity of water, which is to be found by subtracting 13·8 from the weight of barium sulphate obtained in the experiment, and multiplying the difference by 726. The resulting number is the number of septems of water to be added.

"If these operations have been accurately performed, a second precipitation and weighing of the barium sulphate obtainable from 50 septems of the test acid will give nearly the correct number of 13·8 grains. If the weight exceeds 13·9 grains, or falls below 13·7 grains, more water or sulphuric acid must be added, and fresh trials made, until the weight falls within these limits. The test acid thus prepared should be transferred at once to stoppered bottles which have been well drained and are duly labelled.

"To prepare the standard solution of ammonia, measure out as before a gallon of distilled water, and mix with it 50 septems of strong solution of ammonia (sp. gr. 0·88). Try whether 100 septems of the test alkali thus prepared will neutralize 25 of the test acid, proceeding according to the directions given subsequently as to the mode of testing. If the acid is just neutralized by the last few drops, the test alkali is of the required strength. But if not, small additional quantities of water, or of strong ammonia solution, must be added, and fresh trials made, until the proper strength has been attained. The bottles in which the solution is stored should be filled nearly full and well stoppered.

"The mode of testing is as follows: Take 50 septems of the test acid (which is greatly in excess of any quantity of ammonia likely to be found in the gas), and pour it into the glass cylinder, so as to well wet the whole interior surface, and also the glass beads. Connect one terminal tube of the cylinder with the gas supply, and the other with the meter, and make the gas pass at the rate of about half a cubic foot per hour. Any ammonia that is in the gas will be arrested by the sulphuric acid, and a portion of the acid (varying with the quantity of ammonia in the gas) will be neutralized thereby. At the end of each period of testing, the glass cylinder and its contents are to be well washed out with distilled water, and the washings collected in a glass vessel. One-half of this liquid is to be taken in a separate glass vessel, and a quantity of solution of hæmatoxylin (which is turned pink by alkalies) added, just sufficient to cause the liquid to assume a yellowish colour. Then pour into the burette 100 septems of the test alkali, and gradually drop this solution into the measured quantity of the washings collected, stirring constantly; as soon as the colour changes to a pink tint, which remains permanent for five minutes (indicating that the whole of the sulphuric acid has been neutralized), read off the quantity of liquid remaining in the burette. To find the number of grains of ammonia in 100 cubic feet of the gas, multiply by 2 the number of septems of test alkali remaining in the burette, and move the decimal point one place to the left. The remaining half of the liquid is to be preserved in a bottle duly labelled for a week.

"III.—SULPHUR COMPOUNDS OTHER THAN SULPHURETTED HYDROGEN.

"The testing shall be made in a room where no gas is burnt other than that which is being tested for sulphur and ammonia.

"The gas for this test shall be burnt at the rate of about half a cubic foot per hour, until ten feet are consumed, when the supply is stopped by a self-acting movement.

"The apparatus to be employed is represented below by a diagram, and is of the following description: The gas is burnt in a small Bunsen burner with steatite top, which is mounted on a short cylindrical stand, perforated with holes for the admission of air, and having on its upper surface a deep circular channel to receive the wide end of a glass trumpet-tube. On the top of the stand, between the narrow stem of the burner and the surrounding glass trumpet-tube, are to be placed pieces of commercial sesqui-carbonate of ammonia weighing in all about two ounces.

"The products both of the combustion of the gas and of the gradual volatilization of the ammonia salt go upwards through the trumpet tube into a vertical glass cylinder, packed with balls of glass, to break up the current and promote condensation. From the top of the cylinder there proceeds a long glass pipe or chimney, serving to effect some further condensation, as well as to regulate the draught and afford an exit for the uncondensable gases. In the bottom of the cylinder is fixed a small glass tube, through which the liquid formed during the testing drops into a beaker placed beneath.

"The following cautions are to be observed in selecting and setting up the apparatus:

"See that the inlet pipe fits gas-tight to the burner, and that the holes in the circular stand are clear. If the burner gives a luminous flame remove the top piece, and having hammered down gently the nozzle of soft metal perforate it afresh, making as small a hole as will give passage to half a cubic foot of gas per hour at a convenient pressure.

"See that the tubulure of the condenser has an internal diameter of not less than ¾ inch, and that its outside is smooth and of the same size as the small end of the trumpet tube.

"See that the short piece of indiarubber pipe fits tightly both to the trumpet tube and to the tubulure of the condenser.

"The small tube at the bottom of the condenser should have its lower end contracted, so that when in use it may be closed by a drop of water.

"The indiarubber pipe at the lower end of the chimney tube should fit into, and not simply rest upon, the mouth of the condenser, and the upper extremity of this tube may with advantage be given a downward curvature.

FIG. 51.

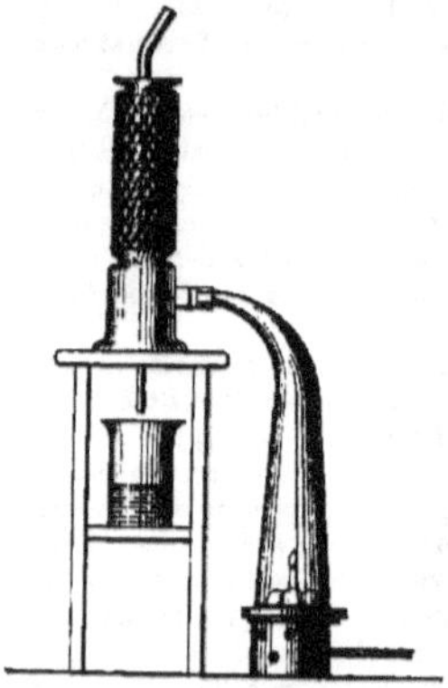

"At the end of each period of testing, the cylinder and trumpet tube are to be well washed out with distilled water. Fresh pieces of sesqui-carbonate of ammonia are to be used each day.

"The gas examiner shall then proceed as follows:—

"The liquid in the beaker and the water used in washing out the apparatus shall be put into the same vessel, well mixed, and measured. One-half of the liquid so obtained is to be set aside, and preserved for a week, properly labelled, in case it should be desirable to verify the correctness of the testing.

"The remaining half of the liquid is to be put into a flask or beaker, covered with a large watch-glass,—treated with hydrochloric acid, sufficient in quantity to leave an excess of acid in the solution,—and then raised to the boiling point, but not so violently as to occasion loss by spirting. An excess of a solution of barium chloride is now to be added, and the liquid again boiled for five minutes. The vessel and its contents are to be allowed to stand till the barium sulphate settles at the bottom of the vessel, after which the clear liquid is to be as far as possible poured off through a paper filter. The remaining liquid and barium sulphate are then to be poured on to the filter, and the latter well washed with hot distilled water. (In order to ascertain whether every trace of barium chloride and ammonium chloride has been removed, a small quantity of the washings from the filter should be placed in a test tube, and a drop of a solution of silver nitrate added; should the liquid, instead of remaining perfectly clear, become cloudy, the washing must be continued until on repeating the test no cloudiness is produced). Dry the filter with its contents, and transfer it into a weighed platinum crucible. Heat the crucible over a lamp, increasing the temperature gradually, from the point at which the paper begins to char, up to bright redness. When no black particles remain, allow the crucible to cool; place it when nearly cold in a desiccator over strong sulphuric acid (to prevent absorption of moisture), and again weigh it. The difference between the first and second weighings of the crucible will give the number of grains of barium sulphate. Multiply this number by 11, and divide by 4: the result is the number of grains of sulphur in 100 cubic feet of gas.

"This number is to be corrected for the variations of temperature and atmospheric pressure in the manner indicated under the head of illuminating power; with this difference, that the readings of the barometer and thermometer are to be taken for the day on which the testing commenced and also the day on which it closed; and the mean of the two is to be used.

"This correction may be made most simply and with sufficient accuracy in the following manner:—

"When the tabular number is between* 955–965, 966–975, 976–985, 986–995; increase the number of grains of sulphur by $\frac{4}{100}$ths, $\frac{3}{100}$ths, $\frac{2}{100}$ths, $\frac{1}{100}$th.

"When the tabular number is between 996–1005, no correction need be made.

"When the tabular number is between 1006–1015, 1016–1025, 1026–1035; diminish the number of grains of sulphur by $\frac{1}{100}$th, $\frac{2}{100}$ths, $\frac{3}{100}$ths.

Examples:—

Grains of barium sulphate from 5 cub. ft. of gas .. 4·3
Multiply by 11, and divide by 4.

4)47·3

Grains of sulphur in 100 cub. ft. of gas (uncorrected) 11·82
Add 11·8 × $\frac{2}{100}$ = ... ·24

Grains of sulphur in 100 cub. ft. of gas (corrected) 12·06

Barometer (mean) .. .. 29·2
Thermometer (mean) 54
Tabular number 985

Result :•
12·1

* These Tables are to be found under the head of Photometer, page 246.

PURIFICATION.

"As to the TIMES AND MODE OF TESTING FOR PRESSURE.

"BAROMETER AND THERMOMETER.

"Readings of the barometer and thermometer shall be made and recorded daily at 8 P.M.

"METERS.

"The meters used for measuring the gas consumed in making the various testings, having been certified by the Referees, shall, at periods of not less than seven days, be proved by the gas examiners by means of the Referees' cubic foot measure.

"No meter other than a wet one shall be used in testing the gas under these instructions.

"The results of the daily testings for illuminating power and purity shall be recorded in the proper form, and delivered as provided in the Acts of Parliament.

"As to the MAXIMUM AMOUNTS OF IMPURITY in each form with which the gas shall be allowed to be charged.

"SULPHURETTED HYDROGEN.

"By the Acts of Parliament all gas supplied must be wholly free from this impurity.

"AMMONIA.

"The maximum amount of this impurity shall be $2\frac{1}{4}$ grains per 100 cubic feet.

"SULPHUR COMPOUNDS OTHER THAN SULPHURETTED HYDROGEN.

"For gas made at Beckton, Bow, and Bromley, the maximum amount of sulphur allowable shall be 20 grains per 100 cubic feet of gas.

"For gas made at other works of the Gas Light and Coke Company, and the works of the South Metropolitan Gas Company, it shall be 25 grains.

"For gas made at the works of the Commercial Gas Company, no maximum is yet fixed, and meanwhile the amount shall simply be recorded."

From the preceding we can understand the extreme delicacy observed in the modern method of gas testing; and although, so far as we are aware, the various details are not adopted by any other gas companies than those under the control of the Board of Trade, it was considered necessary to introduce these instructions in order to illustrate the most advanced operations in connection with our subject. That considerable expense is incurred to arrive at this degree of perfection cannot be denied, and that some companies are deterred from applying to Parliament in consequence of anticipated restrictions is also true, but whether the public are benefited by this state of purity of the gas, or whether that portion of the inhabitants of the metropolis who consume gas so purified are capable of appreciating any difference between it and that as usually furnished by companies, is open to considerable doubt. These questions are, however, in a manner answered by the universal satisfaction experienced in all localities throughout the world where gas is, as we have stated, commercially pure, and which leaves nothing to be desired.

HISTORY OF THE GASHOLDER.

IN our historical sketch we have endeavoured to trace the gradual development of the knowledge of the various gases, from the first vague ideas of Van Helmont, to the wondrous discoveries of Black, Cavendish, Priestley, Lavoisier, Scheele, and others.

This knowledge of the existence of gases being acquired, it then became necessary, for experimental purposes, to employ means of storing and transmitting them; for their application to any commercial or industrial use had probably never then entered into the mind of man. Among the earliest arrangements for such manipulation is that mentioned in some French works on chemistry, published about 1780, which is represented in Fig. 52, and may be described as follows:

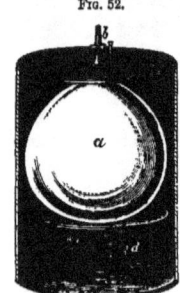

FIG. 52.

Within the cylindrical vessel, which is partly shown in section, and is closed at the top and bottom, was placed the bladder a, its mouth being attached to the tap b, situated in the centre of the top of the vessel. Beneath the bladder was a disc, of rather smaller diameter than the case, which disc was attached to, and actuated by the spiral spring d. Thus, when the gas was admitted into the bladder, the disc and the spiral spring were depressed, and the bladder filled; and whenever requisite, the gas was expelled by the action of the spring raising the disc. It is almost needless to say that the apparatus was very clumsy and ineffective, that no regular supply could be obtained therefrom, nor could any correct idea of its cubical contents be ascertained; in addition to these defects, the capacity was very limited.

Prior to the period referred to, the master mind of Priestley was directed to the construction of apparatus and means of storing and transmitting gas, which resulted in his invention of the pneumatic trough, the jars and sliding trays identical with those used in the chemical laboratory at the present day. He also adopted the method of passing the gas into the jars by means of a pipe.

Here, we may observe, were the three requirements for the construction of the gasholder—namely, the trough or tank, the jar or holder, and the pipe for the admission of the gas. So far, a most important step was made by Priestley towards the invention of that indispensable part of a gasworks; but the completion was reserved for his celebrated *collaborateur*, M. Lavoisier, to whom is due the great merit of being the inventor of the gasholder.

During their important investigations and experiments, a regular correspondence was maintained between Dr. Priestley and M. Lavoisier, a free interchange of ideas passing between them, with all the candour that sincere friendship could dictate. Hence we can understand that what had been accomplished by the former was communicated to the latter, who, probably animated by the achievements of Priestley, applied himself to the study of the means of supplying the great want experienced by chemists in gas manipulation, in which he succeeded, as will be now shown, beyond all expectation.

In the year 1782, M. Lavoisier presented a communication to the Académie des Sciences, "On the Manner of augmenting considerably the Action of Fire and Heat in Chemical Operations." In his communication and drawings, he describes a square gasholder (as now termed) working in a square tank, filled with water, and provided with two pipes for the inlet and outlet, the holder being suspended by a chain, which passed over the curved end of a balance lever, and was counterbalanced by ordinary weights placed in a scale, as will be hereafter represented. But at the time of the invention, the apparatus was intended simply for the purpose of containing oxygen gas, and termed the ("*soufflet hydrostatique*") hydrostatic bellows, and employed for fusing substances by means of the "*chalumeau à gaz oxygène*," or oxygen gas blow-pipe. Beyond this, its inventor appears at the time to have had no other idea of the application of the apparatus.

The same year, however, M. Meusnier, who, in all his various important experiments, was the constant assistant of Lavoisier, communicated to the Académie already mentioned "a description of an apparatus suitable for the manufacture of different kinds of air" (for the term gas had not yet become generally accepted), "in experiments where considerable quantities are necessary, by which a perfectly continuous uniform supply is obtained, capable of being varied at will, and giving at each moment the quantity of air employed with all the precision that can be desired."

In the description he says, "The hydrostatic bellows described by M. Lavoisier has given me the idea to make it an apparatus which would be of the greatest utility in the laboratory, where experiments on air are conducted, when it becomes of the most vital importance to know, with all possible precision, the quantity of these fluids." The drawings which accompanied Meusnier's communication were very similar to those of Lavoisier, the main object of the communication of the first-named gentleman being to represent the desirability of an apparatus for storing gas in large quantities, at the same time to be enabled to appreciate the quantity the vessel contained. The author also described how the pressure may be controlled by variable counterbalance weights.

It is generally, but erroneously, stated that Lavoisier invented the gasholder in 1789. The circumstance which may have led to this error is probably the fact that his work, 'Opuscules Chymiques et Physiques,' was published that year, wherein is presented, we believe for the first time, by the author, a description and drawing of the apparatus, and it may have been assumed that the invention was then of recent date. At the risk of being charged with repetition, we refer to the work mentioned, and give a copy (Fig. 53) of the drawing alluded to.

FIG. 53.

In this the author, after rendering just homage to Dr. Priestley for his various inventions in connection with pneumatic chemistry, and relating the serious impediments experienced when conducting experiments with gases, through the absence of suitable means for their gradual transmission, Lavoisier describes his own admirable invention in the following words :

"I have given the name of GASOMETER to an instrument of which I am the inventor, and had made with the object of producing a continuous and uniform supply of oxygen for experiments of fusion. Since then, M. Meusnier and myself have made considerable corrections and additions to our first trials, and we have transformed it, in a manner of speaking, to be a universal instrument, which it would be difficult to dispense with whenever it may be desirable to make experiments of rigid exactitude.

"The name of the instrument sufficiently indicates that it is destined to measure the volume of gas. It consists of a balance lever 3 feet long, supported by a pillar; to each end of the lever is fixed a portion of an arc." After describing the great care necessary in order to avoid any friction at the centre of the balance, he says: "At the extremity of one of the balance arms is suspended a balance, for the purpose of receiving weights; the chain, which is flat, adapts itself to the curvature of the arc. At the other end of the balance arm is another chain, to which is suspended a (*grande cloche*) great bell of sheet copper of 18 inches diameter, and 20 inches deep, which is partly immersed in another tank of larger dimensions, containing water, and having pipes, shown in dotted lines, for the ingress and egress of the gas." He also describes how the holder, in rising, is filled with gas, and how, by the action of the water, the gas is expelled, and the holder descends. The drawing represented is substantially the same as that in Lavoisier's first communication to the Académie in 1782, the only difference consisting in the fact that the gasholder and tank are cylindrical, instead of square as first proposed by Lavoisier. All is so clear as to render any further description unnecessary.

Thus was produced the first gasholder, which has, indeed, become a "universal instrument," as termed by its great inventor, employed as it is in every civilized town throughout the world, not only for gas lighting, but also for numerous other industrial purposes. And hence the origin of the word "gasometer," a name retained during many years for the gasholder, which is likewise now applied in some Continental languages to signify a gasworks. To repeat Lavoisier's words, "It would be difficult," indeed, "to dispense with" the gasholder, since nothing that ingenuity could devise could possibly replace this important apparatus, particularly at the present time, when its magnitude has reached such gigantic proportions as to contain in some instances upwards of $2\frac{1}{2}$ million cubic feet of gas, whilst the money value of the invisible fluid entering and issuing from these vessels in the metropolis alone in the depth of winter, may be estimated at about 15,000l. per diem.

And here the thought occurs that the first idea of Murdoch to employ gas for the purpose of illumination might have been suggested by the description of the simple means for storing it, as only three years elapsed between the publication of Lavoisier's work describing his invention and Murdoch's first experiments. It is however certain, that had the practicability of gas lighting been discovered before the invention of the gasholder, its application would have been attended with many extraordinary difficulties, and its progress very slow. But on the other hand, as we learn by experience, one great invention leads to another which is subservient to the first, hence we may infer that had the use of gas been known prior to its date, the invention of the gasholder would have speedily followed.

The gasholder consists essentially of two parts, the tank containing the water, which water serves three distinct purposes—viz. it prevents the gas escaping; it is the point of resistance to the action of the gas so causing the holder to rise; and is the means of expelling the gas as the holder descends. The other part, which is sometimes called the bell, is in reality the holder, and rises and descends according to the quantity of gas contained therein. The whole serving as a store or magazine for the gas produced, whence it is delivered as may be required.

A principal consideration in the construction of the bell, apart from its dimensions, is that it shall be of such weight as to give the desired pressure to the gas in the mains with which it is connected, but any extraordinary weight therein which causes excess of pressure is prejudicial, particularly in small works unprovided with an exhauster, for as already observed it increases the deposit of carbon within the retorts, and the loss by leakage on the works.

On the first practical application of gas lighting, the holders were made rectangular in form, in this respect following the first ideas of Lavoisier. Their capacity, to judge from the drawings in Accum's 'Treatise,' was about 500 cubic feet each, and they were supported in the interior by strong iron and wooden framing capable of resisting against enormous external pressure, and yet water seal was only a few inches in depth. From this we may infer that our predecessors, the first engineers, were but imperfectly acquainted with the nature of atmospheric pressure.

The rectangular gasholder was exclusively retained until about 1816, when the cylindrical form was first adopted. A rectangular holder of 19 feet square at the Manchester Works exploded in 1819, causing considerable consternation in the neighbourhood, and for a short time prejudiced the advancement of the new art. This we believe was the first of several accidents of a like nature which have unfortunately happened in some of the large towns of the United Kingdom and the Continent. Happily, such catastrophes are now less frequent than formerly.

Cylindrical holders coming into operation, it then became necessary, in order to meet the requirements of the extended use of gas lighting, to construct them of much larger dimensions than had ever been attempted with the square form. Whether arising from the cost of construction or the want of ability is not clear, but at that period the greatest difficulty was experienced in erecting suitable tanks for the holders. This led to the adoption of wooden vats for the purpose, similar to those used by brewers, indeed a second-hand brewer's vat was purchased by the Chartered Company, and fixed as a gasholder tank at one of their stations, which circumstance is mentioned with the view of comparing the past with the present position of that company. The durability of these vats was very limited, for the slightest leakage of water speedily destroyed the hoops, when the vessel would sometimes suddenly burst and cause considerable damage and annoyance, misfortunes which were admitted by George Lowe in his evidence before a Committee of the House of Commons in 1823. The last of these tanks or vats was removed from the Brick Lane station in 1843.

Among the greatest impediments experienced by Samuel Clegg in his early gasworks, were the difficulties referred to in constructing the tanks of gasholders. At that time cast iron was at a fabulous price, the advantages of hydraulic cements were not known; and these obstacles were naturally augmented in proportion to the depths of the tanks. He, therefore, devoted his great abilities to find remedies for this evil, by inventing descriptions of gasholders which should require tanks of very limited depth. The first of these was the "rotary holder," which, in addition to its intended purpose, possesses considerable interest on account of it being supposed to have suggested to its inventor the first idea of the gas meter.

This rotary holder, represented in section in Fig. 54, is thus described.

The tank t, which was constructed of brickwork or masonry, was filled with water, and provided with a wooden frame ff, for the purpose of carrying the annular vessel or gasholder v, which in cross section was rectangular and closed at the end e, and open to the atmosphere at the end marked a. By means of suitable framing, $g\,g\,g$, the holder v was attached to a central hollow shaft, which worked in a corresponding stuffing box. The pipe p connected the hollow shaft with the end, e, of the holder, and served to convey the gas into and from that vessel, the arrows indicating the passage of the gas thereto. One end of the chain h passed round and was attached to the drum s, which was fixed to the hollow shaft. The other part of the chain passed over a pulley, and to it were suspended counterbalance weights, for the purpose of keeping the holder in equilibrium in all its varied positions. This, however, as the writer was informed many years ago, by a gentleman who was in the service of the Chartered Company at the period, was in practice quite impossible, and in this consisted the main objection to the use of the system, leading probably to the invention of the collapsing holder.

Fig. 54.

The action of the apparatus was as follows:
In the first instance, a quantity of water was placed within the holder to the height indicated in the drawing. The outlet being then opened, the vessel was turned round until the end e touched the water; when on the gas being admitted it pressed between the surface of the water and the end, acting as in an ordinary gasholder, so causing the vessel v to revolve in the direction of the arrows, until the part a would be immersed in the tank, when the holder having made about half a revolution would be full. Then, by changing the valve communicating with the outlet pipe, the holder would, by its weight, revolve in the opposite direction and expel the gas.

BRICK TANKS CONSTRUCTED.

One of these holders was erected at the Peter Street station; they were also fixed at a few other works, but were not successful. Subsequently, in 1818, Clegg invented his collapsing gasholder, which has already been described.

Hence, Clegg's efforts to improve the simplest of all mechanical contrivances ended in complete failure, and since that period the only change that it has undergone has been by the adoption of the telescopic system.

According to Peckstone's 'Theory and Practice of Gas Lighting,' published in 1819, the gasholders generally employed at large works varied from 15,000 to 20,000 feet in capacity. The author recommended either cast-iron or wooden vats for the tanks, but suggested the greatest care in the construction of the latter. We learn from this work that John Malam contributed largely towards the improvements in the construction of cylindrical gasholders, by dispensing with the heavy internal framing, in applying counterbalance weights and chains for relieving the pressure, and guiding the holder by a centre rod and tube. The last system was extensively adopted in gasholders for small works, many of which are still in existence. Referring to the construction of holders, Peckstone says: "Experience has now taught the manufacturer that he cannot construct the gasholder too light. Instead of the wooden frame or weighty iron stays, that vessel consists now of nothing save the plates riveted together, and an angle-iron at the bottom, and another angle-iron inside at the top, for keeping it in form, together with six or eight small rods, which project from the eyebolts, by which the gasholder is suspended. Under this arrangement the gasholder is light, and consequently costs much less in the first instance; it requires less balance weights, lighter friction wheels, and, in short, under all its bearings, it is attended with benefits."

In direct opposition to Peckstone's opinion, the author of a standard work on gas lighting, when detailing the iron necessary to be employed in a gasholder, allows less than one-half of the total weight for the sides and top sheets (which really constitute the holder), the greater portion of the metal being distributed as stays, diagonals, and trusses. This is beyond all question erroneous, for all that is requisite under any circumstances is to have sufficient trussing to support the weight of the roof, and maintain the form of the top curb; and when the holders are of great depth, or of large diameter, suitable vertical stays are desirable to support the sides. For these combined purposes about one-eighth or one-tenth part of the gross weight of the holder will generally be sufficient.

The opinion expressed by Peckstone is shared by some of our most eminent engineers of the present day, the more particularly where strict economy is observed, or where the pressure of the holder is required to be limited. In those cases, the material composing the trussing of the roof is dispensed with, and suitable framing or props are placed within the tank to support the crown when the holder grounds, of which examples will be given.

Brick tanks were first constructed about 1818, since which period other materials, as stone of various kinds, and Portland cement concrete, have been introduced with very marked success; and such is the state of perfection now arrived at in the erection of the vessels under consideration, that there are many tanks built, each of which contains tens of thousands of tons of water without the slightest loss or leakage; hence, in this respect, greater perfection cannot be attained.

Vast improvements have been gradually introduced in the construction of gasholders, by the introduction of the telescopic principle, of which some excellent specimens will be presented; whilst the magnitude of these vessels has far exceeded the most sanguine expectations of Samuel Clegg, when he asserted the possibility of them being constructed equal in diameter to the dome of St. Paul's Cathedral.

GASHOLDER TANKS.

SITES FOR GASHOLDER TANKS.

THE choice of the site for a gasworks must necessarily be controlled by a variety of circumstances—such as the price of land, the facilities of obtaining building and other materials, but more especially the means of obtaining coal, both as regards its price and permanent supply. The site should also be in proximity to the district to be lighted, for the twofold purpose of avoiding useless expenditure in the distributing apparatus, and affording facilities for the transport and disposal of the residual products. And yet the position chosen should be sufficiently distant from a populous neighbourhood to avoid the complaints, just or unjust, that might be raised on sanitary grounds.

Amongst other important considerations in the selection, is the nature of the soil, its solidity and suitability for the necessary buildings; whether expenses are likely to be incurred, either in carting earth to raise, or, in excavating, to lower its level; or whether walls to protect it from inundations may be considered necessary.

There are various other contingencies which must govern the determination; but the choice once made, it then becomes the duty of the engineer to decide on the best, and at the same time the most economical, means of constructing the various buildings and apparatus constituting the works. Of these, so far as regards the nature of the soil, unquestionably the most important are the gasholder tanks.

The first step to be adopted is to ascertain the nature of the substratum. For this purpose, in ordinary ground, three or more pits or dry wells may be sunk in the locality, to the depth of the foundations of the proposed tanks, due care being observed that, when constructed of bricks, masonry, or concrete, the tank shall project a short distance above the ordinary level of the ground, for the sake of preventing, in a great measure, any substance accidentally falling therein. These trial pits being dug, furnish sections of the ground where the buildings are to be erected, and thus a fair idea of the work to be executed can be obtained, and the necessary plans and specifications prepared.

Beyond all question, the best soil for the class of construction under consideration is the plastic clay, as found in abundance in London. A remarkable example of the excellence of this site, for constructing gasholder tanks, exists at the Pancras station of the Imperial Gas Company, where the substratum is of London clay. The engineer, cognizant of this, when specifying the construction of a brick tank of 145 feet in diameter and 55 feet deep, stipulated, in the first instance, that the clay in the centre should be left as the frustum of a cone, relying upon the capacity of that material for retaining the water. But in the progress of the excavation it was ascertained, however, that at about 35 feet from the surface of the ground the clay became somewhat loamy, and it was considered necessary to lay three or four courses of brickwork over the lower portion of the cone to the height of about 22 feet, all the rest being left as excavated. The tank, as thus constructed, was, in every sense of the term, perfect, and when it is considered that a pressure of upwards of 15 lbs. on the square inch was exerted on the clay, without any loss of water by leakage, the extraordinary impermeability of that material is clearly established.

After this may be enumerated compact earths of all denominations—shale, marl, and loam—which permit of the sides of the excavation of tanks of small diameter remaining firm and secure without the aid of shoring; but as excavations increase in diameter, so do they approach nearer to the state of a straight line or wall, and thus lose the retaining form of the arch which exists in those of limited dimensions.

Then follow gravel and sand, demanding much greater angles of repose than any of the preceding soils; or, in other language, the slope of the excavation of these is required to be greatly increased in order to prevent it from slipping or falling after the excavation is made. When sand is encountered, then, on account of its angle of repose, the quantity excavated in the construction of a tank of a given capacity must be very considerable; but, on the other hand, the reduction in the price per yard compensates for the large quantity required to be removed.

In some places the site may be of "made ground," expressed more comprehensively in French as "*terrain transporté*" (transported ground)—that is, ground transported from another locality. This is always very treacherous, and no buildings should be erected thereon unless the foundations are sunk to the firm virgin soil beneath. An instance may here be recorded of the want of care displayed in the study of the substratum which occurred about thirty-three years ago, in the construction of a gasholder for supplying a large town in Germany. The engineer, bearing a name very honourably associated with gas lighting erected a cast-iron tank on the description of ground in question; but after a few days' operation, the whole of the structure, tank, and holder, suddenly gave way, presenting a complete wreck, and putting the town in darkness. The holder, according to our informant, who was an eye-witness after the calamity, assumed the shape of "an old cocked hat"; but making due allowance for some exaggeration in the description, it can be understood that a collapsed holder is by no means an interesting sight.

Or the locality chosen may consist of marsh or peat, and of this class the site of the Gas Light Company's works at Beckton may be considered the worst of all specimens. But as the locality was indispensable on several accounts, the cost of construction of the various buildings became a secondary consideration. This site in appearance is an ordinary average earth without any indication of marsh, but at about 3 feet from the surface for a depth of from 10 to 14 feet thick there exists a stratum of soft earthy peat, of that character that only by the weight of the coal stored thereon the ground is depressed 3 or 4 feet. But at the depth of about 22 feet from the surface a solid gravelly substratum was found.

Hence with such ground it became necessary to sink the whole of the foundations of the buildings to a depth of 24 feet; whilst for each of the numerous pillars which constitute the viaducts hereafter alluded to, a pile of 15 inches diameter and 30 feet long had to be driven to support it, and there is not a single column, whether in the retort house or in the viaduct, but is secured by this or similar means. This, however, is on every account an exceptional case, and probably no works, whether large or small, were ever constructed under such disadvantageous circumstances as regards the nature of the site.

In marshy and muddy localities means must be first adopted to ascertain the depth of the solid substratum. This is accomplished by probing the earth in a vertical direction with an iron rod of about three-quarters of an inch in diameter, which with the strength of a man will sink several feet before reaching the solid; and on this being attained, a judgment may be formed according to its depth, and other circumstances, as to the class of tanks necessary for the gasholders, as also the other buildings, which will be further controlled by the extent of the capital at the command of the engineer.

In some cases it may be necessary to place an iron tank, which will have to be supported by piles placed equidistant from each other, over the whole area of the tank. But when piling cannot be advantageously employed, it is futile to attempt to erect a tank in such a position, for it is better to sustain a positive loss by renouncing the site, rather than expend money on a construction that must of necessity fail.

Again, in some localities chalk or other rock of various kinds is encountered. For the former little skill is requisite, as the excavation can be performed by ordinary labourers; but for hard rock, where blasting becomes necessary, for which a special class of workmen is requisite, the cost becomes a serious consideration, as compared with the construction of a cast-iron or composite tank. The writer once met with a site of this description, and having good miners at hand, the excavation was made to the desired depth, its diameter being about 18 or 20 inches greater than the interior of the tank when completed. The wall, except the parts occupied by the piers for the holding-down bolts, was built of square stones, similar to paving stones, of about 7 inches thick, the space between this and the rock being made good with rubble and hydraulic mortar, and when completed the tank was as tight as could possibly be desired. It is almost needless to say the object of constructing tanks of great thickness in loose ground is to prevent them yielding or bursting; but that when in rock, the slightest wall, if properly built, is sufficient.

There are also sites where water exists on the surface, as in low grounds, or in the proximity of a non-tidal river or sea, as the Mediterranean, which is practically non-tidal, as it only ebbs and flows a few inches each month. Or the locality may be influenced by a tidal river or sea, when the operations can only be carried out at intervals. Or in sinking the trial pits, water may be found at a few feet from the surface. Lastly, the works may be inland, where ground is very valuable, and it is absolutely necessary to construct the buildings, when, water being encountered, it must be contended with. An example of this kind of site exists at the South Metropolitan Works, where water was found in abundance at 11 feet below the surface, the excavation being about 42 feet deep; and in order to carry out the work, it was constantly pumped for months at the rate of from 60,000 to 70,000 gallons per hour, or the enormous quantity of 240,000 cubic feet per diem.

In situations like that described, it is usual to sink a wide well in immediate proximity to the tank, a depth of 3 or 4 feet below the foundation of the intended structure, and in sinking this, one or more veins of water may be traversed; or in loose soil the water often arises from the general drainage of the neighbourhood. The well being sunk to the desired depth, and ample means provided for withdrawing the water as the excavation and construction of the tank proceed, constant communication with the well exists, so that as the water flows therein, it is pumped therefrom.

There is an old adage that "water is an excellent servant, but a very bad master," and in few instances is this more appropriately illustrated than in the construction of gasholder tanks, in localities where this element presents itself in abundance. The engineer should endeavour, therefore, to retain it in the first-named capacity, for when water obtains the mastery, the limit to the expenses, in many cases, is beyond conception. The writer remembers an instance in which the contractor for the construction of two large tanks sustained a serious loss by obstinately persisting in carrying out the plans contrary to the wishes of the engineer, by attempting to remove the water by pumping, and this by manual labour—for thirty years ago steam power was little used for such purposes; but after several weeks of ineffectual operations, working day and night, the futility of the course pursued being evident, it was decided to build the tanks about 2 feet higher than originally intended. The foundations, of hydraulic concrete, were then formed in the water, the work was proceeded with, and no further difficulty was experienced. But by sinking a pit to the desired depth, and the presence of water in a loose sandy soil having been ascertained, the tank would have been designed accordingly, and the steps ultimately necessitated adopted in the first instance, and all the loss and inconvenience would have been avoided. In localities therefore where, in the trial excavations, water is found to exist, the utmost caution must be observed in conducting the works.

Under all these varied conditions the judgment of the engineer is called into requisition to decide on the best means of carrying out the work, so that it may be perfectly executed, and at moderate cost, for any useless expenditure must result in a proportionate and permanent reduction of dividends to the shareholders. On the other hand, cheapness alone should never be a consideration; there should always be a proper exercise of judgment and skill, in order to determine the mode of construction. Whether the tank or tanks should be of brick, masonry, concrete, or iron, or a combination of two or more of these, are questions dependent on numerous circumstances, such as the nature of the site, the cost of suitable building material, the class of skilled labour in the locality, the price of iron, expenses of transport, and customs duties, where such exist, are all subjects for serious consideration.

In England, where the best advice can be obtained at a few days' notice, where building materials of all kinds are ready at hand, where the best of workmen can be procured, and, if desired, contractors to execute the work, the engineer need not incur much responsibility; but the case is widely different with many of our countrymen who fulfil the offices of managers of gasworks in towns in remote parts of the world, and who, in their isolated positions, are thrown entirely on their own resources, and are often required to carry out the execution of works with labour and material of a very inferior description, and under numerous other disadvantages. To this class especially we hope that some of our observations on gasholder tanks may be acceptable.

Having briefly referred to the different kinds of sites to be met with, we now proceed to a description of the various methods of constructing holder tanks.

GENERAL CONSTRUCTION.

The tanks of gasholders, as used in establishments for supplying cities, towns, villages, factories, and other places, are usually constructed either of cast iron, brick, stone of various kinds, or a combination of two or more of these materials, and in some few instances they have been formed of wrought-iron plates, riveted together, and of thickness proportioned to the depth and diameter of the vessel. Within the last few years an important step, on economical grounds, has been made in the construction of these vessels, by employing a thin brick lining for their interior, which is backed to the desired thickness with cement concrete, and still more recently gasholder tanks have been composed entirely of concrete, with unquestionable success in all points with reference to economy, strength, durability, and soundness.

We have already alluded to the difficulties experienced in the early days of gas lighting, in the formation of tanks. These probably arose from the employment of ordinary lime mortar, which, in addition to being slightly soluble by the action of water, was also porous, whilst experience had not then demonstrated the extreme care essential to proper construction. As a remedy for these evils, the use of puddle was resorted to, by placing a layer of plastic clay over the whole area of the intended tank, as well as against the exterior of its wall. The impermeability of this has already been noticed, it can therefore be understood that by its aid, and with only the most moderate care, a sound tank may be ensured.

Up to the present time the system of puddling has, regardless of expense, been very generally retained in these structures, and some engineers of eminence still maintain that, for the perfect construction of the vessels under consideration, it is indispensable; which opinion, however, is not confirmed by general practice.

When deciding on the construction of tanks of brick, masonry, or otherwise, the engineer must be guided by the most suitable and economical building material at hand, for different localities have, of course, their peculiar materials for the object in question. Thus, in London, and other towns and cities where clay is abundant, brickwork is very generally employed. There are towns in the North where the universal building material consists of stones of various kinds. Again, there are places where, in a manner of speaking, bricks are unknown; for instance, in some of the islands of the Mediterranean the only material is a soft porous stone, which is used for all kinds of construction, in blocks for walls of buildings, or, sawn into thin slabs, it is employed to cover roofing, or for flooring; and as the joists supporting it are uncovered, it serves as a ceiling to the apartment beneath. In short, this serves all the purposes of bricks, tiles, slates, and flooring.

In such a locality a contractor, without having visited the place, undertook, amongst other work, to construct a tank for a specified sum; but on his arrival, contrary to all expectation, he ascertained that no bricks were to be had in the island, and to have incurred the expense of transporting them from the Continent would have been attended with serious loss. On the other hand, stone was remarkably cheap, so, after due consideration, it was decided to build the tank of 56 feet diameter and 16 feet depth of this porous stone. This was delivered in blocks of about 3 feet long, 15 inches thick, and 10 inches wide, the face being formed to the curvature of the tank. The hydraulic mortar consisted of ordinary lime mortar, with an admixture of " puzzolana "—hereafter mentioned—and for the backing, large boulders and shingle were procured. With these materials and good workmen the tank was commenced.

The foundations for the wall were of concrete, about 15 inches thick, but this was reduced over the dome in the centre to about 12 inches in thickness. The first course was then laid, the stones being well wetted and embedded in the mortar, and as each course was carried up, it was backed with boulders, shingle, and hydraulic mortar to the desired thickness, in order to give the necessary strength to ensure safety and render it impermeable. During the progress of the work, the earth was well rammed into the excavation, and the wall, when complete and dry, received two good coats of hot pitch, the dome in the centre being rendered or coated with hydraulic mortar.

On filling the tank, about three months after its completion, unexpectedly, it lost considerably during the first few days; but at the end of a fortnight it was as sound as could be desired. The loss may be explained by the porous stone absorbing the water through some pores which were uncovered by the pitch; but once these filled, this action no longer existed, whilst the backing and concrete prevented the water escaping. This tank, as well as others built in a similar manner in the same locality, has now been in operation several years without showing the least signs of leakage. Therefore, with such an example, we may ask, where is the utility of incurring heavy expenses to procure puddle, when with ordinary care it can be dispensed with?—and this with the worst material possible, as, in this instance, the facing stone was so porous as to be employed for filters. The circumstance is mentioned simply with the view of showing the necessity of avoiding any established theory, and applying means for adapting the material at command.

According to the old-fashioned methods of building tanks, they were made perfectly level over their entire internal area, and sometimes covered with Yorkshire flag paving; by these means expense was incurred without attaining any good object. Indeed, the contrary of this was the case; for as the whole area was exposed to the full column of water, any weakness in the substratum was by this means tested. It was also frequently stipulated in specifications that the concrete for the foundations of tanks should be thrown from a great height, generally 30 feet, by which means solidity was supposed to be given to the mass; but, as we shall have occasion to observe when referring to concrete tanks, the very opposite course should be adopted at least to secure impermeability.

For the purpose of comparing the past with the present, we may state that in a standard work the general instructions are, that any tank constructed otherwise than in the solid rock should be secured by puddle over its whole area, and against the outer part of the wall forming it. Probably this was a wise precaution in those days, with the imperfect knowledge and means at command, but it is now no longer necessary. In his estimates, the author gives the cost of puddle at 1s. 6d. per cubic yard, and this is repeated so often that it cannot arise from an error; but as this price would hardly pay for excavation, working, and treading down, we conclude that under such conditions the clay for the purpose must have been obtained in the excavation. But if this were really the case, the puddle at the lower part, at least, could be dispensed with, and following the example given at the St. Pancras station of the Imperial Gas Company, the centre might have been left as excavated.

The author further recommended grouting every fourth course, which certainly is wrong, for if this is employed it should be at every course; but we believe that the best way to build a tank properly is to embed each brick in the cement, and fill up the joints at the same operation, when, of course, grouting will be unnecessary. Besides which the author referred to recommended large blocks of cut stone, in which holes had to be drilled, for the purpose of securing the holding-down bolts, together with other small stones as rests for supporting the holder when down, also coping stones for the top of the tank. All of which are in most localities very costly, and in some places they cannot be procured at any moderate price; moreover, they answer no useful purpose, as the same durability and strength can be obtained whether by employing brick, stone, or rubble.

But it should also be borne in mind that at the period the only materials employed for uniting brick and stone in constructing holder tanks were Roman cement, mortar, and hydraulic lime. The first, as experience has proved, possesses serious defects which render it by no means desirable for the purpose. The second, in addition to being porous, is also soluble in water; and the quality of the third is often so doubtful as to create considerable distrust in its adoption, for all ground limes are not hydraulic; so that the last is frequently more deceptive than the others.

Although Portland cement was invented or discovered, we believe, at the commencement of this century, it was not until 1851, when its remarkable strength was displayed at the Great International Exhibition, that particular attention was directed to it. Since that period its application for hydraulic work has been of the most extensive nature, and each year it has become more and more appreciated, on account of the facility with which it is worked, its remarkable tensile strength and impermeability, the length of time it may be stored in barrels without material deterioration, and its moderate cost, which are all strong recommendations for its employment. It possesses, moreover, still another great advantage, that is, instead of decaying with time, it actually becomes harder until after the lapse of many years it assumes the nature of the toughest rock. We can refer to many instances in London and other places, where the pavement has been set in Portland cement, and where the stones being worn into recesses by the incessant traffic, the cement in the adjacent joints projects in polished ridges; clearly establishing the great degree of hardness this material acquires after the lapse of several years.

According to Mr. Faija, "in colour Portland cement should be of a dull bluish-grey, and should have a sharp, clear, almost floury, feel in the hand; it should weigh from 112 lbs. to 118 lbs. per striked bushel, and when moulded into a 'briquette,' or small testing-block, and soaked in water for seven days, should be capable of resisting a tensile strain of from 300 to 400 lbs. per square inch. The cement during the process of setting should show neither expansion nor contraction."

With material of such extraordinary strength, together with its other excellent qualities, the construction of gasholder tanks should offer no difficulty, all that is requisite is care in the manipulation. If we further consider that in the deepest gasholder tank hitherto made, the pressure of the water is hardly equal to 30 lbs. on the square inch, and that the cement in question possesses upwards of ten times that force, its adaptability for the object under consideration and its superiority over all other similar materials are fully established. According to our every-day experience, the mortar of ordinary walls has often only the cohesive force of a few ounces to the square inch, arising from the bricks being laid dry, or the inferior quality of the mortar.

Possessing, as we thus do, cement of such vast superiority to anything employed in the construction of the earliest gasholder tanks, it may be expected that engineers will eventually discontinue the use of puddle, except in certain cases where it is found in the excavations, or, as in the Rotherhithe tank hereafter referred to, where it was the best and cheapest material that could be possibly employed for the purpose, as by its adoption a large quantity of brickwork was dispensed with; yet it may be remarked that for years after the excellence of Portland cement was established its employment was but limited for the purpose under consideration. This state of things has, however, been gradually changed, although each progressive step was slowly taken, the cement being first employed for the foundations, and afterwards for the joints and rendering of tanks, which was followed by the introduction of the "composite" tank, the name of which is derived from the fact that the tank is composed of a thin brick wall, or lining, forming the interior, and backed with concrete to the desired thickness to ensure strength. This method of construction presents many advantages in an economical point of view, particularly in localities where gravel abounds, in which case a saving of one-half of the entire cost of the building material will be effected, whilst its strength is considerably greater than any stone or brick tank, equal thicknesses of material being assumed. But more recently Mr. George Livesey, encouraged by his former success with a composite tank, entertained the bold idea of constructing a tank entirely of Portland cement concrete, to which we will hereafter refer.

An excellent substance for intermixing with ordinary lime mortar to render it hydraulic is that called "puzzolana," a volcanic production having the appearance of a brown earth. This, although comparatively unknown in England, is much employed in the neighbourhood of the Mediterranean, and is but little inferior to Portland cement; possessing, moreover, the great recommendation of being one-third its price. This material is ground moderately fine and intermixed with good mortar free from earthy matter, in the proportions of two of lime to one of puzzolana. Whether for concrete or whatever purpose to which Portland cement is applied, ordinary lime mortar rendered hydraulic, as stated, can be advantageously employed, and although it requires several days to become hard, yet at the end of a month it attains such a degree of strength and cohesion, that in order to level concrete made with it, the hammer and chisel are necessary, and frequently, such is the tenacity of the material, the pebbles entering into its formation are frequently split in two before becoming detached.

Puzzolana was employed by Smeaton, the celebrated engineer, in the construction of the Eddystone lighthouse.

The first operation in the construction of a tank is the excavation. This, in vessels of limited dimensions, is practised at the present day precisely in the same manner as forty years ago. But with tanks suitable for large holders, similar to those hereafter referred to, which sometimes contain upwards of 20,000 tons of water, the steam crane, trams, derricks, caissons, and other extraordinary appliances become indispensable, and without these means the gigantic undertakings of modern times could hardly be successfully carried out.

Formerly the excavation was made to the desired depth over the whole area, but the modern practice in erecting large tanks is to excavate an annular space, which is securely shored, to the depth of the required foundations for the wall. The foundations and footings are then laid, and the wall is carried up to the desired height. This accomplished, the ground in the centre is dug out, leaving a dome or the frustum of a cone, the dimensions of which are controlled by the nature of the site; for if it be of gravel or sand, then, on account of their angles of repose, the cone or dome will be limited in height, but if in good hard ground, the cone can be carried up to any desired elevation. In sandy or soft soil, and indeed in all uncertain ground, it is necessary to concrete over the whole of the area, and to "render" the centre with cement, or to lay one or two courses of flat tiles or bricks, in order to prevent leakage. Sand is never to be relied on as a foundation, unless when existing under water, and this entirely free from a current of any kind. The concrete foundations being properly prepared with good cement or hydraulic mortar, become in the course of time like a solid rock, and ensure perfect stability; and if the ballast and gravel be wetted, before they are mixed with an ample quantity of cement, so that the whole of the interstices are filled with the liquid cement, impermeability is assured. Again, there are other sites where the ground is so firm and solid as only to require care in the building, and, as already stated, when the excavation is in good plastic clay it is sufficient in itself to retain the water.

The cost of excavating a tank will, of course, be controlled by the nature of the soil in which it is to be constructed; but speaking generally, this, with the filling in of the earth at the back of the wall as the work advances, may be estimated at 1s. 3d. per cubic yard. The cost of removal of the surplus ground from the excavation will be from 1s. 6d. per yard, according to the distance it has to be transported. Frequently, however, there are irregularities in the site of the works, which permit of the earth being advantageously deposited thereon.

The excavation and foundation made, the circle of the wall is defined by wooden templates, formed according to the curvature of the tank, and as one part is built, the template is carried round from place

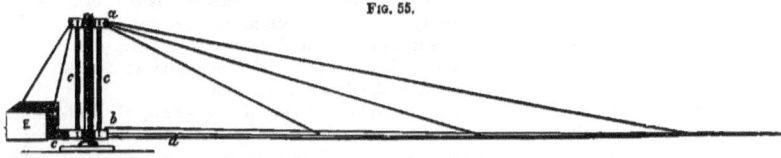

FIG. 55.

to place until the circle is completed. The plan sometimes adopted for such work is to place in the centre of the proposed tank what the workmen term a "compass," which purpose it really serves, as represented in the diagram. This consists of a pipe, placed in a vertical position, and leaded into a socket, which is

z

attached to a wide base, secured to stakes driven into the ground. Two collars, a and b, are provided with two broad flanges, to which are bolted the bars $c\ c$, of corresponding length. The collars and frame revolve freely on the vertical pipe, but have as little play as possible. To one of the flanges of the collar b is bolted a light wooden bar d, tapering towards the end, of a length equal to the radius of the tank, and in section representing the letter T. This bar is suspended by three or more rods to the flange of collar a, so as to retain it always in a horizontal position, and is counterbalanced by weights placed in the box E, which is attached to flange of collar b, and suspended to flange a. The apparatus as arranged can be turned round with the greatest facility to any desired point, and sometimes, when it is likely to interfere with the scaffolding, at some distance from the end there is a hinge, so that a portion of the bar can be folded up out of the way when required. Exactly at the distance of the radius of the tank a hole is pierced in the wooden bar, and from this is suspended a good pointed plumb, which can be raised to any desired point as the work progresses. By this means the cylinder can be assured even to the eighth of an inch, for the apparatus not only serves as a compass, but also that of a plumb, by which means one man being employed in setting the work at various points, the others follow with the templates. Nor is it limited alone to this purpose, but in the formation of the cone or dome in the centre, with a wooden template corresponding to the dome, one end of which is fastened to the flange, and the other resting on the footings of the wall, the excavations can be made with the most unerring exactitude. The simplicity and certainty of the apparatus, its moderate cost, the saving of time and labour affected by it, are all strong recommendations to its use; and whether employed for tanks of the most limited dimensions, or those of the greatest magnitude, it is alike suitable.

The cost of concrete will vary materially, according to the facilities of obtaining the ballast, gravel, or shingle, the first consisting of the large stones found in gravel, of about the size of a hen's egg, or broken stones of the same dimensions; the gravel being the mass of pebbles, stones, and sand as taken from the pit; whilst shingle exists of various dimensions, from large boulders to the smallest pebbles. To produce good concrete, the gravel should be free from clay or loam, and composed of from six to ten parts of gravel, ballast, and sand, to one of Portland cement, dependent on the nature of the former. When the ballast in the mass preponderates, and there is an absence of the various-sized pebbles to fill the interstices, the quantity of cement required will be great; but with a proper admixture of pebbles, so as to fill the spaces, this will be diminished.

The following method is recommended in some works on building, to ascertain the quantity of sand and cement necessary to intermix with the ballast and shingle:—" Fill a bushel measure with a fair sample of the shingle and ballast to be used. Having a known quantity of dry sand at hand, sprinkle a portion of this over the top, and well shake the measure, by which process the sand will fall down between the ballast, and fill up the spaces between the small particles of materials entering into the composition; continue the operation until the measure is filled with sand, and, on ascertaining the quantity used, the required proportion of sand to shingle will be given. Then, with a known quantity of water, pour a portion of this into the measure containing the sand and shingle until full, when the volume of the water employed will give the proportion of cement required." This rule is, however, only applicable to a mixture of ballast and shingle; for wherever gravel is employed, the sand exists therewith, and frequently in excess of the desired quantity, in which case ballast or other materials, as burnt clay or patent ballast, may be intermixed; but it is always indispensable that the concrete, when laid, should have the whole of its interstices filled with the liquid cement. The average price of cement concrete is about 12s. per cubic yard. Lime concrete averages about 8s. per cubic yard.

In situations such as the immediate neighbourhood of a non-tidal river or sea, where the ground is so loose as to render all attempts to remove the water by pumping impracticable, and where it is desirable to obtain as great a depth as possible, the foundations of tanks are sometimes constructed within the water, for which purpose caissons are advantageously employed, particularly in localities where the site is composed of sand.

A tank of 102 feet in diameter and 30 feet deep, built according to the circumstances mentioned, once came under our notice, of which we give a short description.

The site of this tank was ground reclaimed from the Mediterranean, and in its construction the necessary excavation being made to the water level, it became desirable to form within the water an annular foundation of 120 feet external, and 94 feet internal diameter, and 3 feet 6 inches deep. For this purpose, three sheet-iron caissons or boxes, without top or bottom, each 13 feet long, 6 feet wide at one end, and 4 feet 9 inches at the other, corresponding with a portion of the circle, and 3 feet 6 inches deep, were provided, which were slightly tapered so as to permit of them being lifted with facility when the concrete was set therein. During the progress of the work the caissons were placed in their respective positions, as represented in the diagram, with intervening spaces equal to their width. The ground was then excavated from the interior of one of the vessels, at the same time it was driven down by heavy mallets until its upper edge was at the surface of the water. All the earth being cleared out to the depth of the caisson, the operation of filling with stones and cement concrete was then commenced. This was accomplished first by throwing into the centre large blocks, which were readily procured, and afterwards the caisson was filled with concrete. In this operation, of course, the water was expelled, and percolated through the surrounding ground.

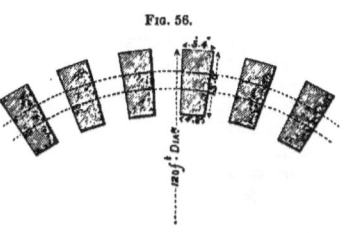

Fig. 56.

At the end of three or four days, the cement being set, the caisson was raised, leaving a block of concrete or stone of about 9 cubic yards, which, like the whole of the foundation, in the course of a short time acquired an extraordinary degree of hardness.

In the meantime, the other caissons were employed in a similar manner, and when a range of six blocks of concrete, as shown, was completed, each of the intervening spaces was closed by a sheet of iron sustained by stakes driven into the ground. This done, the earth was then removed, and the spaces filled with stone and concrete, as described, until the circle was completed, and thus the foundation was formed within the water, without having occasion to pump a single gallon.

The wall, which was 4 feet thick at the bottom, and 2 feet 3 inches at top, was built in the centre of the foundation, as represented by the dotted lines, and consisted of a facing of square stones, averaging 2 feet superficial, set in regular courses, and backed with rubble, the top being finished by a coping. In the centre of the tank a dome was formed by the earth, and over this was placed a layer of concrete, 15 inches thick, the base of which abutted against the annular foundation. The dome was further covered by a course of square tiles set in cement. As the wall projected 5 feet above the surface of the ground, by way of precaution an iron hoop, 5 inches by half an inch, was built into the top just under the coping.

This tank was as perfect as possibly could be desired, and any rainfall was carried off by a pipe inserted for the purpose. On account of the low price of the building material employed, its cost was barely 10*l*. per 1000 feet tank capacity. On the Continent, in small and medium-sized gas-works, the holder tanks, when constructed of masonry, often project 5 or 6 feet above the level of the ground, whilst in England the general practice is to construct them level with the surface. Certainly this latter system has some objections, particularly when, by sinking an extra depth of a few feet, water is met with, which could be avoided by building a portion of the vessel above ground. An objection which has been raised against this method is, that the water arising from a leakage, on becoming frozen would expand and crack the tank, which reasoning is perfectly correct, supposing such a leakage; but in a well-constructed tank this contingency should never exist.

Where stone can be procured at a moderate cost, very excellent tanks are constructed with it, but to employ puddle in such a case would simply be an unnecessary expenditure of money, without the slightest cause or utility; for with the most moderate care in building with ordinary close-grained stone, and even, as we have stated, with porous material well backed with rubble, a perfectly sound tank can be assured.

z 2

GASHOLDER TANKS.

Brick Tanks.

Here it may be remarked that, in various countries, bricks of different dimensions are employed; thus in Belgium they are usually $6 \times 3 \times 1\frac{1}{2}$ inches; in Spain, Portugal, and in countries which were or are dependencies of these nations, they are generally $11 \times 5\frac{1}{2} \times 1\frac{3}{4}$ inches; but wherever the term brick is used in this work in defining the thickness of the walls of tanks, the English brick of $9 \times 4\frac{1}{2} \times 3$ inches, including joints, is understood.

When intended for the construction of holder tanks, the bricks should be thoroughly well burnt; and it can hardly be repeated too often that when employed with Portland or indeed any other cement or mortar, they should be saturated with water at the moment of laying them, and the joints should not be too close. When bricks are laid in a dry state, or only slightly wetted, no cohesion takes place between them and the cement, and on the latter setting, the former becomes detached with the slightest blow.

Brickwork in England is generally estimated by the rod, which is 272 feet superficial, of a thickness of $13\frac{1}{2}$ inches, or $1\frac{1}{2}$ brick, called the standard thickness, by which it is always computed.

A rod of brickwork is calculated to be equal to 306 cubic feet, and contains from 4300 to 4500 bricks, according to the manner the work is executed, and will require about 184 hods of mortar or cement, and contain 235 feet cube of bricks, and 70 cubic feet of cement or mortar, and weighs on the average about 15 tons. A rod of brickwork, when the cement and sand are used in equal proportions, requires about 36 bushels of cement and 36 bushels of sharp clean sand.

The average price of brickwork, set in Portland cement and sand, in equal proportions, for tanks is about 20*l.* per rod, or nearly 35*s.* the cubic yard.

The cost of rendering in neat Portland cement about $\frac{1}{4}$ inch thick is about 3*s.* 6*d.* per yard superficial; this at the same time is flattened with the trowel, so that a smooth surface is produced, which increases its impermeability; but for rendering or coating the dome or cone in the centre of the tank, this price will be exceeded. A bushel of Portland cement, if used carefully, will be sufficient to render from $1\frac{1}{2}$ to 2 yards superficial of wall, but in addition to this the brickwork itself should be built in such a manner as to be relied on to retain the water.

In estimating the capacity of a tank and its corresponding holder, due allowance must be made for the height of the dip or seal, which will depend on the pressure given by the vessel. But it is unsafe, in an economical sense, to work a holder—at least in the daytime, when the temperature of the atmosphere is liable to continuous fluctuations—without sufficient seal, and to losses from that part of the gas apparatus some of the unaccounted for gas is undoubtedly due.

The following is a specification for the construction of a tank of 81 feet 6 inches in diameter and 20 feet deep, and in addition it may be regarded as a sample of the conditions and terms sometimes arranged by gas companies in the erection of these vessels.

"*Specification of work to be done, and the materials to be employed in the construction of a Gasholder Tank and Dry Well for the ——— Gas Company, at their works situated at ———. The exact site will be indicated to intending contractors.*

"*Excavation.* The ground to be dug out to the depth of 20 feet from the surface line, leaving the frustum of a cone in the centre of 12 feet high in vertical section, its base being 70 feet and the top 35 feet in diameter. The excavation being properly shored in order to prevent the earth falling in, and to be kept clear of water until the tank is finished. All the superfluous soil to be carted away.

SPECIFICATION OF A BRICK TANK.

"*Puddle.* A layer of puddle of good plastic clay free from vegetable fibres, stones, or rubbish, and properly tempered, to be laid a thickness of 12 inches over the whole area at the bottom of the tank. The exterior of the wall to be puddled entirely round a thickness of 18 inches at the bottom and diminishing to 1 foot at the top. To be worked in regular courses of not more than one foot, and to be well trodden, each course to be left clean at the surface. The earth to be filled in and well rammed as the work progresses.

"*Bricklayer.* The bricks to be sound, well-burnt hard stock bricks, and to be laid old English bond, in mortar composed of best blue lias lime one part, and clean river sand two parts, well incorporated together, and worked thin in the joint, and well flushed round and grouted every third course. The two lower courses of footings to be of a thickness of 5 bricks in length, or 45 inches; and on these other two courses of $4\frac{1}{2}$ bricks, on which the wall of $3\frac{1}{2}$ bricks, or $31\frac{1}{2}$ inches thick, will be commenced, and carried up of that thickness to a height of 8 feet. At that height the wall will be diminished to 3 bricks for 6 feet in height, and from this height to the top of the brickwork it will be $2\frac{1}{2}$ bricks thick.

"There will be ten buttresses, to be $1\frac{1}{2}$ brick thicker than the wall, to the height of 6 feet $4\frac{1}{2}$ inches, and then to be carried up 4 feet in thickness by 3 feet in width, to be properly bonded with the wall. Three courses of bricks, set in blue lias lime and sand, will be laid over the whole area of the cone in concentric rings, the bricks to be properly embedded in the mortar.

"*Ironwork.* The contractor will have to insert in the inner face of the brickwork 10 cast-iron troughs for the guide rollers to work upon. Also 40 wrought-iron bolts built into the piers, to hold down the columns which will be placed thereon. This ironwork will be provided by the gas company, and must be built in according to the instructions of the company's engineer.

"*Stonework.* The 10 piers or buttresses to be covered with Derbyshire blocks, 4 feet by 3 feet 6 inches and 1 foot thick, to be provided by the contractor, properly squared and dressed, and drilled for holding-down bolts. The contractor also to provide 20 blocks of stone, 18 inches by 18 inches by 12 inches, to be inserted in the wall in front of the piers. Two holes will have to be made in each stone, to receive jagged-headed bolts, which bolts will be run in with lead for securing troughing before mentioned. To provide also ten stone rests, 2 feet by 1 foot 6 inches by 3 inches, for the holder to rest on when down; these must be kept perfectly level with each other. To provide also one block of stone, 3 feet by 1 foot 6 inches by 6 inches, for inlet and outlet pipes to bed upon.

"The remainder of the top of tank to have a coping of 20 inches in width by 6 inches in depth, properly set in cement, and cramped together on the outside.

"The whole of the stone above described to be of the Derbyshire quarries.

"The excavation of the dry well to be 28 feet deep, and 14 feet diameter at bottom. The brickwork to have two footing courses of two bricks each. The walling immediately above this to be 18 inches brickwork to the height of 15 feet, from which point it is to be finished off in 9-inch work, and to be puddled entirely through the whole height a thickness of not less than 12 inches, as also over the area of excavation a similar thickness.

"Two trenches to be made from the centre of the wall towards the middle of the tank, for the inlet and outlet pipes to be laid in. The tank, when completed, to be perfectly cylindrical, and to be 81 feet 6 inches internal diameter by 20 feet deep.

"The dry well, when finished, to be 6 feet 6 inches internal diameter by 25 feet deep.

"*General Conditions.* The contractor shall perform the whole of the work indicated in drawing and specification, and also such other work (if any), although the same may be omitted in the drawing and specification, as shall be necessary to render the tank efficient, sound, and water-tight.

"If any discrepancy exists in or between the said drawings and specification, the contractor shall point out the same in sufficient time to the company's engineer, so as to cause no delay in the execution of the work, and upon such discrepancy (if any) being pointed out, the company's engineer shall rectify the same.

"The contractor shall, at his own cost and charge, find and provide all and every kind of material whatsoever, and all labour, carriage, freight, barrows, planks, scaffolding, tackle, tools, implements, templates, and all and every other matter or thing that may or shall be requisite for the due and complete execution of all the several works to be performed under this contract, and to be at the whole and sole cost and expense of completing the same in every respect.

"The whole of the workmanship, whether or not described, as to the mode of execution, shall nevertheless be done and performed in the most substantial and workmanlike manner, both as to strength of construction and finish.

"The contractor shall proceed with the work contracted for with all due diligence and despatch, until the whole shall be completed.

"The contractor shall deliver up the tank and dry well in a complete state to the company within three months from the date of signing the contract, and shall maintain it in the same order and condition for twelve calendar months at his own expense. From the commencement of the work to its completion, all the risk and responsibility, of whatever nature, appertaining thereto, shall rest with the contractor.

"*Payment.* The contractor shall receive 50 per cent. of the amount of tender, upon the company's engineer certifying that work to that amount and one-fourth more has been done. Another sum, equal to 25 per cent., shall be paid when the tank and dry well are completed; and the remaining 25 per cent. to be retained in the hands of the company, as security for the stability and soundness of the work, for three months after the tank has been filled with water.

"*Fines.* In case the contractor fails to complete the work in the time specified, he shall be subjected to the following penalties—viz. 10*l*. for the first week, 20*l*. for the second week, and 30*l*. for the third week that the work remains unfinished. Such sums to be deducted from the balance due to the contractor as liquidated damages. Should the contractor then have failed to complete the work, the company's engineer shall have power to employ another contractor to finish the tank, and all moneys that may be due to the first contractor shall be paid to the second for completing the work. The tank will be filled with water by the gas company. The contractor shall enter into a bond for 500*l*., in two efficient securities, as guarantee for the due execution of the work specified.

"Tenders to state a lump sum for the completion of the whole, subject to the conditions hereinbefore mentioned, and be addressed to, &c., &c."

For the annexed form of terms and conditions of contract between a gas company and a contractor, we are indebted to Mr. Longworth, of Dukinfield; which may be considered a model of this class of documents, and its insertion will render repeated reference to the general terms of agreement in the specifications of tanks unnecessary.

"*Terms and conditions of contract entered into between* —— *Gas Company and* —— *contractor.*

"The contractor shall, and must at his own expense in all things, provide, deliver, and fix all necessary materials, and supply all carts, carriages, and labour, and all tools, tackling, scaffolding, planks, ladders, ropes, chains, blocks, pumps, and other necessary implements and things for the due execution of the said works and the completion of his contract, in a proper and workmanlike manner, according to the true intent and meaning of the plans, drawings, and specifications.

"The contractor must, also at his own expense, from time to time remove and cart away to such place as he may be enabled to procure as a deposit for the same, the superfluous earth taken out of the said excavations, except such as may be required by the company, and on the completion of the works restore the land upon which the same was temporarily placed to its original level.

"That the whole of the said works must be done and completed according to the true intent and meaning of this agreement, and of the said plans and specification, in a good and substantial workmanlike manner, and to the satisfaction of ——, the company's engineer (or such other person as may be appointed by the said company or the directors thereof for the time being, to superintend the completion of the said works), on or before the —— day of —— next now ensuing.

"And that if it should happen that the whole of the said works should not be completed and finished as aforesaid within the time aforesaid, then the contractor shall forfeit and pay unto the said company as and by way of liquidation, and ascertained damages, and not by way of penalty, the sum of —— pounds, of lawful British money, for each week, and every week which shall be taken up in finishing or completing the said works after the said —— of —— next, such forfeitures to be retained and deducted by the company out of the amount which should otherwise be payable to the contractor under this agreement for the said work.

"That no extras, nor alterations, nor deductions shall invalidate the contract, but in case any alteration shall be deemed by or on behalf of the company necessary, or advisable during the progress of the work, the contractor shall follow the directions of the said engineer, or other the persons to be appointed as aforesaid, given to him in writing expressing the nature of such alterations. And a sum shall be added to, or deducted from, the amount of the original contract, according to the effect of such alteration, such sum to be calculated at and after the rate or price to be fixed upon at the time by some competent and respectable person duly appointed by the contractor, and the said engineer, or other the person to be appointed as aforesaid on behalf of the company, and if no such rate or price shall be so then fixed upon, then the amount to be estimated by the said engineer or such other person so appointed as aforesaid whose decision alone shall be conclusive.

"That if any damage happen to the work before the same shall have been completed to the satisfaction of and certified by the said engineer, or other the person appointed as aforesaid, or if any slips or injury occur in making the said excavation, or any damage of any kind be done or occasioned to any land or buildings adjoining, or near to the same works, or to the works or other the buildings belonging to the company, by reason or on account of the said works so contracted to be done by the contractor, such damage or injury shall be made good by the contractor at his own expense without any charges whatsoever to the company in respect thereof or of removing the same. And the company shall be held harmless and indemnified by the contractor, of and from the same and all costs and expenses in any manner occasioned thereby.

"The plan and drawings referred to in the specifications are to be signed by the contractor and to remain in the custody of the company's engineer during the progress of the works. The contractor being allowed access to them at all reasonable times for the purpose of taking copies or extracts of the whole or any portion thereof he may require.

"It is hereby expressly understood and agreed on the part of the contractor, that the whole of the materials to be used and the workmen to be employed upon or about the said work, or any part thereof, shall be of the best of their respective kind and class. And that the said engineer or other person to be appointed on behalf of the company have power to reject any material which they or either of them shall consider unsound or unfit to be used in the work, and also to dismiss therefrom and discharge any careless, inefficient, or unskilled workman employed thereon (who after such discharge shall not be employed again); that if at any time the contractor should neglect or refuse to exchange or replace any material, or to make good and amend any work which the said engineer, or other the person to be appointed as aforesaid, may object to, or if he should not proceed with and execute the said work hereby contracted for, or should refuse, or omit to comply with the terms of the specification and drawings to the satisfaction of the said engineer or other such person as may be appointed as aforesaid, or in case the contractor shall in any wise delay or neglect the execution of the work contracted for, then and in every other such case it shall be lawful for the directors of the company, by giving seven days' notice in writing under the hand of the clerk of the company, to annul and determine the contract and make the same void; and the company shall have power to employ any other person or persons to make good and amend the work or material objected to, likewise to finish the whole of the remaining work according to the said plans and specification, and the true intent and meaning of this contract. And to retain the costs and charges for so doing and extra price or damage consequent upon the contractor's default, out of the amount hereby agreed to be paid to the contractor.

"All notices to the contractor are to be given in writing, and the delivery of them to his foreman or to the contractor personally, or at any of his usual places of business, or residence, is to be deemed sufficient service.

"That in case any doubt, question, difference, or dispute, shall arise at any time during the progress, or after the completion of the works, as to the meaning of any part of these presents, or of the said plans or specification respectively, or as to any other matter or thing connected with the contract, the same are to be referred to the company's engineer, or manager for the time being, whose decision shall be final and binding both parties.

"And it is further agreed that in consideration of the proper execution of the works stated, that the company shall pay to the contractor, his executors, and administrators, the sum of —— in the following manner, that is to say, such proportion thereof (subject to conditions already mentioned) as shall from time to time be certified in writing by the engineer, or other person to be appointed to superintend and inspect the said works by the directors of the company for the time being, to have been fairly earned according to the proportion of work actually done, shall be paid on the production of such certificate, provided that there shall be reserved in hand by the company until the completion of the work at least one-fifth part of the value of the work for the time being executed, until the said engineer or other the person to be appointed as aforesaid shall have certified to the company that the whole of the works have been completed to his satisfaction and according to the true intent and meaning of this contract."

The following is an abstract of a specification for the construction of a tank 145 feet in diameter and 55 feet deep, at the Pancras station of the Imperial Gas Company already referred to, designed by the late Mr. David Methven, and merits particular attention.

"The tank when finished to be a perfect cylinder of 145 feet in diameter and 55 feet deep; the top of coping to be 4 feet above the present surface level.

"*Excavation.* The earth to be excavated around the entire circumference of the tank to a depth of 53 feet from the present level, forming an annular space in which the wall is to be built; the ground to be properly shored as the excavation proceeds by waling pieces and struts.

"*Brickwork.* When the excavation reaches the desired depth, and the ground properly levelled, the footings wil be commenced. The first eight courses of tank wall and piers to be set in Roman cement. The first, second, third and fourth double courses of footings to be respectively 5 feet, 4 feet 8 inches, 4 feet 4½ inches, and 4 feet in width corresponding with the thickness of the wall. The seventh course to commence the wall, which will be 3 foot 7 inches thick to the height of 10 feet above the footings. The remaining height of the wall, to be reduced as follows, viz. from 10 feet to 20 feet from the bottom of net brickwork to be 3 feet thick; from 20 feet to 30 feet in height, 2 feet 8 inches thick; from 30 feet to 45 feet, to be 2 feet 3 inches thick; from 45 feet to 50 feet, 1 foot 10½ inches thick. The remainder exclusive of coping to be two bricks thick.

"There will be sixteen piers, 4 feet 3 inches wide, for the support of the cast-iron columns, built up with the wall.

"Six courses of brickwork in every 5 feet in height, as also the three finishing courses, and the corresponding courses in piers to be set in cement.

"The brickwork at no part to be carried higher than 5 feet, until the circle up to that level shall have been completed, and properly tempered clay rammed in behind the wall.

"A cast-iron plate and holding-down bolts to be built into each of the piers 10 feet below the level of coping. The stand-pipe well, to be a true cylinder, the foundation of which will be 63 feet deep from the level of the coping of the tank, the bottom to be perfectly level and paved with three courses of bricks on edge set in cement. The footings of the wall to start from the same level as the deepest portion of the wall of tank.

"There are to be thirty-two piers of brickwork built in the bottom of the tank for the foundations of the rest stones, the same to form part of the general footings of the tank, each pier to be capped with stone.

"*Stonework.* In addition to the stones mentioned there will be required blocks of stone, 18 inches by 12 inches by 12 inches, to be inserted in the wall opposite to and at points equidistant from each pier as indicated in the drawing.

"There are also to be sixteen stones, 5 feet square and 18 inches thick, to cap piers for the support of the guide standards of the holder. In each of these four holes are to be pierced for the holding-down bolts; the wall to be surmounted by a coping 12 inches thick, to be bedded in cement."

In the specification provision was made to support the crown of the holder, which was untrussed.

"*Material.* The bricks employed on the work to be the best hard-burnt stocks. The mortar to be composed of one part of fresh-burnt lias lime to three parts of clear sharp river-sand. The Roman cement to be fresh burnt and gauged in equal proportions with sand as above, the cement to be mixed as it is used, and no more mortar than is sufficient for one day's work to be made at one time.

"The contractor to supply all materials of every description here specified, and to provide all scaffolding, struts, and shoring which may be required, the same to remain his property when the work is terminated.

"*Pumping.* The contractor, during the progress of the work, to find the necessary pumps and labour to keep the tank dry, to employ competent men to watch at all times, and strut and shore up any signs of slip or yielding of the earth, and to make good at his own cost any damage whatever which may occur.

"*Excavation.* When the wall shall have been completed, the earth in the centre of the tank to be removed, leaving the frustum of a cone, the base of which shall be 139 feet in diameter, its sectional vertical height 40 feet, and its diameter at the top 30 feet.

"All superfluous earth or clay not required for filling in or puddling to be carted away at the expense of the contractor. The puddle to be kept constantly well moistened and not to be less than 18 inches in thickness, and the earth to be firmly rammed in behind as the work proceeds.

"*Extra Work.* No extra work to be allowed but such as shall be ordered in writing by the company's engineer and addressed either to the contractor or his clerk of the works.

"*Terms of Tender.* The proposed contractor to state the sum for which he will execute the work herein specified, according to its true intent and meaning connected with the drawings."

This specification is presented as an extraordinary example of construction, as the only brickwork originally intended was that specified, namely, the outer wall, whilst the whole of the rest of the vast space was to have been left as excavated, the soil (London clay) being relied on to retain the water, and this at a depth of 55 feet. In the course of construction, however, as already observed, it was discovered that at 35 feet below the surface, the character of the soil became slightly changed, containing a small portion of loam, when it was considered prudent to alter the original plan by laying three or four courses of bricks in cement over the annular space, also around the base of the cone to a height of about 22 feet, and thus the tank was terminated, and which to all intents and purposes has since proved a perfect success. The method of building is particularly instructive, as it demonstrates the great impermeability of plastic clay, and suggests the most economical method of construction in such descriptions of soil.

The cost of this tank was upwards of 11,000*l.*, but if from this we subtract the expense of timber frame for supporting the roof of gasholder, the price will be reduced very considerably below 10*l.* per 1000 feet capacity, and had it been carried out as originally proposed it would have been much cheaper.

The following specification as well as other interesting matter and drawings have been kindly furnished by Mr. Horatio Brothers, engineer of the Western Station of the Gas Light Company.

"*Specification of a Gasholder Tank, erected at the Commercial Gas Company's Works, Croydon.*

"*Brickwork.* The whole of the brickwork is to be built of good, sound, hard, well-shaped, stock bricks, set in mortar in old English bond, that is to say, in alternate courses of headers and stretchers.

"In the walls every fourth and fifth courses are to be laid in herring-bone fashion. All the beds and joints are to be set as close as possible and flushed perfectly full of soft mortar. The face of all brickwork to be neatly flat-pointed.

"*Mortar.* The mortar is to be composed of three parts of Greaves's lias lime, or Ellis's barrow lime (Leicestershire), two parts of clean, fine, sharp sand, and one part of coarse ashes. The lime must be brought to the works in an unground state and be kept free from moisture until required for use, when it must be ground together with the above-named proportions of sand and ashes, and the requisite quantity of water, by means of a proper edgestone or pug-mill. It must be ground as required, and no portion which has been allowed to set or become hard will be allowed to be used in the works.

"*Stone.* The ashlar work must all be of the best 'Bramley Fall,' free from flakes or flaws of any kind, and in all cases must be laid on its natural bed.

"The bottom of the dry well is to be laid with 3-inch flagging or paving, properly tooled and laid in cement.

"*Puddle.* The puddle must be made of the best clay excavated. The clay must be such as the engineer or clerk of the works shall approve, and must be thoroughly tempered and worked with water so as to form a perfectly homogeneous mass, and then wheeled and tipped into the place where it is required, and pounded solid so as to be perfectly water-tight.

"*Concrete.* The concrete is to be made of the best Portland cement and ballast, in the proportions of one of cement to six of ballast.

"*Filling-in.* Any space which may be left between the solid ground and the puddle must be filled in with dry material and pounded solid. The space indicated on the plan, marked '*A*,' around the tank and above the red line showing the ground level, must be filled up to the level of the top of the brickwork with the material excavated from the tank, and pounded solid with such a batter as the engineer may direct.

"*Excavation.* The whole of the excavations must be cut out as nearly as possible to the required sizes and depths, as shown in the drawings. The whole of the earth or other material not required for the works must be wheeled and laid on such part of the ground in such form as the engineer shall from time to time direct.

"*Description of Tank.*

"The tank, when finished, is to be 130 feet diameter and 27 feet deep, to the finished level of the bottom around the bearing stones, being about 20 feet below the present surface of the ground. The bottom is to be coneshaped, rising 10 feet in the centre as shown. The walls are to be built of brickwork, of the form and of the dimensions following, namely: the footings to be constructed at the depth shown in the drawing, seven and a half bricks wide, diminishing by offsets of one brick every 6 inches for a height of 2 feet, where the footings will be four and a half bricks wide. On this will be commenced the wall of the tank, of a thickness of three bricks for a height of 9 feet, where it will be diminished to two and a half bricks, and so carried up for another 9 feet, above which it will be diminished to two bricks to the under part of coping, which will be 2 feet 6 inches wide and 6 inches thick.

"There are sixteen piers built equidistant, each 4 feet 6 inches square at the top, increasing in thickness in the same proportion as the wall, the footings at those points being extended accordingly. The wall to be perfectly plumb and of an exact circle.

"The dry well to be three bricks thick at the bottom for a height of 13 feet 6 inches, and above that point two bricks thick; to be finished by coping 2 inches wide and 6 inches thick.

"The bottom footings of pedestal in centre to be nine bricks square, with offsets every 6 inches to the level of the concrete, where the pedestal will be six bricks square, and thus carried up a height of 2 feet 6 inches.

"Good sound elm planking, 4 inches thick and of the lengths of the cross sections of the footings of the walls, to be placed under the footings. No elm planking shall be used which is not satisfactory to the engineer or clerk of works.

"Sixteen ashlar pillar-stones are to be built in the walls in the positions shown; they must be well boasted on the beds and joints, and fair tooled on the faces, and have four 4-inch holes cut in each to receive the holding-down bolts. Each of these stones (sixteen in number) must be 5 feet × 5 feet × 1 foot 6 inches thick.

"The holding-down bolts, 2 inches diameter, are to be found by the contractor, and must be carefully fixed through the pillar-stones and cast-iron plates, as shown, and secured by washers and bolts underneath the said plates.

"The brickwork underneath the cast-iron plates must be cut to receive the plates and bolt heads. The bolts must have a play at the top of 1 inch in every direction.

"An ashlar coping, 6 inches thick, is to be laid on the top of the walls, fair tooled on the top and edges, the joints boasted and truly radiated to the curve of the wall and joggled, the whole being set in mortar.

"Thirty-two bearing stones, 2 feet × 1 foot × 1 foot thick, are to be fixed at equal distances round the bottom of the stand; these stones must be fair tooled.

"Thirty-two stones, fair tooled, 15 inches × 12 inches × 12 inches are to be built in the walls in the positions to be pointed out by the engineer, to which are to be fixed the cast-iron grooves or guides.

"The bottom of the tank (and underneath the foundations of all walls) is to be well puddled, of the thickness shown.

"A bed of concrete, made as previously described, is to be laid on the puddle in the bottom of the tank.

"The dry well must be about 4 feet 11 inches deeper than the tank and 8 feet diameter when finished. It will be built, coped, and finished in a similar manner to the tank. The arches for the pipes must be built with great care, so as to prevent any settlement in the walls.

"A puddle wall, as shown, must be carried up outside the tank and between it and the dry well, so as to ensure the tank being water-tight, and the contractor must distinctly understand that he is held responsible for making the tank capable of holding water without leakage, and the work will not be certified as complete until the tank has been proved to be water-tight and the dry well perfectly free from water.

"The pipes, valves, and tar receivers are to be fixed as shown on the drawings, the joints are to be well caulked with plaited hemp and made tight with lead 2 inches deep and properly set, and when the pipes are fixed, the same are to be well rammed in with good puddle and the openings built up and made perfectly water-tight.

"The inlet and outlet pipes (24 inches diameter) and two 24-inch 'Donkin' valves, syphon wells, &c., are to be found and fixed by the contractor. The pipes, &c., are to be full three-fourths of an inch thick throughout, and good sound castings, free from scoria, sand cracks, or other imperfections, and are to be such as the engineer shall approve, and are to be fixed to his satisfaction as per drawings.

"The estimate is to include all the works which may reasonably be inferred as belonging to, or be necessary for, the due completion of any of the works described in this specification, although the details thereof may be wholly or partially omitted in the specification and drawings, as the contractor will not be allowed any extra sum excepting for absolute additions which may be ordered by the engineer in writing."

The preceding specifications describe gasholder tanks as generally constructed in England, where plastic clay suitable for puddle exists in abundance. There are, however, other localities which are not placed in such favourable conditions, where the proper clay, or indeed clay of any description, cannot be obtained, when the tank must be constructed without its aid. For this object, with good bricks or stone and hydraulic cement or mortar and ordinary good workmen, no difficulty whatever need be apprehended, either in the erection or in rendering the tank impermeable without puddle, and, as will be hereafter shown, a tank of the largest magnitude is thus constructed.

This opinion of the inutility of puddle for the purpose was forced upon the writer by circumstances, for having to construct a brick tank in a locality where the requisite clay was not to be had, he with some degree of misgiving decided on dispensing with it, and the success which attended the experiment was sufficient to prove the use of puddle absolutely unnecessary. From that period, although we have built tanks of varied capacities, some of large dimensions, with different kinds of stone and bricks, yet from the date of the first experiment puddle has never been employed by us, nor have we had occasion in one single instance to regret its absence.

We have no intention of expressing opinions in opposition to the general practice of some of our eminent engineers, who invariably resort to the use of the material in question for the reasons that they can obtain it with facility, that a sound tank is more easily secured by its use, and they have been accustomed to employ it; but our object is to explain how it may be dispensed with, and thus guide the manager, who perhaps is thrown entirely on his own resources, and probably has to execute the work under difficulties, but amongst these the absence of puddle is not to be reckoned.

Before proceeding farther, however, it may be remarked that the elements of success in building tanks without puddle are: to have good hydraulic cement or mortar; that the bricks employed should be well burned; that the material, whether bricks or stone, should be thoroughly wetted at the moment of being laid, and that they should be completely embedded in the cement, so that the whole of the joints of brickwork are flushed at the operation; above all, that there should be an abundance of cement or hydraulic mortar, and the joints not too thin, as from a deficiency of cement and a want of ordinary care arise the imperfect structures which are sometimes heard of; and at the same time it may be stated that no extraordinary skill is requisite in the construction of tanks of small and medium capacity, but for the purpose the greatest care is indispensable. We will now proceed to indicate the usual method of building these vessels.

The excavation being made to the desired depth and the solidity of the soil being ascertained, then, in order to guard against any weak place in the ground, as well as to obtain a regular foundation for the structure, a layer of concrete corresponding with the dimensions of the tank is then placed over the whole area. Should the tank be of medium or small capacity, say 80 feet in diameter and under, the whole is usually excavated at one operation, leaving the centre, according to the judgment of the engineer, either as a dome, or frustum of a cone, which in uncertain ground gives greater solidity to the structure.

In ordinary good ground for a tank of from 35 to 50 feet in diameter, on the annular space carrying the wall, the concrete will be required of about 12 inches thick, and if only rendered with cement, the same thickness over the dome or cone; but should a course of square tiles be laid over the dome, the concrete may be diminished to 9 inches. This thickness of concrete is gradually increased until tanks attain the greatest diameter, when it is about 18 inches thick.

The foundations being levelled, the footings of the wall are then commenced, and the wall and piers carried up in regular courses, observing during the operation of building that a difference beyond 3 or 4 feet at any point should not exist, or, in other words, it should be carried up approximately level throughout. At the height of receiving the holding-down bolts, these bolts should be placed and secured in their position to receive the columns by wooden templates, having holes corresponding with those in the base of the columns, and when the wall attains the necessary height, should there be any difficulty in procuring stone, the top of the tank can be terminated with a course of bricks on edge.

Portland cement can be used in the proportion of one to three of clean sand, but by any excess of sand much beyond this, its strength is very materially impaired; and for "rendering" or coating, pure cement should always be employed. The concrete foundation in the centre, if properly prepared, will require only to be rendered, and for this purpose the irregularities are first levelled with the ordinary material used in the building, and over this is laid a thickness of about three-eighths of an inch of cement. The wall, if properly built, will be sound without rendering, but it is always safer to place a layer of cement of about a quarter inch thick over the whole area.

When first filling a brick tank, sometimes the water will be absorbed during the first few days, and convey the impression that a leakage exists. This, however, gradually ceases, and after a short time, if only built with moderate care and good material, brick tanks and without puddle can be made as sound as can be desired.

With these preliminary remarks we proceed to give the following specification :

" *Abstract of a Specification of a Gasholder Tank 33 feet 4 inches in diameter and 12 feet 3 inches in height, suitable for a holder 32 feet in diameter and 12 feet deep at the sides, capable of containing 9500 cubic feet of gas, including seal of 6 inches.*

" The ground to be excavated in the place indicated of a diameter of 38 feet, forming an annular space of 4 feet wide, perfectly level, and 12 feet deep, leaving a dome in the centre of 30 feet diameter at base and 7 feet high in

SPECIFICATION OF A BRICK TANK.

vertical section. The excavation to have sufficient slope to prevent the ground slipping, or, if necessary, to be shored in such a manner as not to interfere with the progress of the building. The excavation for the dry well to be made at the same time, also a communication between this and the tank for the purpose of placing the inlet and outlet pipes which will be done before the concrete of the centre is laid. The excavation to be kept clear of water during the construction.

"*Concrete.* A bed of concrete, 12 inches thick, to be laid around the outer part of the excavation and levelled, and to be continued over the whole of the dome a like uniform thickness throughout. The concrete to be formed of seven parts of clean ballast and gravel with one of Portland cement, the whole to be thoroughly mixed and placed alike on the annular space and dome of uniform thickness.

"*Brickwork.* All the bricks to be hard, well-burnt stocks, and each brick to be thoroughly wetted just previous to laying and embedded in the cement, so that all the joints are flushed or filled, to be laid in Portland cement and clean sand mortar in the proportion of one of cement and three of sand. The concrete of the annular space being levelled, a course of footings for the wall of the tank of three bricks wide, also for the four piers of three and a half bricks square, will then be laid, the internal diameter of this being 32 feet 8 inches. On these footings the wall of the tank is then commenced, of an internal diameter of 33 feet 4 inches, and carried up two bricks thick to a height of 7 feet 6 inches. At this point the wall is diminished to one and a half brick thick for a height of 12 feet, on which a course of bricks on edge, 16 inches in width and projecting 2 inches beyond the outside of the tank, is to be laid perfectly level over the whole of the top of the wall. The four piers to be each three bricks square at base, tapering to two and a half bricks at the top of the wall, to be bonded with the wall from bottom to top. At the height of 7 feet 3 inches four cast-iron washers with the holding-down bolts will be built in each of the piers, the holding-down bolts being supported by wooden templates, having holes corresponding with those of the bases of the guide columns, care being taken that they are exactly equidistant and at the proper height and level with each other. A course of flat tiles set in cement over the annular space around the dome. On the centre of dome there will be a dwarf pier of 3 feet high, two bricks square at base, and one and a half brick square at top, for the purpose of supporting the centre pipe of holder when down.

"The ground to be filled in behind the wall and rammed as the work progresses, care being taken not to injure the brickwork. The tank, when completed, to be a true cylinder, and perfectly plumb at the sides. The walls, as well as the centre dome, to be afterwards rendered with neat cement to a thickness of not less than one-fourth of an inch, and to be worked with the trowel so as to form a smooth surface.

"To construct a dry well adjoining the tank of 3 feet 6 inches internal diameter to the depth of 13 feet, the footings of wall to be two courses of one and a half brick, and on this to be built the wall, of one brick thick, set in mortar to the height of the ground level, where it will be finished by a course of bricks on edge, set in Portland cement and sand.

"All surplus earth to be carted away, and the ground left level as before the commencement of the work."

The tank here described is intended for ground of an ordinary character, which renders the concrete foundation necessary. It is simple and cheap in its construction, and if built with care will be perfectly sound, and cost on the average about 140*l.*

In small works sometimes the dry well is dispensed with, and a recess is built in the wall of the tank, in which are placed the inlet and outlet pipes. This method, although economical, possesses the disadvantage that, in the event of an obstruction, the inlet and outlet are not readily cleared.

The same kind of construction as that suggested—with cement concrete foundations, the wall built as described, the centre and walls being rendered with cement, and the prices of materials as stated—a tank, 36 feet 6 inches in diameter and 14 feet 3 inches deep at the sides, with its corresponding dry well suitable for a holder 35 feet in diameter and 14 feet deep, of the capacity of 13,000 feet, will cost about 220*l.* A similar tank, of 41 feet 6 inches in diameter and 14 feet 3 inches deep, for a holder of 17,000 feet, will average 300*l.* If of 46 feet 6 inches in diameter and 16 feet 3 inches deep, for a holder of 25,000 feet, the cost will be about 430*l.*; and of 55 feet 6 inches in diameter and 18 feet 3 inches deep, it will average 645*l.*

A tank, built in ordinary good ground, of 61 feet 6 inches in diameter and 18 feet 3 inches deep, in the manner described, suitable for a holder of a capacity of nearly 50,000 feet, will require a bed of concrete over the whole area of 15 inches thick. Two courses of footings, of four bricks or 36 inches wide, are placed on the concrete, on which the wall is commenced. This to the height of 4 feet should be three bricks thick, from that point it is diminished to two and a half bricks to a height of 12 feet from the footings; from this point to the coping it is two bricks or 18 inches thick. If the concrete in the centre is close and compact, it may be rendered, as already indicated, with cement, or if any doubt exists a course of flat tiles, set in cement in concentric courses, may be laid. A tank of this description, with its corresponding dry well, will cost about 720*l.*; another, of the same construction, of 65 feet 6 inches in diameter and 20 feet 6 inches deep, suitable for a holder of 62,000 feet, will average 860*l.* Some of the foregoing are actually the prices paid according to contract for the work executed; others are calculated according to their quantities and the prices of material already adopted, taken as an average. For a holder of this capacity six columns for its support will be necessary, therefore six piers or buttresses are required, which, however, will be controlled greatly by the height of the holder, as when these are of the largest dimensions, frequently the columns are placed at a distance of only 20 or 22 feet from each other.

A tank, erected in the vicinity of London by one of the most eminent contractors, of 65 feet in diameter and 22 feet deep, built in the best possible manner, entirely in Portland cement and puddled at the bottom and sides, cost 1030*l.*; the difference between this and the price last quoted being due principally to the expense of puddle which in this case had to be procured at a considerable distance from the site of the works where it was built.

As gasholder tanks increase in diameter and depth, so is the surface of their walls proportionably augmented, and this increased surface and depth require greater strength of material and thickness of wall to support it. The increase of depth, and consequently pressure of water within the tank, demands additional care in the execution of the work, in order to ensure soundness. Thus, from the specification of tanks already given, a slightly gradual and progressive increase of strength in all the parts of larger tanks will be necessary.

STONE TANKS.

Stone tanks are very frequently constructed on the Continent, invariably without the aid of puddle, and are rendered impermeable simply by the stone and cement or hydraulic mortar. Sometimes these are built so as to project 5 or 6 feet above the surface of the ground, which system, in addition to the economy in excavating, permits of a deeper vessel being constructed in any locality. In such cases a hoop similar to those used in cast-iron tanks, is built into the stone or brickwork, just beneath the coping.

When erecting these tanks the stones are dressed on five of their sides, similar to those used for paving, but are generally of much larger area. The foundation of the wall being properly laid and levelled, the wall is commenced, and in the operation the stones are embedded in cement or hydraulic mortar, each course being formed of stones of equal thickness; at the same time all the joints are well filled with small stones and cement at the back, in order to prevent the escape of water at these parts. This accomplished, the wall together with the piers are built to the desired thickness with rubble, which is intended to give strength more than for any other purpose, as the soundness is assured when forming the facing, nevertheless it is necessary that the whole of the work should be well filled with cement.

The following is an abstract from a specification for building a stone tank of 102 feet in diameter and 30 feet deep, the upper part projecting 5 feet above the level of the ground; the site, already referred to, being situated immediately adjoining the sea, in sandy soil, with an abundance of water at 25 feet below the surface.

"To excavate the ground to a depth of 25 feet over the whole annular space where the foundations of the wall are to be constructed, the internal area to be left in the form of a dome, its base being 92 feet in diameter and in vertical section 15 feet in height. At the depth of 25 feet, the ground will be further excavated a depth of 3 feet 6 inches, so as to form an annular foundation, the inner diameter of which will be 94 feet, and its outer diameter 120 feet, and 3 feet 6 inches in depth, thus forming an annular foundation 13 feet wide. This foundation to be formed within the water and composed of Portland cement concrete and to be of uniform thickness and width and level throughout. The inlet and outlet pipes being placed within the foundation, in communication with the syphons in dry well, so as just to clear the interior of the tank.

"The concrete of the foundation being levelled, the walls are then commenced, the facing being of square stones each of about a superficial foot area, roughly dressed on five sides and embedded in mortar composed of cement and sand, in the proportion of one of the former to three of the latter, all the stones of each course to be of equal thickness; at the same operation all the joints at the back to be carefully filled in with cement and small stones, and backed with rubble masonry in cement to the required thickness of the wall.

"The diameter of the tank between the walls to be 102 feet, and carried up perfectly plumb and cylindrical to a height of 29 feet 9 inches. The wall will be 4 feet 6 inches thick at the bottom, including rubble and facing stones, gradually tapering to 2 feet thick at the top, at which point an iron hoop, of 6 inches $\times \frac{5}{8}$ inch flat iron, with corresponding jaws secured with cotters, will be built in with the wall around.

"There will be twelve piers, equidistant from each other, each of a width of 4 feet, which at the bottom will project 2 feet beyond the wall, and carried up with the same batter as the wall. Each of the piers to be finished with a square-dressed stone, fine tooled, 4 feet 8 inches $\times$ 4 feet 4 inches and 6 inches thick embedded in cement.

"At the height of 22 feet 6 inches cast-iron plates with the holding-down bolts will be built in with the piers, the bolts being supported by wooden templates having holes corresponding with those in the bases of the guide-standards. The wall to be terminated with a stone coping 2 feet 3 inches wide and 6 inches thick, fine tooled, and formed to the curvature of the tank and properly embedded in cement.

"Opposite the centre of each pier within the tank will be built into the wall at the points indicated six iron lugs, for the purpose of securing the roller guide-plates.

"A layer of concrete, of the thickness of 15 inches, to be placed over the whole area of dome, except at the part which abuts against the annular foundation, where it will be increased to 18 inches in thickness and gradually diminishing in a length of 3 feet. On the whole of the annular space as well as over the dome will be laid two courses of flat tiles, 8 inches square and 1 inch thick, in concentric rings, set in Portland cement and sand as stated.

"In the centre of the dome will be built a square brick pier, of 7 feet high, 4 feet square at the bottom, and tapering to 3 feet at the top, for supporting the king-post of the holder when that grounds.

"The ground to be filled and rammed as the work proceeds, and all superfluous earth to be carted away.

"The dry well to be sunk to a depth of 28 feet from the surface, where there will be a thickness of 15 inches of concrete, extending over the whole area of the centre and walls a circular space of 12 feet in diameter, and on this will be laid three courses of footings, of two and a half bricks wide; from this point the dry well, of one and a half brick, will be carried up to the surface of the ground, where it will be finished with bricks on edge of 16 inches. All the bricks to be set in cement and sand."

The foundations were built in the manner described when referring to caissons, and on its completion the tank was as sound as could possibly be desired; indeed, as a matter of precaution, as the work was "green" it was supplied with water in the first instance only to within 3 feet of the top, but in the course of two or three years the tank was entirely filled by the rain. This can be readily understood, when we reflect that the rain fell over the whole surface, whilst the evaporation was confined to the ring of water which surrounds the holder.

Some years ago an engineer when building a stone tank had every stone in each course keyed and tongued together, which must have caused an immense amount of labour and expense, but the cohesive power of Portland cement being known, such precautions are entirely unnecessary. Moreover, the weight of the materials forming the wall, together with the pressure of the external earth against it, counteract, in a great measure, the pressure of the water within the vessel.

Composite Tanks.

Among the innovations introduced into the construction of holder tanks is the use of concrete for the formation of the walls in combination with brickwork. This system was first introduced with a view to economy, we believe, by Mr. Wyatt, who formed the wall by an inner and outer ring of brickwork, and gradually, as these were built, the intervening space was filled with concrete. It appears that the principal object in having the outer wall was to avoid the difficulty of fitting a mould to the exterior of the tank; however, the success which attended this structure induced that gentleman subsequently to build with an inner facing only, the desired thickness of the wall being formed with concrete, which system is now generally employed, particularly in tanks of the largest capacity, and termed "composite tanks."

A tank of this description was constructed by Mr. Livesey, at the South Metropolitan Works, London, and is 153 feet in diameter and 38 feet deep in the clear. The wall of this is 4 feet 3 inches thick at the bottom, diminishing to 2 feet 3 inches at the top, including bricks and concrete. The whole area of the bottom as well as beneath and outside the wall is puddled. The saving effected in building this tank, by using concrete instead of brickwork, was 1700*l.*, and there is every reason to believe that in addition to the economy, the former is stronger than the latter.

We have been favoured by Mr. Wyatt, constructing engineer of the Gas Light Company, with a copy of the following specification of a composite tank, designed by that gentleman, and constructed at the Redheugh Works, at Newcastle-on-Tyne, of which a half-section is given in Plate 15. As the document possesses general interest, we present it in detail.

"Specification of Gasholder Tank at Redheugh.

" Conditions.

" The contract is to include the provision of all materials and labour for completing the works shown on the drawings, and described in the specification.

" The contractor must set out the works, and must satisfy himself as to the accuracy of the same, as he will be responsible for any error in the form of the works.

" The works are to be performed under the control and supervision of the engineer, or such person or persons as may be appointed for such purpose, and whose decision upon all matters will be final. All orders for new or altered works will be made in writing by the engineer.

" All materials delivered by the contractors upon the site of works to become the property of the company, and shall not be removable without the engineer's consent. All materials disapproved of by the engineer shall be removed from the site of the works within forty-eight hours of the contractor receiving notice thereof from the engineer.

" In case the company shall consider it desirable to alter any portion of the work, adding to or diminishing the amount thereof, or changing one kind of material for another, the same shall be executed under the conditions of this specification, and measured up and paid for at the schedule of prices annexed hereto, or where no price is affixed for the work, the same shall be valued by the engineer, whose opinion thereon shall be final. The value of such works shall be accordingly added to, or deducted from, the amount of the contract.

" The contractor will be required to furnish a monthly statement to the engineer of all extras or departures from the contract, and a monthly stock account of all materials delivered by the contractor upon the works.

" The contractor must guarantee the company harmless against all accident or damage to the company's property or works or the properties and works of others, during the progress of the contract, and incident thereto.

" Should the progress of the works not be deemed satisfactory at any time by the engineer, he may take the works in hand and use the contractor's plant and materials on the ground, and deduct the cost of such proceedings from any balance due to the contractor. Such operation on the part of the engineer shall not relieve the contractor from any liabilities under the contract.

" The contractor shall maintain and uphold the works in an efficient state for twelve months from the time of completion, as certified by the engineer, and shall then hand them over to the company sound and efficient, making good all defects, accidents, and damages during such term. Any reparation performed by the company during the maintenance, in consequence of the contractor failing to do so, shall be charged to the contractor.

" All measurements for quantities and estimates will be based upon the net dimensions shown on the drawings.

" Payments will be made monthly upon the amount of work performed during the preceding month, equal to 80 per cent. of such work done and certified by the engineer, and the remainder will be retained as a drawback or security for the performance of the works. Half the drawback will be paid within three months, and the remaining half within twelve months of the certified completion of the works by the engineer.

" The contractor to find two sureties for the proper performance of the contract in the sum of 1000*l.*

" The work to be completed by the —— and in default thereof the contractor will allow a discount off the gross amount of the contract of 1 per cent. for each month the works remain unfinished.

SPECIFICATION OF A COMPOSITE TANK.

"All wrought and cast iron work (excepting hoop iron) shown on the drawings of tank, and built in same, will be provided and delivered under the gasholder contract, but the fixing of same to be done under the tank contract.

"A sum equal to half per cent. on the contract amount to be included in the tender for the taking out of the quantities attached in this specification."

"*Specification for Tank.*

"*Description.* The tank will be 152 feet in diameter, between the walls, and 31 feet 9 inches deep, from the floor to the top of coping.

"*Excavation.* The whole of the earth to be excavated to the depth shown, and timbered in the trenches of such form to admit of building the side walls to the dimensions given on drawings, including the inlet and outlet well. The price of such excavation to include the necessary plant, timbering engines, pumping (until the completion of gasholder) materials, and labour to complete the work to the satisfaction of the engineer, and also the spreading loading, and forming the earth on any part of the site of the works, as may be directed by the engineer.

"*Surplus Earth.* The whole of the surplus material excavated will remain the property of the company, but such portions of the same as may be fit for the concrete, mortar, or puddle of the tank works, in the opinion of the engineer, shall be at the disposal of the contractor free of expense. Should there not be sufficient of any or either of these materials to complete the work, such deficiency shall be provided by the contractor from other sources at his own expense.

"*Puddle.* The puddle throughout to be 2 feet thick at least, and to be worked up, chopped up with tools, well watered and passed through a pug-mill, driven by engine power, before being used. It is then to be thrown into place from an elevation, and trodden in 9-inch layers uniformly at the back of the walls. Each layer to be chopped into preceding one which is to be kept clean. A large quantity of puddle to be in course of preparation at one time, so as to ensure sufficient working and tempering to same.

"*Brickwork.* The whole of the brickwork to be composed of the best hard-burnt bricks, sound, square, and of an even colour, and laid in Portland cement mortar. The work to be laid English bond, grouted up solidly at every course, and pointed up neatly either as the work proceeds, or at such time after completion as the engineer may direct. At every 4 feet 6 inches in height of the walls, there will be hoop-iron bond in the proportion of one strip to each $4\frac{1}{2}$ inches in thickness of walls $1\frac{1}{2}$ inch wide, $\frac{1}{16}$ inch thick, and well tarred and sanded. It is to be lapped over 2 inches, and rivetted up with two $\frac{3}{8}$-inch diameter rivets at each joint; formed in continuous lengths for the entire circumference of tank and well, and passed through the concrete pocket of piers and wells, and intersected by other bands of hoop iron, as shown on plans. The walls of tank to be of the following dimensions, including the concrete backing; footings to piers 13 feet $2\frac{1}{2}$ inches × 4 feet 6 inches; breadth at wall panels 8 feet $6\frac{1}{2}$ inches, depth of footing 2 feet 6 inches; the walls from top of footings, for a height of 13 feet 9 inches, to be 3 foot $4\frac{1}{2}$ inches thick, next 8 feet 6 inches × 3 feet thick, next 8 feet 6 inches × 2 feet 8 inches thick, to the under side of the coping. The piers sixteen in number, at 29 feet $10\frac{1}{4}$ centres round the inner circumference of tank, to be 8 feet × 4 feet 6 inches, formed of 9-inch walls on three sides, and filled with Portland cement concrete. In each of the piers there will be built the holding-down plates and bolts hereinafter described.

"*Cement Mortar.* The whole of the mortar to be composed of one part of Portland cement, from a first-class maker, to three parts of sand, measured dry and net. The cement not to weigh less than 112 lbs. per imperial bushel, and to be tested upon from time to time by those in charge of the work for cohesive strength and setting properties. There must be a cohesive strength in the pure cement of more than 200 lbs. per square inch, after seven days' immersion in water, and 500 lbs. after one year's immersion.

"*Concrete.* The whole of the concrete to be composed of one part of Portland cement as above, and seven parts of gravel measured dry and net. The gravel to be composed of four parts of coarse screened stones, up to $1\frac{1}{4}$ inch diameter, and three parts of sand. The concrete to the pockets of the piers and walls of tank, to be deposited in 9-inch layers, not exceeding 2 feet 3 inches in depth each day, and then levelled off and grouted. The concrete on cone to be 1 foot thick.

"*Stone.* On the floor of tank and under each post of holders there will be two stones, each 4 feet × 1 foot 6 inches × 9 inches. There will be three stones, each 2 feet × 1 foot × 9 inches, built in tank walls to the vertical line, of each of the guides, at the positions shown on drawings, having two holes in each to receive two $1\frac{1}{4}$-inch diameter bolts. The coping stones to be 1 foot 10 inches × 1 foot in section, in stones not less than 3 feet long, and worked truly to the radius of tank. There will be a stone 6 feet diameter and 1 foot thick at the top of centre pier on cone, which will have eight $\frac{3}{4}$-inch diameter bolts 4 feet long through the same. The stones under standards to framing to be each 8 feet × 4 feet × 1 foot. The coping to wells to be 1 foot 2 inches × 1 foot in stones not less than 3 feet long worked to the radius of same. There will be two 4-inch landings under inlet and outlet pipes in well, of an area of 5 feet × 4 feet 6 inches each; and two 4-inch ditto, 5 feet × 4 feet each under pipes in tank, placed in pairs and jointed to each other. The whole of the stone, except landings, to be of best Yorkshire stone, or stone equal thereto, of sound quality, dressed and fair tooled to beds, joints, and surfaces, and laid on the natural bed on a thin layer of cement, without packing.

"*Well for Inlet and Outlet Pipes.* The well to tank to be 11 feet diameter inside, and 40 feet deep to the floor of same. The platform at bottom will be 2 feet 3 inches thick, with a 1 foot 2-inch invert formed in same. The first 22 feet of wall above the floor to be 1 foot 6 inches thick and the next 17 feet to under side of coping to be 1 foot 2 inches thick. Leading from the bottom of the well to the interior of the tank there will be an arched chamber 9 feet wide, 5 feet high, and 14 feet 1½ inch long from the inside of wall of well, arched 18 inches thick, side walls 3 feet thick, and inverted with an arch 1 foot 2 inches thick. The whole to be rendered on the inside faces of walls with a coating of Portland cement $\frac{3}{4}$ inch thick.

"*Centre Pier.* There will be a brick pier on concrete on the centre of cone. The concrete foundation to be 10 feet square and 8 feet deep. On this will be erected the circular pier, with a diameter at the top of 6 feet 6 inches, and batter sides of 1 in 40. At the level of 2 feet 9 inches below the top of gasholder dome there will be the stone previously described 6 feet diameter and 1 foot thick. On this stone there will be a double layer of 4-inch oak planking, formed of planks, spiked together, about 7 inches wide, of an area 5 feet 6 inches in diameter and laid at right angles to each other. There will be built into the pier from the top of planking eight $\frac{3}{4}$-inch diameter holding-down bolts, 4 feet long, with proper nuts and washers, and held in the brickwork at bottom by four strips of wrought iron, each 4 feet × 4 inches × $\frac{3}{8}$ inch, intersecting each other. The upper surface of planking must be faced and wrought to a true level to receive the bearing plate attached to the under side of rafters. The planking to be thoroughly tarred twice before being placed in position.

"*Cone.* When the cone is dressed off to the form shown on drawings there shall be placed over the entire area of same one foot of Portland cement concrete.

"*Soundness of Tank.* The contractor must undertake to deliver over the tank to the company at the time stipulated, sound, water-tight, in good condition, and to the approval of the engineer. He must also fill the same with water, and should the tank be found defective or leaky, and require to be emptied for remedying defects at any time during the year's maintenance, the same shall be done, the defects made good, and the tank refilled with water, with all other necessary works, at the expense of the contractor. The contractor for the tank will have to perform the necessary pumping to keep the works dry during the construction of the gasholder."

Concrete Tanks.

Encouraged by the success which attended the formation of composite tanks, Mr. J. Douglas, of Portsea, entertained the idea of making one of these entirely of concrete, which, however, when carried into execution, was lined with brickwork, in order to ensure a circular even surface. This differed materially from the "composite," inasmuch as the concrete wall was erected separately from the brick lining, this being added afterwards, as that gentleman observes, "solely for the sake of appearance." The means of construction adopted were similar to those employed for building ordinary concrete walls, or the clay walls of cottages as often practised on the Continent, that is, by a mould or framing, formed by two stout boards of any convenient length and width, placed in a vertical position, and parallel to each other; these being separated a distance equal to the width of the desired wall, but secured from further separation by bolts and nuts, thus forming a mould into which the concrete or clay is placed, so as to fill it, and on the former setting or the latter drying, the moulds are detached by means of the screws, and removed to another part to continue the building.

From Mr. Douglas's description of the construction of this tank it appears that several of these moulds were prepared, some to suit the thickness of the walls and others for the piers, all of which were about 2 feet 6 inches deep. The foundation of the wall being terminated and levelled, a number of moulds were placed exactly in their positions to form the circle, when they were filled with concrete, and on this setting, the moulds were removed and placed in another part, in their respective positions, and the process continued until completing the whole of the circumference of the wall with the corresponding

2 B

piers. By these means a course of 2 feet 6 inches thick, equal to the depth of the mould, was built; on this a second course was laid, and the other courses above were all built as described, thus forming a cylinder of concrete, of the diameter and depth desired, which, according to the experience of Mr. Douglas, is stronger than a similar thickness of brick set in Portland cement, and indeed, in that gentleman's opinion, walls of tanks when thus erected need only be two-thirds the thickness of those of brickwork. This tank, in addition to being lined as stated, was also puddled over the whole of its area, and the exterior of the wall.

The proportions of material were one of Portland cement to seven of shingle. For mixing, a deal platform of about 20 feet square, firmly embedded in the ground, was employed for the facility of shovelling when turning over the material. The ballast was measured, and on the cement being added, it was well mixed in a dry state, after which water was supplied, and the whole thoroughly intermixed, and well moistened throughout. Mr. Douglas refers to the importance of having good cement, which under all conditions should always be sought, and it is the worst of economy to endeavour to obtain this article in the cheapest market, when probably it may be worthless.

Unfortunately we are not in possession of the dimensions of this vessel, therefore any reference to the thickness of the walls will answer no purpose; all that is necessary to observe is that it fulfilled all the expectations of the engineer. When entering into the description Mr. Douglas gives some details, in which he says, "throwing the concrete from a height should be avoided, because the elements of the concrete, having different specific gravities, fall in a heap together, and the cement is thus separated from the other material." We are of opinion that the defect of this arises from the liquid cement being ejected from the mass when falling on the ground, by which the concrete is rendered porous; we also differ on another point from that gentleman, where he asserts: "It must be borne in mind that though neat Portland cement is quite impermeable to water, when mixed with ever so small a quantity of sand it is not so, and concrete will leak like a colander."

In our opinion, the degree of impermeability of a mixture of cement and sand will depend on the freedom of the latter from any foreign matter; for it being conceded that pure cement is not porous, it follows that, if intermixed with pounded glass, both on account of the cohesive affinity existing between the two substances, as well as their nature, the compound will be as impermeable as the pure cement. If, again, the cement be intermixed with perfectly clean sand (and if granulated all the better), the mixture will be found equally desirable for the object. This is not only correct in theory, but is confirmed by experiments made with the view of determining a point upon which much diversity of opinion exists.

But the presence of loam or earth in any material quantity in combination with ballast, gravel, or sand, will have a tendency to render the compound porous, as in this case each particle of the foreign substance surrounds and attaches itself to the stones and pebbles forming the gravel, thus preventing the cement adhering thereto, and subsequently when the loam or earth is washed out by the action of the water, the concrete may be rendered porous as described. From the experiments referred to, together with practical experience, we are firmly convinced that, if due care be observed in the choice of clean sand, ballast, and shingle, together with the proper admixture of the varied sizes of these materials so as to avoid large interstices, accompanied by a sufficiency of cement, and the employment of ordinary intelligence in the manipulation, concrete will be found quite impermeable for the purpose under consideration, but for this purpose the whole should be well wetted with the liquid cement.

A great progressive step in the construction of holder tanks was subsequently made by Mr. George Livesey, remarkable alike for the boldness of the undertaking, the magnitude of the structure, the difficulties occasioned by the presence of an immense quantity of water in the excavation, and lastly in dispensing entirely with puddle; and although, as we shall have occasion to observe, complete and perfect success did not attend the structure in the first instance, this arose from no defect either in the building or in the material employed, but simply from an excess of confidence of its strength entertained by the engineer.

CONCRETE TANKS.

This remarkable structure, represented in section, Fig. 57, is built at the South Metropolitan Company's works, and is 184 feet in diameter, and 47 feet deep at the sides in the clear. In the centre is left the frustum of a cone of about 170 feet diameter at its base and 90 feet diameter at its summit, its height in vertical section being about 38 feet; the whole of the cone, as well as the annular space around it, having a layer of concrete 18 inches thick. On the cone is built a circular wall of 87 feet in

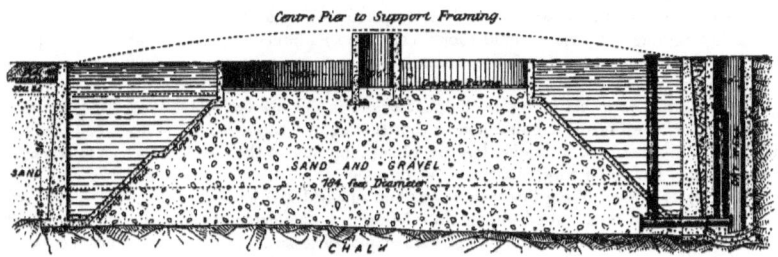

Fig. 57.

diameter and 12 feet high, extending slightly above the level of the coping. In the centre of this is another wall or tower of 10 feet in diameter and 18 feet high, carrying the wooden framing which is extended nearly over the whole of the area of the holder, in order to support its roof when it grounds, the framing being supported by vertical props. The great depth of the holder, together with its luting, and plates of a thickness essential to ensure durability, rendered it necessary to dispense with all trussing in the roof, this being, so far as we are aware, the largest untrussed holder in existence.

The wall of the tank is 5 feet thick at the bottom, diminishing to 2 feet 3 inches at the top. There are 20 piers in the circumference, in each of which is built six 2½-inch holding-down bolts, for the purpose of securing the guide standards. A great advantage in the construction of this tank is the formation of a large cone, with a circular wall above it, thus confining the water space to narrow limits, and in consequence the full pressure of the water is restricted to a comparatively small area, and the liability to leaking is diminished in a corresponding manner.

The dry well is also constructed entirely of concrete, cemented underneath, and on the outside for the purpose of preventing the ingress of the water. The inlet and outlet pipes are of wrought iron.

The concrete is formed principally of gravel and sand from the excavation, intermixed with Thames ballast, burnt clay ballast, old retort rubbish, and ordinary clinkers accumulating from the furnaces. The proportion of the two last-mentioned materials is undoubtedly small, their principal advantage being their irregular angular form, which appears to facilitate the cementation of the compound. From Mr. Livesey's practical experience in the formation of concrete he finds the best proportions suitable for the purpose are seven parts in bulk of the various materials mentioned, in combination with one part of Portland cement. He is, moreover, of opinion that ten to one would answer the object; but, under the extraordinary circumstances, it was considered advisable, both by the contractors, Messrs. Docwra and Sons, and by himself, to adopt the proportions mentioned.

When building this tank, in consequence of the very treacherous nature of the soil and the immense quantity of water encountered, the excavation was attended with some difficulties. This was accomplished in the manner usually adopted in structures of this kind when of great magnitude, as described in the preceding specifications, that is, by excavating an annular trench of sufficient width and depth to permit of the outer wall and piers being erected, and afterwards taking out the earth, leaving the cone in the centre, the height of which is controlled by the angle of repose of the earth. In excavating the trench, the ground was supported by 9-inch deals, placed close together, and supported by struts and walings; and when we consider the great depth, the nature of the ground, and the weight of the surrounding material, it must be regarded as a skilful operation to execute a work of such magnitude without the slightest accident. At a distance of 11 feet from the surface, water was met with in abundance, which naturally augmented as the excavation increased in depth, and, as already observed, the quantity pumped daily for months together was no less than 240,000 cubic feet.

Thus the whole of this stupendous tank is composed entirely of concrete, and as if its author had resolved that nothing but that material should enter into its formation, there is not a single brick nor stone, as generally termed, in the whole structure, and even the rests and the coping are of concrete. But the boldest undertaking in the work, and one contrary to all accepted ideas of the generality of English engineers, was to dispense with the puddle, and in this was undoubtedly achieved the greatest success. The saving estimated to be effected by constructing of concrete instead of brick and puddle, was about 5000*l*., but had it been in another locality, where the whole of the ballast existed in the excavation, this saving would have been very considerably augmented.

It appears that when filling this tank, it was first supplied with water to a height of 35 feet, which was as much as could be admitted without interfering with the construction of the gasholder then in progress, and during the operation the pumps were employed to keep the springs down outside the tank, thus putting it to a most severe test. "The result," as stated by Mr. Livesey, "surpassed expectations, for the most careful measurement taken from day to day over an extended period showed no sign of leakage, the tank being so sound that an hour's rain could be measured, and from rain alone a considerable rise of the water took place." However, on the completion of the gasholder in the depth of winter, the tank was filled during a sharp frost, when, as the last foot of water was being admitted, a slight crack of about the sixteenth part of an inch wide appeared at the point at which the water entered, and from this a leakage of some importance at first occurred. This has since been gradually diminishing, and means have been recently adopted by which the tank is rendered sound. Thus although a positive success was not obtained at first, it was eventually achieved.

In our opinion Mr. Livesey was too confident of success and submitted the tank to such an extraordinary test, that few, if any, similar structures, either of brick or stone, would have been able to withstand, for on pumping the water from the exterior when filling, was to deprive the vessel of its natural support at that point. Again, as Portland cement increases in strength for months and years after its application, by filling the vessel to the full height whilst the work was still fresh or green, another obstacle to success was presented.

However, in the construction of this tank, two most important practical lessons have been learned. The first, that puddle is not essential in these structures, even when of the magnitude to contain upwards of 20,000 tons of water, and of a depth of 47 feet. This is more particularly confirmed in the dry well of the tank in question, which is the only one on the premises free from water, although all the others are puddled. We also learn the important fact that concrete alone can be successfully applied for the construction of tanks, but whether it may be better to employ this in preference to brick or stone is a question to be decided by various circumstances, and particularly the cost of ballast.

The cost of tanks of all kinds, as well as gasholders, is usually estimated according to their cubical capacity, on supposing the former to be flat in the centre, as at one time they were generally constructed. The prices of brick and composite tanks are very variable, depending on the nature of the site and the difficulties encountered either by the presence of water or the proximity to buildings, the kind of structure, the cost of building materials, and other circumstances. Thus we learn from Mr. Douglas that the cost of a brick tank of 158 feet in diameter at Portsea was 28*l*. per 1000 feet capacity, and others at the South Metropolitan Works were about 25*l*. and 27*l*., and the price of a composite tank was 18*l*. for a similar capacity. The average prices of the composite tanks at Beckton were from 13*l*. 4*s*. to 14*l*. 16*s*., and that constructed at Redheugh, by Mr. Wyatt, as we are informed, cost under 10*l*. per 1000 feet; whilst

Mr. Longworth, of Dukinfield, lately erected a brick tank of 82 feet in diameter and 22 feet deep at the low price of 8*l*. for the same capacity. Within our knowledge the sites of the two last mentioned are in clay, therefore offering considerable advantages in the economy of building, notwithstanding they merit attention, particularly as that by Mr. Longworth is probably the cheapest tank ever constructed.

Hence, with these great variations, it is obvious that no approximation to the cost of a tank can be made without first possessing a knowledge of all the various circumstances by which the building would be influenced. However, we believe that the average may be taken at 14*l*. per 1000 feet, which opinion is confirmed by the prices of those at Beckton, as well as of the smaller tanks given, and notably that of 65 feet 6 inches in diameter which was built by a first-rate London firm, in concrete and puddle, the price being at the rate of 13*l*. 10s. for the capacity referred to, and we may assume that the high prices mentioned arose from extraordinary adverse circumstances, which in undertakings of this description are sometimes numerous and unexpected, therefore a contractor must always leave a large margin for contingencies.

There are instances where these average prices may be materially reduced, but they should be rarely exceeded, and it must be remembered that our observations apply to the present time, when a marked progress has been made in building tanks. The various new methods adopted in the construction of these vessels further encourage this opinion, and there is reason to believe that considerable economy is yet to be effected in these expensive apparatus. Under like conditions of nature of ground, price of materials, and manner of construction, there will be but little difference between the cost per 1000 feet capacity of a tank of the largest and one of ordinary dimensions.

Cast-iron Tanks.

As already shown, the earliest gasholders and tanks were rectangular (the latter being formed of cast iron), which method of construction was retained for years. The first explosion of these vessels recorded happened at the Manchester Works in 1819, which accidents have, unfortunately, since been repeated at several towns in the United Kingdom and elsewhere, but now are happily almost unknown. These calamities invariably occurred on the completion of the holder, sometimes by an escape of gas through an unsound valve, thus producing an explosive compound within the vessel, which ignited by the light employed by the workmen. In other cases, on first filling, the holder on descending has become set or bound, so that no pressure existed, when on attempting to ascertain the quality of the gas a light has been applied to an open tap, and there being a large accumulation of air mixed with the gas, an explosion has occurred. But had the trial been made with an argand or other burner having small holes, then, in consequence of the orifices being too small to permit of the passage of the flame, the calamity would have been avoided.

During the early period of gas lighting, when its ultimate success was exceedingly doubtful, and its immediate results by no means remunerative to the shareholders, and when, although only in its first infancy, it was supposed almost to have attained its utmost importance, the directors of gas companies evinced the greatest caution in incurring any extra expenditure. Thus the augmentation of their gas stores or holders, like the rest of the apparatus, was generally gradual in the extreme; commencing in the first instance with a holder of 10,000 feet, which in the course of time was replaced by another of double the capacity, and this in its turn was substituted by one of still larger magnitude according to the requirements of the business; and as the site occupied by gasworks was invariably very restricted, these changes of gasholders were usually made in the same locality, and it may be asserted that frequently in old works, no less than five different gasholders have been erected at successive periods on or adjoining the same site.

For these numerous changes cast-iron tanks were admirably adapted, as by the sale of the old vessel a considerable portion of its cost was realized, whilst by the gradual growth of gasworks, a ready market was always open for the purchase of tanks and holders which, by extended business, had become too limited for other establishments. There are many of these still in existence that can be traced to have been employed at no less than three different towns at distant parts of the kingdom; and only recently in the neighbourhood of London, a tank of about 6000 feet capacity came under our notice, which, according to a cast tablet thereon, was constructed nearly fifty years ago for a city in the north of England. From there it passed to a town in the west, and ultimately was purchased for the works where it is now in active service.

About thirty-five years ago, the telescopic holder of 60 feet in diameter, having three guide standards each formed by two columns attached to the same base and entablature, was very generally adopted in England, many of which were fixed abroad, and for a period it appears to have been considered that a greater capacity was not desirable, and some of our readers will remember the plan of a works, designed by a gentleman of eminence, in which there were no less than ten telescopic holders with cast-iron tanks, the united contents of which did not reach one million cubic feet, and at the present time there still exists at one of the storage stations of the metropolis a similar collection of small holders, some of which may probably have been in use almost since the first formation of the company.

Cast-iron tanks are, however, very desirable under peculiar conditions and in certain localities, such as where the site is composed of rock, or a want of solidity of the site does not admit of the construction of any other kind, or where water exists within a few feet of the surface, or in places where the requisite skilled labour for the construction of those of brick or masonry is not available. Under these and other contingencies cast-iron holder tanks are sometimes indispensable.

The advantages of cast-iron tanks consist in the facility and promptitude with which they can be erected, together with the limited skilled labour necessary for that purpose; and as the material entering into their formation is very considerably lighter than that employed in other descriptions, less solidity in their foundations is requisite; indeed, in some instances when they are erected in a gravelly site, all that is necessary is to level the ground, and then lay the bottom plates of the tank. Usually, however, baulks of timber extending the full width of the vessel are placed perfectly level parallel to each other, at intervals of two or three yards, and on these the tank is erected. Hence in many instances the base of the tank can be placed on a level with the surface, so that the whole projects above the ground, and thus no excavation whatever is requisite.

When cast-iron tanks are erected on marshy ground, then, in order to ensure their solidity, extraordinary precautions have to be observed by piling over the whole area of the site of the tank. Under these circumstances the length and thickness of the piles will be controlled by the number employed, the capacity of the tank, and the depth of the solid substratum. The tops of the piles being perfectly level with each other, the bottom plates are then laid, but as the whole of the weight of the tank, holder, and water, is entirely supported by these, every possible precaution is necessary by having a sufficiency of piles well driven into the solid substratum.

Cast-iron tanks are constructed of flanged plates bolted together, the joints being afterwards caulked with cement in order to render them sound; their bases are universally flat, similar to brick tanks as formerly constructed. The centre of the base of small tanks is circular, and in one plate, but in those of larger dimensions it is formed of two or more plates. Around this are concentric rings, each ring being about 5 feet wide, the plates forming it being of course of that length, and averaging about 12 square feet area. The flanges projecting upwards average 3 or $3\frac{1}{4}$ inches wide, and about $\frac{5}{8}$ inch or $\frac{3}{4}$ inch thick, having bolt holes cast therein 6 inches apart, and are further strengthened by brackets placed midway between the bolt holes. The side or wall of the tank is composed of a number of tiers of plates, averaging 3 feet 6 inches or 4 feet in height, also provided with flanges, bolt holes, and strengthening brackets, as described. The flanges are cast on the outside of the tank, which facilitates its erection, also offering means for stopping any leakage that may occur.

When erecting a tank all the bottom plates are first temporarily placed as near as possible in their position; the first tier of the side plates is then fixed, when some "humouring" of the whole will be necessary so as to arrange all the joints throughout alike over the base as well as at the sides of an average thickness of three-eighths of an inch. This accomplished, the joints are then caulked with cement; in the meantime the other tiers are being fixed ready for cementing.

But in order to render a cast-iron tank of sufficient strength to sustain the pressure of the water, a wrought-iron hoop is necessary for each tier of plates. These hoops are of flat bar iron, and for a tank of 62 feet in diameter a bar 5 inches × $\frac{1}{2}$ inch will be necessary, put together with jaws and tightened with cotters, each plate being further secured by a wooden block firmly wedged between it and the hoop. To show the importance of these hoops, we may mention that on one occasion a small tank of this kind came under our notice, in which the engineer, supposing the vessel to be sufficiently strong without them, dispensed with their use; hardly, however, had three feet of water been pumped into the vessel, when it began to leak on all sides, but on placing the necessary hoops, properly wedged, it became as sound as could be desired. It must be stated that in this case the flanges were placed within the tank, and naturally the strength of the vertical joints was weakened materially.

With cast-iron tanks the columns or guide standards of the holder are generally built on piers of brickwork detached from the tank, and having suitable foundations, but in small tanks of 30 feet diameter or less, brackets are sometimes cast at the desired points on to the plates of the upper tier of the tank, and on these brackets are fixed the columns.

Similar to the construction of those of brick and other material, the thickness of the plates of an iron tank is augmented in proportion to the depth and diameter of the vessel. Thus in a tank of 36 feet 6 inches diameter and 12 feet deep, the bottom central plates will be required $\frac{5}{8}$ inch, and the outer ring $1\frac{1}{4}$ inch thick; the two lower tiers of plates will be $1\frac{1}{8}$ inch, and the upper tiers $\frac{1}{2}$ inch thick, and the hoops will be required $3\frac{1}{2}$ inches wide and $\frac{1}{2}$ inch thick, with their corresponding jaws and cotters; whilst a tank of 71 feet 6 inches diameter and 22 feet deep will require the bottom plates $1\frac{1}{4}$ inch thick, and the two lower tiers of the same strength, the third tier being 1 inch, the fourth $\frac{7}{8}$ inch, and the fifth $\frac{3}{4}$ inch thick. One hoop, 5 inches wide and $\frac{3}{4}$ thick, drawn up by means of cotters and wooden wedge blocks, will be necessary for each tier of plates.

The following specification will more clearly describe the manner of constructing the vessels under consideration.

"*Specification of a Cast-iron Tank for a Gasholder 41 feet in diameter and 12 feet deep.*

"The tank to be 41 feet in diameter and 12 foot deep to top of internal flanges on the bottom plates, internal measure, and to be constructed as follows:—

"The bottom to be formed of four rings of plates, with a centre plate 3 feet in diameter by $\frac{3}{4}$ inch thick. The first row next to the centre plate to be 4 feet 4 inches long by 3 feet 1 inch wide at the outer end, by 10 inches at the inner end, and $1\frac{1}{4}$ inch thick of metal. The second row to be 4 feet $10\frac{1}{2}$ inches long, by 2 feet 11 inches wide at the outer end, by 1 foot 7 inches wide at the inner end, and $1\frac{1}{8}$ inch thick of metal. The third row to be 4 feet $10\frac{1}{2}$ inches long, by 3 feet wide at the outer end, by 2 feet $\frac{1}{2}$ inch at the inner end, and $1\frac{1}{8}$ inch thick of metal; and the fourth row to be 3 feet wide at the outer end by 2 feet 3 inches wide at the inner end, and $\frac{3}{4}$ inch thick, and to be 4 feet $10\frac{1}{2}$ inches long from the inner end to the inside of flange on the outer end; this outer flange to project 3 inches above the plate and to be curved so as to coincide with the radius of the tank. On the top of this flange is to be formed the flange for bolting to the bottom ring of side plates, thus:

"The whole of the bottom plates to have flanges $\frac{5}{8}$ inch thick cast on them, and projecting 3 inches above their upper surface, and provided with $\frac{3}{4}$-inch square holes for bolts placed 6 inches apart from centre to centre, also brackets $\frac{5}{8}$ inch thick between the bolt holes. Each flange to be set back $\frac{7}{16}$ inch from the edge of the plate so as to leave a space of $\frac{7}{8}$ inch for jointing.

"The plates to be bolted together by bolts $\frac{5}{8}$ inch thick, square under the head, with strong hexagonal heads and nuts, and provided with $1\frac{1}{8}$-inch washers, the joints to be made with ordinary iron cement, well caulked in from the inside.

"The sides of the tank to be formed of three tiers of plates, each tier consisting of forty-three plates 4 feet high and 3 feet wide, and truly curved on the inner side to a radius of 20 feet 6 inches.

"The bottom tier to be $\frac{3}{4}$ inch thick, the middle tier $1\frac{1}{4}$ thick, and the top tier $\frac{5}{8}$ inch thick. Each plate to have a flange $\frac{5}{8}$ inch thick, cast on each of its four sides, and projecting 3 inches beyond its outer surface, with 1¼ square holes for bolts placed 6 inches apart from centre to centre, and intervening brackets $\frac{5}{8}$ inch thick between all the bolt holes. The outer extremity of these flanges to have chipping pieces cast on them $\frac{1}{10}$ inch thick the thickness of the plate, so as to leave a space of $\frac{3}{4}$ inch for jointing.

"These, like the bottom plates, to be bolted together by bolts $\frac{5}{8}$ inch thick square under the heads, with strong hexagon heads and nuts, and provided with 1¼ washers. The joints to be made with ordinary cement, well caulked from the outside.

"The bolt holes must be cast on the flanges in such a manner as to coincide exactly with each other, without chipping or enlarging when the plates are fitted in their respective places. On the outside of the bottom and middle tier of the side plates, and at a height of 2 feet above their lower edges, a rib or flange is to be cast 3 inches wide and $\frac{5}{8}$ inch thick, with brackets on each side $\frac{5}{8}$ inch thick and 6 inches apart. These ribs are intended to form a continuous and level bearing around the outside of tank for the wrought-iron hoops. The vertical flanges at the points where these ribs intersect them to have projections cast thereon, for the purpose of supporting the wrought-iron hoops.

"These hoops or bolts to be three in number, one 4 inches broad and $\frac{3}{4}$ inch thick, for the bottom tier of the plates, and the others 3 inches broad and $\frac{5}{8}$ inch thick for the remaining tiers. Each bolt to be provided with three joints or couplings, in no part of less strength than the body of the bolt, and with cotters of the same thickness. Each section of these bolts to be in two pieces, which are to be connected by a fish-plate 18 inches long, and of the width and thickness of the respective bolts, and riveted thereto by $\frac{5}{8}$-inch rivets, at 2 inches apart from each other.

"In the centre of one of the plates, forming the centre row of bottom plates, a hole 5 inches in diameter to be cast for the inlets, and in the centre of the adjoining plate a hole 6 inches in diameter for the outlet pipe. Each plate for a space of 3 inches round the hole to be made $\frac{3}{4}$ inch thicker than the rest of the plate.

"The whole of the plates to be fitted together and plainly marked for fixing in their respective places, and to be painted with a coat of metallic oxide paint, previous to delivery. The whole of the material herein specified to constitute a tank, which, when erected, shall be truly cylindrical and of the dimensions given."

The total weight of the tank thus specified is 57 tons 7 cwt., of which about 52 tons are cast iron; the rest, consisting of the hoops, bolts, and nuts, are of wrought iron. On supposing, therefore, the former to be 8*l*. and the latter 18*l*. per ton, the price delivered would be 512*l*. To this must be added carriage, say 20s. per ton, and, estimating the expenses of preparing the ground and the erection of the tank to be 60*l*., a tank of this kind under ordinary circumstances would cost about 630*l*. To this, again, must be added the expense of foundation and piers, which would make a total of 710*l*., or nearly 45*l*. per 1000 feet capacity of tank, or about three times the cost of one of similar dimensions in brick or masonry.

The preceding specification describes the usual manner of constructing cast-iron tanks. In some instances the plates are strengthened by diagonal ribs, cast therewith, and entering into their formation; and, in our opinion, all plates, whether for the bottom or sides, but more particularly the former, should be provided with these ribs, which usually extend from the four corners to the centre of the plate, being at that point 2 or 3 inches deep and tapering to the angles, and of a thickness of $\frac{3}{4}$ to 1 inch, according to the dimensions of the vessel constructed.

Generally the outer part of the plates forming the wall of the tank is ornamented with beadings, representing panels, as shown in the Plate hereafter referred to ; but where economy is necessary, ornamentation should yield to utility, by the adoption of diagonal ribs, whereby, with a given quantity of metal, the greatest strength is obtained.

As tanks increase in diameter and height, the same area of plates as those mentioned is approximatively retained, the augmented magnitude of the vessel being obtained by increasing the number of rings forming the base, as well as the tiers of plates composing the side. The thickness of these, as well as that of the flanges and hoops, are all augmented according to the dimensions of the tank of which they enter into the formation. In like manner, as tanks diminish in magnitude below that specified, all the respective parts are made proportionately lighter. Thus, in a tank of 21 feet diameter and 10 feet depth, suitable for a holder of 20 feet diameter and the same depth as the tank, of a capacity of about 3000 cubic feet, the plates will be half an inch thick.

CAST-IRON TANKS.

The following details of the weight of two gasholder tanks have been furnished by an extensive manufacturer of gas apparatus.

A cast-iron tank, 23 feet in diameter and 10 feet 3 inches deep, together with the brackets on which the guide standards or columns are fixed, weighs 18 tons. To this must be added the expense of columns and girders, which latter in holders of this capacity are merely T-iron; together with the transport and erection, in order to ascertain its total cost complete. The plates forming this vessel are $\frac{1}{2}$ inch thick at the inner rings or upper tiers, but increased to $\frac{11}{16}$ths at the outer circle and lower tier.

A similar tank to the foregoing, but 31 feet in diameter and 10 feet 3 inches deep, suitable for a holder 30 feet diameter and 10 feet deep, weighed 29 tons 5 cwt. 3 qr.

The following are the details of the weights of the two kinds of tanks referred to:

TANK, 23 feet by 10 feet 3 inches.

	Tons	Cwt.	Qr.
24 Side plates $\frac{9}{16}$ths thick, 362 feet	3	11	0
24 " $\frac{11}{16}$ths " 379 "	4	8	0
14 Bottom plates $\frac{5}{8}$ths " 120 "	1	6	0
26 " $\frac{11}{16}$ths " 296 "	3	8	3
Flanges 3 inches by $\frac{5}{8}$ths 333 "	3	11	2
Brackets .. .. 35 "	0	8	0
1260 $\frac{5}{8}$th bolts	0	6	0
Cement	0	15	2
2 Hoops	0	5	1
Total weight	18	0	0

TANK, 31 feet by 10 feet 3 inches.

	Tons	Cwt.	Qr.
33 Side plates $\frac{5}{8}$ths thick, 512 feet	5	10	0
33 " $\frac{3}{4}$ " 487 "	6	2	0
34 Bottom plates $\frac{5}{8}$ths " 352 "	3	15	2
34 " $\frac{11}{16}$ths " 403 "	4	13	2
Flanges 3 inches by $\frac{5}{8}$ths	4	15	2
Brackets	0	15	2
Ribs	1	13	0
Bolts and nuts, 1956	0	8	3
Cement	1	4	0
Hoops	0	8	0
Total weight	29	5	3

Both these tanks are remarkable for the strength of their plates, but are given as constructed. We have seen a tank 24 feet in diameter, of which no plates exceeded $\frac{1}{2}$ inch thick, and have erected another of 41 feet 6 inches diameter, the stoutest plates of which did not exceed $\frac{5}{8}$ inch thick. It should be borne in mind, however, that whenever the thinner plates are employed for the bottom of the tank, the greater caution is necessary, so that they may be well supported beneath, particularly when piling is employed. The above details also furnish the method of calculating the weight of the vessels in question.

Another tank, without pipes or columns, of 71 feet 6 inches in diameter and 22 feet deep, weighs about 195 tons; a still larger tank, of 102 feet in diameter and 24 feet deep, complete with columns, girders, and pipes, weighs about 345 tons, and, erected, cost approximately 4500*l*. From these data the reader is enabled to form an opinion of the cost of such vessels, which will be found to average from 30*l*. to 40*l*. per 1000 feet capacity of tank, which, as already stated, is considerably more than the average cost of tanks in brickwork or masonry; therefore only imperative necessity can induce the engineer to make choice of these expensive structures.

Plate 12 represents a gasholder of 84 feet in diameter, and 22 feet deep, in its corresponding cast-iron tank of 87 feet in diameter, and of a like depth as the holder, one half being drawn in elevation, and the other half in section.

The tank, as shown, is of greater diameter than is necessary for a single holder, it being so constructed for the purpose of converting the apparatus to the telescopic system, whenever that may be desirable by the extended business of the company, which precaution has been sometimes observed by companies of limited capacity, when desirous of restricting their expenditure to their more immediate wants, at the same time making provision for future increased consumption.

For such a purpose the tank is increased in diameter, and the capitals of the columns so arranged to receive the upper columns at a future period. Beyond this no extra expense is incurred at the time of erection, whereas the capacity of the holder may be doubled at any future period, whenever it may become necessary. Although this plan is frequently adopted, the economy derived therefrom must often be questionable, inasmuch as the expense of labour and inconvenience of emptying the tank, the extra expense incurred by cupping the holder, and the erection of the outer holder, under disadvantage, all increase the cost materially beyond that of making a telescopic holder in the first instance. According to our opinion, therefore, no advantage can be derived from such an arrangement, unless a very long period intervenes between the first construction and the alteration, in which case the interest of capital would compensate for the extra expense.

The engraving, in elevation, shows the method of constructing cast-iron tanks, and, as represented, there are five tiers of plates forming the sides, with the corrresponding wrought-iron hoops already referred to. The bottom of the tank is composed of a large centre-piece, which may be of one or more plates, and nine concentric rings of plates, the lowest tier of the side being bolted to a vertical flange, cast on the outer ring of plates forming the bottom. The standards, or guide columns, are cylindrical, the description very generally adopted, whether for large or small gasholders. These columns, when of great length, are cast in two or more pieces, having an internal collar at one end, turned at the points of junction, and secured together with cotters; as in the present case, where the ultimate intention is to render the holder telescopic, the standards are arranged to receive another tier of columns to be superposed thereon, with their corresponding girders, chairs for pulleys, and counterbalance weights, should the holder be of limited diameter.

The girders of holders are of varied designs, some exceedingly ornamental, and frequently constructed of cast iron. Those represented are of wrought iron, which possesses the advantage of freedom from breakage in transport, and more particularly in shipment.

The holder, represented partly in section, in addition to the roof trussing, is supported by trussed stays at the sides. One of these stays is placed opposite to each column, and consists of a 5-inch T-iron "upright," bolted to the bottom and top curbs, with a strut and tie-rods as shown, the strut being tightened by the two screws. Frequently, in holders of 50 feet diameter, the stay is simply an angle iron, whilst in those of smaller dimensions it is dispensed with. This stay, however, like that of the roof trussing, is only necessary to sustain the sides when the holder is down, and deprived of the support of the gas within; but it may be stated that the weight of the sides, in consequence of the increased pressure thereby given when in action, aids in supporting the roof, when this is untrussed.

The drawings are represented as the holder was actually executed, and in this we find a deviation from the established rule, by attaching the tension rods to the main bars, instead of to the crown plate, showing another instance of the diversity of opinion existing amongst manufacturers of these apparatus.

The other engraving represents the half plan of the holder, and showing a portion of the roof covered with the sheets; the other portion showing the trussing of roof, which is similar to that described, but having, in addition to the main secondary bars, others which extend from the third ring of purlins to the curb.

The principal parts of the holder in question are as follows:

The crown plate, 7 feet in diameter, and $\frac{1}{4}$ inch thick.

The top and bottom curbs, 5-inch by $\frac{1}{2}$-inch angle iron, strengthened by the outer row of plates or roof, 3 feet 9 inches wide, of No. 7 gauge.

The top and bottom tiers of plates, as well as those riveted to centre plate, are of No. 8 gauge. All the rest of the plates are No. 12 gauge.

The main bars are 5 inches by 3 inches by $\frac{1}{2}$-inch T-iron.

The secondary bars, $3\frac{1}{4}$ inches by $2\frac{1}{4}$ inches by $\frac{3}{8}$-inch T-iron. All the purlins are of angle iron.

The outer ring and fifth ring of these being 3 inches by $2\frac{1}{4}$ inches by $\frac{3}{8}$ inch.

The second and sixth rings are 3 inches by 2 inches by $\frac{3}{8}$ inch.

The third ring being 4 inches by $2\frac{1}{4}$ inches by $\frac{3}{8}$ inch, and the sixth row, 2 inches by 2 inches by $\frac{1}{4}$ inch.

The centre pipe or king-post is 12 inches in diameter and 20 feet long; and the truss cup, 5 feet in diameter at the bottom.

The suspension rods are $1\frac{1}{2}$-inch round iron.

The inlet and outlet pipes, 18 inches in diameter.

The estimated price for this gasholder and tank, delivered at railway station, was 4650*l.*

At the present time the progress of gas companies from year to year may be regarded as definite, although variable according to the magnitude or description of towns, their kinds of industry, or other causes; and this on the average may be estimated at from 8 to 10 per cent., following progressively year after year, and in some cases this is exceeded. With this datum to proceed upon, the requirements of a gasworks, and, consequently, the proper dimensions of the gasholders, may be readily estimated for ten or twenty years to come.

A difficulty attending the use of iron tanks in cold climates, when they project above the level of the ground, is the facility with which the water within them freezes. The same observation would apparently apply to the luting of telescopic gasholders; but whether arising from the heat combined with the gas, when entering the holder, neutralizing the effects of the frost, or from other causes, it is certain that the water forming the luting of holders is not so susceptible of freezing as the surface of the water in the tank. The effect of frost, in such cases, if disregarded, would result in the holder becoming fixed, and thus, deprived of its pressure, the district supplied by it would be put in darkness. In most works, however, there is a portable steam boiler employed; and, to counteract the evil alluded to, a pipe should be taken from this to the tank, to supply sufficient steam to keep the temperature of the water above the freezing point. In some works fires are suspended to the tank, and sometimes salt is placed in the lute cup to accomplish this object; and in many parts of Russia, on account of the intense cold, the gasholders are enclosed within buildings constructed for the purpose of preventing the water-lutes freezing, and the mains are laid at an unusual depth, in order to avoid the action of the severity of the weather on the gas.

Annular Tanks.

Another kind of tank which has sometimes been constructed of brickwork, is the annular tank. This is formed by two circular walls, the one being built concentric to the other, the annular space between them forming the tank. One of these vessels, of very large dimensions for the period, was erected in Paris some years ago by Mr. John Manby; but, as the writer was informed by that gentleman, the expense and trouble experienced were far beyond his expectations, and sufficient to induce him to renounce any repetition of such a system of construction. Other engineers have also built annular brick tanks at various times; we believe, however, that no one has ever entertained this class of structure a second time; we may assume, therefore, that it is not accompanied with any advantages.

Annular tanks have been constructed, with more or less success, of cast and wrought iron; of the first kind, that formerly existing at the works of the late City of London Company's works merits particular mention. This tank was erected under very peculiar circumstances, and attended with considerable difficulties.

The business of the company having increased in a remarkable manner, and the site of their premises being very limited, it was desirable to build the largest gasholder possible in the confined locality allotted for that purpose. Under these conditions, a cast-iron tank was decided on by Mr. Mann, the engineer of the company, as presenting the advantage of being built of much larger capacity than a brick tank, on account of the thickness of the walls of the latter, thus profiting to the utmost of the limited space at his disposal for the structure. That gentleman intended at first to have erected a double-lift telescope holder, rising to a height of 98 feet, and of course requiring a tank of 50 feet in depth; but on sinking a well in order to ascertain the nature of the ground, it was found that a greater depth than 33 feet was impracticable. This being known, and as large capacity was of paramount importance, Mr. Mann then resolved on building the holder with three lifts, instead of two, so obtaining the desired storage; and we may assume that, for the facility of erection, as the site was closely surrounded by dwellings, an annular tank was determined on.

This was accordingly constructed, its outer cylinder being 84 feet and the inner cylinder 77 feet in diameter, both 33 feet deep, thus forming an annular tank of 3 feet 6 inches wide and of the depth stated. The foundation of concrete being laid, the lower plates were then placed, all the flanges of those as well as the wall of the tank were inside, by which means the excavation was limited in dimensions, a larger area obtained, and greater facilities were afforded for the erection.

However, the great merit did not consist alone in erecting this tank under the various difficulties which attended it, but in the construction of a three-lift holder, which rose to a height of 98 feet, whilst its mean diameter was 80 feet; and when we remember that telescopic holders are often built which only rise to a height equal to one-fourth of the diameter, and seldom exceed one-half of that, the extraordinary height of the vessel under consideration, as compared with its diameter, can be well imagined. This holder was in operation during several years, and was in every sense of the term successful, and to those of the gas profession was one of the most interesting apparatus of the kind ever constructed. It is only fair to state, that although the design is due to Mr. Mann, it was executed by the firm of Westwood and Wright.

This apparatus has been recently removed, in consequence of the station being no longer required, through the amalgamation of the City with the Gas Light Company.

There are, however, certain localities, where a tank is shallow as compared with its diameter, in which the annular form shows some degree of economy; and to illustrate this we will give as an instance one of 41 feet in diameter by 12 feet deep. The area of the sides and bottom of an ordinary tank of this size would be 2900 square feet, whilst the area of an annular one would be 3300 square feet, showing a difference in favour of a tank as ordinarily made of about 14 per cent. in amount of surface. But in a tank of 127 feet in diameter by 20 feet deep, the annular form has the advantage. Thus the sides and bottom of an ordinary tank contain 22,000 square feet, those of an annular tank contain 17,300 square feet, leaving 4700 square feet in favour of the latter, or 21 per cent. less surface than in the ordinary mode of construction.

As a rule, it may be stated, that it is only when the diameter of a tank is not less than four times its depth, that the annular system will bear comparison, as regards cost, with the other, or ordinary form.

Wrought-iron tanks, made of $\frac{3}{8}$-inch or $\frac{1}{2}$-inch plates riveted together similar to an ordinary boiler, have come under our notice at some small works; but their great expense has been an impediment to their more general use.

The application of a wrought-iron annular tank has been highly successful at the works of the Surrey Consumers' Company at Rotherhithe. This class of structure was not, however, one of choice, but of necessity, occasioned by the very unstable nature of the ground whereon it is erected.

The site of these works immediately adjoins the river Thames, the ground being a loose sand, through which the water percolates with all facility at a depth of 12 feet from the surface, which rendered the construction of a brick tank of the dimensions required out of all question, since no contractor under any conditions would undertake such a structure. The question then arose of erecting a cast-iron tank; but this, made in the ordinary manner, would necessarily involve great expense and difficulty, on account of the large area of the operations and the impossibility of clearing the water, so as to enable the men to work. This being obvious, Mr. Finlay, the engineer of the company, decided on building an annular tank, the foundations of which, as well as the erection, could be made in portions or segments, thus presenting a comparatively small area for operation, and permitting of the water being pumped from the excavation and enabling the work to be carried out with facility. The expense of wrought against cast iron, together with the unstable nature of the ground, and the flexibility of the former in the event of any settlement, all of these points being duly considered, Mr. Finlay decided on constructing the tank of wrought iron, and, as stated, annular.

The outer cylinder of this is 153 feet in diameter and 25 feet deep, the inner cylinder being 142 feet diameter and 14 feet deep, which is surmounted by the frustum of a cone hereafter mentioned of 11 feet 6 inches in sectional height, thus forming an annular tank 5 feet 6 inches in width; the tank extending about 11 feet above the surface of the ground.

The foundation is annular, of a width of 7 feet 6 inches, and 5 feet deep, the base of this being about 7 feet below the ordinary water level existing in the ground; the foundations of the eighteen brick piers were made of greater width.

The bottom of the tank is formed of wrought-iron plates $\frac{3}{8}$ inch thick, which, like the whole of the tank throughout, is put together with butt joints and lapping pieces, all riveted with double rows of rivets. The angle-iron curbs for attaching the sides to the bottom are 5 inches × 5 inches × $\frac{3}{4}$ inch, also butt jointed and double riveted.

The sides are formed by seven tiers of plates, the lowest being $\frac{5}{8}$ inch thick, and diminishing at the top to $\frac{7}{16}$ inch thick, the top of both cylinders being terminated by angle iron 5 inches × 5 inches × $\frac{3}{4}$ inch. Each of these cylinders is further strengthened by five curbs of angle iron of the same dimensions placed equidistant in their height.

The eighteen piers are carried up in brickwork from the annular foundation to 6 feet above the ground level, on each of which is placed an ornamental pedestal, with its respective column.

The centre of the tank is formed like the frustum of a cone, puddled throughout, the base of which commences 18 inches below the top of the inner cylinder, its summit being about 6 inches above the level of the water within the tank, which puddle alone retains the water.

The outer cylinder of the holder within the tank is 150 feet in diameter and 25 feet deep. The inner holder is 147 feet in diameter, and 25 feet 6 inches deep at the sides, the dome rising 6 feet 6 inches. The plates entering into the formation of the holder vary in thickness from No. 5 to No. 10 Birmingham wire gauge.

The holder is guided by eighteen cast-iron standards, with two tiers of ornamental girders and thirty-six rollers within the tank, the whole being of elegant design and admirably finished; and we may observe with all confidence that there are very few specimens of ironwork of this description and magnitude which can be compared with this wrought-iron annular tank. We may further add, that during several years that they have been in operation, both holder and tank have given the most complete satisfaction.

The weight of this annular tank, together with its wrought-iron inlet and outlet pipes, is about 400 tons, and cost 9200*l.*, or less than 20*l.* per 1000 feet capacity, to which, however, must be added the expenses of the foundation, excavation, and puddle. The tank and holder were constructed by Messrs. Thomas Pigott and Co., of Birmingham.

Compound Tanks.

At a period in the annals of gas lighting, not very remote, but few engaged in that art were disposed to depart from the beaten track, or to adopt methods whether in construction, manufacture, or distribution, other than those previously employed. As examples of this, we may state that tanks of holders were built, during several years, perfectly level in the centre; and although there could be no advantage gained by this, still the system was universally adopted. Again, with the settings of retorts, for a considerable period no general attempts at improvements were made, and it was sufficient that a plan was employed at a particular establishment, or by a certain engineer, for the majority of superintendents of gasworks, as then termed, to entertain the same system. But a more remarkable proof of this spirit of conservatism was, that although the lighting rod for the public lights had been employed on the Continent almost from the first establishment of gas, yet the use of the ladder for lighting and extinguishing was obstinately continued in the United Kingdom until within the last few years.

In this respect, however, a marked change has taken place, for at the present day engineers now think, judge, and act for themselves; and although at times, by this exercise of individual judgment, errors may be committed, yet, as in all other matters, the progress of gas lighting must be advanced by this combination of intelligence.

These observations are considered necessary on introducing a system of construction which has hitherto been but of limited application; but, as hereafter shown, the compound tank has been most successfully applied by an engineer of eminence, thus offering an inducement, under certain conditions, for its more general adoption; and impressed as we are that the method in question must in certain cases be alike economical and present great facilities in erection, we enter on its description.

The compound tank consists of an external wall of cast iron, similar to that of ordinary cast-iron tanks, the internal area being either concrete, brickwork, masonry, or puddle. The desirability of such a structure will depend on various circumstances, such as the nature of the soil where erected, the cost of building materials, the facility of obtaining the requisite skilled labour for the construction of brick or stone tanks, and other contingencies already enumerated: but whenever it may be requisite to build a cast-iron tank, and supposing the ground to possess sufficient solidity for the purpose, the compound tank may, in our opinion, be adopted with great advantage; but to ensure success, proper care must be observed in its formation, and more particularly as relates to the junction of the iron with the cement or other material employed.

A tank of the description in question, of 100 feet diameter and 20 feet deep, was constructed by Mr. G. W. Stevenson, at Halifax, some years ago, and was built of four tiers of plates, the lowest of which was bolted to a flanged plate of a width of 2 feet, extending around and projecting within the circumference of the tank. The foundation consisted of a circle of about 2 feet 6 inches wide, formed of dressed stones properly levelled, and in which was cut a circular channel 96 feet in diameter, of a depth and width of 3 inches, for the purpose of receiving a rib, cast on, and forming part of, the flanged plate referred to. Preparatory to laying each plate on the foundation, a bed of Portland cement was placed over the surface of the foundation stone as well as in the channel, and on this the plates were laid. The circle of the flange plates being completed and properly bolted and cemented, the first tier of the side plates was fixed, after which the others were erected in the usual way, each of the three lower tiers being bound with a 6-inch × $\frac{3}{8}$-inch wrought-iron hoop. The centre of the tank was simply puddled, which, as in the instances referred to at the Pancras station, as also in the annular tank at the Rotherhithe works, is the only means adopted to retain the water in the central area, and, as thus constructed, the tank was perfectly successful.

It should be observed, that the vessel thus built is embedded in the earth to the greater part of its height, which, however, is not material to the point in view, namely, the question of combining iron with cement or puddle in such a manner as to ensure a sound joint and prevent the water escaping, and this, as stated in the tank referred to, has been accomplished. Hence, with this example before us, it is obvious that the system is capable of extended application in some instances with marked economy, and in other cases offering considerable facilities for construction; and if we reflect on the limited thickness of puddle requisite to retain water, even under heavy pressure, when this presses upon the puddle, and by reason of the clay being forced into the interstices, and preventing leakage, the facilities for the class of construction in question are evident.

In order to appreciate more fully the value of this method of building holder tanks, let us assume that in a certain locality it is required to build a tank, the whole or the greater portion of which shall project above the level of the ground, which may be occasioned by any of the various circumstances already referred to. Now as the average areas of the interior of tanks are equal to the surface of their walls, and as the plates for the former are thicker than the latter, it follows that the centre contains a greater weight of metal than the walls; and when we come to tanks of great diameter and shallow depth, this variation is very important, as we find in a tank of 150 feet in diameter and 22 feet deep, that the surface of its central area is about 70 per cent. more extensive than the walls. Hence, under these conditions, it becomes simply a question of the price of iron against the other building material which may be employed to replace it; and if we suppose this to be Portland cement concrete, adapted to a tank of large dimensions, say 150 feet in diameter, it would have a thickness of 15 inches, and when rendered with pure cement, would cost about 9s. per square yard. On the other hand, the weight of a square yard of plate for a similar description of tank would be about $3\frac{1}{2}$ cwt., costing about six times the price of the concrete. Thus under these circumstances, by the adoption of the compound tank of the diameter stated, instead of forming the whole of cast iron, an economy will be effected under average conditions of upwards of 1700*l.*; and in addition to this there is the other consideration, that the whole of the concrete work could be executed by ordinary labourers, which, in certain localities abroad, where skilled labour is difficult to obtain, is a further recommendation.

Another site still more economical for the tank under consideration, is that composed of plastic clay, and capable of retaining water, such as we have had occasion to refer to. In such a locality, all that is requisite is to sink the excavation to the desired depth, and render it perfectly level throughout the circumference where the plates are to be laid; then a thin layer of puddle of one or two inches thick is placed thereon, and the plates, previously wetted and smeared with clay, are laid in their positions; which effected, it will be necessary to carefully puddle at the point of junction where the clay and iron meet. This kind of tank, so far as our knowledge extends, has never been carried into execution; it must, therefore, be regarded as completely theoretical; but in our opinion, if carried out it would be numbered among the progressive achievements of gas lighting; and as we commenced by observing the deviation that has been made in recent years from old established methods, in like manner our suggestions may lead to the practical adoption of the principle referred to. And, with the example of its application mentioned in view, whenever this may be accomplished, it will be infinitely less hazardous than was the construction of the Imperial Company's tank described in page 175, or the concrete tank of Mr. Livesey, which was designed in direct opposition to all previous practice, at least so far as regards the exclusive employment of concrete for that purpose. Yet both of these, like the compound tank, have been successful, resulting in considerable economy; and by the introduction of these innovations, and by the establishment of these several methods of forming tanks according to the nature of the site or the material at command, the engineer's art is materially advanced.

GASHOLDERS.

As represented in the illustration of the gasworks erected on the premises of Mr. Akermann, page 21, the earliest gasholders were rectangular, in this respect repeating the form of the vessel as first invented by the illustrious Lavoisier. This description seems to have been retained until about 1816, when the cylindrical shape began to be adopted; and, as observed, to John Malam is undoubtedly due the merit of improving these apparatus, by dispensing with the massive wood and iron framework which had hitherto been regarded as essential to their construction, as we find in Peckston's work a description of gasholders so constructed by Malam in the simplest manner possible, guided with merely a central pipe and tube, some of which holders made in the earliest years of gas lighting are actually in use at the present day.

On the other hand, Clegg seems to have retained much of the strength of the primitive holders; for, according to the specifications for the formation of these vessels given in his son's 'Treatise,' the trussing and diagonals are described as equal in weight to the whole of the rest of the holder put together, but, as now usually made, the trussing forms but a fractional part of the vessel, and in some cases it is altogether dispensed with.

If we may judge from the designs of various engineers, the best method of constructing gasholders, combining efficiency with economy, appears to be a matter of uncertainty, as these vessels are erected by some of our ablest men without any trussing whatever to the roof; whilst, on the other hand, there are other gentlemen equally eminent who adopt the very opposite plan; therefore, with these contending views, we will confine our observations principally to the description of the various kinds of holders, venturing on suggestions as we proceed.

For the purpose of inquiring into the necessity for the trussing of the roof of a holder, we must first consider the different circumstances which influence that vessel. Thus, when in action, it contains gas having a resisting force or pressure somewhat greater than the atmosphere. Then we may suppose the plates forming the roof to weigh 5 lbs. per square foot, and the pressure given to be equal to 4 inches, or $20 \cdot 83$ lbs. on the square foot as hereafter observed, occasioning a displacement of water of the same weight as the holder. Under these conditions, with the force of the gas considerably greater than the weight of the top plates, we find that the trussing is practically useless, since in its absence, in the event of the roof being unsupported, it would be raised whenever the holder commenced to rise; and in all trussed roofs, those parts of the plates midway between the curb and the crown plates, are invariably lifted away from the trussing. Therefore, when in action, so far as regards the support to the roof-plates, the trussing of the holder serves no purpose.

Another contingency to which a holder is subjected, is the influence of high winds, which, in the event of weakness in the top curb, might cause it to yield at that point, and its rollers to leave their guides, when it would be a matter of difficulty to restore the vessel to its proper position. This evil is, however, not likely to occur in a trussed roof, as the principals are always placed directly opposite to the guide standards or columns, and thus they support the curb at those points; and, as will be shown, the system adopted by some engineers is to make the principals of the trussing of suitable strength to sustain the curb. Moreover, we find in large gasholders, that, in the absence of the trussing, the curb is invariably required to be increased in strength; it therefore becomes a question whether the metal is best employed in strengthening the whole of the curb, or in forming the principals to support the roof and sustain the curb at the weakest points; and although less metal may be required in the former case, this is not all economy, as, in the absence of trussing, considerable expense is incurred in erecting scaffolding to place the crown plates, besides the permanent props and framing which support the roof when the holder grounds.

TABLES OF WEIGHT OF IRON. 201

We are disposed to believe, however, that the untrussed roof for small holders of 50 feet and under is not only considerably cheaper, but can be easily made of ample strength, as frequently proved by experience, whilst the pressure given by a holder of this kind, with equal thickness of sheet entering into its formation, is of necessity much less than with the other description.

When of small capacity, intended to supply factories, villages, or small towns, holders are usually provided with counterbalance weights suspended by chains at two or more points. By this means the weight, and consequently the pressure given by them, can be readily adjusted to any required degree; and in small works, holders should be so counterbalanced as to give little more than the maximum pressure necessary to supply the district; and this in any town, moderately level, should not exceed 1½ or 2 inches, which will, however, depend upon the size of the mains and the position of the works.

The pressure of a holder is determined by its weight as compared with its area: thus, when of large diameter and shallow depth, with average strength of sheets and trussing (where this is employed), the pressure is limited accordingly; but when the holder is of small diameter and great depth, the pressure is always excessive. In defining the dimensions of single holders, that is, the depth of the sides as compared with their diameters, engineers appear to have no defined rule, as no uniformity exists between the various structures; but the average proportions are, that the height of the holder is about one-third its diameter. With telescopic holders, however, a more definite law appears to be generally observed, or at least with those of large magnitude: in which the height of the two lifts is usually equal to two-fifths the diameter of the holder, and in most cases, alike in single as well as telescopic holders, the rise of the dome is equal to $\frac{1}{70}$th part of the diameter.

When deciding on the construction of a holder of any given dimensions, in order to ascertain its approximative cost, as well as the pressure it will give, it is necessary to know its weight; for this purpose, the area of the top and sides, the thickness of the sheets of which it is formed, and the various parts of the internal trussing being known, it then becomes a matter of simple calculation to determine the weight, and from that the pressure it will give is ascertained.

For this object, and for making other estimates in connection with the apparatus of a gasworks, the following tables of the weight of iron—as sheet, T, angle, round, square, and flat bars—are given. The first two relate to plates or sheets, their thickness being recognized by the Birmingham wire gauge, which is usually adopted for sheet iron, copper, and brass:

TABLE OF THE THICKNESS OF SHEETS, WITH THE WEIGHT PER SUPERFICIAL FOOT, AND THEIR NUMBERS CORRESPONDING WITH THE BIRMINGHAM WIRE GAUGE.

Thickness in Parts of Inch.	Weight per Square Foot in Pounds and Parts.	Thickness corresponding with Number on B. W. G.	Thickness in Parts of Inch.	Weight per Square Foot in Pounds and Parts.	Thickness corresponding with Number on B. W. G.
$\frac{1}{32}$	1·25	22	$\frac{7}{16}$	17·5	—
$\frac{1}{16}$	2·5	16	$\frac{1}{2}$	20·0	—
$\frac{1}{8}$	5·0	11	$\frac{9}{16}$	22·5	—
$\frac{3}{16}$	7·5	7	$\frac{3}{4}$	30·0	—
$\frac{1}{4}$	10·0	4	$\frac{7}{8}$	35·0	—
$\frac{5}{16}$	12·5	1	Inch	40·0	—
$\frac{3}{8}$	15·0	—			

The foregoing table may also be applied for ascertaining the weight of cast iron, either for the tanks of gasholders, boxes of purifiers, scrubbers, cases of station meters, pipes, and other purposes; for the area and thickness being known, the rest is readily calculated, due allowance being made for the difference between the weight of wrought and cast iron, the former being about 6 per cent. heavier than the latter.

A more comprehensive table for all the various sheets used in the construction of gasholders and

2 D

other apparatus, is the following, which gives the weight of a superficial foot of sheet iron in pounds and fractions, with its corresponding number of the Birmingham wire gauge, and its thickness in fractions of an inch; the second gives the weight of angle iron.

WEIGHT OF A SUPERFICIAL FOOT OF SHEET IRON IN POUNDS AND FRACTIONS, WITH CORRESPONDING NUMBER OF BIRMINGHAM WIRE GAUGE.

No. of Birmingham Wire Gauge.	Thickness in Parts of Inch.	Weigh per Square Foot, lbs.	No. of Birmingham Wire Gauge.	Thickness in Parts of Inch.	Weight per Square Foot, lbs.
1	·312	12·5	13	·094	3·75
2	·290	12·0	14	·080	3·13
3	·273	11·0	15	·071	2·82
4	·250	10·0	16	·064	2·50
5	·230	8·74	17	·056	2·18
6	·202	8·12	18	·048	1·86
7	·187	7·50	19	·044	1·70
8	·168	6·86	20	·040	1·54
9	·152	6·24	21	·036	1·41
10	·134	5·62	22	·032	1·25
11	·122	5·00	23	·028	1·12
12	·108	4·38	24	·025	1·00

TABLE OF THE WEIGHT IN POUNDS OF A LINEAL FOOT OF ANGLE IRON OF VARIED SIZES AND THICKNESSES.

Breadth in Inches and Parts.	Thickness at Throat in Inches and Parts.	Thickness in Middle of Web in Inches and Parts.	Weight of 1 Foot.		Breadth in Inches and Parts.	Thickness at Throat in Inches and Parts.	Thickness in Middle of Web in Inches and Parts.	Weight of 1 Foot.	
			lbs.	oz.				lbs.	oz.
1¼ × 1¼	3/16	1/8	2	0	3¼ × 3¼	1¼	1/4	11	12
1¾ × 1¾	1/4	3/16	3	1	3½ × 3½	1¼	5/16	15	0
2 × 2	1/4	3/16	3	0	3¾ × 3¾	1	1/4	14	0
2¼ × 2¼	5/16	1/4	5	10	4 × 2¼	3/4	3/16	7	10
2¼ × 2¼	5/16	1/4	5	8	4 × 3	3/4	3/16	9	0
2½ × 2½	5/16	1/4	6	0	4½ × 3	1	1/4	10	0
2¾ × 2¾	3/8	5/16	6	8	5 × 3	7/8	3/16	11	0
2¾ × 2¾	3/8	5/16	7	0	6 × 3	1	1/4	13	0
3 × 3	3/8	5/16	8	8	6 × 3¼	1	1/4	13	12
3¼ × 3¼	3/8	5/16	8	0					

To apply the two preceding tables, let it be supposed that the weight of a holder, without trussing to the roof, made in the simplest way possible, is required, the diameter of which is 30 feet and its height at the sides 12 feet. Here it may be observed, that the plates of the roof-holders are invariably stouter than the others, on account of the more rapid oxidation in their interior at that part. These we will imagine to be of No. 14 gauge, and those of the sides of No. 16 gauge sheets, and the top and bottom curbs of 2¾ × 2¾ angle iron weighing 7 lbs. per foot run. Then, to allow for the weight of the rivets and laps, together with the increased area caused by the dome of the holder, 10 per cent. may be added to the sides and curbs, and 12 per cent. to the area of the roof, when we have:

Area of roof 790 feet, No. 14 gauge	2472 lbs.
„ sides, 1232 feet, No. 16 gauge	3080
Top and bottom curbs, 206 feet, 7 lbs. per foot run	1442
Then assuming the weight of the 3 top guides and bottom rollers with their carriages	380
The total weight of holder will be ..	7374 lbs.

TABLE OF THE WEIGHT OF A LINEAL FOOT OF T-IRON.

Breadth in Inches and Parts.	Thickness in Inches and Parts.	Weight in Pounds and Ounces per Foot run. lbs.	oz.	Breadth in Inches and Parts.	Thickness in Inches and Parts.	Weight in Pounds and Ounces per Foot run. lbs.	oz.
1¼ × 1¼	3/16	1	11	3½ × 3½	⅝	15	0
1½ × 1½	3/16	2	1	3¼ × 3¾	⅝	14	6
2 × 2	¼	3	3	4 × 2½	⅜	8	0
2¼ × 2¼	5/16	5	6	4 × 3	7/16	9	2
2½ × 2½	⅜	5	10	4½ × 3	½	10	5
2¾ × 2¾	⅜	6	14	5 × 3	½	11	8
3 × 3	7/16	6	8	6 × 3	7/16	13	6
3½ × 3½	7/16	8	4				

TABLE OF THE WEIGHT OF A LINEAL FOOT OF ROUND AND SQUARE BARS IN POUNDS.

Diameter or Side, in Inches or Parts.	Round.	Square.	Diameter or Side, in Inches or Parts.	Round.	Square.
½	·653	0·830	1⅞	9·18	11·08
⅝	1·02	1·30	2	10·44	13·28
¾	1·47	1·87	2⅛	11·80	15·00
⅞	2·00	2·55	2¼	13·23	16·81
1	2·61	3·32	2⅜	14·73	18·74
1⅛	3·31	4·21	2½	16·32	20·80
1¼	4·08	5·20	2⅝	18·00	22·89
1⅜	4·94	6·28	2¾	19·76	25·12
1½	5·88	7·48	2⅞	21·59	27·46
1⅝	6·90	8·78	3	23·52	29·92
1¾	8·00	10·20			

The preceding tables are useful for estimating the weights of principals of roofing for retort houses or coal stores, or for the trussing of gasholders, for lattice girders, for tie and suspension rods, fire bars, hoops for cast-iron tanks, in short for round and square bar iron of all kinds, as employed in the construction of all the various apparatus of a gasworks.

TABLE OF THE WEIGHT OF A LINEAL FOOT OF FLAT BAR IRON IN POUNDS.

Breadth in Inches and Fractions.	Thickness in Parts of Inch.								
	⅛	3/16	¼	⅜	½	⅝	¾	⅞	1 Inch.
½	·21	·31	·42	·63	..	..	..	..	..
¾	·31	·47	·63	·94	1·26	1·57	..	..	..
1	·42	·63	·84	1·26	1·68	2·10	2·52	2·94	..
1¼	·52	·78	1·05	1·57	2·10	2·62	3·15	3·67	4·20
1⅜	·57	·86	1·18	1·73	2·31	2·88	3·46	4·04	4·62
1½	·63	·94	1·26	1·89	2·52	3·15	3·78	4·41	5·04
1¾	·73	1·10	1·47	2·20	2·94	3·67	4·41	5·14	5·87
2	·84	1·26	1·68	2·52	3·36	4·20	5·06	5·88	6·72
2¼	·96	1·41	1·89	2·83	3·78	4·72	5·66	6·61	7·56
2½	1·05	1·57	2·10	3·15	4·20	5·25	6·30	7·35	8·40
2¾	1·15	1·73	2·31	3·46	4·62	5·77	6·93	8·08	9·24
3	1·26	1·89	2·52	3·78	5·04	6·30	7·56	8·82	10·08
3¼	1·36	2·04	2·73	4·09	5·46	6·82	8·19	9·55	10·92
3½	1·47	2·20	2·94	4·41	5·88	7·35	8·82	10·29	11·76
3¾	1·57	2·36	3·15	4·72	6·30	7·87	9·45	11·02	12·60
4	1·68	2·52	3·36	5·04	6·72	8 40	10·08	11·76	13·44
4½	1·89	2·83	3·73	5·67	7·56	9·45	11·34	13·23	15·12
5	2·10	3·15	4·12	6·30	8·40	10·50	12·60	16·70	17·80
6	2·52	3·78	5·04	7·56	10·08	12·60	15·12	17·64	20·16

By the tables combined, taking into consideration the difference remarked between the weight of wrought and cast iron, the weights of tanks, either of wrought or cast iron, suitable for gasholders or for water, or the weight of scrubbers, washers, purifiers, girders, columns, and telescopic holders are readily determined. For that purpose, if we suppose the total area and lengths of the various parts forming the structure to be obtained with the greatest detail, the respective parts are then multiplied by their corresponding weight, when the total will give the gross weight.

TABLES OF PRESSURES.

Gasholders of small dimensions are generally constructed with their sides of No. 16 gauge sheets, and the top or crown of No. 14 or 15 gauge; but as these vessels increase in magnitude, their sheets are increased in thickness, according to circumstances. Where capital is limited, holders of 60 or 70 feet in diameter are often made of the thickness indicated, except in the sheets riveted to the bottom and top angle-iron, which are augmented in strength by one or two numbers of the gauge. The durability of these vessels under average conditions, constructed of No. 16 gauge, may be twenty-five years; whereas by the addition of $\frac{1}{4}$th more to the thickness of the metal, their duration can be fairly estimated at double that period. It is therefore very false economy to construct holders of light material.

The rule for ascertaining approximatively the weight of a single gasholder, when not counterbalanced, of a given area and pressure; or, the area and pressure being known, to find the weight; or, the weight and pressure given, to find the area, is very simple.

For this rule, the weight of a cubic foot of water is the basis, which weighs 62½ lbs., or 1000 ounces; a cubic foot of air weighs 1¼ ounce; consequently, if we disregard the small fraction, it follows that a cubic foot of air possesses sufficient buoyancy to float a weight of 62½ lbs.; and as the weight of water displaced is equal to the weight of the holder displacing it, hence the weight divided by the area gives the column, or pressure; and it is obvious that, two points being known, the third can be readily ascertained. For the application of this rule the following tables are given:

TABLE OF THE PRESSURE PER SQUARE FOOT, CORRESPONDING WITH VARIOUS COLUMNS OF WATER EXPRESSED IN TENTHS OF AN INCH.

		Tenths of Inch.			Lbs. per Square Foot.
Column of water or pressure		½	equal to pressure or weight of		·260
,,	,,	1	,,	,,	·520
,,	,,	2	,,	,,	1·041
,,	,,	3	,,	,,	1·562
,,	,,	4	,,	,,	2·083
,,	,,	5	,,	,,	2·604
,,	,,	6	,,	,,	3·124
,,	,,	7	,,	,,	3·645
,,	,,	8	,,	,,	4·166
,,	,,	9	,,	,,	4·687
,,	,,	10	,,	,,	5·208

TABLE OF THE PRESSURE PER SQUARE FOOT, CORRESPONDING WITH VARIOUS COLUMNS OF WATER EXPRESSED IN INCHES.

		Inches.			Lbs. per Square Foot.
Column of water or pressure		1	equal to pressure or weight of		5·208
,,	,,	2	,,	,,	10·416
,,	,,	3	,,	,,	15·624
,,	,,	4	,,	,,	20·833
,,	,,	5	,,	,,	26·040
,,	,,	6	,,	,,	31·248
,,	,,	9	,,	,,	46·872
,,	,,	12	,,	,,	62·500

The first table is given more especially for calculating instruments of precision as governors, and pressure registers for future reference, whilst the two combined give any ordinary pressure. To apply the rule in question, let us suppose the weight of a holder to be required, the diameter of which is 50 feet and the pressure given by it to be 4 inches. Then the area of this will be 1963·5 feet; and on referring to the preceding table, we find the weight of each square foot corresponding with the pressure mentioned to be 20·833 lbs., when the area multiplied by the weight per foot gives 40,904 lbs. as the weight of the holder.

Reversing the proposition, on having the weight of the holder and its diameter or area, to find the pressure, then on applying the same figures, the weight 40,904 lbs. divided by the area 1963·5 gives 20,8332 lbs., which on reference to table corresponds with 4 inches as the pressure required.

Or, the weight and pressure being known, the area of the holder may be ascertained by dividing its weight by the column of the pressure. Thus 40,904 lbs. divided by 20·8332 lbs., the weight of a column of water 4 inches high, gives 1963·5 feet as the area of the top of the holder. This rule, which is of the greatest simplicity, is also useful for constructing experimental gasholders, governors, regulators, and for ascertaining the weight of counterbalance for holders, &c.

When ascertaining the approximative weight of gasholders by this method, the ascending power of the gas must be taken into consideration, and this may be averaged at 40 lbs. for each 1000 feet of gas in the vessel; hence, if this be 50 feet in diameter and 16 feet deep, containing 30,000 feet, it would act with an ascending power of 1200 lbs., which would have to be added to the weight.

The pressure of a holder varies slightly according to its position, or the degree to which the sides are immersed in the tank, in consequence of the increase of the weight of the iron when out of the water; this, with single holders of ordinary construction, is insignificant, and practically need not be taken into consideration.

The accompanying engraving, Fig. 58, represents a holder, 45 feet in diameter and 15 feet deep at the sides, partly in section and partly in elevation, with its corresponding brick tank and dry well. The tank is shown, as built without puddle, as already described, the centre being domed, having a layer of concrete, which is afterwards rendered with Portland cement. In the centre is built a brick pier, and on this reposes the truss cup when the holder grounds, by which means the strength of the roof trussing may be reduced to a very considerable extent. The form of the centre of the tank, although somewhat more costly than the flat bottom, is decidedly stronger, and should the ground be hard, the cost of building will be compensated by the saving effected in the excavation, and whenever rock is met with, the mound in the centre can be materially increased. The frustum of a cone is more generally adopted than the dome for the centre of a tank; but as the latter has no angle, we believe that greater facilities are offered in its formation and the layer of concrete is more uniform.

The trussing is represented in elevation by the main bar formed to the curvature of the dome, one end of which is attached to the top curb, and the other to the crown plate in the centre of the holder. The tension rod is bolted to the curb and the truss cup, where there are regulating screws by which the crown can be raised to the necessary height. The suspension rods are sometimes bolted to the crown plate, at others to the main bar, and the struts and bracket bars hereafter mentioned complete the trussing. Fig. 59 is a half plan of the same holder, with a portion of its plates removed in order to show the parts: $a\,a$ is the curb, $b\,b$ two of the main bars, c one of the secondary bars, and $d\,d\,d$ the bracket bars.

With these general observations, we proceed to give a specification of a small holder, suitable for a factory or other similar establishment, or village, or small town.

"*Specification of a Gasholder 45 feet diameter and 15 feet deep at the sides.*

"*Curbs.* The two curbs top and bottom to be of angle iron, 3 inches by 3 inches by $\frac{3}{8}$ inch, curved to a radius of 22 feet 6 inches, to be butted at the joints with proper lapping pieces, securely riveted with $\frac{7}{16}$-inch rivets.

"*Sheets.* The sheets employed to be of best Staffordshire quality, all properly flatted before being punched. The sheets of roof and the top and bottom rows of the sides to be of the thickness of No. 14 Birmingham wire gauge, all the others to be No. 16 B. W. G. The sheets of the sides not to exceed 2 feet in width and 4 feet in length, from centre to centre of rivet holes. To be riveted with $\frac{1}{2}$-inch rivets, one inch from centre to centre, a hempen cord, dipped in paint, being placed between the plates at the joints at the time of riveting.

"*Crown Plate.* The roof of holder to be domed, and to rise to a height of 2 feet 3 inches; the centre or crown plate of roof to be 3 feet 6 inches in diameter and $\frac{3}{8}$ inch thick, having four bolt-holes for centre pipe, twelve $\frac{3}{4}$-inch bolt-holes for securing the ends of main bars, and six bolt-holes for suspension rods. To have a wrought-iron ring, 3 feet 6 inches in diameter and 4 inches wide and $\frac{3}{8}$ inch thick, with holes corresponding with those in the crown plate for securing the main bars.

"*Centre Pipe.* The centre pipe to be of cast iron, 6 feet long 4 inches in diameter and $\frac{1}{2}$ inch thick, to have two flanges, each 3 inches wide in the clear. The upper flange to be bolted to the crown plate with four $\frac{1}{2}$-inch bolts, the lower flange to be bolted in like manner to the cast-iron truss-cup, which is provided with six holes to receive the tension rods.

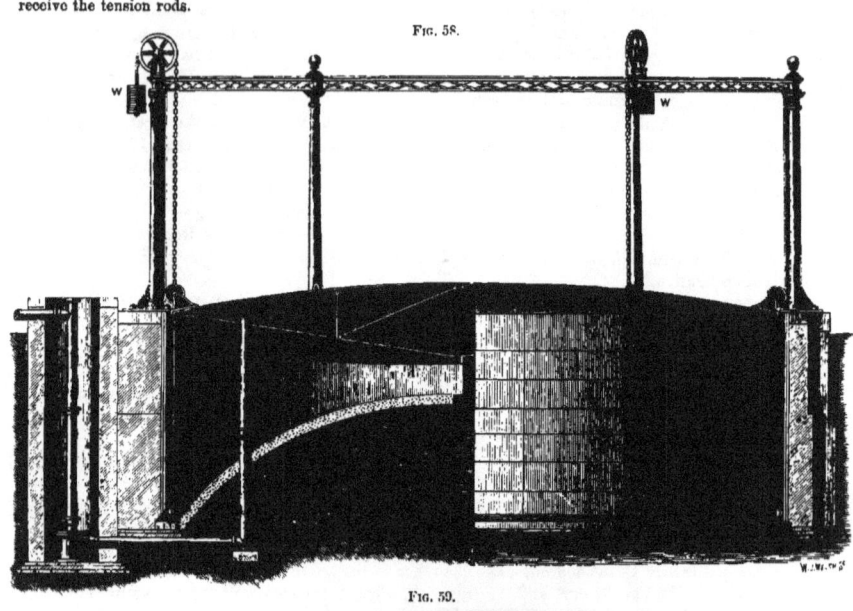

Fig. 58.

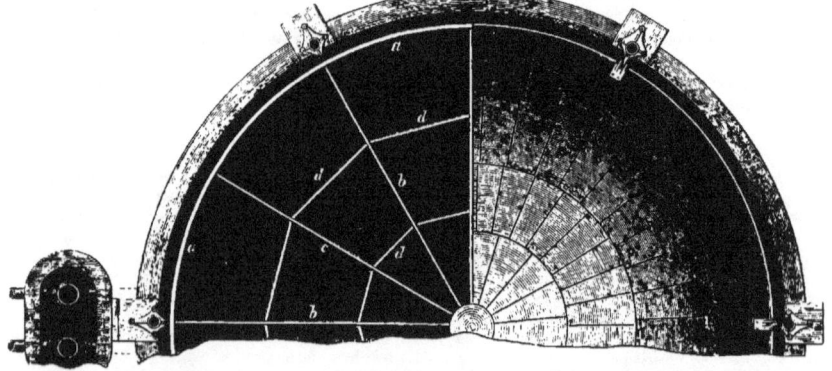

Fig. 59.

"*Main Bars.* To have six main bars and six secondary bars of T-iron, $2\frac{1}{2}$ inches by 2 inches by $\frac{1}{4}$ inch. The upper ends to have clips and eyes, to be secured with $\frac{3}{4}$-inch bolts to crown plate and ring, their lower ends being in like manner secured to the curb.

"*Trussing.* To have six tension rods each $1\frac{1}{2}$-inch round iron, provided with long screws and two nuts at one end, the other end being formed with an eye, and secured by a $\frac{3}{4}$-inch bolt to the main bar, curb, and sheet of roof, the bolt passing through the four pieces. To have six suspension rods each $\frac{3}{4}$-inch round iron, secured at one end by eye and $\frac{5}{8}$-inch bolts to crown plate, the other end having an eye through which the strut passes. The strut to be $1\frac{1}{2}$-inch round iron, the upper end being jawed and secured to main bar by $\frac{1}{2}$-inch bolts, the lower end to have a long screw, which passes through tension and suspension rods and is secured by two nuts, the one above and the other beneath. To have two rows of bracket bars or purlins, the outer row to be 2 inches by $\frac{1}{4}$ inch angle iron, the inner row to be $1\frac{3}{4}$-inch $\times \frac{7}{16}$-inch angle iron, all riveted with $\frac{3}{8}$-inch rivets.

"*Manhole.* To have a manhole cover adjusted to manhole ring of 21×14 inches $\times \frac{3}{4}$ inch flat bar, internal measure, to be riveted, and manhole cut where desired, the cover and ring to be put together with $\frac{3}{8}$-inch screws.

"*Rivets.* The rivets for the sheets to be $\frac{1}{4}$-inch snap head, placed 1 inch apart from centre to centre. The rivets for top and bottom curbs to be $\frac{5}{8}$-inch, and in the former to be placed $1\frac{1}{2}$ inch apart from centre to centre, those of the bottom curb being 3 inches apart.

"*Chairs and Rollers.* To have six chairs with corresponding grooved rollers for guiding the holder in the columns; each of the chairs to be placed on a plate of $\frac{1}{4}$ inch thick, corresponding with its dimensions, and attached to the roof opposite to their respective columns. To have six rollers and frames, to be fixed to the bottom curb for guiding the holder at that point.

"To have six wrought-iron guides, each 15 feet 3 inches long, 6 inches wide, and $\frac{1}{4}$ inch thick, with necessary lugs to build into brickwork.

"*Columns.* To have six cast-iron columns, each 16 feet 6 inches long, $7\frac{1}{4}$ inches diameter at bottom and $5\frac{1}{4}$ inches at top, the base plate to be $1\frac{1}{2}$ inch thick, of the form represented in drawing, and to have three 2-inch holes cast therein for the holding-down bolts. A vertical guide-rib to be cast in the front of each column, of not less than $\frac{1}{2}$ inch thick. The top of the column to have two wings cast therewith, to which are bolted the wrought-iron girders, and the capitals of three of the columns to be surmounted by an ornament as represented. No part of the column to be less than $\frac{3}{4}$ inch thick.

"*Holding-down Bolts.* To have eighteen holding-down bolts, each of $1\frac{1}{2}$-inch round iron, 6 foot 6 inches long, with head at one end, the other end being screwed and provided with a nut. To supply eighteen cast-iron plates, each 6 inches square and $\frac{1}{2}$ inch thick, having a $1\frac{3}{4}$-inch hole cast in the centre.

"*Pulleys and Fixings.* On three of the columns to be placed a set of fixings or frames, into which work the axles of the pulleys, and which work the balance chains.

"*Chains.* To supply three $\frac{5}{8}$-inch chains and with corresponding shackles and balance-weight rods, one end of the chain to be attached to the shackle which is bolted to the holder, the other end being secured to balance-weight rod.

"*Balance Weights.* To supply the two tons of balance weights.

"*Girders.* To have six wrought-iron lattice girders, each formed of two bars of 2 inches $\times$ 2 inches T-iron, placed 5 inches apart in the clear, the latter being formed as in drawing, with $1\frac{1}{2}$ inch $\times \frac{1}{4}$ inch flat bar iron. The girders to be bolted to the wings cast on the top of columns.

"*Pipes.* To supply inlet and outlet pipes of 5 inches in diameter with the corresponding T-pipes complete as shown.

"*Valves.* To supply two 5-inch valves, together with all the nuts and bolts and all necessary material.

"The whole of the wrought ironwork to have two coats of paint before leaving the works of the contractor, and the gasholder to be completed in the true sense of the term, and anything that may have been omitted in this specification shall be supplied by the contractor.

"The contractor to supply all material, labour, tackle, scaffolding, and other apparatus necessary for the construction and erection of the gasholder, together with the columns and girders in a good substantial manner, to the satisfaction of the company's engineer. The holder to be afterwards tested with air at the expense of the company, but any defect to be made good by the contractor."

The cost of a holder like that described, erected in a locality where no extraordinary expense is incurred, either in transport or labour, would be about 380*l*.

When erecting a gasholder, a rude scaffolding is first fitted up in the interior of the tank (leaving room for the sides of the holder to descend) in such a manner that planks may be placed at the desired point to enable the men to rivet the plates of the roof. Planks are then placed around the top of the tank, and on these the parts forming the bottom curb of the holder are adjusted, and riveted or bolted together.

This effected, the bottom row or two rows of plates are placed in their positions and the circle riveted, when the bottom frames with their rollers are securely bolted, exactly opposite to their respective guides. A number of long screws corresponding with the columns are provided, each of about 7 feet long and 1½ inch diameter, having a hook at one end. A nut works freely on the screw, and beneath this nut is a swivel, so arranged that it can be attached to the guide standards.

The swivels of the long screws being bolted to their respective standards, then by turning the nuts the screws are raised to nearly their highest point, when each of them by means of a chain and hook grapples the bottom curb; then on raising the screws still higher, that portion of the holder is suspended. The planks are then removed, and by the action of the nuts, the whole is gradually lowered to the desired depth ready to place another row of plates. Thus all the side plates are riveted and lowered as required, when the top curb is also riveted.

The crown plate with the centre pipe are then placed in their position, and the main and secondary bars are bolted to the curb and crown plate. The tension rods, struts, and suspension rods are then fixed, and the bracket bars riveted or bolted, thus forming the trussing of the roof. The plates covering the roof are then temporarily placed over the whole area, and when accurately arranged they are then riveted, when the holder is lowered to its bearings and the long screws removed. This effected, the upper guide rollers are then fixed accurately in their positions. A manhole is always left in the roof, so that after the interior of the holder is painted, the scaffolding can be removed at that point; it is also intended for the purpose of clearing out any obstruction that may occur in the pipe.

At this stage of the operations two important points have to be observed; first, not to close the manhole before the tank is filled with water and its soundness assured; for, as we once witnessed, through inattention to this a remarkably strong holder of 75 feet in diameter was completely collapsed, which arose from the circumstance that after admitting water into the tank the manhole lid was secured, and the absorption of the brickwork, or perhaps leakage, was sufficient to cause a partial vacuum of some feet, when, of course, the holder gave way and presented a complete wreck.

The other precaution is, not to connect the gas with the holder until the whole is completed and ready for use, as many serious accidents have arisen by acting otherwise, either occasioned by a leaky valve, or perhaps some thoughtless person meddling with a valve when gas has passed into the holder, whilst the men are still at work therein; and on the approach of a light a terrible explosion has happened, with a serious loss of life and property. Severe accidents arising from this cause have occurred in some of the towns in the United Kingdom as well as on the Continent, but happily now they are never heard of.

Small holders up to 50 or 60 feet in diameter are always made single lift, and they may be estimated at from 12*l.* to 16*l.* per 1000 feet capacity, delivered. At the present day few gasholders are made telescopic of a less diameter than 80 or 90 feet, and, according to our first engineers, the cost of gasholders for a given capacity gradually diminishes until reaching 2,000,000 feet; beyond which it appears that they begin to increase in price.

Telescopic Holders.

These apparatus derive their name from the fact of one holder being concentric to the other, and on the inner vessel, or single lift, filling, it raises the outer holder, or, as generally termed, the second lift; and in some rare instances they are made in three parts, and called "treble-lift" holders.

As we learn from Creighton, telescopic holders were invented and in use about 1824; but no importance was attached to them until about nine years afterwards, when they were first erected at the works of a Metropolitan Company then established, the engineer of which laid claim to and patented the invention. From that period, on account of the extended application of gas lighting and the limited sites of works, telescopic gasholders were rapidly introduced.

Telescopic holders were first brought into requisition in consequence of the extremely limited space

TELESCOPIC HOLDERS.

occupied by the early gasworks, situated, as they were sometimes, in the heart of a city, in the midst of populous neighbourhoods, and where ground had to be purchased at fabulous prices. But the increasing demand for gas, and the necessary enlargement of these vessels, have rendered the telescopic system indispensable in an economical point of view, as by it the cost in the construction for a given capacity is very materially decreased, and it is therefore adopted in the most extensive and modern works. Our general observations of weight and area apply equally to telescopic holders, but in these the hydraulic cup and grip, together with the weight of the water forming the lute, all tend to increase considerably the pressure when the lower lift is brought into action; and all these must be included in the collective weight in order to define the pressure that the holder will give. In telescopic holders of medium capacity, the lower lift is counterbalanced in order to reduce the pressure, but in holders exceeding 100 feet in diameter the counterbalance is generally dispensed with.

Formerly, when the sites of some of our metropolitan gasworks were exceeded in extent by the manufacturing premises of many private firms, the main object in employing the vessels in question was to extend the operations of the works within their limited sphere. But at the present day, when the full importance of gasworks is well understood, and sites for their erection are chosen which are practically ample for all time, or at least for many years, the question of space is no longer an inducement for the construction of such holders, the great consideration being economy; and when we reflect on the immense sums expended in this description of apparatus, any saving in their cost must be desirable.

The telescopic holder is comprised of a single vessel of the ordinary construction, guided at the top by rollers working against the guide standards, and surrounded at its base by an annular cup c, of from 12 inches to 18 inches in depth and from 6 to 9 inches in width, represented in section in the annexed diagram. This inner holder is enclosed by a cylinder or outer "lift" concentric to it, of somewhat larger diameter, but about the same depth as the former. To the top of this outer cylinder is attached an annular "grip" $b\ b$, corresponding with the cup as shown. Thus, each time that the inner holder is filled by means of the grip, when ascending, it lifts the lower holder, and as the cup is always filled with water when rising, on the grip being immersed therein an hydraulic joint extending the circumference of the holder is formed. By these means, so long as the holder retains its level position, and in the absence of a pressure superior to that of the dip, the gas is retained. The rollers d and a are for the purpose of preventing any unnecessary play, which would be fatal to the action of the apparatus.

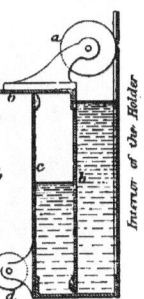

To demonstrate the advantages of telescopic holders in an economical point of view, let us assume the average cost of a stone or brick tank to be 15*l*. and that of a single holder to be 12*l*., making together 27*l*. per 1000 feet capacity. To render the dimensions of the tank suitable for a telescopic holder, a slight increase of expenditure only would be necessary; the price of which may, therefore, be estimated at 16*l*., and if we suppose the outer lift to cost an additional 12*l*. per 1000 cubic feet capacity, then, as the holder is doubled in storeage, an economy of about 35 per cent. is effected by the adoption of the telescopic principle, instead of erecting two separate gasholders.

There are, however, certain localities where the limits of the gasworks have been so restricted, or the site of such a nature, as to compel the engineer to resort to three-lift telescopic holders. These are constructed in a similar manner to those described; but having two concentric cylinders with their corresponding cups and grips, with chains and counterbalance-weights; the holder, when full, rising to three times the height of the single holder. To the excellent specimen of this class of structure formerly existing at the Blackfriars works we have already referred.

The three-lift holder, however, is not to be recommended as economical in erection, except, perhaps, when of great diameter, and shallow depth, in consequence of the extensive framing and columns necessary to prevent the wind from tilting it when at its highest elevation.

2 E

210 TELESCOPIC HOLDERS.

An argument raised against telescopic gasholders is the liability of the water forming the lute, or seal, to freeze, in which case their action would be impeded. For this objection, when this class of holders is erected in cold climates, it must be admitted that there are some grounds, but as the period of frosty weather is limited, and the means of removing the ice very simple, either by breaking it as it forms, or preventing its formation by admitting steam or placing salt into the cup, the question therefore arises whether the little trouble attending these operations bears any material proportion to the economy effected in the construction; in our opinion the former is insignificant as compared with the latter.

Modern works of the kind under consideration demand infinitely more intelligence than the single-lift holder, where, if an excess of play is disregarded, or if it should be out of level to that extent as to be disagreeable to the eye, no ill effects are experienced—at least, so long as the holder remains sealed—but the telescopic holder demands the most rigid care in its design, construction, and erection.

This can be readily understood when it is explained that they frequently exceed 200 feet in diameter (one at the Fulham station of the Imperial Company being 233 feet), and that their cups seldom

FIG. 60.

ELEVATION OF GASHOLDER AT CROYDON.

exceed 18 inches in depth, one half the water of which may be displaced by the pressure. Under these conditions, therefore, the utmost care is indispensable in erection, in order to maintain the vessel perfectly level at all the varied elevations and prevent the water escaping from the luting, and yet to avoid any binding that would add to the pressure, and might cause the holder to blow. For such constructions, the rule of thumb operations of the past are quite unsuitable.

In telescopic holders of 90 feet in diameter and under, the upper lift is generally counterbalanced, but when of larger dimensions their great area renders this no longer necessary, at least that is the generally accepted opinion, which, however, is a question that should be decided by the degree of pressure requisite to supply the town; for if 3 inches be ample for that purpose, then by having a pressure of 8 or 10 inches on the holders, so much extra work is thrown on the exhausters, accompanied with a great increase of fuel, and we are of opinion that if this subject were to be more closely investigated a more perfect method of counterbalancing gasholders would be adopted.

We have been favoured by Mr. Horatio Brothers with the following specification and drawings of a telescopic gasholder 128 feet in diameter, rising 52 feet, erected at the Commercial Gas Company's works, Croydon, represented in elevation in Fig. 60, which is remarkable for its elegant appearance.

Specification of Telescopic Gasholder.

"*Works.* The tank is 130 feet internal diameter.

"The gasholder is to be constructed upon the telescope principle, the outer or lower portion being 128 feet diameter and 26 feet deep. The inner gasholder is to be 126 feet diameter and 26 feet 6 inches deep at the sides.

"*Framework of Inner Gasholder.* The inner gasholder is to be a true circle, 126 feet diameter and 26 feet 6 inches deep at the sides, and 32 feet 6 inches deep when measured from the centre of the top, which will have to rise in a dome or spherical form 6 feet above the top curbs of angle iron forming the top of the sides of the said inner gasholder. The centre crown plate is to be 7 feet diameter and $\frac{5}{8}$ inch thick, and the under plate is to be 6 feet diameter by $\frac{5}{8}$ inch thick, with the necessary holes prepared for the main bearing bars and centre tube.

"The interior framework of the above gasholder for the support of the wrought-iron sheeting is to consist of thirty-two main bearing bars or principals, consisting of wrought-iron T-iron 5 inches × 6 inches × $\frac{1}{2}$ inch thick each, with strong eyes at each end and double fish-plates $\frac{3}{8}$ inch thick (the fish-plates are to be firmly riveted to the main bars with $\frac{4}{5}$ rivets). One end of the main bars is to be secured to the two wrought-iron centre crown plates as before described, and the outer ends are to be secured to the two top curbs of angle iron with 1$\frac{1}{2}$-inch screw bolts and nuts, and the curvature of such principals is to be so set as to rise in a spherical form as before mentioned, and to be trussed as hereafter described.

"Intermediate between the aforesaid main bearing principals must be inserted thirty-two secondary bars, consisting of T iron 4 inches × 4 inches × $\frac{1}{2}$ inch thick each, with eyes and fish plates as before described secured to the fifth rows of purlin bracket bars, and the aforesaid top curbs of angle iron with 1-inch screw bolts and nuts. The main and secondary bearing bars are to be stayed laterally by eight rings or purlins of bracket bars. The fifth row is to consist of T iron 4 inches × 4 inches × $\frac{1}{2}$ inch thick, and all the intermediate rows of such purlin bracket bars are to consist of angle iron 2$\frac{1}{2}$ inches × 2$\frac{1}{2}$ inches × $\frac{3}{8}$ inch thick, and the whole to be well and satisfactorily secured to the main bearing bars with $\frac{3}{4}$-inch screw bolts, and to the secondary bars with $\frac{5}{8}$-inch screw bolts and nuts, and set to the true curvature of the top of the inner gasholder.

"The thirty-two principals are to be well trussed and supported inside by a wrought-iron centre tube and truss cup. The tube is to be about 22 feet long by 2 feet 3 inches diameter, made of boiler plates $\frac{3}{8}$ inch thick, with strong curbs or rings of angle iron at each end 4 inches × 4 inches × $\frac{1}{2}$ inch thick, the upper end to be secured to the under-crown plate with twelve 1-inch bolts and nuts.

"About 2 feet 6 inches from the lower end there must be a strong wrought-iron truss cup 6 feet diameter and 2 feet deep, made of three rings of angle iron 4 inches × 4 inches × $\frac{1}{2}$ inch, and boiler plates $\frac{1}{2}$ inch thick, all of which are to be firmly riveted and welded together with holes to receive the principal tension rods.

"The thirty-two principals are to be trussed with thirty-two tension rods 1$\frac{1}{4}$ inch diameter, and each is to be about 61 feet long with adjusting screw. Each end is to be secured to the main principal and to the top curb of angle iron with a pair of strong jaws and 1$\frac{1}{2}$-inch screw bolts. Each principal is to be trussed at four places between the centre tube and the outer end. The sixty-four middle truss bars are to be 1$\frac{1}{2}$ inch diameter. The sixty-four other truss bars are to be 1$\frac{1}{4}$ inch diameter, with strong jaws and adjusting screws and double nuts.

"The thirty-two main diagonal tension rods are to be 1$\frac{1}{2}$ inch diameter, with strong jaws and eyes and fish plates with adjusting screws and double nuts as before described, and are to be fixed so as to keep the top in its proper shape without interior pressure. All the screws in these rods are to be 1$\frac{3}{4}$ inch diameter.

"The top curb is to consist of two rings of angle iron 5 inches × 5 inches × $\frac{1}{2}$ inch thick, one ring to be placed above the main bars to connect the top and side sheets together, and the other to be close under the main bars, and connected to the side sheets and main bars with screw bolts and rivets, and each ring is to be set true and properly jointed with strong back and front corner plates $\frac{1}{2}$ inch thick, 3 feet long, and well riveted with $\frac{3}{4}$-inch rivets, and all the joints are to be crossed.

"The sides are to be stayed and vertically supported by thirty-two vertical bars of T iron 4 inch × 4 inch × $\frac{1}{2}$ inch thick, secured at the top end to the top curb of angle iron, and at the bottom to the hydraulic cup with one-inch screw bolts and nuts, and to each tier of side sheets with $\frac{3}{8}$-inch rivets.

"The thirty-two vertical supports are to be trussed on the bow-string principle with tension rods 1 inch diameter, with strong eyes and jaws and also sixty-four truss bars 1$\frac{1}{2}$ inch diameter, with strong jaws and adjusting screws passing through the eyes of the tension rods.

"The secondary bar fixed to the fifth row of purlins in the top frame is to be trussed on the bow-string principle, with $\frac{3}{4}$-inch tension rods, eyes, and jaws, and 1$\frac{1}{4}$-inch truss bars, jaws, &c.

"*Hydraulic Cup.* On the strong row of sheets at the bottom of the sides of the inner gasholder is to be secured a hydraulic cup 18 inches deep by 8 inches wide, which is to be made of a channel of rolled iron, as shown, about 4 inches deep and $\frac{1}{4}$ inch thick at the bottom and $\frac{3}{8}$ths at the sides, to which the sheets are to be well riveted. And a stiffening curb of half-round bead iron, 2 inches wide by $\frac{3}{4}$ inch thick, is to be riveted to the outer edge of the cup. The side sheets of the cup must be $\frac{1}{4}$ inch thick.

"*Sheeting of Inner Gasholder.* The first row of sheets on the top which join the centre or crown plates, and the outer row of sheets next the curb of angle iron are each to be $\frac{1}{4}$ inch thick, except sixteen plates on the outer row, on which the sixteen sliding carriages are to be fixed, which are each to be $\frac{3}{8}$ths thick.

"All the intermediate rows of sheets on the top are to be full No. 10 Birmingham wire gauge.

"The top and bottom rows of side sheets are to be $\frac{1}{4}$ inch thick, and the next row to these is to be No. 11 Birmingham wire gauge, and all the intermediate rows are to be No. 12 wire gauge.

"There are to be three oval manholes in the top of the gasholder, each about 2 feet 6 inches by 2 feet, fitted with removable covers with $\frac{3}{4}$-inch screws and bolts.

"The sheets both of the inner and outer gasholders are to be well and securely riveted with $\frac{5}{16}$ths rivets made of the best iron, and the said rivets are not to be more than an inch apart from centre to centre, a hempen tape being inserted between the joints of the plates.

"*Outer Gasholder.* The outer gasholder is to be 128 feet diameter and 26 feet deep. The sides are to be supported by thirty-two vertical bars of flat iron, 6 inches wide by $\frac{1}{2}$ inch thick, and secured with two $\frac{3}{4}$-inch screw-bolts and nuts at each end to the bottom curb of the angle iron and the hydraulic cup, and to each row of side sheets with counter-sunk bolts and nuts.

"The lower row of strong sheets is to be secured to the bottom curb, which is to consist of two strong rings of angle iron 5 inches × 5 inches × $\frac{1}{2}$ inch thick, one ring to be above and the other below the friction rollers and carriages. The curbs are to be set true and properly jointed with strong corner plates, &c., as before described. On the upper row of strong sheets is to be fixed an inverted hydraulic cup or dip 18 inches deep by 8 inches wide, and made the same as described for the inner gasholder, of which this is to be a counterpart.

"*Sheeting.* The top and bottom rows of sheets are to be $\frac{1}{4}$ inch thick, the next rows are each to be No. 11 Birmingham wire gauge, and all the intermediate rows of sheets are to be No. 12 wire gauge.

"*The Suspension Frame.* The suspension frame is to consist of sixteen columns, girders, guides, entablatures, &c.

"Each column is to be 55 feet high. The base is to be 4 feet 6 inches square and 12 inches deep. The shaft of the column is to be 36 inches inside diameter at the base and 27 inches at the top. The cap of the column is to be 3 feet 9 inches square underneath the top plate by 5 inches deep, and the column is to be made with proper top and bottom beads and mouldings, as shown in the drawings.

"The entablature is to be about 6 feet 6 inches high by 2 feet 6 inches wide at the narrowest part, and $\frac{5}{8}$ inch thick, and to be firmly secured to the top of the column. Each column is to be cast in six lengths and secured together with inside flanges with $\frac{3}{4}$-inch brackets, to be turned and fitted at each joint in a lathe, so that when put together they will have the appearance of one piece.

"The metal of the columns is to be full $1\frac{3}{8}$ inch thick for the base and three bottom lengths, and full $1\frac{1}{4}$ inch for the remaining lengths.

"Each section or part of the said columns is to be of an equal thickness throughout, perfectly cylindrical in bore, and to be cast vertically, of good iron, second running, and when cast to be free from scoria, sand holes or other imperfections whatever, and no portion will be received which is in any part less than the specified thickness.

"The inside flanges are to be of the same thickness as described for the portion of the column of which they form part, and the sections are to be bolted together with at least twelve 1-inch bolts at each flange, and the joints are to be properly made.

"To each column is to be fitted a cast-iron openwork guide, about 6 inches deep at the bottom and about 11 inches deep at the top; it is to be 4 inches wide at the front on which the guide-pulleys work and 9 inches wide on the part bolted to the column, and is to be secured to the main column about every 2 feet 9 inches with 1-inch screw-bolts and nuts.

"Each guide is to be cast in six lengths, and to be $1\frac{1}{2}$ inch thick of metal throughout.

"*Wrought-iron Girders.* The columns are to be bound together at the top by sixteen wrought-iron trellis-work girders, with an ornamented boss or star of cast iron secured to the centre of the diagonal bars with a 1-inch bolt and nut, as shown in the drawings. A pattern of the stars shall be submitted to the engineer before casting the same.

"Each girder is to be about 26 feet 4 inches long by 3 feet 5 inches deep, and is to be formed of a framework of double angle iron, top and bottom, each 3 inches × 3 inches × $\frac{3}{8}$ inch thick; the bottom flange is to be of boiler plate, 12 inches wide by $\frac{1}{4}$ inch thick, and the top flange is to be 10 inches by $\frac{3}{8}$ inch thick. The diagonal bars are to be 4 inches wide by ½ inch thick, and the vertical bars are to be 8 inches wide and ⅜ inch thick, firmly riveted to the angle irons, and each end of the girder is to be well secured to the top of the column with three 1-inch screw bolts and nuts, and to each other with the same number.

"Each column is to be secured to the foundation stones at the top of the tank with four holding-down bolts, 2 inches diameter and about 14 feet long, with proper hexagonal nuts and washers, but these are included in the contract for the tank.

"To be provided and fixed thirty-two cast-iron guides in the tank, each 26 feet long, 16 inches wide at the flange, and about 3½ inches deep, and each guide is to consist of three sections, which are to be bolted with 1-inch bolts to the stones in the tank walls prepared to receive the same.

"The contractor will have to drill the stone work in the tank to receive the said bolts, and to caulk the same with lead in a firm and secure manner.

"*Guide Carriages, Pulleys, &c.* On the rim of the inner gasholder at the top are to be fixed sixteen wrought-iron carriages and adjusting plates, the pulleys are to be of cast iron, and to be about 18 inches diameter, with a 3-inch flange, and 5 inches wide on the face to work against the guides of the columns, to keep the inner vessel true and steady.

"The carriages are to be made as per drawings, and the iron is not to be less than the thickness shown in the drawings. The framework of the carriages is to be composed of angle iron 3½ inches × 3 inches × ½ inch welded to the form required. The bed plates are to be ⅝ths thick. The whole to be securely riveted and bolted, as shown in the drawings.

"On the bottom of the inner gasholder and secured underneath the hydraulic cup are to be fixed thirty-two friction rollers and wrought-iron carriages, to work inside the outer vessel against the flat bars prepared for the purpose.

"On the hydraulic cup of the outer gasholder are to be fixed sixteen guide-pulleys and wrought-iron carriages and adjusting plates, as shown, to work against the column guides. The wrought-iron carriage is to be made, as before described, of welded angle iron 3 inches × 2½ inches × ⅜ inch.

"The cast-iron pulley is to be 12 inches diameter, with 2½-inch flange. The whole to be securely riveted and bolted, as shown.

"On the bottom curb of the outer gasholder are to be fixed thirty-two rollers and wrought-iron carriages, to work in the guide troughs in the tank, so as to keep the outer gasholder true and steady.

"The whole of the guide-pulleys, rollers, and carriages are to be properly fitted together, and turned and bored so as to work freely with strong spindles or axles.

"The whole of the sheets and wrought-iron work must have one coat of boiled linseed oil before they leave the contractor's yard, and the gasholder, when completed, shall be painted at the joints inside and all over outside with a coat of the best oxide of iron paint. The cast-iron columns and girders and carriages are to be painted with a coat of good oil paint, stone colour.

"The whole of the materials and work herein specified are to be of the best quality for the purpose intended, and the workmanship to be of the very best description, and the gasholder, when completed, shall work perfectly true and free at all points, and shall be perfectly gas-tight, and symmetrical in form and even in the sheeting.

"When the gasholder is completed, the Gas Company will fill the tank (at their own expense) with water, under the superintendence and at the risk of the contractor, and the contractor shall then, at his own expense, prove the soundness of the holder and its working by filling the same with atmospheric air, and shall let it remain so filled as long as the engineer may direct, and shall then gradually let out the air. If the gasholder is not air-tight, or if it does not work easily, freely, and evenly, the contractor shall make it air-tight and work satisfactorily, and shall prove the same to the satisfaction of the engineer, and when once the Company has filled the tank with water, in case it may be necessary to empty the tank and to fill it again with water, the expense thereof, and any expense to which the Gas Company may be put in consequence thereof, shall be borne by the contractor.

"The contractor shall be responsible for any loss of gas by leakage or imperfect working of the holder during the period of maintenance, and the engineer shall be judge of such loss and the amount of the same, and the Company may deduct the estimated amount of such loss from any money due or owing to the contractor.

"The Company will provide and fix the holding-down bolts and plates, and prepare the foundations ready to receive the columns."

In Plate 13 at the end of the volume are represented the centre pipe, together with the method of trussing the roof, the stays for the sides, the guides and columns; and Plate 14 shows a fourth part of the plan of the holder, of which a portion is covered with the sheets; in the other portion are shown the top curb, the principals, secondary bars, and the purlins described in the specification.

GASHOLDER AT REDHEUGH.

The following is an abstract of specification of a gasholder of 150 feet in diameter, and rising to a height of 66 feet 6 inches, capable of containing one million cubic feet of gas, designed by Mr. Wyatt for the Newcastle Gas Company, which is represented half in section in Plate 15, wherein is shown one of the "main cross rafters" hereafter mentioned. The conditions being similar to those on page 183 are consequently omitted.

Specification of a Gasholder and Framing.

"*Description.*—The gasholder will consist of two parts, of a telescopic form, and will be, when raised to its full height, 66 feet 6 inches from the underside of curb of lower vessel to top of curb of upper vessel. The lower vessel will be 31 feet in height, over all, and 150 feet in diameter. The upper vessel will be 31 feet in height at sides, and 147 feet 4 inches in diameter. The dome to the upper lift will be built with its trussing to a segment curve, having the span of 147 feet 4 inches, and 7 feet 3 inches rise at centre. The gasholder will be wholly composed of wrought iron.

"*Quality of Iron and Workmanship.*—The wrought iron to be equal to the best Staffordshire iron, and of such strength that any bar shall bear a tension strain before breaking of not less than twenty tons per *square* inch of section for plates, angle irons, T-irons and bars; but the rivets and bolts uniting these parts must take a strain of twenty tons per *circular* inch of section. There must be no permanent set in the iron at a strain of *ten tons* per square inch in tension. Sample bars, girders, and detached members of the structures will be selected by the engineer from time to time to be tested by direct loading or otherwise in the manufacturer's yard, such testing including all the necessary labour and materials incident thereto, to be borne by the contractor at his expense. The rivets to fit the holes accurately, which must be carefully punched to true centres, rather smaller than the diameters of rivets, and afterwards rhymered out to the exact sizes of rivets with rhymers by hand. No drifting of the holes will be permitted, and the holes of two or more plates placed on each other to come truly together. The engineer may direct that any or all of the rivet holes be drilled instead of punched, and the bolts, where used instead of rivets, to be turned; and he may substitute bolts for rivets, or *vice versa*, with such other changes in the details of manufacture as may appear to him to be necessary, without extra charge. The contractor must not, without permission in writing from the engineer, substitute bolts for rivets where shown on the drawings, or alter the details of any of the parts without express permission in writing by the engineer.

"*Plates, Bolts, and Rivets.*—The screw threads to bolts to be of the Whitworth pattern, and the heads to be forged in one with the bolt. The nuts and heads of bolts to be of the thickness of one diameter, and of the width of two diameters over the narrowest part. The rivets generally for the trussing, curbs, posts, and large members, where same do not form gas joints, to be $\frac{3}{4}$-inch diameter, spaced with 4-inch centres, where the plates are $\frac{3}{8}$-inch thick and upwards; and $\frac{5}{8}$-inch diameter rivets for $\frac{1}{4}$-inch, and $\frac{5}{16}$-inch plates with similar centres. Where the plates and irons have gas joints they will be formed as follows: $\frac{5}{8}$-inch and $\frac{1}{2}$-inch plates and irons to have $\frac{3}{4}$-inch rivets, $2\frac{1}{4}$ inches centres and $2\frac{1}{2}$-inches laps; $\frac{3}{8}$-inch and $\frac{7}{16}$-inch plates and irons to have $\frac{5}{8}$-inch rivets, 2-inches centres and $2\frac{1}{4}$-inches laps; $\frac{1}{4}$-inch plates, $\frac{1}{2}$-inch rivets, $1\frac{1}{2}$-inch centres, and $1\frac{3}{4}$-inch laps; and $\frac{1}{8}$-inch plates, $\frac{3}{8}$-inch rivets, $1\frac{1}{4}$-inch centres, and $1\frac{1}{2}$-inch laps. Where butt plates are used with cover strips for double riveting, the latter to be zigzagged, the rivets having a pitch of five times their diameter. The joints and laps to have inserted in same a broad tape saturated with oil and red lead, and riveted up solidly with the joint. Before one plate is laid on another in the process of riveting to form a joint, the same must be well painted with red lead or anti-corrosive paint. All rivets above $\frac{3}{8}$-inch diameter to be riveted hot, and, wherever possible, to break joint with each other, and not to be on same line of cross section. The plates and irons to curbs, posts, and principal members to have butt joints covered with plates 2 feet 8 inches long, of the same sectional area as the irons joined, and with as many rivets on each side of joint as will equal the section of plate in the line of punched holes. The joints of the several irons to curbs, trusses, and main members to break joint, and not come in line nearer than 3 feet of each other, and no plate or irons to be less in length than 20 feet, unless specially shown on drawings. The edges of all plates and irons abutting against each other to be planed and faced to a true fit and cut square, and the outside edges also to be lined up. The rivets to underside of central part of trussing, and such other bearings as may be necessary, to be flush riveted.

Lower Gasholder.

"*Curbs.*—This gasholder will be, as above described, 150 feet diameter, and 31 feet high over all. It will be composed of a bottom curb arranged with the following plates and irons; two plates placed horizontally, 1 foot asunder, each 7 inches × $\frac{1}{2}$ inch, and connected with the lower plates of sheeting by two angle irons, each $2\frac{1}{2}$ inches × $2\frac{1}{2}$ inches × $\frac{1}{2}$ inch, forming an external trough in section. The top curb will be formed of one horizontal plate 1 foot 4 inches × $\frac{5}{16}$ inch, three angle irons $2\frac{1}{2}$ inches × $2\frac{1}{2}$ inches × $\frac{5}{16}$ inch, which connect the previous plate to side plates, pendent plate for grip 18 inches × $\frac{1}{4}$ inch, with a rounded iron 3 inches × 1 inch at bottom of same, covered at joints with covers 1 foot 6 inches × 3 inches × $\frac{3}{8}$ inch inside, and riveted thereto. The curbs will be built with butt jointed plates and irons, having covers at all joints 2 feet 8 inches long, of the sectional area of the plates joined, with the requisite number of rivets each side of joint. The cover plates to pendent plates to grip to be 1 foot 6 inches × 1 foot 6 inches × $\frac{1}{4}$ inch and placed on the inside thereof.

"*Posts.*—There will be round the circumference of the lower vessel and on the outside of sheeting plates sixteen posts, spaced equidistant apart, centre to centre, each composed of two angle irons, each 5 inches × 3 inches × $\frac{3}{8}$ inch, riveted to sheeting. There will be two rollers in the height of the lower vessel, exclusive of those on the top and bottom curbs, carried by two jaw-plates 9 inches × 8 inches × $\frac{1}{2}$ inch each, riveted by four $\frac{3}{4}$-inch rivets to the vertical plates of posts, and in which the axis of the roller will operate. The rollers will be 6 inches diameter and $4\frac{1}{2}$ inches broad, of wrought iron turned and fitted.

"*Guide Rails.*—On the inside of the vessel, and centred with each post, there will be sixteen guide rails running the entire height of the lower vessel, and for 8 inches below lower curb, weighing 40 lbs. per lineal yard, rivetted to the $\frac{5}{16}$-inch plate at the bottom flange, with $\frac{5}{8}$-inch rivets, 6 inches centres on both sides of web, the rivets breaking joint on either side to form a continuous guide.

"*Plates for Sheeting.* The plates on the side of lower lift to be for the top and bottom rows next to the curbs 3 feet wide and $\frac{1}{4}$ inch thick, and the remaining plates to average about 2 feet 9 inches wide and $\frac{1}{8}$ inch thick, excepting the middle tier of plates, which will be $\frac{1}{4}$ inch thick, and the vertical strip riveted at back of each post, which will be 1 foot 6 inches wide, $\frac{5}{16}$ inch thick, and of the entire height of the vessel, in two plates, the joint being a butt one, and covered by an external cover strip 1 foot 6 inches × 1 foot 6 inches × $\frac{5}{16}$ inch. The thinner plates not to be less than 10 superficial feet in area, and the thicker plates not less than 15 superficial feet.

Upper Gasholder.

"*Curbs.* The upper gasholder will be 147 feet 4 inches in diameter and 31 feet high at the sides. It will be composed of a cupped bottom curb, having a horizontal plate 1 foot 4 inches × $\frac{5}{16}$ inch, three angle irons 3 inches × 3 inches × $\frac{3}{8}$ inch connecting same to the lower tier of plates and to the vertical plate of cup. The vertical plate to be 1 foot 6 inches deep, $\frac{1}{4}$ inch thick, covered at joints with plates 18 inches × 18 inches × $\frac{1}{4}$ inch, at the top of which there will be riveted a rounded iron 3 inches × 1 inch. The upper curb to be formed of a vertical plate 12 inches × $\frac{3}{8}$ inch, and one angle iron 6 inches × 6 inches × $\frac{5}{8}$ inch. There will be an external belt also 9 inches × $\frac{5}{16}$ inch round the gasholder at the first joint of plates, 3 feet below the top, which will be double riveted, and covered at the joints by plates 2 feet 8 inches long, to form a continuous ring. The several plates and irons, forming the curbs, to break joint with each other more than 3 feet, and to be in lengths not less than 20 feet, being similar as to joints and covers to those described for the curbs to lower gasholder previously.

"*Posts.* The circumference of the upper gasholder will be stiffened by sixteen posts, or vertical supports, equally spaced centre to centre, and 18 feet 6 inches high to the springing of arched rafters.

"*Main Rafters.* The dome of the upper gasholder will be trussed with eight main rafters, as shown on the plan of framing, stretching from the upper curb to the centre of dome. Four of these rafters, named "main cross rafters" (shewn in Plate 15), will cross each other at the centre of dome, and be secured together, and between the two centre plates. The main rafters will spring from eight of the posts at the level of 18 feet 6 inches above the underside of lower curb, and be of the form shown on the drawings.

"*Secondary Rafters.* The secondary rafters will be eight in number, and will span from the outer curb of the upper gasholder to the concentric girder.

"*Tertiary Rafters.* The tertiary rafters will be T irons 5 inches × 4 inches × $\frac{3}{8}$ inch, and secured to the ring girders, A, B, and C.

"*Side Plates.* The sheeting plates to the side of upper vessel to be for the top row next to the top curb, 3 feet wide by $\frac{5}{16}$ inch thick. This row of plates to be formed with a but joint, and secured to next row of plates below by an external belt 9 inches × $\frac{5}{16}$ inch, double riveted at $2\frac{3}{4}$ inches centres diagonally. The bottom row of plates to be also 3 feet wide and $\frac{1}{4}$ inch thick. The remaining side plates to average 2 feet 9 inches wide and $\frac{1}{8}$ inch thick, excepting the middle tier, which will also be $\frac{1}{4}$ inch thick, and the verticals trip, riveted behind each of the posts, which will be 1 foot 6 inches wide by $\frac{5}{16}$ inch thick, and in two lengths, with a butt joint and cover strip over same externally, 1 foot 6 inches × 1 foot 6 inches × $\frac{5}{16}$ inch. The sizes of plates to be similar to those described for the lower vessel previously, and all sheeting plates to be as to laps, centres, and diameters of rivets according to the rules set forth previously in this specification.

"*Top Plates.* The first row of plates for the top of the gasholder, next to the curb, to be 3 feet wide by $\frac{5}{16}$ inch thick, riveted to curb, main and tertiary rafters over the entire width, with butt joints, cover strips 7 inches × $\frac{7}{16}$ inch, and double zigzag rivets. The second row of plates from the curb to be 3 feet wide by $\frac{1}{4}$ inch thick, and the third row of plates to be also 3 feet wide by $\frac{3}{16}$ inch thick, lap-jointed. The centre plate to be 8 feet diameter and $\frac{1}{2}$ inch thick, attached to the next row of plates by a butt-joint, and a cover-strip 9 inches × $\frac{1}{2}$ inch double riveted, and zigzagged ; and also secured and riveted to Γ irons of main rafters. The first row of plates from the centre plate to be about 3 feet wide by $\frac{3}{8}$ inch thick, the second row also about 3 feet wide by $\frac{1}{4}$ inch thick, the third row 3 feet wide and $\frac{3}{16}$ inch thick, and the remaining plates, not previously described, between the curb and centre, $\frac{1}{8}$ inch thick, of an area of not less than 10 superficial feet each. The whole of these plates, not otherwise specially defined, to be lap-jointed, and to be in accordance with the previously-described stipulations as to plates, laps, riveting, &c., and to be cut to radial lines following the true curve of the dome.

"*Rollers.* There will be the wrought-iron rollers, 5 inches diameter, and $2\frac{1}{2}$ inches broad, with wrought-iron carriages, secured to the bottom curb of upper vessel of gasholder. They will be placed under each side-post, and adjusted to the wrought-iron guide-rail, attached to the inside of lower vessel. The carriage and bed-plate to be secured to the underside of curb by four 1-inch diameter bolts and nuts as per detail. There will also be sixteen cast-iron rollers 2 feet diameter and 7 inches broad over all, having a flange on one side $3\frac{1}{2}$ inches deep placed alternately to the guide-rails, right and left, with wrought-iron carriages, &c., as per detail drawings, secured to the top curb of upper lift by bolts and rivets, as shown, and to have bed-plates capable of adjustment with the rail-guides in front of standards. There will also be sixteen cast-iron rollers 10 inches diameter and $6\frac{1}{4}$ inches broad to paths, with wrought-iron carriages, to top curb of lower vessel, as per details; and also forty-eight wrought-iron rollers 6 inches diameter and $4\frac{1}{2}$ inches broad, in wrought-iron jaws, attached to outer side of lower vessel to work on tank guide-plates and rail-guides. The rollers to be turned to a true circle working truly on the guides.

"*Proving Gasholder.* The gasholder to be proved in the first instance by atmospheric air, by means of a blowing-machine forcing up the two vessels to their maximum lift, to test the accuracy of their movement, level, and working, and also the soundness of joints. After all defects are made good the gasholder will then be proved a second time by filling the two vessels with air, and the inlet and outlet pipes with water, in which state they will remain under test for three days. When these tests are completed to the satisfaction of the engineer, and all defects remedied, then the gasholder will be filled with gas and tested again for any further defects. The expenses of testing, repairs, water, gas, and other matters incidental thereto to be borne by the contractor.

Wrought-iron Bracing to Framing.

"*Lower Bracing to Framing.* The cast-iron standards forming the framing to gasholder will be connected together, by bracing twice in the height. The centre of first tier of bracing will be 30 feet $1\frac{1}{2}$ inch above top of coping to tank. It will be 2 feet 6 inches deep, composed of wrought-iron flanges, each formed of two angle irons 4 inches × 4 inches × $\frac{3}{8}$ inch, with lattices 5 inches × $\frac{7}{16}$ inch, secured at each end by two $\frac{7}{8}$-inch diameter rivets. The Γ irons will be cranked down the standards at each end 1 foot 3 inches, and covered at joint by a double cover strip 1 foot 4 inches × $3\frac{1}{2}$ inches × $\frac{3}{8}$ inch. At each end of bracing-girders there will be No. 8 $\frac{3}{4}$-inch diameter bolts, four on each side of web, as shown on the details, passing through the standards and the two ends of bracing-girders.

"*Upper Bracing to Framing.* The upper tier of bracing will be at its centre 61 feet 6 inches above the top of coping to tank, and be formed of flanges each composed of two angle irons 5 inches × 4 inches × $\frac{3}{8}$ inch, riveted side to side, with 4 inch centres, and $\frac{3}{4}$-inch diameter rivets, and lattices 6 inches × $\frac{7}{16}$ inch, riveted with two $\frac{7}{8}$-inch rivets at each end. The bracing-girders will be made continuous all round the circumference of framing and pass through slots formed in the caps of columns.

"*Holding down Bolts and Plates to Framing, &c.* Under the base of each column there will be four holding down bolts, each 11 feet 2 inches long, exclusive of head and nut, and $2\frac{1}{4}$ inches diameter built in with the concrete and brickwork of tank piers, and secured to four horizontal wrought-iron anchor plates. The nuts, heads, and washers of bolts to be as shown on details. These bolts to be provided and delivered under the gasholder contract, but they will be fixed under the tank contract.

Cast-iron Framing and Guides.

"*Quality of Cast Iron.* The whole of the cast iron at the foundry for this work to be composed of a mixture of at least two different kinds of pig iron, added to 10 per cent. of scrap iron, melted in the cupola furnace, and run from the same, free of all cinder and other impurities. The castings must be free from any imperfections,

SPECIFICATION OF GASHOLDER AND FRAMING.

such as honeycombs, "cold shuts," cracks, or flaws, and to be cast, wherever at all practicable, in a vertical form with the heaviest part of the casting downwards. The cast iron to be of such a quality that a bar, run from the cupola from which the castings are prepared, 3 feet 6 inches long, 2 inches × 1 inch in section, placed with a clear bearing between the supports of 3 feet, shall not break with a less weight at the centre than 30 cwt. The engineer will test bars of this section at the foundry from time to time at the expense of the contractor. The ends and fitting-parts of joints of all castings to be machined, faced, and accurately fitted. The holes for all bolts and taps to be drilled to true centres, when the parts are fitted together *in situ*. The facing joints to be smeared over with boiled oil and red lead before being fitted together. Where cast iron bears upon stone, or upon wrought iron, 5 lbs. sheet lead to be used in the joint; or the same packed up and run solid with lead.

"*Framing.* The framing for guiding the gasholder will be made up of sixteen cast-iron standards, united by two horizontal tiers of wrought-iron bracing. Each standard will be 60 feet 3 inches high, from coping to underside of cap, formed of five castings. The metal will be 1 inch thick throughout, except to lower part of base, which will be 1¼ inch.

"*Cap to Standards.* The cap to be 4 feet 6 inches high, and 3 foot × 2 feet square, cast in a separate piece of ⅝-inch metal, and secured by ¾-inch diameter bolts.

"*Guides to Tank.* There will be sixteen wrought-iron guides or roller-paths 12 inches × ¾ inch, secured to the inside of tank-wall at every wall stone to tank provided for the purpose. They will be fixed by two 1¼ inch diameter bolts and nuts, 1 foot 2 inches long, of the form shown on details. These are to be provided and delivered under the gasholder contract, but will be fixed under the tank contract.

"*Wrought-iron Guide-rails in front of Standards.* There will also be sixteen wrought-iron guide-rails, weighing 75 lbs. per lineal yard, fixed as per details, attached to the inside of standards, in line with those in the tank. These rail-guides will be secured to the standards by ⅝-inch bolts, spaced to 9 inch centres, on alternate sides of web. Each guide-rail will be secured to tank guide at the coping level, as shewn on detail.

"*Inlet and Outlet Pipes.* The contractor to provide and deliver the wrought-iron inlet and outlet pipes 30 inches internal diameter, with the necessary flanges, bends, &c. The pipes to well hole to be of cast-iron, and the metal in flanges to be 1¼ inch, and in the body of pipes 1 inch thick; the flanges to be 4½ inches wide over all, and each flange will be secured with 1⅛-inch diameter bolts at an average of 7 inches from centre to centre. The joints to be formed with a ring of lead bolted up inside the joint and between the bolts and inner face, and the remainder of the joint to the outside of flange to be caulked with iron cement, or the joint may be a mill board joint, or formed in such other manner as may be directed by the engineer.

"*Painting.* The whole of the cast and wrought iron work to be immersed in a bath of hot boiled linseed oil at the manufacturer's yard for an hour at least immediately after being prepared, and before any rust has been formed on same. The iron can then be delivered for fixing, and when this is completed the whole to have inside and outside three coats of approved preservative paint, mixed with boiled and plain linseed oil, the last or finishing coats to be of such tints and varied in such manner as the engineer may direct. All holding-down bolts and iron built in the walls of tank to be dipped in hot tar before being placed in the work."

QUANTITIES.

			Rate.	£	s.	d.
Lower vessel	34¾	Tons wrought-iron curbs, stiffening-irons, &c., to lower vessel				
	50¼	Tons wrought-iron sheeting plates to lower vessel				
Upper vessel	88½	Tons wrought-iron curbs, ribs, posts, trussing, &c., to upper vessel				
	106½	Tons wrought-iron sheeting-plates to upper vessel				
		(*Note.*—The lower vessel weighs nett 85 tons, the upper vessel 195 tons. Total floating weight of gasholder = 280 tons of wrought iron				
	65½	Tons wrought-iron bracing to framing				
	10¾	Tons wrought-iron inlet and outlet pipes				
	10	Tons cast-iron ditto ditto				
	148	Tons cast-iron standards to framing				

GASHOLDER AT BROMLEY.

The preceding specification illustrates very forcibly the care observed in the execution, and the requirements of modern structures of this description, and contrasts in a marked manner with the system of erecting gasholders some years ago. Undoubtedly it would have been desirable to give the specification *in extenso*, but, in order to render it intelligible, several drawings would have been necessary. The holder with its ornamental standards and girders has a handsome appearance, and is another example of the skilful employment of material by which excellence of design is obtained at a minimum of expenditure, as the holder of a capacity of one million feet cost 10,000*l*., or the low sum of 10*l*. per 1000 feet capacity.

The largest holder hitherto constructed exists at the Fulham station of the Gas Light Company, and was designed by Mr. N. F. Kirkham. This is telescopic, erected in a brick tank, and is 233 feet in diameter, and rises to a height of 66 feet, and was constructed by Westwood and Wright.

To Mr. Kirkham, we believe, is also due the merit of designing the holder represented in plate 11, which may be regarded as one of the best specimens of gas holders in existence. This we have termed the "Imperial holder," as they have been exclusively constructed by the late Imperial Company, of which there are several in use at the new works at Bromley, as well as at the Hackney station.

The tanks of these holders are built of brick and puddled, each of a diameter of 203 feet in the clear and 38 feet deep from the bottom of the footing to the top of the curb. The outer holder is 200 feet diameter and 35 feet deep, and is guided in its tank by 56 wrought-iron guides. The inner holder is 197 feet in diameter and 35 feet 4 inches deep, and is guided above by twenty-eight piers of cast-iron columns, with the two tiers or ornamental girders. The columns are surmounted by their respective entablatures, the whole forming perhaps the most elegant specimen of gasholders in existence, and except that at the Fulham station they are of about the largest capacity, as each holder is capable of containing two millions cubic feet of gas.

The roof of the inner holder is trussed with twenty-eight main bars, a like number of secondary bars, and seven rings of bracket bars or purlins. The main bars extend from the top curb to the centre cup, and represents in section a web 12 inches wide by $\frac{1}{2}$ inch thick, which we may assume to be necessary to support the curb at the points where the upper guide roller exists.

The secondary bars are formed of 5 inches × 4 inches × $\frac{7}{16}$-inch T iron, trussed by struts and tension rods.

The centre pipe, together with the trussing cup, is 30 feet long, and to the latter the suspension rods are attached in the usual manner. The roof is domed to a height of 9 feet. The crown plate is 12 feet in diameter and $\frac{7}{16}$ inch thick, made in four segments, the plates being butted together with lapping pieces.

The plates forming the top of the holder are as follows: On the outer ring nearest the curb are twenty-eight plates, situated at equal distances apart, each 4 feet wide and $\frac{7}{16}$ inch thick, and on these repose the upper guide carriages. All the rest of the ring is composed of plates $\frac{3}{8}$ inch thick. The plates of the second ring from the curb, as well as these attached to the crown plate, are $\frac{1}{4}$ inch thick. In the third ring from the curb, and the second from the crown plate, the sheets are $\frac{7}{16}$ inch thick, and the next ring No. 8 gauge, all the rest of the plates forming the roof being No. 9 gauge.

The plates of the side are of the same strength as those of the lower lift, that is, there are seventeen tiers of sheets, the lowest being 3 feet deep, the bottom and top sheets are $\frac{7}{16}$ inch thick, the next tiers are of No. 6 gauge, and all the rest are of No. 10 gauge. Both the upper and lower lifts have vertical stays both inside and outside, and are provided with their corresponding rollers on the grips as also at the bottom. All the seams of the holder are double rivetted.

The top curb is formed by an upper and lower curb, the former being a simple ring of 5 inches × 5 inches × $\frac{7}{16}$-inch obtuse angle iron. The under curb is formed by two rings of 5 inches × 4 inches × $\frac{7}{16}$-inch angle iron, the one being smaller than the other, rivetted to an annular plate 16 inches wide and $\frac{3}{8}$ inch thick, corresponding with the dimensions of the two curbs.

The hydraulic cup is 18 inches deep, and is formed by a bottom plate 20 inches wide and $\frac{3}{8}$ inch thick, and of three angle irons 5 inches × 5 inches × $\frac{7}{16}$ inch, two of these being riveted to the bottom annular plate.

The vertical stays are a combination of the lattice and box girders; of these there are fifty-six in the interior and twenty-eight flat stays on the exterior. The holder at the top is guided by twenty-eight rollers running in wrought-iron jib carriages. The bottom curb of the outer holder is formed in a similar manner to that of the inner holder. The grip is 15 inches deep.

Each of the twenty-eight pairs of columns with its pedestal is 75 feet high, $2\frac{1}{4}$ inches thick at the bottom, and $1\frac{1}{2}$ inch thick at the top. The girders are composite, that is, composed of wrought and cast iron. A handrail is placed around the top to prevent accidents to workmen on the roof.

The inlet and outlet pipes are of wrought iron 36 inches in diameter, each of which is enclosed by a cap attached to the crown of the holder, by which means when the holder grounds and the water is at its full height in the tank, the pipes are sealed, and thus admit of the manhole cover being taken off to remove any obstruction that may occur.

The first of these holders built at the Hackney station, together with the brick tank, cost 47,000*l.*, or 23*l.* 10*s.* per 1000 cubic feet capacity, this being less than the cost of the tank alone in some establishments. But there is reason to believe that as gasholders increase in dimensions up to that in question, this price per 1000 feet is diminished.

To convey an idea of the magnitude of these holders, it may be stated that to fill one of them with gas about 220 tons of coals would be required. The weight of the gas contained in each, assuming its specific gravity about 480, will be no less than 33 tons, and yet possessing an ascensive power of about 40 tons, in other words 2,000,000 feet of air, weigh 73 tons, therefore the adage, "light as air," is not so appropriate as is generally considered. These estimates are, however, only approximative, and are influenced by every change of temperature and pressure of the atmosphere, which will have our attention hereafter.

Hence, in the daily operation of filling one of these holders, the quantity of coal carbonized, the gas produced, and the rental derived therefrom, will be about equal to the yearly operations of many works for supplying a town of 1500 inhabitants.

THE STATION METER.

For our present purpose it will be sufficient to limit our observations to the meter as the "station-meter," which, in its general construction, is almost identical with the wet meter employed by consumers. In describing the former, the principle of the latter will therefore be understood, and hereafter, when treating on "distribution," the progressive history of consumers' meters, both wet and dry, will have our attention. We may here observe the singular circumstance that, although we are indebted to the French for the word meter, or measure, yet in that language the instrument is called *compteur*, or counter, a name which it has received in other Continental languages.

The importance of the station-meter in a gasworks, but more particularly in small establishments, where the utmost care and intelligence are necessary in order to earn a dividend, is hardly susceptible of exaggeration, for by this instrument all the gas produced is measured and recorded hourly and daily; it serves as a check on the quantity and quality of coals carbonized; it shows whether the operations in the retort-house are being successfully conducted or otherwise; and, so far as regards the production and sale of gas, it affords unerring means of ascertaining the operations of the company. In addition to these advantages, the station-meter, when properly constructed, is amongst the most durable of all the apparatus in a gas manufactory. Once placed, beyond being properly supplied with water, it requires no attention, and during a series of years is a faithful recorder of all the gas produced.

On the other hand, when this instrument is dispensed with, it is practically impossible to ascertain, with any strict degree of accuracy, the quantity of gas manufactured and sold. To determine this in its absence, constant reference has to be made to the gasholders, the contents of which are continuously influenced by every change of temperature of the atmosphere, perhaps at one hour, in consequence of the gas being expanded by the heat of the sun, showing an extraordinary fabulous make, and at another period, with reduced temperature, no production whatever. Moreover, when a holder receives and delivers the gas simultaneously, it is utterly impossible to ascertain the quantity produced. It is obvious, therefore, on these accounts, that, for economical working, the instrument in question is indispensable.

For the station-meter, that known as the "wet meter" is universally adopted. In some few instances the "dry meter" has been tried for the purpose, but has invariably resulted in failure, occasioned by the large accumulation of condensed liquid therein, which impedes its action; therefore, however desirable, under certain conditions, that instrument may be for consumers, it is quite unsuitable as a station-meter for a gasworks.

When treating on the history of gasholders a drawing and description of Clegg's rotary holder was given, from which it is generally supposed he derived his first idea of making a rotary meter. This supposition is further favoured by our knowledge of the fact that this holder was erected and in use at the Peter Street station a considerable period before Clegg left the Chartered Company, which happened early in 1817. And, if we reflect on the time that would be required in those days to construct a vessel of such complication and magnitude, and bear in mind the date of his patent, we may fairly infer that the idea of the rotary gasholder first occurred to him, and from this the invention of the gas-meter followed. However, as we shall have occasion to show hereafter, the invention of Clegg differed very materially from the instrument now in use.

Considering the most important and delicate functions of the instrument in question, it may be classed among the simplest of mechanical contrivances, yet, in consequence of its peculiar formation, it is one of the most difficult to explain; however, by the aid of engravings, we hope to be enabled to make its operations well understood.

THE STATION METER.

A station-meter may be simply described as a cast-iron case, containing a cylindrical vessel, usually termed the "wheel," or "drum," which revolves freely in a horizontal position in suitable bearings. This wheel is divided into four equal compartments, each having its corresponding inlet and outlet. The case is charged with water to a short distance above the centre of the shaft of the wheel. The gas entering one of these compartments, by the force or pressure it possesses, expels at the same time a corresponding gas from another compartment. Thus at each revolution the wheel receives and expels a quantity of gas equal to its capacity, and this being known, then by suitable wheel-work the number of revolutions of the wheel, or the quantity of gas passing through the meter, is indicated on the dial or index.

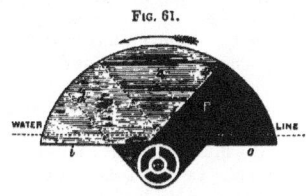

Fig. 61.

Fig. 61 represents the elevation of one of the measuring compartments of the wheel, attached to an axle, A, this being in reality a gasholder in the form of the fourth part of a cylinder, having its sides, P P, placed, for reasons hereafter stated, at slight angles to the axle. This is similar to Clegg's rotary holder, with the difference that whereas the holder revolves in one direction when receiving, and in the opposite when discharging the gas, the vessel in question, whether receiving or delivering, rotates always in the same direction as indicated by the arrow. One of the radiating partitions, P, is shown, the other being represented by dotted lines. The inlet, i, and outlet, o, are long narrow passages seen in section in Fig. 63, and in elevation are of a triangular form, comprised between the letters i and d, and are sometimes called " hoods."

For the purpose of illustration we will suppose this vessel to be supported by suitable bearings, within a tank in which it can revolve freely, the tank being charged with water to the point marked "water line." As represented, both the inlet, i, in front and the outlet, o, at the back of the compartment, are immersed or sealed, consequently the air within cannot escape. But if now the vessel be turned slowly by the hand in the direction of the arrow, the outlet, o, will become unsealed, when the air contained in the chamber will be gradually expelled as that revolves, until eventually the vessel is immersed.

Fig. 62.

By continuing the operation, the inlet, i, will first issue from the water, the air entering thereby, and gradually the whole of the vessel will emerge, and when attaining the position represented it will be again filled with air, with its inlet and outlet both sealed as before. Thus, in revolving, the vessel in rising from the water is filled with air, which is expelled by continuing the rotary motion.

This is precisely the action which takes place with each chamber of the meter wheel, each of its divisions being filled with gas and emptied in succession.

Fig. 62 is an elevation of a meter wheel, either the back, or the front part with the "hollow cover" removed; this being really four divisions represented in the preceding figure attached to one axle, A. The inlets are indicated by the letters $i i i i$, the outlets being at the opposite side of the wheel. The axle is attached to the wheel by means of the central stud, at which point a small portion of each of the partitions, P, is shown. The inlet and outlet passages are separated by the "hollow cover," and the gas conveyed therein by the "spout," hereafter mentioned.

Fig. 63 represents a small station-meter partly in section, C C C C being the outer case, D the inlet, and E the outlet, the dial of the index being situated at the point K. The axle, a, carrying the wheel, has one of its bearings fixed to the centre of the back of the case, which is hidden by the spout, the other bearing being attached to the index box as shown. A "dog" or carrier, not represented, is attached to the end of the axle, and gives motion to its counterpart, which passes through a stuffing box and actuates the index.

A fourth part of the periphery of the wheel is represented as cut away, in order to show its interior The openings, m m, are for the purpose of permitting the water to flow with all freedom to and from the various compartments as the wheel revolves, thus avoiding the "drag" or friction that contracted passages would occasion.

The hollow cover is a concave disc, Q Q, shown in section, soldered to the wheel already described, having a circular orifice, at its centre, as represented. Attached to the back of the case, C C, and passing through the centre orifice of the hollow cover, is the bent pipe or "spout," h, h, shown in section, and so arranged that on the wheel revolving the spout is quite isolated therefrom.

Previous to being put in operation the meter is charged with water to within a short distance of the top of the spout, as indicated by the dotted lines; by this means the openings, m m, and the orifice in the hollow cover, are sealed, thus separating the inlet from the outlet passages, and leaving nearly the half of the wheel above the level of the water, as represented in the following figure.

FIG. 63. FIG. 64.

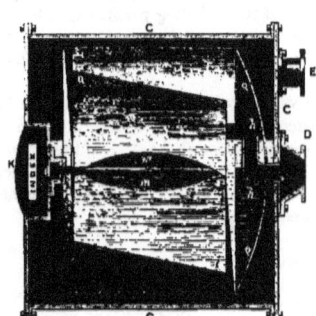 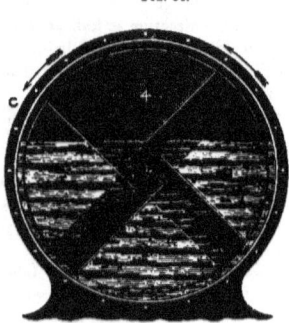

Fig. 64 is a section of the same meter direct through the centre of the wheel; the case, C C, being charged with water to the required height, as indicated. The partitions, P P P P, as stated, are placed at slight angles with the line of axle for the purpose of diminishing and equalizing their resistance when passing through the water, for if they were placed straight with the axle, the obstruction would render the instrument entirely useless for private consumers, on account of the excessive oscillation of the lights supplied thereby.

The measuring chambers are numbered 1, 2, 3, and 4, and as already represented in Fig. 61, the chamber when in the position of that marked 4 is filled with gas, its inlet and outlet both being sealed by the water. If we now suppose the meter to be set in action, the gas will enter the chamber 1, and by its force will expel that from chamber 3; at the same time, by the rotation of the wheel in the direction of the arrows, the outlet of chamber 4 will become unsealed, when the gas is expelled therefrom. In the meantime, chamber 1 gradually rises, increasing in capacity, expelling the gas from 4, and eventually chamber 2 comes into operation, when chamber 1 is filled, and its inlet and outlet sealed. This action is continuous, all the chambers being filled in rotation, at the same time expelling the gas from those preceding them. Here the method of giving motion to the wheel by hand, as proposed when describing Fig. 61, is superseded by the force or pressure of the gas; and so long as this is maintained, and any supply issues from the meter, its action is continued.

THE STATION METER.

Hence the capacity of one revolution of the wheel being known—say, 10 cubic feet—it can be readily understood how, by simple gearwork, such as wheels and pinions, the quantity of gas passing will be recorded on the index, with the most unerring accuracy, during a period of many years, depending on the duration of the instrument, and the maintenance of the water-line at the proper height.

When constructing meter-wheels, their approximate capacity is obtained by calculation; but the exact position of the "water-line" is ascertained by experiment. It is then carefully marked on the front of the case, so that, in the event of the gauge being broken or lost, the mark is always retained. With the same object, some manufacturers attach a brass plate to the case, having the words "water-line" engraved thereon; others have it marked on a brass tube which surrounds the glass of the gauge, a space being left for observation.

The causes of the friction or pressure required to work a meter are not generally understood, hence we find that various absurdities have, from time to time, been patented in order to diminish this. Amongst these, glass bearings for the axles have been proposed; but it must be evident, from the great facility with which a meter-wheel revolves in its case when empty, and the difficulty with which it moves when charged with water, that the cause of obstruction must arise from that alone, concerning which we will make a few observations.

FIG. 65.

As observed, the partitions are placed at angles; therefore, in revolving, they form a kind of vane or screw, and whilst in action draw continuously a small portion of water through the centre opening, which water is expelled at the back of the wheel. A similar result, although to a very slight degree, is effected by the inlet and outlet hoods. Under these circumstances the orifice in the hollow cover should be, therefore, of ample capacity, in order that the water may pass therethrough with all freedom. Further, that a meter may work with a minimum of pressure, care should be observed to make the passages, $m\ m$, at the centre of its wheel, communicating between the various chambers, sufficiently large, so that the water may flow freely into and from the chambers during the revolution of the wheel. By inattention to these points, the friction or pressure required to work the meter is materially increased, and by due regard thereto it may in like manner be diminished.

The accompanying engraving (Fig. 65) represents a station-meter, with cast-iron cylindrical case, suitable for a small works. It is mounted on a pedestal of brickwork or masonry, with the object of bringing the index to the level of the eye of the observer, also for the sake of ornamentation, which is at all times desirable in a gasworks, when attainable at moderate cost, as it creates a feeling of care, cleanliness, and order, among the men who have charge of the apparatus.

The motto on the pedestal, "out of smoke to give light," was adopted by Winsor from Horace, and has since been retained and attached to the various apparatus, and sometimes affixed to the entrance of gasworks.

It may be here observed that the general appearance of a gasworks is invariably a good indication of the operations of the company, for wherever there exists untidiness or disorder, there is sure to be waste, bad management, and loss. The same observation applies to any other business, but there are few kinds of industry that offer such facilities for waste and loss as a gasworks. In establishments of magnitude, hundreds or thousands of tons of coke may pass uselessly and imperceptibly away from the furnaces annually, according to the settings of the retorts, without showing the least indications of the loss, except by the returns of the sales. An increase of only 5 per cent. in the fuel account at the Beckton Works would represent about 4000 tons of coke per annum. In small works, losses from this cause are more likely to occur, in consequence of the want of experience in retort setting.

The capacity of a station meter is dependent on the maximum production of gas, the general calculation being that the speed of the wheel shall not exceed 120 revolutions per hour. The principal defects arising from any increased velocity beyond that stipulated; consisting in the extra friction or pressure necessary to work the meter; also, as ordinarily made, by the increased pressure, the water becomes depressed in the receiving chambers of the wheel, and consequently their capacity is diminished and the meter registers "slow," that is, indicates less gas than has actually passed. Therefore adopting as basis 120 revolutions per hour and the maximum make for a works of 7200 feet per hour, the wheel of the meter should be of such dimensions as to deliver 60 feet at each revolution.

Subjoined is a list of the maximum production per hour for which station meters are calculated, the corresponding capacity of each revolution of their respective wheels, and the average prices of well-known firms for these instruments delivered at a railway station:—

Maximum Hourly Production of	Capacity of Revolution of Wheel.	Price of Meter, including stamping.
1200 Cubic feet.	10 Cubic feet	£38
1500 ,,	12½ ,,	45
1800 ,,	15 ,,	51
2400 ,,	20 ,,	56
3000 ,,	25 ,,	62
3600 ,,	30 ,,	75
4800 ,,	40 ,,	94
7200 ,,	60 ,,	140
9600 ,,	80 ,,	168
12,000 ,,	100 ,,	196
24,000 ,,	200 ,,	295

Beyond these sizes, the prices are arranged by special agreement; for the tell-tale apparatus, an addition of 10l. 10s. must be made.

The annexed engraving represents an index for a station meter, provided with that most indispensable machine in a gasworks—a clock—also a "tell-tale" apparatus, which records the production of gas during each successive hour. For this ingenious contrivance we are indebted to Samuel Crosley, and it was applied for the first time by Lowe, at the Chartered Gasworks, in 1823.

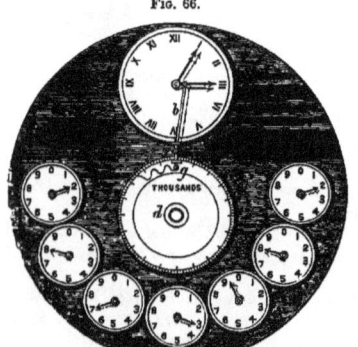
FIG. 66.

The dials of station, unlike those of consumers' meters, are made so that their figures are placed in the same direction, and, of course, their hands revolve in like manner—a very desirable arrangement, as it facilitates the reading of the index, and diminishes the liability to error. In the centre is a circular disc, fixed to an arbor which gears with the general index, of which it forms part.

A circular card or paper of the same size as the disc is attached thereto by means of a small plate, d, and thumb-screw, so that a fresh card can be placed daily with facility. The outer ring of this card is divided according to the maximum production for which the meter is estimated, that in question being 200,000 feet; of course having a corresponding number of divisions.

The minute hand of the clock, at a short distance from its axle, is provided with a projecting stud or pin, which is connected to, and actuates, the rod *b*, and thus gives motion to the small rod *e*, having guides to keep it always in the same position. At the end of this rod is a pencil holder *g*, containing a soft pencil, pressed by a light spring, which marks the card. Thus, by the action of the minute hand of the clock, the pencil is caused to move to and fro in a vertical direction, and if the card were stationary, the pencil would make simply a vertical line, but as it rotates, so, in direct proportion to its speed, is formed a curve of greater or less length, which, corresponding with the divisions on the card, records the quantity of gas passing through the meter each hour during the day and night.

The "tell-tale" is both ingenious and useful, and in our opinion its application should be extended to small works where the number of men employed is limited to two or three at each "shift," for in such places often a spirit of rivalry or jealousy exists amongst them, when at times it is difficult to ascertain the result of these respective operations. This difficulty the tell-tale is well calculated to remove.

In order to render the meter operative, it must be charged with water to a certain height, called the "water-line," and whenever the water within the case is above this point, the capacity of the wheel is diminished, and the instrument indicates a quantity greater than that which has passed therethrough, when it is technically termed "fast." If, on the other hand, the level of the water is below its proper position, the wheel is increased in capacity, when the meter indicates less than the quantity passed, and is "slow."

FIG. 67.

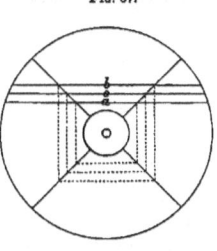

To illustrate this variation of measurement, we will imagine the annexed diagram to represent a wheel with its four measuring chambers, and, for the facility of explanation, we will suppose this to be about 13¼ inches in diameter, its superficial area being 136 square inches, and its depth 1 inch, thus making 136 cubic inches. For it is obvious that, whatever may be the depth, whether 1 inch or 6 inches, the capacity of the wheel will be alike affected by any alteration of the water-line; we have therefore, for the sake of simplicity, imagined the wheel of that limited depth.

Now, if we suppose the proper water-line to be marked *o*, and situated exactly 3 inches above the centre of the axle, then, in the revolution of the four chambers, the space occupied by the water within the wheel would represent a square, as shown by the corresponding dotted lines, each side of which would be 6 inches. Consequently, in the revolution of the wheel, 36 cubic inches would have to be deducted from its capacity, thus making the quantity of gas received and expelled 100 cubic inches.

But if the level of the water be raised 3½ inches, to the point marked *b*, then a square of 49 inches would have to be subtracted from the capacity of the wheel, thus rendering the meter 15 per cent. "fast." On the other hand, if we suppose the water to be at the point *a*, 2½ inches from the centre of the axle, then only 25 cubic inches would have to be deducted from the total capacity of the wheel, under which conditions the meter would only register 100, whereas 111 cubic inches of gas would have passed therethrough, or in other words it would be 11 per cent. "slow." On further abstracting the water below the centre ring, which corresponds with the orifice in the hollow cover, the meter is rendered inoperative, the gas passing freely without being measured.

When describing the results of different water-levels in the meter, no attempt has been made at mathematical accuracy; the angular position of the partitions of the wheel, and the quantity of gas contained in the inlet-hoods or passages, would render the calculation too complicated for our object; but these are mentioned for the purpose of calling attention to the necessity of maintaining the water-level at the proper height, for only by this means can correct measurement be assured.

This is a weak point which exists in some station meters, a circumstance at times taken advantage of by unscrupulous persons, in order to show a pretended increase in the production of gas per ton, and their own feigned superior abilities. These instruments, however, are constructed by the best makers so

2 G

as to prevent such deception, by placing an overflow pipe, the mouth of which is at the water-line, so that any excess of water flows off from the meter to a drain; and, indeed, in order to preserve the meter, as well as to prevent the water therein becoming contaminated with ammonia, which under some conditions would be communicated to the gas in its passage, it is desirable to have a small stream of clean water, proportioned to the make of gas, passing continuously through the instrument. This plan is adopted by some manufacturers, and should be applied by all.

It is essential that the station meter should be perfectly level from back to front, and the centre punch-marks made across the front by the manufacturer should also be horizontal. Neglect in either of these points will affect its accuracy.

When station meters are made of larger capacity than 20,000 feet per hour, their cases are generally rectangular, the base and sides being formed of flat plates, or with slight ridges or recesses, according to the style of ornamentation. The joints are planed and put together with bolts, their surfaces being previously smeared with "red mastic," composed of red and white lead, well intermixed.

Various manufacturers of station meters have, regardless of expense, displayed considerable ability in their construction, as may be witnessed in the majority of our large establishments, where they exist of elegant architectural design, sometimes ornamented with allegorical figures, and having mottoes in Latin or Greek. Indeed, the instrument in question, alike in small as in large works, is the principal ornament to the establishment, and an important object to present to casual visitors.

The largest station meter hitherto constructed has been recently erected at the Bromley Works by Messrs. Parkinson and Co., successors to Samuel Crosley, the first manufacturer of gas meters, which is represented in Plate 19 at the end of the volume. This meter is made to deliver 200,000 feet per hour, each revolution of the wheel being 2000 feet. Its outer case is 20 feet long, 18 feet wide, and 16 feet high, and weighs 53 tons. The measuring wheel is 16 feet long and 15 feet diameter, its weight being 7 tons. Thus its periphery travels, when in full action, at the rate of $1\frac{1}{4}$ mile per hour.

This apparatus is as near as possible of the same magnitude as the retort house described by Accum, already illustrated, as suitable for "lighting a town or large district." Its cost is equal to the whole of the capital invested in some towns of the United Kingdom having upwards of 1000 inhabitants, for the entire works, mains, lamp-posts, lanterns, complete; whilst the capacity of one revolution of its wheel would be sufficient for an average night's consumption of such a town. These facts convey a faint idea of the gigantic extent of gas manufacture in the metropolis, and the rapid progress it has made from the date of Accum's writing, hardly sixty years ago.

In large station meters the connecting pipes, inlet and outlet, are always provided with valves; there is also a by-pass valve on the main, or a combined arrangement for the purpose. Thus, in the event of an accident occurring to the meter, it can be thrown out of action immediately and disconnected for repairs, without interfering with any other part of the works.

This arrangement is very desirable for large works, but in very small establishments, where everything that tends to increase the capital must be avoided, these valves and by-pass for the meter can be consistently dispensed with. Interruptions to the proper action of station meters are so exceedingly rare, that the writer, during a long experience, never saw or heard of one; but he has more than once experienced the inconvenience of a plurality of valves in a small work, for, in the absence of the men in charge, officious meddling persons have sometimes shut a valve, and caused some little confusion. Moreover, suppose in a small works a meter to break down, all that is required is to open the tap or plug attached thereto, and allow a portion of water to flow off, when the gas will pass freely, and it will be no inconvenient matter to stop the charges for a short time whilst the instrument is disconnected. For these reasons we consider the valves and by-pass to the meters of small works neither necessary nor desirable.

The meter patented by Messrs. Warner and Cowan, to which we shall have occasion to refer hereafter, possesses the peculiar advantage that, under all the varied water-lines within a given range, its measurement is always practically correct; for this reason it must be admirably adapted for the purpose in question.

In concluding this subject, we may observe that, considering the simplicity of the construction of the meter, and the important functions it performs; that by it every particle of gas, in passing, is recorded with exactitude during an indefinite period; that its action neither impedes the flow of gas, nor affects the steadiness of the lights; that its work is performed in silence; that it requires little or no attention for months, and even years, except the necessary observations for taking its registration; that, when properly constructed and fairly used, its durability is almost unlimited—we say that, taking all these combined advantages together, the meter may be classed among the most beautiful and useful of all mechanical inventions; and so long as gas is used and measured by it, the names of its inventors—Samuel Clegg, John Malam, and Samuel Crosley—must ever be associated therewith.

GOVERNOR.

For the purpose of delivering the gas from the works to supply a town or district, the governor is indispensable, as by this instrument all the irregularities of pressure of the various gasholders are "governed," and the gas is supplied from the works at a uniform pressure according to the requirements, irrespective of any variation in the consumption; and, once the pressure adjusted, it is by this instrument maintained at the exit of the works with the most rigid accuracy during any indefinite period. But whatever this pressure may be at the exit of the works, it is controlled in the district by the various circumstances of the size of the mains supplying, the quantity of gas being consumed, and the varied elevations of different localities in relation to the source of supply.

An instrument of the description in question was amongst Clegg's earliest inventions, and patented by him in conjunction with his rotative meter, wet-lime purifier, and rotary retort, to which we have already referred. The general impression is that the governor first invented by Clegg was identical with that now in general use; this, however, was not the case, for the apparatus described in his specification, in addition to being very complicated, could not have been a practical machine, in consequence of the valve being formed by a plate or disc dipping into a tank charged with water; and thus, when delivering small quantities, the gas would have a tendency to "blow," and render the instrument useless.

The evils attending the use of this description of apparatus appear to have been at once recognized; but the first idea obtained, and its utility or necessity being evident, the instrument speedily assumed a practical form, as we find it described in Accum's 'Treatise,' published in 1819, very similar in construction to that of the consumers' wet regulator hereafter mentioned. At that early period it was manufactured by Clegg and Crosley, for the purpose of regulating or obviating the oscillations in the lights produced by their first consumers' meter, for which object a governor was always fixed in conjunction with its corresponding meter.

In 1825 Samuel Crosley obtained a patent for "improvements in the construction of gas regulators or governors," which was dissimilar to any apparatus of the kind previously made, as it dispensed with the water, which was essential to their action, and, in short, was what is now termed a "dry governor." This vessel is represented in section in Fig. 68, and consists of a cylindrical vessel A A, provided with a diaphragm or flexible division B, made of thin leather or other air-tight material, having the inlets and outlets at the bottom. From the centre of the diaphragm is suspended a rod, at the lower part of which is a conical plug E, adjusted to the flange above, the two forming a valve. The gas is admitted by the pipe h, and is emitted by that marked g; thus, by any increase of pressure beyond that to which it is adjusted, the diaphragm becomes elevated, raises the cone, and so controls the size of the orifice of the valve according to the supply of gas. As represented, the governor is supposed to be out of action, but on gas being admitted thereto the diaphragm will be elevated proportionably to the number of burners supplied, the pressure on the inlet, and the weight or pressure on the diaphragm.

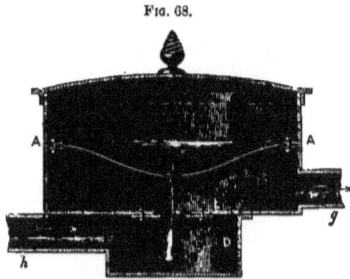

Fig. 68.

We may here observe, in order to show the limited knowledge of several inventors in connection with gas lighting, that since the date of Crosley's invention innumerable different patents have been obtained for dry "governors" or "regulators"; for the terms are synonymous, as by the former is understood the instrument for controlling the pressure as the gas issues from a works to the town (although for this purpose the dry governor is seldom used); and by the latter when it is fixed on the premises of the consumers, or on public lamps. Yet amongst all these patents no material change nor alteration from Crosley's governor has ever been produced, arising unquestionably from the circumstance that the instrument first invented embraced all the elements of simplicity, with perfection of action. It is, moreover, a singular fact that the value of this apparatus for the purpose of controlling the consumption of the public lights was not appreciated for nearly forty years after its invention.

Fig. 69 represents a governor as usually made, the apparatus being partly in section, for the purpose of facilitating the explanation. As ordinarily constructed, the outer case as well as the standards, bridge, and pipes, are of cast iron. The holder or bell is made of tinned sheet-iron, guided at the bottom by rollers, and at the top by a central vertical flat bar, passing between two rollers attached to the bridge, which is supported by the standards, and in some instruments on this flat bar is a scale which indicates the extent of the opening of the valve. At the lower part of the bell is an annular float, as seen in section of sufficient buoyancy to float the holder with the conical plug suspended therein. The inner pipe is the inlet, on the top of which is bolted a flange, corresponding with the cone, and adjusted with such precision that when the bell is elevated to its greatest height, they form a valve, which completely shuts off the gas, when there is no draught from the governor, or when all the weights are removed. But in proportion as the quantity of gas required for the town increases, so does the bell descend, and opens the orifice of the valve for its passage accordingly. The discs on the top of the bell are weights placed to give the desired pressure, which pressure is dependent on the area of the holder together with its weight. By this simple apparatus the pressure of the gas at the outlet of the governor is always controlled.

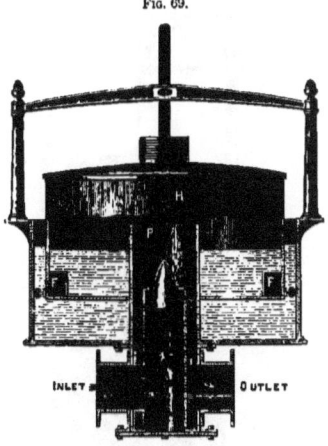

Fig. 69.

A drawing of a station governor manufactured by Messrs Parkinson and Co. for a main 36 inches in diameter is given in Plate 18, and is presented on account of its magnitude. The action of this is identical with that described; T is the tank, H the holder, S S the standards, F the float, C the cone, and V the valve. Some manufacturers make the top of the bell to serve the purpose of a tank, into which water is allowed to flow from a reservoir, by this means gradually giving the desired pressure, and when requisite to diminish this, the water is caused to flow therefrom by means of a flexible tube. This method has the advantage of avoiding the sudden irregularity occasioned by placing or removing the weights on the holder.

Another form of governor is represented by Fig. 70, in which the float is dispensed with and the holder is suspended by a chain passing over pulleys, and counterbalanced by sufficient weights to raise it and close the valve according to the requirements of the town or district supplied.

These weights, like those already referred to, correspond with a fraction of an inch of water pressure, as a tenth, or two-tenths; and with this apparatus, unlike the last instrument represented, in order to give the desired pressure the weights have to be removed instead of being added. The advantage derived by this form of governor consists in the absence of the float, which in the course of time is liable to decay, when it becomes exceedingly troublesome, in consequence of the water entering therein; this evil, however, is capable of being remedied by making the float of white metal, similar to that employed in the construction of wet-meter wheels. On the other hand, as hereafter shown, the absence of the float is attended with a corresponding inconvenience.

Some years ago an accident of a most alarming nature occurred at one of the Metropolitan works by the use of a governor of the construction in question, and was stated to have arisen from a workman—engaged in the governor-room—placing his foot upon the bell in such a manner that it tilted over and so got jammed ; and (whether through ignorance or fright was never explained) the vessel not being replaced in its proper position, the gas escaped therefrom in abundance, and was ignited by some neighbouring fire, producing a terrible explosion, which resulted in the death of several men, and rendering two gasholders at the station useless for some time.

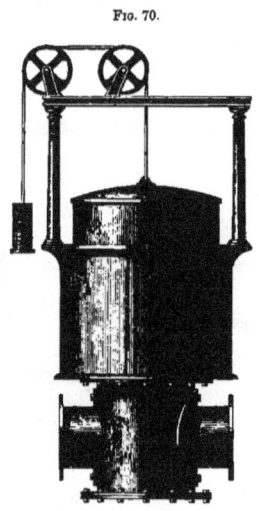

Fig. 70.

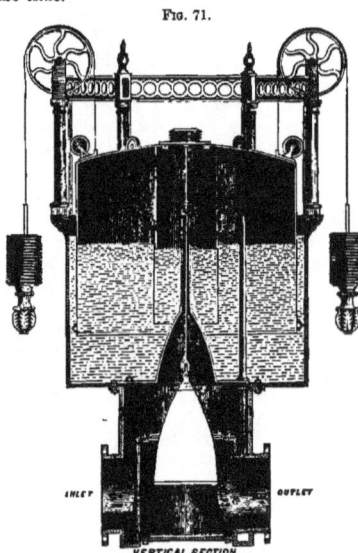

Fig. 71.

This calamity produced considerable alarm concerning the use of governors, and several means were proposed in order to avoid any repetition of such a misfortune. Most conspicuous of these were the inventions of Mr. Pritchard, of London, Mr. Everist, of Kidderminster, and Mr. Key, of Accrington, all of which inventions possess merit and fulfil the desired object.

The most recent apparatus introduced for a like purpose is that patented by Messrs. Braddock, of Oldham, and represented in section in the annexed figure, 71. This governor differs from the others in having two counterbalance weights, and reversing the position of the inlet, which delivers the gas on the top of, instead of beneath, the cone as usually adopted. The cone is suspended to the bell by the small rod which passes through the central pipe; there is also another small pipe in connection with the outer part of the bell. Thus, in the event of any accident, either through extraordinary pressure or other causes, only a very small portion of gas can possibly escape, which in any ordinary ventilated locality could not possibly be productive of mischief. In addition to this advantage, by the chamber within the bell having the same area as the cone, the pressure is always counterbalanced, and the disagreeable oscillations hereafter referred to, either by the heavy pressure acting on the base of the cone or the small area of the bell, are avoided.

All cone-valve governors, as usually made, have a tendency to jump or oscillate on the admission of any sudden heavy pressure, which, acting beneath the base of the cone, produces that annoying effect, and this is particularly the case with governors having small bells, and suspended; but when provided with floats this defect is not so likely to occur. Another evil arises from the increased weight of the bell when emerging from the water, and consequently augmenting the pressure, both of which contingencies are avoided by having the holders of large area.

The small governors of the present day have bells of much greater area relatively than the larger ones, in order to avoid the jumping action, as well as to render the instruments of sufficient importance in appearance. A good general working rule concerning the size of the valve is that its area shall be one-half the area of the connections or the main for which the governor is suitable, and its length equal to double its diameter. This rule is by no means absolute, for much depends upon the pressure given by the gasholders, the relative levels of the district to be supplied, and the maximum quantity of gas to be delivered within a certain time. However, accepting the rule given, then for a governor for a 4-inch main we have a valve-opening of $2\frac{3}{4}$ inches diameter, and as the cone terminates with a conical seating, its diameter at that point will be about $3\frac{1}{4}$ inches. The bell for such a governor will be 18 inches in diameter, or about twenty-two times greater in area than the cone-base, and the tank will be the required 20 inches in diameter.

A 30-inch governor made in the same proportion will have a $21\frac{1}{2}$-inch valve-opening, and with 20 as a multiplier would require a bell 8 feet 2 inches in diameter, which is about the proportions most universally adopted; but there are localities where, on account of limited space, or from motives of economy, smaller bells and tanks may be desirable, and to meet these requirements manufacturers have displayed considerable ingenuity.

In order to be enabled to prevent the oscillation in governors, and to construct them of small capacity without interfering with their efficiency, Mr. F. W. Hartley causes the principal part of the gas to pass from the valve-opening directly to the outlet pipe, and permits a small portion only to enter the bell through a bushed hole in the centre of the cover plate fixed on the upright outlet pipe P, in Fig. 69. The cone is suspended by a turned rod attached to the bell, which rod passes through the bushed hole in the cover, or, as that gentleman terms it, the "safety plate." By this arrangement oscillations of the bell are avoided, and accidents from the escape of gas similar to that referred to rendered impossible. When constructing governors with bells of limited capacity, Mr. Hartley employs a governor with a counterpoised bell dipping into a water tank, such bell being of the same diameter as the cone-base. The water tank is fitted internally with a stand pipe of suitable size, and connection is made between it and the inlet of the governor, so that any variation in the inlet pressure operates equally on the cone and on the counterpoise, just as the addition or subtraction of an equal weight from each side of a balance will leave the conditions of equilibrium undisturbed.

This method of compensation renders it practicable to reduce most materially the dimensions of the bells of large governors. Thus, in a 36-inch governor erected by Alexander Wright and Co. at the Commercial Gas Company's works, London, the area of the bell is barely $5\frac{1}{4}$ times the area of the cone-base, yet, owing to the compensator, there is no sensible variation under considerable changes of pressure, and the action of the instrument is perfectly steady.

It should be stated that independently of Mr. Hartley, and without any knowledge of his proceedings, a very similar contrivance to that just mentioned was adopted by Mr. Warner, both gentlemen being engaged at the same time in solving the difficulty connected with the use of small bells for station governors.

According to the invention of Mr. Peebles, of Edinburgh, a large station governor is controlled by an ordinary dry consumers' regulator in communication therewith. We have not had the opportunity of examining this apparatus, which appears paradoxical, but should it fulfil its duty effectively, which we have no reason to doubt, such a governor would be highly desirable in small works, where the manager or foreman could have a regulator situated in the most eligible position, either in his office or anywhere else, so that he could control the pressure without leaving the place.

There is also a governor, manufactured by Messrs. D. Hulett and Co., of London, which is alike applicable for gasworks and as a regulator for consumers' premises and street lamps. In this apparatus mercury is substituted for the water; and as the former is about fourteen times heavier than the latter, the dip, or seal, is diminished in a corresponding proportion. Thus, supposing a maximum pressure of 18 inches upon the works, and demanding that height of dip or seal in the governor, as ordinarily manufactured, with the mercurial governors only a seal of about 1¼ inch would be requisite. Hence, one advantage of these instruments is their extreme compactness.

There are various other methods of making the instruments under consideration, but differing so slightly from those mentioned that reference to them is unnecessary.

The rule for calculating the construction of the governor as usually made is simple. Thus, taking the cone as described as one-half the area of the connections, and the area of the holder twenty times that of the orifice of the cone, or ten times the area of the connections, then the two principal points are determined. If, now, we refer to the table of the pressure, corresponding with various columns of water, page 204, when referring to the action of gasholders, we have the means of ascertaining the influence of the weight and area of the bell together with the weight or float necessary for a certain pressure. To employ the first table let us suppose that it is required to know the weight to give a certain pressure on a holder of a given area, the former we will assume to be $\frac{1}{10}$ths and the latter 20 square feet. Then 1·562 lb. the weight per square foot corresponding with $\frac{1}{10}$ths pressure, multiplied by the area, gives 31·24 lbs. as the weight required. Or, if we suppose the weight and pressure are given, then, to find the area, divide the total weight by the number corresponding with the pressure; thus, 31·24 lbs. ÷ 1·562 lb. = 20 feet area; Or, the weight and area given, to find the pressure, then divide the weight by the area, hence 31·5 ÷ 20 = 1·575, corresponding with the weight per square foot for $\frac{1}{10}$ths pressure.

The capacity of the float for a holder of a governor is ascertained by allowing at the rate of a cubic foot of float for each 62·5 lbs. of the weight of holder, including the float and cone plug; hence, supposing the holder and cone, together with the estimated weight of the float, to be 2 cwt. or 224 lbs., this divided by 62·5 would give 3·568 cubic feet as the capacity of the float. In practice the smallest fraction is often disregarded, as it is hardly possible to work with such a degree of accuracy as that mentioned.

The depth of the bell will be controlled principally by the pressure on the works.

The cone, as stated, is in length usually twice that of its diameter at the base above the flange, and should be so formed that as the bell rises the orifice of the valve should be closed uniformly throughout its length, so that the scale on the vertical guide-bar or on the quadrant of the lever beam governor should indicate the extent of the opening in inches. This, although disregarded by manufacturers generally, is desirable in order to know the quantity of gas passing at any time, which is readily accomplished by reference to a table corresponding with the various degrees of opening and pressure.

In most works the governor is provided with a by-pass, and in all such cases its valve should be secured by lock and key, or other means, against the action of any mischievous or malicious person. To anticipate danger is to suggest means to avoid it; and if we reflect on the amount of evil that might occur, perhaps in the middle of the night, by any malicious person opening the by-pass valve, and allowing the full pressure of heavy holders to issue to the district or town, with all the numerous hydraulic joints of chandeliers in dwellings, of which some would of necessity become unsealed, and permit the gas to escape, then we realize the danger that would attend such an act. To prevent such a calamity the by-pass valve should be beyond the control of anyone except the foreman in charge of the works, and who alone should be responsible. The same kind of accident might happen with a few kinds of governors by the plug of the valve becoming detached from the holder, but from the nature of their construction the chances of this are very remote indeed.

Governors are also employed for controlling the pressure in localities where considerable variation exists in the levels of the district or town lighted. We believe that these were first introduced at Halifax, by Mr. G. W. Stevenson, but since then their use has been much extended, and with a view of explaining the advantages arising from them, the following observations may be necessary.

It is well known that when gas is enclosed within a main, with little or no draught therefrom, its pressure is increased about one-tenth of an inch for every augmentation of ten feet of elevation; thus, in a locality situated 200 feet above the level of the works, whilst the day pressure at the works would be $\frac{7}{10}$ths, it would be increased at the higher elevation to $\frac{27}{10}$ths. Hence, as the loss by leakage is increased in proportion to the square root of the pressure, and as no advantage is to be gained by any excess of this, it becomes desirable, under these circumstances, to reduce this pressure to within ordinary limits; and for which purpose, in some places, the "district governor" fulfils all requirements.

The most approved apparatus of the kind is that patented by Mr. Cathels, and consists of a dry governor of a peculiar construction, enclosed within a cast-iron case, having its inlet and outlet in a line with the main, provision being made for carrying off any condensation that may occur. The top is covered by a flange with bolts and nuts, so that at any time, should it be desirable, the pressure of the instrument can be altered. Unlike ordinary dry governors, in this the pressure enters upon the top of the cone, by which means the outlet pressure controls the instrument, hence, at whatever degree this is adjusted, that pressure is always maintained.

These governors are fixed on the leading main underground, and regulated to the desired extent, which we may suppose to be $\frac{7}{10}$ths, all that is necessary at the lowest points of a district. Thus, if they be placed at differences of 50 feet elevation on the main, extending through the town, the pressure would never exceed $\frac{12}{10}$ths in any part, whereas without these instruments it will vary according to the respective elevations. Hence, in certain localities, on the grounds of economy, and the prevention of loss of gas by leakage from the mains, the district governor is an important apparatus for a gas company.

However, we must not lose sight of the fact that gas, when in a main as described, on descending a hill loses its pressure in the same ratio that it gains in ascending. In a district, therefore, where great irregularities of elevation exist, when deciding on placing the instruments in question, the position of the lower levels should be well considered.

PRESSURE GAUGE AND REGISTER.

Gas, as generated from the coal during the process of destructive distillation, when subjected to the ordinary pressure existing in the apparatus of a gasworks, assumes a volume about 380 times greater than the coal from which it is extracted; and this volume is capable of augmentation and diminution by the effects of the variable atmospheric pressure, as well as temperature, to which reference is made hereafter. It is almost needless to say, that if the gas were not retained, it would be diffused throughout the atmosphere, and probably by becoming the source of alimenting vegetable creation, would again return to the state from which it sprung.

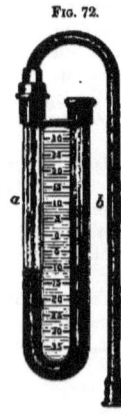

Fig. 72.

For the purpose of appreciating the extent of this pressure in a gasworks, the pressure gauge, represented in Fig. 72, is employed. This, in its simplest form, consists of a glass tube of about five-eighths of an inch internal diameter, bent as represented, half filled with water, and having the scale in the centre divided into inches and tenths of an inch. The part a being in communication with the gas, the water is depressed therein, and elevated in b, the top of which is open to the atmosphere; the distance between the two levels indicating the pressure, which, up to 2 or 3 inches, is generally described in "tenths;" from those points the term "inches" is applied.

In the gauge represented, both tubes have the same area, the depression of the liquid in one tube being equal to its elevation in the other; but when two tubes of unequal diameters are employed, the depression and elevation of the liquid is according to their respective areas.

At one period, when the baneful effects of heavy pressure in a gasworks were not understood, it was by no means unusual to find a pressure of 30, or even 40 inches, in some works, occasioned by small mains, defective purifiers, or stoppages. In these cases, mercurial pressure-gauges were used, which were similar to that described, but with tubes of only about a quarter of an inch in diameter; and as mercury is, as stated, about fourteen times heavier than water, these gauges were made of proportionate length. Thus, a column of 30 inches of water would be substituted by a mercurial gauge of $2\frac{1}{4}$ inches in length; but with our advanced knowledge, such pressure never exists in gasworks at the present time; therefore the mercurial gauge is no longer in requisition.

But if tubes of unequal area are employed, say the one of ten times larger sectional area than the other, the water will be depressed nine-tenths in the smaller, and elevated but one-tenth in the larger tube, and by taking advantage of this circumstance the pressure gauge can be made to indicate with one tube only, which in some cases may be desirable. This is accomplished by having a small enclosed cylindrical vessel, of 99 times the area of the glass tube, which is communicated to the side of the vessel in such a manner that the water level is perceptible in the tube at the point o of the scale. Thus the gas acting on the water within the cylinder depresses that only one ninety-ninth part, and raises it to the full extent of the pressure in the tube, and is appreciated by the scale attached.

To render the pressure gauge not only more ornamental, but more sensitive to the eye, it is sometimes made by combining a dial with pointer in connection with a float enclosed in a case, which float is elevated or depressed by the water in which it is immersed, and this again by the pressure of the gas. An arrangement of this nature is represented in Fig. 73, and known as King's gauge, being the invention of the late Alfred King, of Liverpool. In this instrument the lower part is divided into two concentric compartments, of which the outer is enclosed, but in communication with the gas and connected at the bottom with the central chamber, wherein is placed a metal float attached to a cord which passes over a pulley carrying a counterbalance weight, and on the shaft of the pulley is fixed the dial hand. The effect is, that the float, in rising or descending with the water contained in the compartments, gives motion to the hand of the dial, which dial is graduated according to the pressure, and by these means each tenth is increased to any desired degree, so that the smallest fraction is appreciable. This, together with the fact that the pressure is observed at a glance, and the ornamental character of the apparatus, are strong recommendations to their use.

Fig. 74.

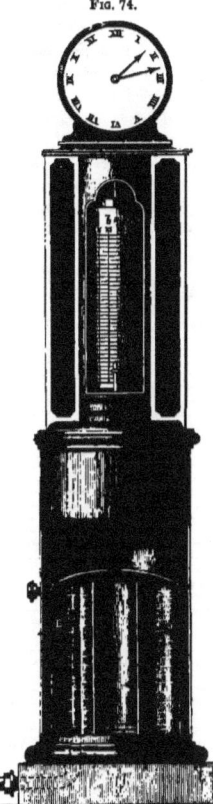

Fig. 73.

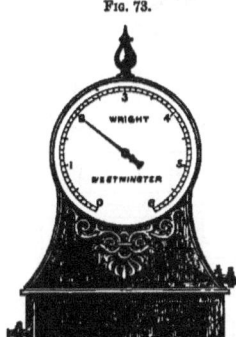

The same principle is applied by attaching a rack to the float, which in rising or descending actuates a pinion carrying the pointer.

In all the various parts of a gas-works, whether in the retort house, yard, or purifiers, the simple tube pressure-gauge is preferable to all others, as there is no likelihood of it being disarranged, whereas with others, if the water is not maintained strictly at its level, or when roughly handled, they are liable to mislead. By means of the gauge we are enabled to know the exact pressure existing in any part of the works, and from this ascertain the existence of any obstruction, either in the connections or in the apparatus; and, as described in the chapter on mains, stoppages in the mains of the town are readily detected by its aid.

We have now to refer to the constantly recording pressure-gauge, or, as generally termed, the pressure register, named by the French *le mouchard*, "the spy." For this simple and effective instrument we are indebted to Samuel Crosley, by whom it was invented in 1824, and the same year it was put in operation at the Brick Lane Station of the Chartered Company.

This instrument, like the governor, is automatic in its action, and serves to record continuously the pressure in the main with which it is in communication; all that is required for the purpose being to place a new sheet or "table" on the drum daily, and to wind up the timepiece regularly. These points attended to, the pressure is faithfully recorded during the twenty-four hours; and even any momentary irregularity or error in the operations committed by the attendant is noted.

The accompanying Fig. 74 represents a pressure register, shown partly in section, for the purpose of explaining its action. The lower case is either of cast or tinned wrought iron. Within this is a small holder, E, of about $13\frac{1}{2}$ inches in diameter, corresponding with a square foot in area. This holder is guided by rollers at the bottom, and by a rod at the top, which rod passes into the chamber above, containing the vertical drum b. In the centre of the holder is the float, F, of such capacity, that when the vessel is charged with water to the proper height, the holder is just buoyed or lifted from its

WRIGHT'S PRESSURE REGISTER.

bearing. The pipe *a* conveys the gas into the vessel, and, according to the pressure, so is the holder more or less elevated. At the top of the apparatus is a timepiece, and from this protrudes a rod into the chamber beneath, to which rod is connected the vertical drum *b*, which, by the action of the clock, makes exactly a revolution in the course of twenty-four hours. The middle chamber is provided with a glass door, so that the instrument can be observed continuously, the door being secured by lock and key, in order to prevent any tampering with the apparatus.

The ruled paper, or "table," Fig. 75, is in length exactly the same as the circumference of the drum, and has twenty-four numbers, with vertical lines corresponding with the hours of the day. There are also any number of horizontal lines, according to the construction of the apparatus, corresponding with the pressure to be indicated. This table is coiled on the drum, and kept in its position by two indiarubber bands; and as the drum revolves by the action of the clock, each of the figures of the hours corresponds with the time indicated by the clock. For the purpose of recording the pressure, there is attached to the top of the rod, in connection with the holder, a lead pencil, pressing with a light spring

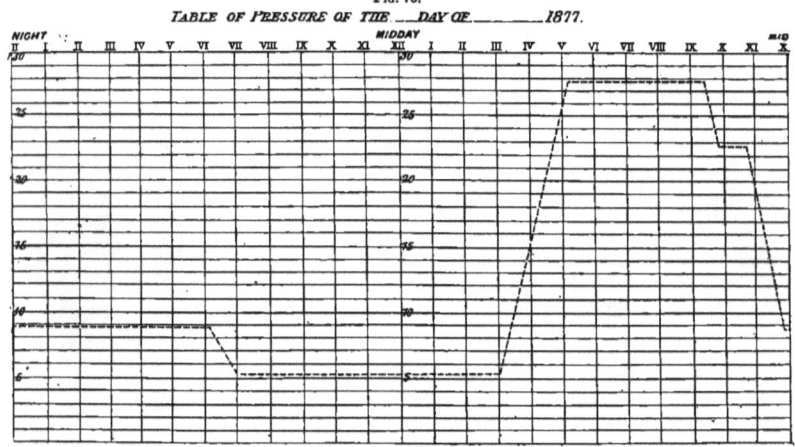

FIG. 75.

against and marking the ruled paper. Thus, on a sudden pressure being communicated to the holder, and withdrawn immediately, the rod with the pencil attached will rise and descend, making a vertical line; but, if the pressure be continued for any protracted period, the time will be indicated by the length of the horizontal line, in relation to the vertical lines.

In some works the table is ruled with pencil, according to the desired pressure, before being attached to the drum, the date being filled in for future reference; it then becomes the duty of the man in charge of the governor to adjust the pressure in accordance with the table. In other establishments the system adopted is to have a sheet on which the pressure during the various periods of the day and night is mentioned, and the attendant has to follow these instructions; the table of the pressure gauge indicating the degree of accuracy observed, one of these tables being required for each day's operations.

Another description of pressure register invented by Wright is represented in Fig. 76, and possesses the advantages of being more compact, and about half the price of the other. This instrument is made with a central circular chamber, open to the atmosphere at the top, but connected at the bottom to an enclosed annular chamber in connection with the gas, and formed by the case A which surrounds it.

Within the central chamber is a float, to which the rod b is attached, and on the instrument being charged with water to the desired height, it is ready for action; when, according to the pressure, the gas depresses the water in the annular chamber and elevates it in the other, in which is enclosed the float, which carries the rod b, provided with a pencil actuated by a spring, and guided by the four rollers. The paper disc c is placed on the dial of a clock-movement, which revolves, instead of the hour hand, once in twenty-four hours. The radiating lines correspond with the hours indicated by the respective numbers, and the concentric rings with the various degrees of pressure. Thus, as the disc revolves according to the degree of pressure, so will the rod b be elevated or depressed, and by means of the pencil the pressure at all the periods during the day and night is recorded. The main difference between Crosley's and Wright's instruments consisting in the fact, that the former represents the degrees of pressure on a magnified scale, whilst by the latter they are shown half their ordinary size.

Fig. 76.

Modifications of these instruments are also applied to register the action of the exhauster on the retorts. For this object, by some makers the holder is floated to that extent, that when in equilibrium with the atmosphere, the pencil is at the top of the drum, which point is marked o, and from this the different degrees of exhaust are indicated on descending the table; but in practice generally the instruments are made to show a few tenths of pressure above zero, which is accomplished by making the float so as to be in equilibrium at the desired point. The importance of the register to record the operations of the exhauster can be well understood, as any momentary stoppage or imperfect action of that instrument is thereby faithfully made known; and certainly no exhauster should be employed without having its corresponding register.

PHOTOMETER.

The photometer, or light measurer, is an instrument by which the relative illuminating power of any two sources of light may be estimated; as, for instance, a gas flame, which may be compared either with a variable number of candles, or with another gas flame, or with a lamp consuming a given quantity of oil in a definite time.

The last-mentioned system is very extensively adopted on the Continent, where the usual obligations of gas companies are to supply gas of such quality that a burner with a maximum consumption of 105 litres, or 3·608 cubic feet per hour, shall give the light of a Carcel lamp burning 42 grammes, or 648 grains, of colza oil during the same period. In these comparisons or testings, the disc, or shadow rod, hereafter referred to, is placed at exactly the same distance from both the gas flame and lamp flame, the question simply being to ascertain whether the stipulated quantity of gas is equal in luminosity to the lamp, or to determine the quantity of gas necessary to give the required amount of light.

The method of testing adopted in the United Kingdom and many other places differs materially from this, as in these cases the value of the gas has to be defined by a variable number of candles, which number is dependent on different circumstances, such as the stipulations of the Acts of Parliament by which the operations of the respective companies are governed; the contracts made with the local authorities, and other causes; and in imitation of the continental system referred to, this testing might be accomplished by placing the required number of candles at one end of the photometer bar, the gas flame being at the other, with the disc or shadow rod midway, in which case the shadows from the rod would be thrown on a horizontal white screen. But as this system would be attended with considerable inconvenience, advantage is therefore taken of the well-known law, which defines that the intensity of light varies inversely as the square of the distance the object illuminated is placed from the source of illumination.

Thus, if a screen of white paper or other material be placed at the distance of 1 foot from the flame, it will be illuminated with a certain degree of luminosity, which we may designate as *one*; but if it be removed to 2 feet, the degree of light will be diminished to *one-fourth*; and if, again, the screen be placed at a distance of 3 feet, the light thrown thereon will be only equal to *one-ninth* part of the intensity of that when the screen is placed at the shortest distance. To illustrate this the following table is given:

	Squares of distance.	Degree of light thrown on screen.
The screen distant from the flame 1 foot	1	1
,, ,, 2 feet	4	$\frac{1}{4}$
,, ,, 3 ,,	9	$\frac{1}{9}$
,, ,, 4 ,,	16	$\frac{1}{16}$
,, ,, 5 ,,	25	$\frac{1}{25}$

Hence it is obvious, that when the screen is placed at a distance of 1 foot from the light, it receives twenty-five times more illumination than if placed at 5 feet therefrom, all the other distances producing corresponding results.

This action is illustrated in a more forcible manner by supposing a hollow sphere of 2 feet diameter, and a light to be placed in the centre, when it would be situated 1 foot distant from every point of the surface, illuminating the whole equally throughout. But if the globe be increased to 4 feet in diameter, its surface would be 2 feet distant from the flame, and as the areas of spheres vary in the direct ratio of their respective diameters, thus the same light would be spread over four times the surface, and hence, the intensity of the light is diminished in a corresponding proportion. These are the principles upon which all photometers are based, as regulated by the Legislature for estimating the illuminating power of coal gas.

On the first establishment of gas lighting, the means adopted for determining the luminosity of gas was by placing a short vertical rod on a sheet of white paper situated between the two lights under investigation, which rod and paper were moved to and fro until attaining that point where the two shadows of the rod became of the same degree of obscurity.

The distances between the two lengths being then measured and squared, the greater divided by the lesser number gave the proportion desired. Thus, if the distance of the gas light be 75 inches, and that of the candle 25 inches, the square of the former divided by the square of the latter gives the illuminating power of the gas as equal to nine candles.

This method was superseded by Count Rumford's photometer, represented in Fig. 77, which consists of a vertical white screen S, placed against the wall or otherwise supported in its position. A rod d, having a suitable stand, is fixed a few inches from the screen, and G and C are the two lights under examination. The gas light G being adjusted to consume the desired quantity, is, for simplicity of calculation, fixed at a given distance from the screen, and by the intervention of the rod d it throws the shadow a. The lighted candle C is then gradually approached towards the rod until its shadow b is of the same degree of obscurity as the other. The distance of G from the screen being known, that of C is measured, and the square of the former divided by the square of the latter gives the value of the gas in candles, as already explained.

FIG. 77.

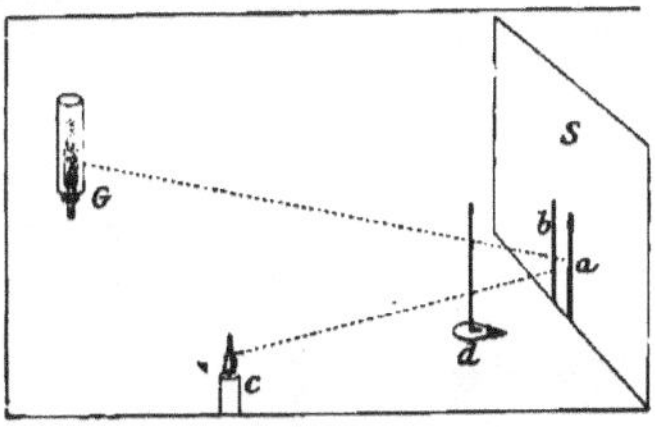

Subsequently Dr. Ritchie constructed a photometer, which consisted of a rectangular box of about 10 inches long and 3 inches square, open at the ends and blackened inside, having an aperture about 3 inches wide at the centre of the top, which was covered by a semi-transparent screen. Within the box, and immediately beneath this, were two small mirrors set at angles of 45°, their reflecting surfaces turned upwards towards the screen on the top of the box, with their upper extremities resting against each other, thus dividing the aperture into two equal parts, which were separated at the top by a slip of black wood to prevent the light from the mirrors combining. When using the instrument, it was placed so that the two openings at the end were on a level with the lights under examination, when it was moved towards the lesser light until the two portions of the screen were equally illuminated, the eyes of the operator during the observation being shaded by his hands to shut out any external light. The means of ascertaining the value of the respective lights was similar to those already mentioned.

Professor Wheatstone devised an instrument for the purpose in question, which consisted of a small circular box, and in the upper part of this was fixed a circular rack, gearing into an eccentrically-placed toothed wheel. A winding apparatus was attached to the bottom of the box, which gave motion to the internal gear-work, causing the eccentric wheel to gyrate rapidly round the circular rack. On top of the wheel was fixed a bright steel bead, placed in such a position that on giving motion to the winder this was caused to define a straight line across the box. All the parts except the bead were painted dull black. Thus, when set in action, the operator moved the instrument between the two lights under examination until the two lines of light reflected from the bead were of equal intensity. The instrument was both ingenious and beautiful, but its use was attended with considerable inconvenience, occasioned by strong vibration, which rendered it a difficult matter to determine by scale measurement the exact position indicating equality in the two lights.

All these instruments which were formerly in general use, like those of the present day, came under the operation of the law explained, where the relative value of two lights of unequal power placed at irregular distances, is ascertained. But, in addition to the difficulty of determining the intensity of the respective shadows, the calculations attending each operation were far too complicated for general use, and the extended requirements of gas lighting demanded a simpler and more sensitive instrument, and capable of universal application, which want was supplied by the photometer invented by Bunsen, and for the first application of which we are indebted to the late Mr. Alfred King, of Liverpool.

To illustrate familiarly the action of the instrument in question, if we take a sheet of writing paper on which a small spot of oil or grease is placed, the spot will be transparent when the paper is held between the observer and the light, but will be dark when in the opposite position; and if held in an apartment at right angles to the window in a vertical direction, at a point where the light on both its sides are of the same intensity, the spot then disappears. Here we may remark, that the light directly opposite a window of an apartment appears over a considerable space of the same uniformity; yet such is the sensitiveness of the simple appliance in question, that it is a matter of some little difficulty to ascertain the exact point where the light is of that equality to cause the spot to disappear. This is the principle of the Bunsen photometer, now universally adopted for determining the illuminating power of coal gas, and teaches us what we may learn by simply observing the effect of light on a piece of greased paper.

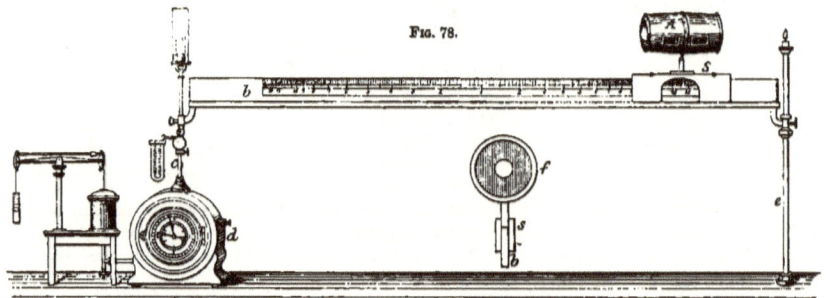

Fig. 78.

BUNSEN'S PHOTOMETER.

The Bunsen photometer, unlike those referred to, is so constructed as to require no calculation nor measurements whatever to determine the relative luminosity of two flames; for by it, the point of equality of light being ascertained, the scale at once affords the necessary information. This instrument in its simplest form is represented in Fig. 78, and consists of a flat wooden bar of any suitable length, of 60, 75, or 100 inches between the gas burner and candle. The bar b is supported at one end by the tube c, which is attached to and in communication with the experimental meter d, furnished with a pressure gauge and tap, and carrying the argand burner consuming the gas under investigation, and at the other end by the tube e carrying the candle. This bar is provided with a scale, the centre of which is marked 1; from this point to the end, where the candle is fixed, it is divided, according to the law already mentioned, into parts corresponding with squares of distances from one to thirty-six candles. To the saddle-piece S, which is provided with a pointer, and capable of being glided freely along the bar, is attached the conical tube or screen A, its centre being on a level with the gas and candle flames, having two orifices in front for the purpose of observing the disc which is placed within, and divides the screen into two equal parts.

This disc, which acts identically like the oiled paper already mentioned, is of white paper of about $4\frac{1}{2}$ inches in diameter, and in its centre is a round spot of about $1\frac{1}{4}$ inch in diameter left in its ordinary condition, the surrounding part being saturated with melted spermaceti. As thus prepared, when seen by a light situated on the same side as the observer, the centre part will be light and the outer ring dark, as shown at f, which represents the disc enclosed within its frame, and supported by the saddle-piece s on the bar b. In this case the disc is seen by reflected light; but if, now, the disc be placed between the observer and the light, the outer ring becomes transparent, consequently the light is transmitted through it. Hence these opposite effects are produced either by reflected or transmitted light, but in placing the disc in such a position that neither the one nor the other predominates, that is, when the light on both sides is exactly of the same uniformity, the centre spot then disappears, and both surfaces of the screen assume the same white appearance throughout, and on this simple contrivance the photometer is based.

When operating with this apparatus, the usual conditions are to employ an argand burner, the number of orifices as well as the other dimensions of which are defined, consuming 5 cubic feet per hour, that commonly used being the Sugg No. 1 argand burner, with a chimney 6 inches long and 2 inches wide for sixteen-candle gas, or 6 inches by 1⅝ for fourteen-candle gas. This burner flame is contrasted with the flames of two sperm candles, instead of one as represented in the drawing, in order to ensure greater regularity of the light given by the candles, each "standard" candle being made to consume as

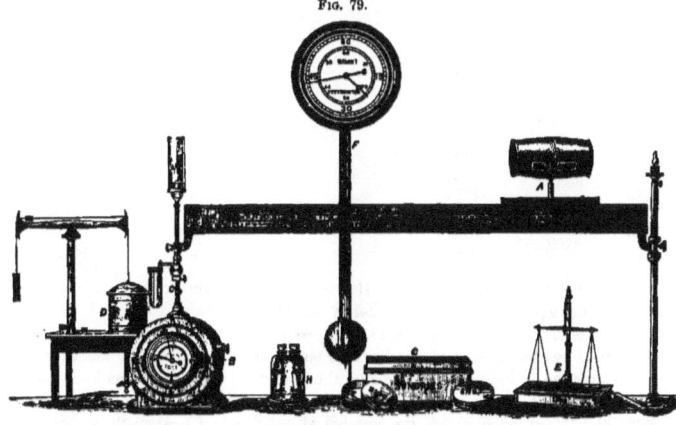

FIG. 79.

near as possible 120 grains of sperm per hour. In addition to the apartment being closed so as to shut out all daylight, its walls and ceiling are blackened or hung with black cloth or calico; if the latter, the glazed side is placed against the wall, the object of this being to avoid any reflected light which would interfere with the investigation.

A good experimental meter indicating the consumption per hour by an observation of a minute, a clock striking the minutes, a small regulator to prevent any variation in the pressure during the operation, the means of weighing the candles, and a thermometer, complete the list of apparatus essential for ordinary photometrical observations; but when the highest degree of accuracy is desired, other more minute details are required to be entered into as described hereafter in the instructions of the Metropolitan Gas Referees for the examiners of the gas supplied by the various companies.

The above figure, 79, represents a set of photometrical apparatus complete, as manufactured by A. Wright and Co. A is the screen containing the disc; B, the experimental meter; C, the micrometer regulating tap and pressure gauge; D, the regulator to control the pressure of the gas under examination; E, scales for weighing the candles; F, a clock striking each minute; G, a box containing sundries; H, a constant test: the whole being ready for operation.

RULE FOR MAKING PHOTOMETER.

For those managers who may be unprovided with a photometer, and who may be disposed to make that instrument, the following distances on the scale up to thirty-six candles are given; the fractional parts may be ascertained with tolerable accuracy by subdividing the intermediate space between any two numbers.

For this purpose, for the longer kind of photometer, the wooden bar will be required 100 inches long from the centre of the burner to the centre of the candle, about 4 inches wide and 1 inch thick, the scale being marked in front. Exactly in the centre, 50 inches from the centre of the burner and candle, is marked 1, and 8·58 inches from this, or 41·42 inches from the candle, 2, and so on throughout, according to the various distances. The saddle piece is about 8 inches long, and is lined with green baize or cloth to prevent friction. The discs can be readily purchased and are better suited for the purpose than any that can be prepared in the ordinary way.

No. of candles.	Distance of division from centre of candle. Inches.	No. of candles.	Distance of division from centre of candle. Inches.
2	41·42	17	19·52
3	36·61	18	19·08
4	33·34	19	18·67
5	30·90	20	18·26
6	28·99	21	17·96
7	27·43	22	17·57
8	26·12	23	17·25
9	25·00	24	16·95
10	24·14	25	16·67
11	23·16	26	16·40
12	22·40	27	16·14
13	21·71	28	15·90
14	21·09	29	15·64
15	20·52	30	15·41
16	20·00	36	14·28

The rule for finding the distances on the scale for the various number of candles is as follows:

First, ascertain the square root of the number of candles, from which subtract 1, and should there be no decimals, two cyphers must be added and the decimal point removed two places to the right. This has now to be divided by the number of candles less one, when the distance in inches will be obtained. For example, it is required to know the distance on the scale for 16 candles, of which number the square root is 4, subtracting 1, leaves 3, to which the cyphers are added, and the decimals removed, when we have 300 inches, and this divided by 15, the number of candles less 1, gives 20 inches as the distance desired. Again, if we suppose the number of candles to be 8, then the square root of this is 2·8284271, from which, on subtracting 1 and removing the decimal point, we have 182·84271, and this divided by the number of candles less 1, equal 7, gives 26·12 as the distance.

In photometers, as now generally made, at the back of the disc are placed two mirrors, cut from the same piece of glass in order to avoid any difference of colour, by which means the operator can compare both sides of the disc at the same time, instead of being obliged as formerly to carry in his memory the shade of one side when comparing the other. This system is now applied to all descriptions of photometers.

To compensate for the varying light given by the gas and candle, a series of observations must be made and the mean of these taken as the illuminating power of the gas; usually the operation will occupy from twenty minutes to half an hour, the gas and candles being lighted about ten minutes before commencing; this precaution is necessary, as burners when first lighted do not pass the same quantity of gas at any given pressure as they do when they become thoroughly heated, nor do candles immediately attain their maximum rate of consumption.

For reading the scale, a small hand looking-glass is sometimes used, and in some photometers, a diminutive mirror is attached to the screen close to the orifice and placed at such an angle as to reflect the light from the gas on to the scale at the pointer. Among other precautions to be observed are to burn out all trace of air in the fittings and meter, the more particularly when this is a dry meter, for, as already stated, a very small quantity of air is sufficient to deteriorate materially the illuminating power of gas. The gas should be burnt as near as possible at the rate of 5 feet per hour, for if it be sensibly below this, the test will be unfavourably conducted for the gas.

The greatest difficulty in connection with photometry is the want of uniformity of the candles both as regards the quality of the material of which they are composed, as well as their rate of consumption. Pure sperm candles are not easily procured, and when otherwise than pure they cannot be relied on; then as to their consumption, it is seldom that they are made to burn 120 grains per hour, but more frequently 130 or 135 grains, when the conditions are changed, as a candle burning the larger quantities will give more light in proportion to the quantity of material consumed than when burning the defined quantity. Another cause of uncertainty in the operations is the irregularity of the candles in burning, which demands a series of experiments. A further source of error arises from the difference of the colour of the two lights, resulting from the gas and candle; therefore under all these conditions, it can hardly be surprising if a variation of half a candle exists in the experiments of two operators, when examining the same gas.

Fig. 80.

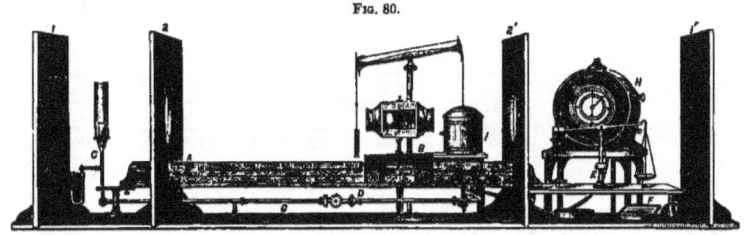

LETHEBY'S PHOTOMETER.

On the other hand, according to Mr. Hartley, who is a good authority on the subject, with candles of first quality, good apparatus, and ordinary skill, a remarkable degree of accuracy can be attained, as in a series of experiments conducted by that gentleman in conjunction with another "by no means practised photometerist" in testing the quality of eight different kinds of gas, made independently of each other, the extreme variation did not amount to $2\frac{1}{2}$ per cent., and the mean was less than one per cent. between the results obtained by the operators. Hence, from this we learn the importance of having proper standard candles.

Another arrangement of the instrument in question has been made by Messrs. Church and Mann, in which the disc is placed in a frame, at a fixed distance of 10 inches from the candle, which frame glides backwards and forwards on a graduated bar; the advantages of this photometer consisting in its limited length (being only 5 feet), and the proximity of the disc to the candle by which the scale is read with greater facility, but on the other hand, the moving of the candle is objectionable.

A further improvement in the photometer introduced by the late Dr. Letheby consists in employing four blackened screens as represented in the annexed Fig. 80, those numbered 2 and 2' having orifices in their centres through which the rays of light from the candle and gas pass, the screens numbered 1 and 1' are entire and prevent any reflected light passing from those points. The advantage this class of apparatus possesses is that the eyes of the operator are protected from the lights under examination, which permits of increased accuracy being obtained. In this photometer, as represented, two candles are employed and placed in a balance E, so that they are weighed during the operation of testing the gas. The regulating tap is at the centre D, under the control of the operator when observing the reflections of the discs on the mirrors, as shown through the orifice.

A modification of the photometer by Mr. F. J. Evans consists in enclosing the burner, candles, screens, and disc within a dark chamber or box, of the length of the photometer, placed on stands, as represented in Fig. 81, so that the sight hole in the centre is brought to the level of the eye.

This box is blackened on its interior and has a slot at the bottom, extending from end to end, and in this slot a socket for carrying the candles is made to move freely backwards and forwards by means of an endless band working over pulleys affixed beneath the instrument. The slot also permits of a proper supply of air entering for the combustion of the gas and candles, whilst provision is also made for the escape of the products of combustion. The front of the box consists of three doors corresponding with the panelling, which permit of the arrangement of candles, gas, and screen, but are closed during the operation; in addition there is a small curtain which encloses the operator during the investigation. Thus the great advantage of this apparatus arises from the circumstance that it can be employed anywhere, without blacking the walls or closing the shutters, or similar precautions observed with all other photometers. At the centre of the box and midway between the gas and candles is fixed the

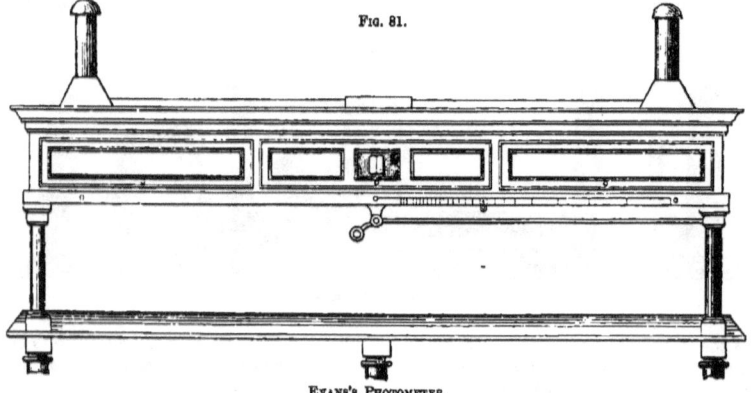

Fig. 81.

Evans's Photometer.

screen and disc, and behind it are two mirrors, so that in observing by the sight hole the reflection of both sides of the disc is seen, as explained in the last illustration. A handle in the centre of the instrument just beneath the sight hole actuates the endless band, and brings the candles to the required position, the same time that the operator is observing the disc.

The action of this instrument is similar to the others described, but as the distance from the disc to the gas is constant, namely, 50 inches, and the distance of the candles variable, the scale is divided in the opposite direction to that mentioned. In the former the lowest numbers are nearest to the centre, whilst in this the reverse is the case.

The rule for ascertaining the distance for any desired number of candles for this class of photometer is as follows:

As the number of candles is to 1, so is the square of 50 to the square of the number of inches required.

Thus if the distance of 25 candles is wanted, then as $25 : 1 :: 50^2 : 100$, the square root of which is 10, the number of inches the scale will be required.

In all photometrical experiments, in order to ensure accuracy, the greatest care is indispensable, and in addition to the other precautions mentioned, the apartment should be at a moderate temperature, and with some apparatus no one besides the operator should be permitted to approach the disc during the

INSTRUCTIONS OF GAS REFEREES. 245

investigation, as the reflection of any article of dress is sufficient to mislead. The gas on being lighted is left burning for some minutes, at the same time its pressure is properly regulated by a governor, in order to avoid any variation in the supply, when its consumption is accurately adjusted to the required degree. As there is always great irregularity in the form of the candle, the lower part being thicker than the other, to avoid errors arising from this cause the candle to be employed is cut in two pieces, and the lower half trimmed so as to burn the two opposite ends together, and when their wicks are incandescent, on the external light being shut out, all is ready for operation.

The observation is then made by advancing the disc gradually to the point where the spot disappears and both sides of the disc are alike or nearly so, and at this part of the experiment the greatest exactitude is requisite, as the eye is easily deceived by the different colours of the lights from the gas and candle. This accomplished, the pointer attached to the saddle indicates on the scale of the bar the illuminating power of the gas, but as photometers are usually made to burn one candle only, and as in the experiments, for reasons stated, two candles are employed, it follows that if we suppose the number indicated on the scale to be eight, this will have to be multiplied by two, thus showing the illuminating power of the gas to be 16 candles.

In the following comprehensive instructions other detail is entered into, which renders it unnecessary to say any more on the matter here.

THE INSTRUCTIONS OF THE METROPOLITAN GAS REFEREES "as to the times and mode of testing the illuminating power of the gas of some of the London companies":

"The testings for illuminating power shall be three in number daily.

"The photometers to be used in the testing stations shall be the improved form of the Bunsen photometer, which shall be certified by the referees.

"The disc in the photometer shall be changed at least once a week.

"The glass chimneys must be cleaned daily at the commencement of each set of observations.

"The candles shall be such as are described in the Metropolitan Gas Act of 1860, namely, sperm candles of six to the pound, each burning 120 grains per hour.

"Two of these candles shall be used together.

"The quantity of gas burnt in each burner shall be 5 cubic feet per hour.

"The gas in the photometer is to be lighted at least fifteen minutes before the testings begin. The gas shall be kept continuously burning from the beginning to the end of the testings.

"Each testing shall include ten observations of the photometer, made at intervals of one minute.

"The consumption of the gas is to be carefully adjusted to 5 feet per hour, which is shown by the long hand of the motor and the seconds hand of the clock travelling together continuously for some minutes.

"The candles are to be lighted at least ten minutes before beginning each testing, so as to arrive at their normal rate of burning, which is shown when the wick is slightly bent and the top glowing. The standard rate of consumption for the candles is 120 grains each per hour. Before and after making each set of ten observations of the photometer, the gas examiner shall weigh the candles, and if the rate of consumption shall have been more or less per candle than 120 grains per hour, he shall make and record in a book to be kept for that purpose the calculations requisite to neutralize the effects of this difference.

"The gas examiner shall observe and record the temperature of the gas as shown by the thermometers attached to the meters, and also the height of the barometer. The volumes of the gas operated upon during the testings may be corrected by these data in the following manner, the standard being, for the barometer, 30 inches, and for the thermometer 60 degrees. Suppose the thermometer stands at 30 inches, and the barometer at 30·5 inches; multiply the quantity of gas consumed by the *Tabular Number* corresponding to the indicatures of the barometer and thermometer as given in the table, pages 246 and 247, the product will be the corrected volume of the gas, i. e. the volume the gas would have occupied at the standard temperature and pressure; thus:

Volume of gas consumed .. .. 5 feet
Tabular number of barometer and thermometer .. 1·025

Then $1·025 \times 5 = 5·125$, the corrected volume.

"The same object may be attained by simply dividing the illuminating power by the tabular number; thus the former being 15·334 candles, divided by 1·025, the correction for temperature and pressure gives 14·9 candles as the corrected illuminating power.

"The average of each set of ten observations is to be taken as representing the illuminating power for that testing. And the average of the three testings is to be taken as representing the illuminating power of the gas for the day.

"The gas examiner shall record his observations and calculations for illuminating power in the proper prescribed form.

"The calculations for working out the corrections, &c., for the illuminating power of the gas proceed in the following manner. Add the observations together and divide the sum by 10 to get the average; then, as two candles are used, multiply by 2 to get the illuminating power of the gas if tried against one candle. Then, as the standard rate of the consumption of the candles (viz. 120 grains) is to the average number of grains consumed by each per hour, so is the above-obtained number to the actual illuminating power. Finally, make the correction for

TABLE TO FACILITATE THE CORRECTION OF THE VOLUME OF GAS AT

BAR.	THER. 32°	34°	36°	38°	40°	42°	44°	46°	48°	50°	52°	54°	56°	58°	60°
28·0	·988	·984	·980	·976	·971	·967	·963	·960	·956	·952	·948	·944	·941	·937	·933
28·1	·991	·987	·983	·979	·974	·970	·967	·963	·959	·955	·952	·948	·945	·941	·937
28·2	·995	·991	·987	·983	·978	·973	·970	·968	·963	·959	·955	·951	·947	·944	·940
28·3	·998	·994	·990	·986	·982	·977	·974	·970	·966	·963	·958	·955	·951	·947	·943
28·4	1·002	·998	·993	·990	·985	·980	·978	·973	·970	·966	·962	·958	·954	·951	·947
28·5	1·005	1·001	·997	·993	·988	·984	·980	·978	·973	·970	·965	·961	·958	·954	·950
28·6	1·009	1·005	1·000	·996	·992	·987	·984	·980	·976	·973	·969	·965	·961	·957	·953
28·7	1·012	1·008	1·004	1·000	·995	·991	·987	·983	·980	·976	·972	·968	·964	·961	·957
28·8	1·016	1·012	1·008	1·003	·998	·995	·991	·987	·983	·980	·976	·971	·968	·964	·960
28·9	1·020	1·015	1·011	1·007	1·002	·998	·994	·991	·987	·983	·979	·975	·971	·968	·963
29·0	1·023	1·019	1·015	1·010	1·006	1·001	·998	·994	·990	·986	·982	·978	·974	·971	·967
29·1	1·027	1·022	1·018	1·014	1·009	1·004	1·001	·998	·993	·989	·985	·982	·978	·974	·970
29·2	1·030	1·026	1·022	1·017	1·012	1·008	1·004	1·001	·997	·993	·989	·987	·981	·977	·973
29·3	1·034	1·029	1·025	1·021	1·016	1·011	1·007	1·004	1·000	·996	·992	·988	·984	·981	·977
29·4	1·037	1·033	1·029	1·024	1·019	1·015	1·011	1·008	1·004	1·000	·996	·992	·988	·984	·980
29·5	1·041	1·036	1·032	1·028	1·023	1·018	1·014	1·011	1·007	1·004	1·000	·995	·991	·987	·983
29·6	1·044	1·040	1·036	1·031	1·026	1·022	1·018	1·015	1·010	1·006	1·002	·999	·994	·991	·987
29·7	1·048	1·043	1·039	1·035	1·030	1·025	1·021	1·018	1·014	1·010	1·006	1·002	·997	·994	·990
29·8	1·051	1·047	1·043	1·038	1·033	1·029	1·025	1·022	1·017	1·013	1·009	1·005	1·001	·997	·993
29·9	1·055	1·050	1·046	1·042	1·037	1·032	1·028	1·025	1·021	1·017	1·013	1·009	1·004	1·001	·997
30·0	1·058	1·054	1·050	1·045	1·040	1·036	1·032	1·028	1·024	1·020	1·016	1·012	1·008	1·004	1·000
30·1	1·062	1·057	1·053	1·049	1·043	1·039	1·035	1·032	1·028	1·023	1·019	1·015	1·011	1·007	1·003
30·2	1·065	1·061	1·057	1·052	1·047	1·043	1·039	1·035	1·031	1·027	1·023	1·019	1·015	1·011	1·007
30·3	1·069	1·064	1·060	1·056	1·051	1·047	1·043	1·039	1·034	1·030	1·026	1·022	1·018	1·014	1·010
30·4	1·072	1·068	1·064	1·059	1·054	1·050	1·046	1·042	1·038	1·034	1·029	1·026	1·021	1·017	1·013
30·5	1·076	1·071	1·067	1·063	1·058	1·053	1·050	1·045	1·041	1·037	1·033	1·029	1·025	1·021	1·017
30·6	1·079	1·075	1·071	1·066	1·061	1·056	1·053	1·049	1·045	1·040	1·036	1·032	1·028	1·024	1·020
30·7	1·083	1·079	1·074	1·070	1·064	1·060	1·056	1·052	1·048	1·044	1·040	1·036	1·031	1·027	1·023
30·8	1·087	1·082	1·078	1·073	1·068	1·063	1·060	1·056	1·051	1·047	1·043	1·039	1·035	1·031	1·027
30·9	1·090	1·086	1·081	1·077	1·072	1·067	1·063	1·059	1·055	1·051	1·046	1·043	1·038	1·034	1·030
31·0	1·094	1·089	1·085	1·080	1·075	1·070	1·067	1·063	1·058	1·054	1·051	1·046	1·042	1·037	1·033

INSTRUCTIONS OF GAS REFEREES.

temperature and pressure by dividing the illuminating power obtained by the average of ten observations, after making the proper corrections for the consumption of the candles, by the tabular number 1·025. Thus, supposing the illuminating power to be 15·334 candles, this divided by 1·025 gives 14·9 = the corrected illuminating power in candles."

The following table, a portion of which forms the appendix of the instructions referred to, is substantially the same as that which has appeared from time to time in several works on gas; some of the higher and lower degrees of temperature are omitted by the Referees, which we have, however, reproduced.

DIFFERENT TEMPERATURES AND UNDER DIFFERENT ATMOSPHERIC PRESSURES.

BAR.	THER. 90°	62°	64°	66°	68°	70°	72°	74°	76°	78°	80°	82°	84°	86°	88°
28·0	·881	·930	·926	·922	·919	·915	·911	·907	·904	·900	·897	·893	·891	·888	·885
28·1	·884	·933	·929	·926	·922	·919	·914	·911	·907	·904	·900	·896	·894	·891	·888
28·2	·888	·936	·933	·929	·925	·922	·917	·914	·911	·907	·903	·899	·898	894	·891
28·3	·891	·940	·936	·932	·929	·925	·921	·917	·914	·911	·906	·902	·901	·897	·894
28·4	·894	·943	·939	·936	·932	·928	·924	·920	·917	·914	·910	·905	·904	·900	·897
28·5	·897	·946	·943	·939	·935	·932	·927	·923	·920	·917	·913	·909	·907	·904	·900
28·6	·900	·950	·946	·942	·939	·935	·931	·927	·923	·919	·916	·912	·910	·907	·903
28·7	·903	·953	·949	·945	·942	·938	·934	·930	·926	·922	·919	·915	·913	·910	·907
28·8	·906	·956	·952	·949	·945	·941	·937	·934	·929	·925	·922	·918	·917	·913	·910
28·9	·910	·960	·956	·952	·948	·944	·940	·937	·933	·928	·925	·921	·920	·916	913
29·0	·913	·963	·959	·955	·952	·948	·943	·940	·936	·902	·928	·925	·923	·919	·916
29·1	·916	·966	·962	·959	·955	·951	·947	·943	·939	·935	·932	·928	·926	·923	·919
29·2	·919	·969	·966	·962	·958	·954	·950	·947	·942	·938	·935	·932	·929	·926	·922
29·3	·922	·973	·969	·965	·961	·957	·953	·950	·946	·941	·938	·935	·933	·929	·926
29·4	·925	·976	·972	·969	·965	·961	·956	·953	·949	·945	·941	·938	·936	·932	·929
29·5	·928	·979	·975	·971	·968	·964	·960	·956	·952	·948	·945	·941	·939	·935	·932
29·6	·932	·982	·979	·975	·971	·967	·963	·959	·955	·952	·948	·945	·942	·939	·935
29·7	·935	·986	·982	·978	·974	·970	·966	·962	·959	·955	·951	·948	·945	·942	·938
29·8	·938	·989	·985	·981	·977	·974	·970	·966	·962	·958	·954	·951	·948	·945	·941
29·9	·941	·992	·989	·985	·981	·977	·973	·969	·965	·961	·957	·954	·952	·948	·944
30·0	·944	·996	·992	·988	·984	·980	·976	·972	·968	·964	·961	·957	·955	·951	·948
30·1	·947	·999	·995	·991	·987	·983	·979	·975	·972	·968	·964	·960	·958	·954	·951
30·2	·950	1·003	·998	·994	·990	·987	·983	·979	·975	·971	·967	·963	·961	·958	·954
30·3	·954	1·006	1·002	·998	·994	·990	·986	·982	·978	·974	·970	·966	·964	·961	·957
30·4	·957	1·009	1·005	1·001	·997	·993	·989	·985	·981	·977	·974	·969	·968	·964	·960
30·5	·960	1·012	1·008	1·004	1·000	·997	·993	·989	·984	·980	·977	·973	·971	·967	·963
30·6	·963	1·016	1·012	1·008	1·004	1·000	·996	·992	·988	·984	·980	·976	·974	·970	·967
30·7	·966	1·019	1·015	1·011	1·007	1·003	·999	·995	·991	·987	·983	·979	·977	·973	·970
30·8	·970	1·022	1·018	1·014	1·010	1·006	1·002	·999	·994	·990	·986	·982	·980	·977	·973
30·9	·973	1·026	1·022	1·017	1·014	1·010	1·006	1·002	·997	·993	·990	·985	·984	·980	·976
31·0	·976	1·029	1·025	1·021	1·017	1·013	1·009	1·005	1·000	·996	·993	·989	·987	·983	·979

The preceding instructions, emanating from such authorities, may be considered the most complete and perfect for the investigations under consideration, and indicate all the necessary precautions to be observed in photometrical experiments.

Whilst speaking of the candle it has been remarked that, if gas lighting had existed prior to the candle, the invention of the latter would have been regarded as a marvel of ingenuity, and in this there is no exaggeration; for in the burning of a common candle we find a number of operations continually in progress; firstly, on the wick being ignited, its flame melts the tallow necessary for its supply, at the same time forming a reservoir of ample capacity, securely walled round to prevent any loss of material; from that reservoir small globules are seen constantly in motion towards the wick, many of them, as a writer observes, "attached thereto, studding it all over like little sparkling diamonds;" this wick in its turn forms a series of capillary tubes conveying the fuel to the flame in the form of hundreds of tiny globules which are seen continually ascending the wick, whilst hundreds of others are every instant exploding and discharging their contents into the flame, there to be decomposed and rendered into gas at the moment it is required for ignition. In short, the whole operation of gas manufacture is conducted in the candle, and this without fear of stopped ascension pipes, strikes of stokers, or Board of Trade restrictions.

FIG. 82.

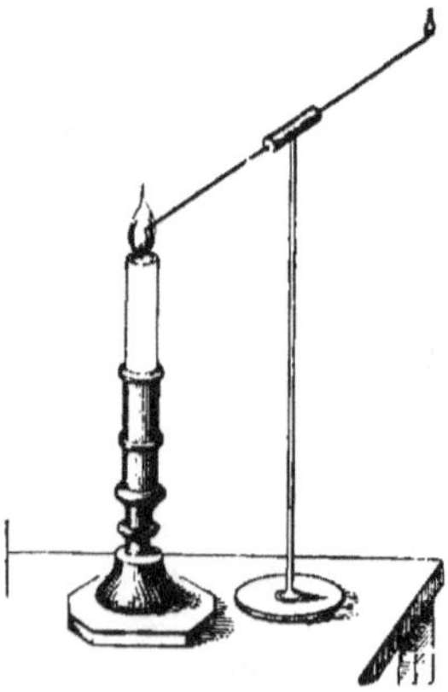

Hence we find, that when comparing gas with a candle it is in fact the comparison of one gas with another, which assertion is readily proved by the following experiment first brought into notice by Sir Humphry Davy. In the annexed diagram is represented a stand supporting a brass tube of about an eighth of an inch in diameter, one end of which is placed in the centre of the flame of the candle where the unignited gas exists. From that point the gas is conveyed through the tube, and may be ignited at the end as shown. It is hardly necessary to observe, that by holding the tube steadily in the candle the same effect is produced.

Among the extraordinary discoveries of the present day bearing on our subject is the radiometer of Mr. Crookes, by which it is demonstrated that motion is obtained by the action of the rays of heat and light, and as that gentleman has proposed to employ the instrument as a photometer it merits our special attention. There are two kinds of these instruments, but the most striking is that represented in Fig. 83, which we describe in its inventor's own words.

"This consists of four arms of some light material suspended on a hard steel point, resting in a cup so that the arms are able to revolve horizontally upon the centre pivot in the same manner as the arms of Dr. Robinson's anemometer revolve. To the extremity of each arm is fastened a thin disc of roasted mica or pith, white on one side and lampblacked on the other, the black surfaces of all the discs facing the same way. The whole is enclosed in a thin glass globe, which is then exhausted to the highest attainable point and hermetically sealed.

"The arms of this instrument rotate with more or less velocity under the action of radiation, the rapidity of revolution being directly proportionate to the intensity of the incident rays. Placed in the sun or exposed to the light of burning magnesium, the rapidity of the revolution is so great that the separate discs are lost in a circle of light. Exposed to a candle 20 inches off, another instrument gave one revolution in 182 seconds; with the same candle placed at a distance of 10 inches off, the result is one revolution in 45 seconds, and at 5 inches off, one revolution was made in 11 seconds. Thus it is seen that the mechanical action of radiation is inversely proportional to the square of the distance. At the same distance two candles give double and three candles give three times the velocity of one candle and so up to twenty-four candles.

CROOKE'S RADIOMETER.

"In all respects, therefore, it is seen that the radiometer gives indications in strict accordance with theory.

"A small radiometer was found to revolve at the velocities shown in the following table when exposed to the radiation of a standard candle 5 inches off.

TIME REQUIRED FOR ONE REVOLUTION.

Source of Radiation.			Time in Seconds.
1 candle 5 inches off, behind green glass			40
"	"	blue "	38
"	"	purple "	28
"	"	orange "	26
"	"	yellow "	21
"	"	light red "	20

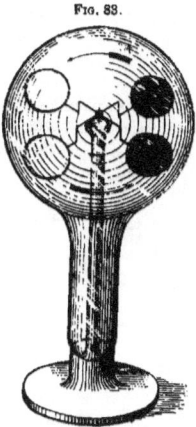

FIG. 83.

CROOKES'S RADIOMETER.

"The position of the light in the horizontal plane of the instrument is of no consequence, provided the distance is not altered. Thus, two candles one foot off give the same number of revolutions per second, whether they are side by side or opposite to each other. From this it follows that if the radiometer is brought into a uniformly lighted space it will continue to revolve.

"In diffused daylight the velocity was one revolution in from 1·7 to 2·3 seconds according to the intensity of the incident rays. In full sunshine at 10 A.M., it revolved once in 0·3 second, and at 2 P.M. it made four revolutions per second.

"When heat is cut off by allowing the radiation to pass through a thick plate of alum, the velocity of rotation is somewhat slower, and when only dark heat is allowed to fall on the arms (as from a vessel of boiling water) no rotation whatever is produced.

"Several radiometers of various constructions as regards details, but all depending on the discovery in question, have been exhibited at the Royal Society, where their novelty and unexpected indications excited a considerable amount of interest.

"By timing the revolutions of the instrument when exposed direct to a source of light, a candle for instance, the total radiation is measured. If a screen of alum is now interposed, the influence of heat is almost entirely cut off, the velocity becomes proportionately less, and the instrument becomes a photometer; by its means photometry becomes much simplified, flames the most diverse may readily be compared between themselves or with any other sources of light; a 'standard candle' can now be defined as one which at x inches off causes the radiometer to perform y revolutions per minute, the values of x and y having been previously determined by comparison with some ascertained standard; and a statement that a gaslight is equal to so many candles may with more accuracy be replaced by saying that it produces so many revolutions."

The engraving is about the full size of the radiometer; the four discs are placed at right angles to each other, and revolve continuously in the direction indicated by the arrows; the speed being, as stated, in direct proportion to the degree of light to which it is subjected, and such is the degree of sensitiveness, that in dull weather when revolving about once in six seconds, on holding a lighted candle close to it, the instrument revolves with rapidity, and on withdrawing the light, it returns to its former sluggish pace. Some of these instruments are exhibited for sale in shops in London, where they are placed close to the window, thus enabling any one to satisfy himself by experiment of the reality of this phenomenon. With the ordinary gas light the radiometer makes a revolution in six seconds, but on shading it with the hand its velocity is reduced to one revolution in fourteen seconds; but on being again exposed to the gas light, the beautiful tiny instrument acquires its former speed, or when submitted to the action of the sun, by intercepting its rays, the speed of the little machine is at once retarded.

2 K

How far this proposed system of photometry is practicable we are unable to express an opinion, and as the invention is of very recent date, sufficient time has not yet transpired to carry this into successful operation. It is, however, within the range of possibility that the radiometer is destined to fulfil a purpose similar to the barometer and thermometer, namely, to record the degree of light, and that its speed will be communicated to a scale on which the various degrees of light will be defined, and thus, in addition to temperature and pressure, the quantity of light may be recorded, and by this means perhaps other secrets of nature may be revealed.

Lowe's Jet Photometer.

The jet photometer is based on the fact that the length of a gas jet issuing from an orifice of a certain size under exactly uniform pressure varies according to the illuminating power of the gas, and the richer the gas, the longer will be the flame. This instrument was brought into use by Mr. Lowe about 1860, and is represented in one of its forms in Fig. 84, which consists of a steatite jet fixed upon a very delicate King's pressure gauge, of that degree of sensitiveness as to be capable of indicating the hundredth part of an inch of pressure. The gas supplied is regulated by a delicately acting dry or wet governor, and the burning jet is enclosed in a chimney having a scale of inches and parts engraved on it, and the height of the flame as shown by the chimney indicates the quality of the gas. In some works the instrument in question is enclosed within a small cupboard, provided with suitable means for carrying off the products of combustion, having a glass door in front on which is marked the scale of inches and parts, and immediately behind the jet is fixed a corresponding scale in porcelain, the lines of which are exactly level with the scale on the glass, which replaces the chimney, so that when making the observation the lines of the two scales are observed.

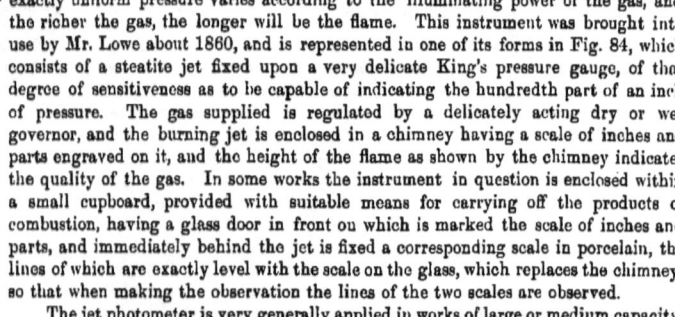

Fig. 84.

The jet photometer is very generally applied in works of large or medium capacity, where the gas is required to be delivered at any particular standard, for which it is accordingly arranged, hence the jet for this being, we will suppose, 6 inches in length, any augmentation or diminution of this will show an increased or diminished quality of gas. Therefore one of these instruments fixed in the engineer's office, burning the gas as produced, indicates at once any error, whether arising from excessive exhaustion, neglecting to blow off the air from the purifiers, or other circumstances which would prejudice the quality of the gas. The name is not applicable, and, as Dr. Bowditch observes, the instrument is not a photometer, but an indicator of *constancy of quality*, for which object it is invariably applied.

The quantity of gas passing through a small orifice at a given pressure, as already remarked, varies according to the density of the gas, and upon this principle attempts have been made to construct photometers, so that by merely observing a flame of gas, an index would give its luminosity without any calculation; and according to Dr. Bowditch, such an instrument was invented by Mr. Alcock, engineer of the Leeds New Gas Company, in 1836, which, however, could not have been reliable, as any admixture of air, although adding to the specific gravity, would at the same time diminish the illuminating power of the gas.

With the jet photometer, although based on the law referred to, the presence of air, whether arising from excessive exhaustion or carelessness in blowing off the air from the purifiers, or other causes, will always be indicated by the diminished flame.

DISTRIBUTION.

Mains.

Prior to the year 1810, the mains of the New River Company, which then supplied a large portion of the metropolis with water, were simply the trunks of elm trees left on their exterior in their rough state, and bored out throughout their length to the desired diameter. One end of the pipe was made conical like a spigot, the other end having a corresponding socket, from which undoubtedly the terms of spigot and socket given to gas and water mains were derived.

At that period, about three years before the first practical application of conducting gas on an extensive scale, cast-iron pipes were introduced by the company mentioned. The first pipes used by Winsor in Pall Mall were leaden, and probably formed of the sheet rolled and soldered together, for within our recollection drawn pipes were limited in diameter; and the continuous pipe made by hydraulic pressure, the metal being highly heated during the operation, had its origin about thirty years ago.

The first mains laid by Clegg for lighting a portion of London were 2 inches in diameter, but these being found too small, were speedily replaced by others of larger capacity, and as we shall have occasion to show, one of the greatest defects during many years consisted in the limited dimensions of this part of the gas apparatus.

Various descriptions of materials have been proposed for the formation of mains for conducting water and gas; amongst these may be enumerated, earthenware, paper, wood, pitch, or a combination of these, and we have heard from reliable authority that a brick main was tried at Cambridge in the early days of gas-lighting; but as can be imagined, on account of expense and inefficiency, it was promptly abandoned. Earthenware mains have been extensively applied by some of the gas companies in France, and within our knowledge by one of those at Lyons, and at a town in the north the loss of gas sustained by the use of this kind of main was so excessive as almost to terminate in the failure of the company who so unwisely adopted them. The leakage, however, did not arise so much from the pipe as from the joints, which were made with cement; thus any settlement, however slight, was sure to break either the junction of the service or the joint of the main.

Some years ago a manufactory was established in Paris for making gas and water mains of paper. For this purpose a mandril was provided of the diameter and length of the intended pipe, and around this was laid a cloth having a thin coating of loam. The paper after being dipped in hot pitch was then coiled round the mandril in continuous layers until acquiring the desired thickness of the pipe, when it was removed to cool. These pipes were made of the same thickness as cast iron, and they probably would have answered the intended purpose, at least for water, had there been any means of connecting them together, but in the absence of this they failed entirely. It was also attempted to make pipes of slips of wood, each slip representing a segment of the pipe, and put together like the former with hot pitch. Of these a considerable number was manufactured, and from analogous causes they shared the fate of the others mentioned.

Mains of bottle glass were also proposed, but never carried into use on an extensive scale, for in addition to the difficulty of connecting them together, there was the further evil of not having proper means of attaching the services.

The only description of main besides the cast iron which has met with any degree of success is that manufactured by Chameroy and Co., of Paris, and called after its inventor, the "Chameroy" pipe. These are formed of leaded sheet iron of a thickness corresponding with the diameter: for the smallest kinds the sheets do not exceed No. 20 B. W. G. in thickness; whilst the largest we have seen (about 18 inches in diameter), are of the thickness of No. 16 gauge. When made the sheet is cut to the desired length and width, and formed cylindrical by passing through rollers; it is then riveted and soldered along the seam, when corresponding metal screws composed of lead and antimony are cast in suitable moulds at each end of the pipe, and are afterwards soldered thereto. Each pipe is then tested by hydraulic pressure in order to detect any leakage, after which it has a coating of hot pitch both inside and outside, and completed by a layer of asphalt intermixed with fine gravel of about ¼ inch thick over its exterior.

When laying these pipes, a hempen washer dipped in tallow is placed on the flange which forms part of the screw; they are then screwed up by means of a short wooden lever with a cord which passes around the pipe, answering the purpose of the ordinary gun-barrel tongs. To attach a service the asphalt is chipped off from the part, a hole is bored in the main, and into this is inserted the end of the lead pipe constituting the service, which is then soldered. This effected, the asphalt is heated and replaced on that portion of the pipe where it was chipped off.

Although from the description, the Chameroy pipes must appear very fragile, yet from our former experience, extending over a period of ten or twelve years, they were all that could be desired, both for soundness and durability, and we regret that our information concerning a more lengthened usage is somewhat conflicting. These mains have, however, been very largely employed on the Continent during nearly forty years, thus giving ample opportunity to discover any defects they may possess.

The description of gas main almost universally adopted throughout the world is the cast iron, and under all the various circumstances of price, efficiency, and durability, they are unquestionably preferable to all others. These when first adopted at the commencement of gas-lighting, were sometimes made with flanges and put together with bolts, but whether arising from the settlement of the ground or whatever cause, they were very susceptible to breakage, which caused them to be abandoned, and the lead joint became universally employed. Some years ago the turned and bored joint was introduced by the late Mr. King, of Liverpool, and although various methods of jointing pipes have been proposed from time to time, only these two systems, the leaded, and the turned and bored joint, are generally adopted.

The lead joints afford greater facilities than the other for laying where there are curves or bends and short lengths, and in addition they possess a degree of elasticity which in some localities may be desirable. The turned and bored jointed pipes, although somewhat higher in price in consequence of the labour in turning, as will be seen hereafter, are much cheaper when laid. For that purpose, all that is required is to clean the turned faces and smear them over with red mastic, when they are connected and driven home either with a mallet, or by a pipe suspended to shear legs, or by means of a screw working against a stake driven into the ground. When laid in a straight line of considerable length these pipes make an excellent job; they are readily put together, and perfectly sound; but under some conditions, as when placed on a dead level or slightly deviating from that, they demand extra care in laying, as the position of the pipe as regards its level is controlled by the manner it is forced home; thus, if struck above the centre, the end of the pipe will have a tendency to rise, and if beneath, it will descend below the proper level.

The effects of expansion and contraction in mains are often much exaggerated, and suggestions have frequently been proposed to introduce appliances to counteract this, which, we have no hesitation in saying, are unnecessary and useless, as in practice no inconvenience or loss is experienced from these causes. In corroboration of this assertion, it may be stated that a line of 6-inch main once came under our observation with turned and bored joints, of about 2¼ miles in length, laid perfectly straight, at a depth of about 18 inches, and every pipe driven home, which of course was eventually subjected to the ordinary variations of temperature. Thus in this case, if there had been any material expansion, the ground immediately over the main would have been elevated at one point or another; this, however, did not occur, so that we may conclude therefore that no important effect of the kind takes place. Besides, at a depth of 2 feet below the surface of the earth even in the extreme seasons, there is but a slight variation of temperature, consequently the effect of expansion and contraction in gas mains, as generally laid, may, in our opinion, be disregarded.

The durability of cast-iron mains is entirely dependent on the soil wherein they are laid. In clay the metal is kept in a high state of preservation, and we have heard that in some localities the mains are always imbedded in that material with a view to their durability; indeed, an experiment of the kind on a small scale was tried some years ago in London. In sandy saline ground an oxide intermixed with sand is formed on the exterior of the pipes, which increases in thickness according to the period they are buried. These we have seen, after being interred for twenty-five years, with a coating of incrustation of about half an inch thick, and on breaking this off, the pipe was found to be materially diminished in thickness, but the iron when cut was in a good state of preservation. In made ground cast-iron pipes become rotten, and when in muddy soil, after the lapse of years, they become of a nature similar to plumbago, and can be cut with the same facility as the crust of bread. But the worst of all ground, as we are informed by a gentleman of the most extensive experience, in which pipes can possibly be laid, is that containing cinders or ashes and moisture; in this the pipes are covered with marks throughout, similar to pock marks, some of which eventually penetrate the metal. This action is probably due to electrical agency; each of the numerous particles of carbon coming in contact with the iron and aided by the moisture, producing the slow but positive decay.

Hence we find that in some instances, as in clay, sandy soil, and ashes, the nature of the metal is preserved, and in the first case in all its integrity; in the second, when in the saline sand, the pipe is rendered thinner by the accumulation of oxide on its exterior; therefore in relaying these, it is always better to place them as they are taken from the ground without any attempts at cleaning. Whilst in ashes an infinity of small holes are gradually but continually being formed in them. In made ground, or muddy soil, the iron seems to undergo a chemical change and becomes very rotten, but whether sufficiently porous to permit of the escape of the gas is hardly certain.

Therefore when the question of the "life" of a gas main arises, it can only be determined approximatively by the knowledge of the kind of soil in which it is laid.

But we must not forget that, when gas mains were laid many years ago, no attempts to preserve them were then thought of. To this important point the attention of engineers, however, has of late years been directed, by coating them with a layer of pitch, and even imbedding the iron service pipes in the same material. In our opinion too much importance can hardly be attached to this, as pitch possesses the power of resisting against many, if not all, the corroding agencies so detrimental to cast and particularly wrought iron, and there can be no doubt that if a good coating of pitch be applied to the mains when slightly heated, their durability in all soils will be increased very considerably.

Undoubtedly one of the causes of the leakage many years ago was the smallness of the mains, and the great pressure required to force the gas to the end of the districts, augmented by the very imperfect manner in which they were laid. In one instance within our knowledge, at a metropolitan works, the gas issued at a pressure of 6 inches, and yet at a distance of two miles there was not a tenth pressure in the main, and, singular to relate, the company at one time refused new customers in a portion of their district on account of their inability to supply. We believe the largest main in the metropolis in 1839 was a 10 inch; in 1840 a main of 14 inches diameter was laid from the Brick Lane station of the Chartered Company, and the following year a "large main" of 18 inches was laid from the Westminster station, in which all the joints were coated with a mixture of pitch and tallow, and afterwards, when tested with a pressure gauge, the main was found perfectly sound. Now we find them in many works of 36 inches diameter, whilst from Beckton there are two 48-inch mains of a joint length of nearly twenty miles, always containing about 1,250,000 cubic feet of gas.

Years ago it was supposed that the illuminating power of gas was materially impaired in its passage through the mains, which by engineers was considered equal to a loss of one candle in the illuminating power during the transit of the gas through the first mile of main; consequently, by the Sales of Gas Act of 1860, the examiner's testing offices were required to be situated not less than one thousand yards from the works where the gas was produced. This supposition, however, is now proved to be erroneous, and it is now known that gas can travel eight or ten miles (how much beyond that experience has not yet demonstrated) without being affected to any important degree. Another asserted cause of the deterioration of gas was by the action of diffusion, and a well-known chemist at one period stated that an admixture of 25 per cent. of air with the gas in the mains produced by this action once came under his notice. A little practice is sometimes more reliable than any amount of theory, and practice teaches us that this diffusive power with gas enclosed within cast-iron mains is not capable of being appreciated, as we find by the reports of the various examiners of the gas of any particular company, whose offices are situated far distant from each other as well as from the works, yet according to these reports the illuminating power of the gas is substantially the same in all the various localities.

It has been often urged that old mains are conducive to excessive leakage, but this does not always appear to be the case, as we find by the returns of the metropolitan companies that the greatest percentage of unaccounted-for gas exists at a comparatively modern establishment, but whether this is due to leakage or other causes it is difficult to state. In addition to the loss by leakage in the distributing plant, unaccounted-for gas may arise in a great measure from a want of proper surveillance of the consumers' meters, or from an excess of supply to the public lamps beyond that contracted for, either in the quantity per burner, or in the hours of lighting.

On the other hand, such a degree of care may be observed, or other means adopted in the consumption of the public lights, as to show no loss whatever. In confirmation of this statement it may be observed that we were once acquainted with an engineer who pretended that he actually sold more gas than he produced, and not being able to assign any other reason for this, and the works being situated in the tropics, he maintained that the gas was permanently expanded by the action of the sun, and "sun gas" always appeared in his reports and accounts, this being "unaccounted-for gas" in another sense. This enigma is explained when it is stated that only $3\frac{1}{2}$ feet per hour were consumed by each public light, and that 4 feet were charged for the same; hence, in addition to the loss by leakage there would be an excess, which was the "sun gas."

The unaccounted-for gas of the metropolitan gas companies averages $7 \cdot 32$ per cent., which includes loss from all sources, whereas thirty years ago it was generally estimated at 30 or 35 per cent. on the production, which was invariably put down to leakage. This low percentage of the metropolitan works, however, can hardly be expected in small towns, where the rental for mileage of mains bears no comparison with the other, but in all cases the intelligence of a manager should be directed to reduce the unaccounted-for gas to a minimum.

The diameters of mains will necessarily be decided according to the quantity of gas required to be delivered by them, together with their length and position as regards level. A main ascending from the works will deliver a greater quantity than another on a level, and still more than another which descends, since for every 10 feet of increased height above a given point, the gas in the main, in consequence of its low specific gravity, acquires an increase of about one-tenth of pressure. Thus if we suppose at the works a pressure of 5-10ths to exist in a main from which there is no draught, then at a point 100 feet above, the pressure in the same main will be 15-10ths, or if we imagine a pressure of 15-10ths at the works, then at a point 100 feet below there will be but 5-10ths pressure. Hence the delivery under the two conditions is according to the mean pressure within the mains, and in certain localities, where the works are much lower than the town supplied, the gas can be delivered in the day time with four or five tenths 'exhaust,' according to the difference of elevation, as we witnessed some years ago at Madrid. The table of the delivery of gas from mains of varied diameters hereafter given is intended to apply to those placed on a level.

Obstructions in gas mains arise from various causes: sometimes by the accumulation of water therein at a point where the pipes have been badly laid, or by naphthaline, which is often collected by a piece of tarred yarn projecting from the joint, on which a few grains of that troublesome compound is deposited, and this in the course of time increases in such a manner as seriously to impede the passage of the gas. The appearance of an accumulation of naphthaline within a main, which once came under our notice, where a 12-inch main was opened close to the obstruction, was similar to a crystal grotto, and was of such tenacity that with difficulty it was dislodged by striking the main above with a sledge hammer. We once observed in a large works, where the mains were being replaced by those of larger capacity, that the old pipes were literally half filled with naphthaline, which of course diminished their capacity; but had the engineer been aware of this obstruction, which he could have ascertained with the pressure gauge, it could have been easily removed, and the expense of replacing them might have been saved, or at least deferred for some time. There have been other very curious obstructions: in one case a basket of tools was left in a 10-inch main and was only discovered twelve years afterwards, when the main was worked to its utmost, and the stoppage indicated by the pressure gauge. In another case a wooden plug nearly the diameter of the pipe was enclosed therein.

Within our knowledge at a small works, then recently constructed, and shortly after the commencement of operations, the pressure to the town nightly diminished, and on investigation, contrary to all expectation, the obstruction was found to be in the governor. This ascertained, the inlet and outlet were detached without discovering the evil, but on taking out the holder, a workman's cap was found on the outlet pipe, from which it appeared that the fitter, in putting the apparatus together, placed his cap in that position and forgot to remove it before placing the holder. In another instance, part of a district supplied by a company abroad, from the commencement of operations was almost in darkness, and continued in that state for years; as the manager observed, "There never has been nor never will be any light there." But on investigation this was found to arise from one of the pipes being choked with coal, which had collected there no doubt during the transport, the coal and pipes forming the same cargo, and the men in laying had neglected to see if they were clear.

In order to detect a stoppage in the main, where its locality cannot be determined by observing the lights, the pressure gauge must be applied.

For this purpose the pressure must be taken on the main during the full lighting, at intervals of 200 or 300 yards, on the public lamps. Formerly "stand pipes" (which were short pieces of ¾ pipe tapped into the main, capped and covered with syphon boxes) were placed at various parts on the mains expressly for the purpose of ascertaining the pressure. The public lamps, however, answer the same object, and have the further advantage that they exist throughout the district.

First, commencing at a point where the full pressure is known to exist, then, in making the observation from lamp to lamp at the distances stated, a gradual reduction of pressure will occur as the investigation proceeds, this reduction depending on the size of the main and the extent of the supply therefrom. But in the event of there being a stoppage between two points, then a marked diminution of some tenths of the pressure will be indicated. This ascertained, the experiment is repeated at shorter distances, that is, from lamp to lamp, between two of which the point of obstruction will be discovered, when four or five stand pipes should then be placed on the main at intermediate distances, and on the pressure being taken at night as before during the full lighting, the position of the stoppage will be known within a few yards. The main is then opened at the two points, and the obstruction, if naphthaline, removed, but should it arise from the imperfect position of the main occasioning the accumulation of water, it will have to be relaid, or a syphon placed to receive the condensation.

From these statements we learn the importance of the pressure gauge for the purpose of ascertaining the presence of obstructions, whether existing on the works or in any part of the mains, as any sudden decrease of pressure between two points indicates a stoppage; whilst any progressive great decrease of pressure is due to the smallness of the mains.

Formerly, the loss arising from leakage was considered an inevitable evil, but at the present time this opinion no longer exists, for mains can be and are laid in such a manner as to be perfectly sound, and in a length of two or three miles will not show any appreciable leakage even with the pressure gauge; that is, on gas being admitted into the main at a certain pressure, and the valve closed, on observing a pressure gauge fixed on the main, no appreciable diminution of pressure is experienced. Moreover, as we observe every day, water mains with a pressure of 200 or 300 feet show no indication of leakage; if therefore this state of excellence with that great pressure exists, it would be discreditable to gas engineers if the results stated were not obtained. Hence, no considerations of rigid but false economy should induce any gas company to risk the soundness and efficiency of this most important part of their plant, but the whole of the mains should be well and efficiently laid, and, when possible, tested for soundness before being used. The importance of sound mains and services for a gas company is not susceptible of exaggeration, for on this principally depends the proper return for the quantity of gas produced.

Another requisite is, that whilst having a sufficiency of syphons to collect the condensation in the mains, any excess of them should be avoided, as with a large number some may be forgotten; besides which the labour of pumping is increased. Further, a good map drawn to a large scale should show every main and syphon in the district or town. In the absence of this in the early years of gas-lighting, through the change of managers or other circumstances, it was by no means uncommon for the main in a particular street to be forgotten, and a second one laid accordingly; and although not likely to occur at the present time, it has been done repeatedly.

In keeping with all other kinds of gas apparatus, cast-iron mains have been greatly improved during the last few years. Formerly they were cast at a slight incline from the horizontal position, when often, by the shifting of the core, one side of the pipe was considerably thicker than the other, besides which serious defect, leakages frequently existed at the core stays. At the present time they are invariably cast in a vertical position, thus ensuring equal thickness throughout, and soundness in the castings. The larger sizes have also been increased from 9 feet to 12 feet in length, by which considerable economy is effected in laying them.

We venture to assert that every gas manager, at least of works situated in small or moderate-size towns, should know the amount of loss by leakage from the mains, whether this be 100 or 1000 feet per hour or diem, and although it may appear to present some difficulty, we believe this more imaginary than real; of course some pains must be taken in order to ascertain such an important point, for this knowledge acquired, then attention can be directed to the real cause of unaccounted-for gas, which may be placed rightly or wrongly to leakage.

With the view of ascertaining this, let us suppose a meter to be placed in conjunction with the main leading to the town by means of a by pass, for which purpose in most towns ordinary gun-barrel will be sufficient, and on the main between the inlet and outlet of the meter, a good sound valve is placed, and if hydraulic the better. The meter may be a 20-light or 50-light, or of other dimensions, according to the magnitude of the town or district.

The trial should be made on a summer's morning some time after daylight, when those who require gas for their occupations during the night will have dispensed with it, and before the population is stirring, when by shutting the valve the gas is allowed to pass through the meter until the pressure in the main attains a certain degree, say one inch, which is regulated by the governor. This accomplished, then by a careful observation of any given time, say fifteen minutes, the leakage will be arrived at, from which the daily loss can be calculated.

The objections to this may be that there will always exist a few lights burning; but these in small towns are very few, and in many localities there will be none, so that the investigation will not be much influenced thereby. On one occasion when examining the mains of a small town extending about eight miles in length, by this system, the small amount of leakage was surprising, and we are led to the belief that gas mains are generally sounder than they are supposed to be.

Should, however, the mains on investigation show any leakage, the question arises whether this is sufficient to justify any expenditure to remedy the evil, and in the event of this being decided in the affirmative, the first step to be adopted is to overhaul or relay the old services, which are often a fruitful source of loss to a gas company.

This effected, it may be desirable to test the soundness of the various districts by means of the meter on the by pass already referred to. For this object the main may be cut and plugged, or if there is a syphon in a suitable locality this may be "logged"; in the first instance taking the district nearest the works and noting the loss, and afterwards continuing throughout the various districts, when probably the leakage may be discovered at some out-of-the-way place where it was least expected.

Some engineers recently have resorted to the system of stripping the whole of their mains and services with the view of repairing such escapes, but the leakage in a town must be excessive indeed to demand such measures.

The annexed is a table of the weights of mains as usually employed, but as the pipes of different manufacturers vary in thickness, the weights must be considered therefore as approximative.

TABLE OF WEIGHTS AND LENGTHS OF CAST-IRON MAINS.

Diameter of Pipe in Inches.	Weight per Pipe.				Weight per Yard.			Length of Pipe in feet.
	tons.	cwt.	qr.	lb.	cwt.	qr.	lb.	
2	0	0	1	16	0	0	22	6
2½	0	0	2	0	0	1	0	6
3	0	0	3	18	0	1	6	9
4	0	1	1	13	0	1	23	9
5	0	1	3	8	0	2	12	9
6	0	2	1	15	0	3	5	9
8	0	3	0	24	1	0	8	9
10	0	4	2	6	1	2	2	9
12	0	7	2	8	1	3	16	12
14	0	8	1	20	2	0	12	12
15	0	11	0	0	2	3	0	12
16	0	12	1	8	3	0	9	12
18	0	15	0	0	3	3	0	12
20	0	18	0	0	4	2	0	12
22	1	0	0	0	5	0	0	12
24	1	4	1	8	5	3	9	12
30	1	10	2	0	7	2	14	12
36	2	0	0	0	10	0	0	12
48	3	0	0	0	15	0	0	12

At one period it was the practice to lay the mains in the centre of the road or street, quite regardless of its width or the length of the services; when the mains and services were so laid, they were frequently disturbed during the alteration of old or the formation of new sewers, and often broken, occasioning serious leakage of gas and loss to a company. In some cases in the Metropolis on the gas escaping from these accidental breakages it would be conveyed through the old dead wooden water-mains a distance of one or two hundred yards, thus rendering it a matter of extreme difficulty to discover the locality where the defect existed.

The present system in all large towns and cities, at least in the most public thoroughfares, where the roadway has a width of 25 feet or more, is to lay a main on each side of and within the footway where practicable, one of which is the leading main, conveying the supply to other localities; the other being a 3 or 4-inch pipe, connected at two or more points with the larger main. This on various accounts is a great improvement on the old method, for, as the mains and services are beyond all interference of the sewer operations, and are protected from the heavy cart traffic, the leakage formerly occasioned by these is avoided, whilst the diminished length of the services afford increased facilities for the execution of the work, without presenting the impediments to street traffic which at one time were so common.

When laying services, all the holes in the mains should be drilled previously to being tapped, by which means the work is done in a proper manner. The clumsy method of gouging the holes has undoubtedly occasioned serious loss to gas companies, for sometimes services have come under our notice where the pipe has been actually loose in the main, the white lead being relied on to prevent the loss of gas.

By the kind permission of the author we are enabled to present the following Tables of the average cost per yard for laying mains and paving, as well as the price of mains per yard, which are copied from Mr. Newbigging's 'Handbook for Gas Engineers and Managers,' and, like the general contents of that work, may be relied on for rigid accuracy:

TABLE showing the AVERAGE COST PER YARD of LAYING MAINS WITH TURNED AND BORED JOINTS, and WITH LEAD JOINTS, including the TOTAL EXPENSE OF MATERIAL (THE PIPES EXCEPTED), EXCAVATING, REINSTATING, AND MAINTAINING THE GROUND FOR SIX MONTHS AFTERWARDS. AVERAGE DEPTH FROM THE SURFACE OF THE GROUND TO THE UPPER SIDE OF THE PIPE, 1 FOOT 9 INCHES.

Diameter in Inches.	2		2½		3		4		5		6		7		8		9	
Description of Joint.	Turn'd and Bored.	Lead.	Turn'd and Bored.	Lead.	Turn'd and Bored.	Lead.	Turn'd and Bored.	Lead.	Turn'd and Bored.	Lead.	Turn'd and Bored.	Lead.	Turn'd and Bored.	Lead.	Turn'd and Bored.	Lead.	Turn'd and Bored.	Lead.
	s. d.	s. d.	s. d.	s. d.	s. d.	s. d.	s. d.	s. d.	s. d.	s. d.	s. d.	s. d.	s. d.	s. d.	s. d.	s. d.	s. d.	s. d.
In ordinary ballast ...	0 7	0 10	0 8	0 11	0 9	1 2	1 0	1 5	1 2	1 7	1 3	1 9	1 6	2 3	1 8	2 5	1 9	2 8
In roads macadamized with Welsh or limestone	0 9	1 0	0 10	1 1	0 11	1 4	1 2	1 7	1 4	1 9	1 5	1 11	1 8	2 5	1 10	2 7	1 11	2 10
In paved roads, ordinary	0 10	1 1	0 11	1 2	1 0	1 5	1 3	1 8	1 5	1 10	1 6	2 0	1 9	2 6	1 11	2 8	2 0	2 11
In footpaths made with ashes, sand, or gravel	0 5	0 8	0 6	0 9	0 7	1 0	0 8	1 1	1 0	1 5	1 1	1 7	1 3	2 1	1 6	2 3	1 7	2 6
In footpaths flagged	0 7	0 10	0 8	0 11	0 9	1 2	1 0	1 5	1 2	1 7	1 3	1 9	1 6	2 3	1 8	2 5	1 9	2 8
In footpaths asphalted	1 8	1 11	1 11	2 2	1 11	2 4	2 1	2 6	2 3	3 2	2 6	2 4	2 10	2 7	3 4	2 0	3 6	2 10	3 0

Diameter in Inches.	10		11		12		14		15		16		18		20		24	
Description of Joint.	Turn'd and Bored.	Lead.	Turn'd and Bored.	Lead.	Turn'd and Bored.	Lead.	Turn'd and Bored.	Lead.	Turn'd and Bored.	Lead.	Turn'd and Bored.	Lead.	Turn'd and Bored.	Lead.	Turn'd and Bored.	Lead.	Turn'd and Bored.	Lead.
	s. d.	s. d.	s. d.	s. d.	s. d.	s. d.	s. d.	s. d.	s. d.	s. d.	s. d.	s. d.	s. d.	s. d.	s. d.	s. d.	s. d.	s. d.
In ordinary ballast ...	2 0	3 0	2 2	3 4	2 4	3 8	2 9	4 5	3 0	4 11	3 6	5 6	4 1	6 0	5 0	7 9	6 1	8 7
In roads macadamized with Welsh or limestone	2 2	3 2	2 4	3 6	2 6	3 10	2 11	4 7	3 2	5 1	3 8	5 8	4 3	6 5	5 2	7 11	6 3	8 9
In paved roads, ordinary	2 3	3 3	2 5	3 7	2 7	3 11	3 0	4 8	3 3	5 2	3 9	5 9	4 4	6 7	5 3	8 0	6 4	8 10
In footpaths made with ashes, sand, or gravel	1 10	2 10	2 0	3 2	2 2	3 6	2 7	4 3	2 10	4 9	3 4	5 4	3 11	6 1	4 10	7 7	5 11	8 5
In footpaths flagged	2 0	3 0	2 2	3 4	2 4	3 8	2 9	4 5	3 0	4 11	3 6	5 6	4 1	6 0	5 0	7 9	6 1	8 7
In footpaths asphalted	3 1	4 1	3 3	4 5	3 5	4 9	3 10	5 6	4 1	6 0	4 7	6 7	5 2	7 5	6 1	8 10	7 2	9 8

One of the greatest evils in laying mains is the system of contracting at an unreasonably low price, which contract is sometimes afterwards sublet. Under such circumstances, it follows that instead of the joints containing their proper quantity of lead, often only one-third of that quantity is employed, when, although the joints may be caulked so as to be rendered sound at the time, yet on the slightest settlement of the ground they will become defective, and a continual source of loss to the company.

A good substitute for the clay belt used for pouring the lead into the joints, and particularly when the mains are of large dimensions or when the men are not expert at the work, is a flat iron ring, of about ¼ inch thick and 1 inch or 1½ inch wide, according to the size of the pipe. This is cut in two and jointed at the centre, so as to resemble a pair of calipers; the interior of which fits closely to the pipe, and when placed ready for making the joint, an opening is left at the top for the lead to enter. When using these with small pipes, all that is necessary is to form the "gate" above, and to smear clay round the ring, to prevent the lead escaping when it is poured into the joint. When the pipes are of large diameter, the ring or calipers will have to be secured at the top; for this purpose the ends are bent at right angles, leaving a space for the gate, and fastened by a screw. This is a very rapid and simple manner of making joints, and does not require a skilled workman for the purpose.

For temporarily stopping the supply of gas when laying mains, a bladder or bag valve is sometimes employed, which consists of a bladder, or by preference an indiarubber spherical bag, which when distended by blowing into it shall fill the pipe; the bladder or bag is attached to a pipe of about $\frac{3}{4}$-inch diameter and one foot long, at the end of which is a tap. At the point desired a hole is drilled on the top of the main, just sufficiently large to permit of the bag entering, care being taken not to injure this in the operation. The bag being carefully placed within the main, is then inflated by a man blowing through the tap, which accomplished, the tap is then closed, when the stoppage or valve is completed. To guard against accidents, which sometimes arise by the breakage of the bag or bladder, whenever required for a considerable time, two of such valves should be placed, the one close to the other.

A liability attending this appliance is the possibility of a slight leakage sufficient to form an explosive compound in a length of main one end of which is open to the atmosphere, when by the approach of a light an explosion would occur. An accident of this kind once happened, and was probably occasioned by a man passing close to the end of the main with a lighted yarn. When laying mains which require to be stopped at repeated intervals, undoubtedly the better system is to place a sheet of iron or tinplate between the spigot ends of two of the pipes, which can be removed at night if desirable, and the aperture closed with roman cement, and the following morning the plate can be again replaced; afterwards this part is made good with a double socket that has been left on the pipe expressly.

By accident at times, either in laying services or otherwise occupied, men are subject to the influence of the issuing gas, and so sudden is its action, that we have seen on more than one occasion a man fall senseless in the trench, when, if he had not been speedily removed, he would have died. The sensations arising from inhaling an excess of gas are particularly disagreeable, but are speedily removed by a good dram of spirits of any kind; in the absence of this remedy, a racking headache for a lengthened period is the result.

On account of the great facility with which small mains are broken, even when embedded in the ground, occasioned by the slightest settlement, only when capital is limited and in the absence of heavy traffic in the locality, should 2-inch mains be employed, and in which holes no larger than sufficient for half-inch gun-barrel should be drilled, and when required for lamp services, short pieces of $\frac{3}{8}$-inch pipe screwed into the main will be ample; and in all cases where small pipes are employed, they can be increased by a socket to the desired size of the service. A reduction in the diameter of a service for a length of a few inches will not affect the quantity of gas delivered, as we find by experiment that a $\frac{3}{8}$-inch gun-barrel 6 inches long with a pressure of $\frac{1}{10}$ths of an inch will deliver 70 feet of gas per hour; another pipe of $\frac{1}{2}$-inch diameter, of the same length and with the same pressure, will deliver 110 feet per hour; and from a third of $\frac{3}{4}$-inch diameter, under like conditions, the supply will be 230 feet per hour. From these the deliveries from larger pipes can be estimated.

The only possible objection that can be raised against small apertures in the main is the chance of obstruction by naphthaline, which contingency, however, under any circumstances, is very remote, and in the event of its occurring the stoppage can be readily removed with the service cleaner. On the other hand, by drilling unnecessarily large holes in small mains, the pipes are weakened at those parts to a very remarkable degree, and from breakages arising from this cause the loss of a large quantity of gas can often be traced.

The following table by Mr. Newbigging, with some addition to the larger sizes, gives the cost price of mains per yard from 2 inches to 48 inches in diameter, varying from 4*l*. 15*s*. to 12*l*. per ton, which, together with the preceding table of the prices for laying mains, will undoubtedly be acceptable to all engaged in the construction of gasworks:

PRICES OF MAINS PER YARD.

THE MAINS OF LONDON. 261

We have endeavoured to obtain some information respecting the lengths of the mains of the various gas companies of the metropolis, in which, however, we have not met with success, and must therefore attempt to arrive at an approximation thereto through other channels.

According to Sir Joseph Bazalgette, the engineer of the Metropolitan Board of Works, the sewers of London are about 1300 miles long; but as there are many miles of streets and roads lighted by gas to which these do not extend, therefore from this source little information is acquired; but on turning to the list of the water mains, something more definite is obtained.

We gather from the returns made of the several water companies appointed by the Board of Trade in 1872, that at that period there were upwards of 2800 miles of main pipes employed to deliver water to the inhabitants of the metropolis; but of these there are some streets and roads where two or more water mains exist, and, besides, portions of the districts of some of the water companies extend beyond the limits supplied by the London gas companies. From the returns of the Metropolitan Police, made in 1850, the streets and roads then included within the inner district were 1700 miles long, whilst those of the City measured 50 miles, or, together, 1750 miles. Since that period twenty-six years have elapsed, and, so far as we are aware, no official return has been made; but, assuming that the streets and roads have been augmented one-fifth, then their present length will be 2100 miles.

Hence, after duly considering these various accounts, the extent of mains in London for the purpose of supplying gas must be a matter of conjecture; but with the facts stated, and the knowledge of the public lamps supplied, we will not be far off the correct length.

We find that the number of public lamps lighted by all the companies in 1874 were 54,000, as follows:

The Imperial	Company supplied	.. ..	..	16,984
,, Chartered	,, ,,	.. ..	..	15,000
,, Phœnix	,, ,,	.. ..	..	5,497
,, London	,, ,,	.. ..	..	4,821
,, Commercial	,, ,,	.. ..	..	3,706
,, South Metropolitan	,, ,,	.. ..	..	3,666
,, Surrey Consumers'	,, ,,	.. ..	..	1,949
,, Independent	,, ,,	.. ..	..	1,507
,, Ratcliff	,, ,,	.. ..	..	870
		Total	.. ..	54,000

But here again a difficulty presents itself, arising from the great irregularity of the distances between the lights, which in some few localities are placed only 30 yards apart, whilst in others they are 70 or 80, and even 90 yards, distant from each other. Assuming, however, that the average distance to be 70 yards (a distance sufficient to include all "independent" mains), this gives 2150 miles as the total length of the gas mains of the metropolis, corresponding with the estimated length of the streets and roads, and cannot be very far from strict accuracy. Therefore, taking this as basis, we conclude that the length of the mains of the various companies, as they existed in 1874, may be estimated as under:

The Imperial	Company	.. ..	..	675 miles of mains.
,, Chartered	,,	.. ..	..	600 ,,
,, Phœnix	,,	.. ..	..	217 ,,
,, London	,,	.. ..	..	190 ,,
,, Commercial	,,	.. ..	..	150 ,,
,, South Metropolitan	,,	.. ..	..	146 ,,
,, Surrey Consumers'	,,	.. ..	..	77 ,,
,, Independent	,,	.. ..	..	60 ,,
,, Ratcliff	,,	.. ..	..	35 ,,
		Total	..	2150 ,,

If this be correct, the mileage of mains of the metropolis is considerably less than the average of that of the majority of provincial towns, where it is considered to be equal to from 1 mile to 1¼ mile per 1000 inhabitants, whereas in London it is little more than half a mile for a similar population. The metropolitan companies are also highly favoured in other respects, but more particularly in the amount of their capital as compared with their annual revenue, the latter being about 33 per cent. of the former, or, in other words, their gross receipts every three years are nearly equal to the whole of their invested capital, whilst the annual gross receipts of many provincial works is not equal to one-sixth of their capital.

But the preceding estimate does not convey a correct impression of the length of gas pipes embedded in the streets and roads of the metropolis, where formerly, in consequence of the competition existing in some localities, the mains of three or four enterprises were laid, and in a few streets there were no less than three or four mains belonging to one company. However, on the amalgamation of the various companies, instead of three mains being required for a street or district, in the majority of places one alone was sufficient, and as the worth of the old pipes embedded in the ground was not equivalent to the expense of removing them and paving, they were consequently disconnected and left "dead." To arrive at an approximation of the extent of these dead mains, we must take into consideration the length of the streets and roads in London in 1850, when competition had attained its utmost limits, which, as stated, was 1750 miles. Of this a large portion of the north was supplied by the Imperial Company, where competition never extended. The Chartered, Independent, and Western companies were similarly situated in portions of their districts at the north, east, and west of London, whilst the uncontested districts south of the Thames were also very extensive, the great opposition being confined principally to the main thoroughfares. The competition existing between the British and the Commercial companies, although very violent, was of short duration, as the former was bought up by its more energetic rival before the latter completed the whole of their mains, thus avoiding a large extension of duplicate pipes. In short, in addition to that already stated, the opposition was principally restricted to a portion of the west central districts and the whole of the city; therefore, taking all these points into consideration, we believe that the whole of the dead mains in London do not exceed 500 miles in length, making a total of 2650 miles of gas mains embedded in the streets of the metropolis.

Dimensions of Mains.

Probably there is no branch of gas engineering which has received so much attention as the flow of gas through mains, and it may be observed that there is no other on which such unanimity of opinion is entertained; for, while engineers possess different views respecting the best method of constructing gasholders and tanks, or the most advantageous means of setting retorts, or the best system of purification, or whether the wet or the dry meter is the most advantageous; yet the questions of the capacity of mains for the delivery of gas, under all the varying conditions of their length and diameter, together with the pressure and specific gravity of the gas, is determined with a degree of certainty sufficient for all practical purposes, as proved by everyday experience, and leaving little or nothing to be desired.

This important knowledge has been acquired by the practical experiments of engineers, aided by the researches of eminent mathematicians, among whom may be mentioned Girard, D'Aubuisson, Clegg, Mayniel, Arson, Hughes, Professor Pole, and particularly the late Thomas Greaves Barlow, to whom we are indebted for a most comprehensive and complete table of the delivery of gas through mains under the various controlling influences of the length and diameter of the main, with the pressure of the gas, which table first appeared about twenty-six years ago in the columns of the 'Journal of Gas Lighting,' and is here reproduced. By means of which, together with the formulæ contributed by Mr. Lewis Thompson, and the tables of square roots of pressures and specific gravities hereafter given, any question connected with the subject under consideration may be readily solved.

TABLE OF THE DISCHARGE OF GAS THROUGH MAINS.

Table by the late Thomas Greaves Barlow of the discharge of gas in cubic feet per hour through pipes varying from ½ inch to 36 inches in diameter:

DISCHARGE OF GAS IN CUBIC FEET PER HOUR THROUGH PIPES OF VARIOUS DIAMETERS AND LENGTHS AT DIFFERENT PRESSURES. THE SPECIFIC GRAVITY OF THE GAS ESTIMATED AT ·400, AIR BEING 1·000.

DIAMETER OF PIPE ·5 INCH.

Length in yards							10	20	30	50	75	100	150
Quantity delivered with	0·1 inch pressure						37·7	26·7	21· 7	16·8	13·8	11·9	9·7
"	"	0·2	"				53·4	37·7	30· 6	23·8	19·5	16·8	13·8
"	"	0·3	"				65·2	46·3	37· 7	29·1	23·8	20·7	16·8
"	"	0·4	"				..	..	..	33·7	27·5	23·8	19·5
"	"	0·5	"				..	..	..	..	..	26·7	21·7

DIAMETER OF PIPE ·75 INCH.

Length in yards							10	20	30	50	75	100	150
Quantity delivered with	0·1 inch pressure						104·3	73·8	60·0	46·6	37·9	32·9	26·9
"	"	0·2	"				147·5	104·3	84·9	65·8	53·7	46·6	37·9
"	"	0·3	"				..	..	104·3	80·9	65·8	57·0	46·6
"	"	0·4	"				..	..	..	93·2	75·9	65·8	53·8
"	"	0·5	"				..	..	..	..	..	73·8	60·0

DIAMETER OF PIPE 1 INCH.

Length in yards							10	20	30	50	75	100	150
Quantity delivered with	0·1 inch pressure						214	151	124	95	78	67	55
"	"	0·2	"				302	214	175	135	110	95	78
"	"	0·3	"				..	..	214	165	135	117	95
"	"	0·4	"				..	..	..	190	156	135	110
"	"	0·5	"				..	..	..	..	..	151	123

DIAMETER OF PIPE 1·25 INCH.

Length in yards							25	50	75	100	150	200	300
Quantity delivered with	0·1 inch pressure						236	167	137	118	96	84	68
"	"	0·2	"				333	236	192	167	137	118	196
"	"	0·3	"				..	289	236	205	167	144	118
"	"	0·4	"				..	..	..	236	192	167	137
"	"	0·5	"				..	..	..	..	..	187	152

DIAMETER OF PIPE 1·5 INCH.

Length in yards							25	50	75	100	150	200	300
Quantity delivered with	0·1 inch pressure						374	264	215	187	152	132	107
"	"	0·2	"				528	374	304	264	215	187	152
"	"	0·3	"				..	456	374	322	264	229	187
"	"	0·4	"				..	..	..	374	304	264	215
"	"	0·5	"				..	..	..	..	..	295	239

DIAMETER OF PIPE 2 INCHES.

Length in yards							50	75	100	150	200	300	500
Quantity delivered with	0·1 inch pressure						540	441	381	311	270	220	170
"	"	0·2	"				763	623	540	441	381	311	241
"	"	0·3	"				..	763	665	540	468	381	296
"	"	0·4	"				..	..	..	623	540	441	341
"	"	0·5	"				..	..	..	..	..	492	381

DIAMETER OF PIPE 2·5 INCHES.

Length in yards							50	75	100	150	200	300	500
Quantity delivered with	0·1 inch pressure						943	770	667	545	471	335	298
"	"	0·2	"				1335	1090	943	770	667	545	421
"	"	0·3	"				..	1335	1172	943	819	667	516
"	"	0·4	"				..	..	..	1090	943	770	596
"	"	0·5	"				..	..	..	..	..	861	667
"	"	0·6	"				..	..	..	..	..	..	731

TABLE OF THE DISCHARGE OF GAS THROUGH MAINS.

Discharge of Gas in Cubic Feet per Hour, &c.—continued.

Diameter of Pipe 3 inches.

Length in yards								100	150	250	500	750	1000	1250
Quantity delivered with	0·1 inch pressure				..	..		1054	850	666	471	384	333	298
,,	,,	0·2	,,		..	..		1440	1214	942	666	543	471	375
,,	,,	0·3	,,		..	..		..	1487	1153	815	666	576	529
,,	,,	0·4	,,		..	..		..	..	1332	942	768	666	500
,,	,,	0·5	,,		..	..		..	..	..	1054	850	744	600
,,	,,	0·6	,,		..	..		..	..	..	..	942	815	730
,,	,,	0·8	,,		..	..		..	..	..	..	..	942	845
,,	,,	1·0	,,		..	..		..	..	..	..	..	..	942

Diameter of Pipe 4 inches.

Length in yards								100	250	500	750	1000	1250	1500
Quantity delivered with	0·1 inch pressure				..	..		2160	1366	966	788	683	611	557
,,	,,	0·2	,,		..	..		3054	1932	1366	1114	966	864	788
,,	,,	0·3	,,		..	..		..	2366	1673	1366	1183	1058	966
,,	,,	0·4	,,		..	..		..	..	1932	1576	1366	1222	1114
,,	,,	0·5	,,		..	..		..	..	..	1761	1526	1366	1245
,,	,,	0·6	,,		..	..		..	..	..	1932	1672	1496	1366
,,	,,	0·8	,,		..	..		..	..	..	..	1932	1728	1576
,,	,,	1·0	,,		..	..		..	..	..	..	..	1932	1761
,,	,,	1·5	,,		..	..		..	..	..	..	..	..	2160

Diameter of Pipe 5 inches.

Length in yards								100	250	500	750	1000	1250	1500
Quantity delivered with	0·1 inch pressure				..	..		3540	2245	1587	1296	1122	1000	910
,,	,,	0·2	,,		..	..		5005	3174	2245	1832	1587	1414	1296
,,	,,	0·3	,,		..	..		..	3888	2748	2245	1943	1732	1575
,,	,,	0·4	,,		..	..		..	..	3174	2592	2245	2000	1820
,,	,,	0·5	,,		..	..		..	..	..	2888	2508	2236	1934
,,	,,	0·6	,,		..	..		..	..	..	3174	2748	2449	2245
,,	,,	0·8	,,		..	..		..	..	..	..	3174	2828	2590
,,	,,	1·0	,,		..	..		..	..	..	..	..	3174	2877
,,	,,	1·5	,,		..	..		..	..	..	..	..	..	3540

Diameter of Pipe 6 inches.

Length in yards								250	500	750	1000	1250	1500	1750
Quantity delivered with	0·1 inch pressure				..	..		3770	2660	2170	1880	1680	1530	1420
,,	,,	0·2	,,		..	..		5320	3770	3130	2660	2370	2170	2010
,,	,,	0·3	,,		..	..		6520	4620	3770	3270	2920	2660	2460
,,	,,	0·4	,,		..	..		7540	5320	4340	3770	3360	3060	2840
,,	,,	0·5	,,		..	..		..	5970	4860	4210	3770	3430	3180
,,	,,	0·6	,,		..	..		..	..	5320	4620	4130	3770	3460
,,	,,	0·8	,,		..	..		..	..	..	5320	4740	4340	4020
,,	,,	1·0	,,		..	..		..	..	..	..	5320	4860	4500
,,	,,	1·5	,,		..	..		..	..	..	..	..	5970	5500
,,	,,	2·0	,,		..	..		..	..	..	..	..	..	6360

Diameter of Pipe 8 inches.

Length in yards								250	500	750	1000	1250	1500	1750
Quantity delivered with	0·1 inch pressure				..	..		7,760	5,470	4,470	3,880	3,400	3,100	2,920
,,	,,	0·2	,,		..	..		10,940	7,760	6,310	5,470	4,880	4,470	4,130
,,	,,	0·3	,,		..	..		13,400	9,450	7,760	6,700	5,980	5,470	5,050
,,	,,	0·4	,,		..	..		15,520	10,940	8,940	7,760	6,920	6,320	5,840
,,	,,	0·5	,,		..	..		..	12,200	9,900	8,640	7,760	7,020	6,520
,,	,,	0·6	,,		..	..		..	..	10,940	9,450	8,480	7,760	7,150
,,	,,	0·8	,,		..	..		..	..	..	10,940	9,780	8,940	8,260
,,	,,	1·0	,,		..	..		..	..	..	..	10,940	9,900	9,230
,,	,,	1·5	,,		..	..		..	..	..	..	..	12,200	11,300
,,	,,	2·0	,,		..	..		..	..	..	..	..	..	13,040

TABLE OF THE DISCHARGE OF GAS THROUGH MAINS.

Discharge of Gas in Cubic Feet per Hour, &c.—continued.

Diameter of Pipe 10 inches.

Length in yards		500	750	1000	1250	1500	1750	2000
Quantity delivered with	0·1 inch pressure	9,560	7,800	6,750	6,050	5,520	5,100	4,780
,, ,,	0·2 ,,	13,500	11,040	9,560	8,520	7,800	7,300	6,750
,, ,,	0·3 ,,	16,500	13,500	11,700	10,520	9,560	8,850	8,250
,, ,,	0·4 ,,	19,120	15,600	13,500	12,100	11,040	10,200	9,560
,, ,,	0·5 ,,	21,300	17,400	15,050	13,500	12,380	11,400	10,650
,, ,,	0·6 ,,	..	19,120	16,500	14,800	13,500	12,500	11,650
,, ,,	0·8 ,,	..	..	19,120	17,050	15,600	14,400	13,500
,, ,,	1·0 ,,	..	..	..	19,120	17,400	16,150	15,050
,, ,,	1·5 ,,	..	..	..	..	21,300	19,600	18,500
,, ,,	2·0 ,,	..	..	..	..	..	22,800	21,300

Diameter of Pipe 12 inches.

Length in yards		500	750	1000	1250	1500	1750	2000
Quantity delivered with	0·1 inch pressure	15,100	12,300	10,700	9,550	8,700	8,050	7,550
,, ,,	0·2 ,,	21,400	17,400	15,100	13,450	12,300	11,350	10,700
,, ,,	0·3 ,,	26,100	21,400	19,500	16,500	15,100	13,880	13,050
,, ,,	0·4 ,,	30,200	24,600	21,400	19,100	17,400	16,100	15,100
,, ,,	0·5 ,,	33,600	27,500	23,800	21,400	19,440	18,050	16,800
,, ,,	0·6 ,,	..	30,200	26,100	23,300	21,400	19,800	19,500
,, ,,	0·8 ,,	..	..	30,200	26,900	24,600	22,700	21,400
,, ,,	1·0 ,,	..	..	..	30,200	27,500	25,450	23,800
,, ,,	1·5 ,,	..	..	..	..	33,600	31,250	29,250
,, ,,	2·0 ,,	..	..	..	..	..	36,100	33,600

Diameter of Pipe 14 inches.

Length in yards		500	750	1000	1250	1500	1750	2000
Quantity delivered with	0·1 inch pressure	22,100	18,100	15,600	13,950	12,750	11,800	11,050
,, ,,	0·2 ,,	31,200	25,500	22,100	19,800	18,100	16,700	15,600
,, ,,	0·3 ,,	38,400	31,200	27,100	24,250	22,100	20,500	19,200
,, ,,	0·4 ,,	44,200	36,200	31,200	27,900	25,500	23,600	22,100
,, ,,	0·5 ,,	49,400	40,400	35,000	31,200	28,500	26,460	24,700
,, ,,	0·6 ,,	..	44,200	38,400	34,300	31,200	28,900	27,100
,, ,,	0·8 ,,	..	..	44,200	39,600	36,200	33,400	31,200
,, ,,	1·0 ,,	..	..	..	44,200	40,400	37,300	35,000
,, ,,	1·5 ,,	..	..	..	..	49,400	45,700	42,600
,, ,,	2·0 ,,	..	..	..	..	..	52,920	49,400

Diameter of Pipe 15 inches.

Length in yards		500	750	1000	1250	1500	1750	2000
Quantity delivered with	0·1 inch pressure	26,300	21,400	18,600	16,600	15,200	14,000	13,150
,, ,,	0·2 ,,	37,200	30,400	26,300	23,500	21,400	19,900	18,600
,, ,,	0·3 ,,	45,500	37,200	32,250	28,750	26,300	24,300	22,750
,, ,,	0·4 ,,	52,600	42,800	37,200	33,200	30,400	28,000	26,300
,, ,,	0·5 ,,	58,700	48,000	41,600	37,200	34,000	31,400	29,350
,, ,,	0·6 ,,	..	52,600	45,500	40,700	37,200	34,450	32,250
,, ,,	0·8 ,,	..	..	52,600	47,000	42,800	39,800	37,200
,, ,,	1·0 ,,	..	..	..	52,600	48,000	44,400	41,600
,, ,,	1·5 ,,	..	..	..	..	58,700	54,300	50,800
,, ,,	2·0 ,,	..	..	..	..	..	..	58,700

Diameter of Pipe 16 inches.

Length in yards		500	750	1000	1250	1500	1750	2000
Quantity delivered with	0·1 inch pressure	31,000	25,250	21,850	19,550	17,850	16,550	15,500
,, ,,	0·2 ,,	43,700	35,700	31,000	27,700	25,250	23,400	21,850
,, ,,	0·3 ,,	53,600	43,700	38,100	34,000	31,000	28,700	26,800
,, ,,	0·4 ,,	62,000	50,500	43,700	39,100	35,700	33,100	31,000
,, ,,	0·5 ,,	69,120	56,600	49,000	43,700	39,900	37,150	34,560
,, ,,	0·6 ,,	..	62,000	53,600	47,900	43,700	38,100	40,700
,, ,,	0·8 ,,	..	..	62,000	55,400	50,500	46,800	43,700
,, ,,	1·0 ,,	..	..	..	62,000	56,600	52,400	49,000
,, ,,	1·5 ,,	..	..	..	..	69,120	63,900	60,100
,, ,,	2·0 ,,	..	..	..	..	..	74,300	69,120

DISCHARGE OF GAS IN CUBIC FEET PER HOUR, &c.—continued.

DIAMETER OF PIPE 18 INCHES.

Length in yards				500	750	1000	1500	2000	2500	3000
Quantity delivered with	0·1 inch pressure			41,400	33,800	29,400	23,900	20,700	18,400	16,900
,,	,,	0·2	,,	58,800	47,800	41,400	33,800	29,400	26,200	23,900
,,	,,	0·3	,,	71,800	58,800	50,800	41,400	35,900	32,100	29,400
,,	,,	0·4	,,	82,800	67,600	58,800	47,800	41,400	36,800	33,800
,,	,,	0·5	,,	92,600	75,700	65,600	53,500	46,300	41,400	37,850
,,	,,	0·6	,,	..	82,800	71,800	58,800	50,800	45,400	41,400
,,	,,	0·8	,,	..	..	82,800	67,600	58,800	52,300	47,500
,,	,,	1·0	,,	..	..	..	75,700	65,600	58,800	53,500
,,	,,	1·5	,,	..	..	..	..	80,000	71,800	65,600
,,	,,	2·0	,,	..	..	..	..	..	82,800	75,700
,,	,,	2·5	,,	..	..	..	..	..	..	84,500

DIAMETER OF PIPE 20 INCHES.

Length in yards				500	750	1000	1500	2000	2500	3000
Quantity delivered with	0·1 inch pressure			54,000	44,000	38,250	31,200	27,000	24,200	22,000
,,	,,	0·2	,,	76,500	62,400	54,000	44,000	38,250	34,200	31,200
,,	,,	0·3	,,	93,500	76,500	66,100	54,000	46,750	41,800	38,250
,,	,,	0·4	,,	108,000	88,000	76,500	62,400	54,000	48,400	44,000
,,	,,	0·5	,,	120,500	98,800	85,300	69,800	62,250	54,000	49,400
,,	,,	0·6	,,	..	108,000	93,500	76,500	66,100	59,100	54,000
,,	,,	0·8	,,	..	..	108,000	88,000	76,500	68,400	62,400
,,	,,	1·0	,,	..	..	..	98,800	85,300	76,500	69,800
,,	,,	1·5	,,	..	..	..	..	102,300	93,500	85,300
,,	,,	2·0	,,	..	..	..	..	..	108,000	98,800
,,	,,	2·5	,,	..	..	..	..	..	..	110,200

DIAMETER OF PIPE 22 INCHES.

Length in yards				500	750	1000	1500	2000	2500	3000
Quantity delivered with	0·1 inch pressure			68,600	56,000	48,400	39,600	34,300	30,700	28,000
,,	,,	0·2	,,	96,800	79,200	68,600	56,000	48,400	43,400	39,000
,,	,,	0·3	,,	118,800	96,800	84,000	68,600	59,400	53,300	48,400
,,	,,	0·4	,,	137,200	112,000	96,800	79,200	68,600	61,400	56,000
,,	,,	0·5	,,	153,500	122,500	108,200	88,600	76,800	68,400	61,200
,,	,,	0·6	,,	..	137,200	118,800	96,800	84,000	75,000	68,600
,,	,,	0·8	,,	..	..	137,200	112,000	96,800	86,500	79,200
,,	,,	1·0	,,	..	..	..	122,500	108,200	96,800	88,600
,,	,,	1·5	,,	..	..	..	..	132,000	118,800	108,200
,,	,,	2·0	,,	..	..	..	..	..	137,200	122,500
,,	,,	2·5	,,	..	..	..	..	..	..	140,000

DIAMETER OF PIPE 24 INCHES.

Length in yards				500	750	1000	1500	2000	2500	3000
Quantity delivered with	0·1 inch pressure			84,000	68,600	59,500	48,500	42,000	37,500	34,300
,,	,,	0·2	,,	119,000	97,000	84,000	68,600	59,500	53,400	48,500
,,	,,	0·3	,,	145,500	119,000	103,000	84,000	72,700	65,200	59,500
,,	,,	0·4	,,	168,000	137,200	119,000	97,000	84,000	75,000	68,600
,,	,,	0·5	,,	187,500	155,000	135,600	108,800	93,800	84,000	77,500
,,	,,	0·6	,,	..	168,000	145,000	119,000	103,000	92,000	84,000
,,	,,	0·8	,,	..	..	168,000	137,200	119,000	106,000	97,000
,,	,,	1·0	,,	..	..	..	155,000	135,600	119,000	108,600
,,	,,	1·5	,,	..	..	..	..	163,000	145,500	135,600
,,	,,	2·0	,,	..	..	..	..	..	168,000	155,000
,,	,,	2·5	,,	..	..	..	..	..	..	172,000

DIAMETER OF PIPE 26 INCHES.

Length in yards				750	1000	1500	2000	2500	3000	4000
Quantity delivered with	0·1 inch pressure			85,000	73,500	60,000	52,000	46,500	42,500	36,750
,,	,,	0·2	,,	120,000	104,000	85,000	73,500	65,800	60,000	52,000
,,	,,	0·3	,,	147,000	127,000	104,000	90,000	80,600	73,500	63,500
,,	,,	0·4	,,	170,000	147,000	120,000	104,000	93,000	85,000	73,500
,,	,,	0·5	,,	189,000	165,000	134,000	116,000	104,000	94,500	82,500
,,	,,	0·6	,,	208,000	180,000	147,000	127,000	114,000	104,000	90,000
,,	,,	0·8	,,	..	208,000	170,000	147,000	132,000	120,000	104,000
,,	,,	1·0	,,	..	..	189,000	165,000	147,000	134,000	116,000
,,	,,	1·5	,,	..	..	..	201,000	180,000	165,000	142,000
,,	,,	2·0	,,	..	..	..	..	208,000	189,000	165,000
,,	,,	2·5	,,	..	..	..	..	..	213,000	184,000
,,	,,	3·0	,,	..	..	..	..	..	..	201,000

TABLE OF THE DISCHARGE OF GAS THROUGH MAINS.

DISCHARGE OF GAS IN CUBIC FEET PER HOUR, &c.—*continued.*

DIAMETER OF PIPE 28 INCHES.

Length in yards			1000	1500	2000	2500	3000	4000	5000
Quantity delivered with	0·5 inch pressure		198,000	161,000	140,000	125,000	114,500	99,000	88,600
,, ,,	1·0 ,,		280,000	229,000	198,000	177,000	161,000	140,000	125,000
,, ,,	1·5 ,,		..	280,000	241,000	216,000	198,000	171,000	153,500
,, ,,	2·0 ,,		..	..	280,000	250,000	229,000	198,000	177,200
,, ,,	2·5 ,,		..	..	..	280,000	255,000	222,000	198,000
,, ,,	3·0 ,,		..	..	..	..	280,000	241,000	216,000

DIAMETER OF PIPE 30 INCHES.

Length in yards			1000	2000	3000	4000	5000	7500	10,000
Quantity delivered with	0·5 inch pressure		234,000	166,000	135,000	117,000	105,000	86,000	74,500
,, ,,	1·0 ,,		332,000	234,000	192,000	166,000	149,000	121,500	105,000
,, ,,	1·5 ,,		..	287,000	234,000	203,000	182,000	149,000	128,500
,, ,,	2·0 ,,		..	..	270,000	234,000	210,000	172,000	149,000
,, ,,	2·5 ,,		..	..	..	263,000	234,000	192,000	166,000
,, ,,	3·0 ,,		..	..	..	..	257,000	210,000	182,000
,, ,,	4·0 ,,		..	..	..	..	..	243,000	210,000

DIAMETER OF PIPE 36 INCHES.

Length in yards			1000	2000	3000	4000	5000	7500	10,000
Quantity delivered with	1·0 inch pressure		530,000	372,000	303,000	265,000	234,000	192,000	166,000
,, ,,	1·5 ,,		..	456,000	372,000	322,000	288,000	234,000	204,000
,, ,,	2·0 ,,		..	..	428,000	372,000	332,000	271,000	234,000
,, ,,	2·5 ,,		..	..	..	416,000	372,000	303,000	265,000
,, ,,	3·0 ,,		..	..	..	..	407,000	332,000	288,000
,, ,,	4·0 ,,		..	..	..	..	..	384,000	332,000

The preceding table has been repeatedly tested in practical operations by various engineers, by which its accuracy has been fully established, it therefore fulfils all requirements so far as relates to the various conditions mentioned of dimensions of pipes and pressures; and, in order to meet other circumstances, the following formulæ for the distribution of gas in mains by Mr. Lewis Thompson will be found of the greatest utility.

The application of the table needs but little explanation, for all that is requisite is to know the quantity of gas to be supplied each hour, the length of the main, and pressure at which it is to be supplied, when, by simply referring to the approximative quantity, the diameter of the pipe is ascertained. But, in order to arrive at a correct knowledge of other facts, such as the varied conditions of the pressure necessary for any augmented or diminished supply of gas other than those mentioned in the table, or the quantity of gas that will be delivered by any alteration in the diameter or in the length of the mains, or the quantity of gas of any specific gravity other than that for which the table is calculated, Mr. Thompson's formulæ in the next page will supply the want.

GAS MATHEMATICS OR FORMULÆ FOR THE DISTRIBUTION OF GAS IN MAINS.

By LEWIS THOMPSON, Esq.

WITH the view of substituting in some measure the use of the preceding tables, we now offer a set of formulæ adapted to every condition of gas distribution, accompanied by rules and examples for their more easy comprehension, and also illustrated by comparison with the practical results set down in the preceding table by Mr. Barlow.

How to accommodate the pressure of gas to the demand.

FORMULA. The present quantity of gas is to the quantity demanded as the square root of the present pressure is to the square root of the pressure required.

LEMMA. The quantity of gas now delivered is 322,000 cubic feet per hour, with $1\frac{1}{2}$ inch pressure, and I desire to deliver, by the same pipe, 416,000 cubic feet per hour, what pressure must I use?

To ascertain this, I divide the quantity required by the present quantity, and multiply the product by the square root of the pressure, by which I obtain the square root of the pressure I am seeking, and this, multiplied into itself, gives me the required pressure, which in Barlow's table is 2·5 inches.

The quantity required, 416,000, divided by present quantity, 322,000, gives 1·292, and this multiplied by 1·225, the square root of $1\frac{1}{2}$, gives 1·582, which, multiplied into itself, gives 2·502 for the pressure required.

To find the quantity of gas that will be passed by change of pressure from one pressure to another.

FORMULA. The square root of previous pressure is to square root of altered pressure as the previous quantity of gas is to the gas that will pass by altered pressure.

LEMMA. Under half inch pressure I was passing 1145 cubic feet per hour, but some one has changed the pressure to 2 inches; what quantity of gas am I now passing in consequence?*

Here the previous pressure was $\frac{5}{10}$ths, the square root of which is ·2236, and the present pressure 2 inches or 20 tenths, the square root of which is 4·472, and the quantity of gas multiplied by this, that is, 1145 × 4·472, gives 5120·440, which, divided by ·2236, produces 2290 cubic feet, and this is exactly the same as Barlow's table.

To ascertain the flow of gas by tubes of different diameters, the pressure remaining unchanged.

From a small to a larger diameter. Rule: divide the large diameter by the small diameter, and multiply the square of the product by the square root of the large diameter, and divide by the square root of the small diameter, then multiply the result by the quantity of gas passing through the smaller tubes, which will furnish the amount of gas that will pass by the larger tube.

LEMMA. A tube 18 inches in diameter is passing under 2 inches pressure, at a distance of 3000 yards, a quantity of gas amounting to 75,700 cubic feet, how much will a tube of 36 inches in diameter pass under the same circumstances?

Here 36, divided by 18, gives 2, and this squared and multiplied by 6 (the square root of the large diameter), gives 24, which divided by 4·243, the square root of the small diameter, gives 5·656, and this, being multiplied into 75,700 (the quantity of gas passing), yields 428,159 cubic feet, which, by Barlow's table, is 428,000.

By reversing this rule we can go from large to small diameters.

---

* The author supposes the consumption in the meantime not to be checked at the burners.

GAS MATHEMATICS.

The pressure and distance being given, to know the amount of gas that will be delivered at any other distance.

RULE. The quantities are in the inverse ratio of the square root of the distances.

LEMMA. I am supplying a main which, under 1 inch pressure, is delivering 530,000 cubic feet per hour, at a distance of 1000 yards, how much will it deliver under the same pressure at 4000 yards?

Here the square root of 1000 is 31·620 and the square root of 4000 is 63·24, consequently if we multiply 530,000 by 31·62 and divide by 63·24 we obtain 265,000 cubic feet, which agrees exactly with Barlow's table.

Q. A certain main delivers 530,000 cubic feet per hour, at 1000 yards, how much under the same pressure will it deliver at a distance of 120 miles? *A.* 36,471 cubic feet per hour.

The quantity of gas delivered of the specific gravity of ·400 being given in the preceding Tables, to ascertain the quantity that will be delivered of any other specific gravity.

RULE. Multiply the quantity of gas delivered by the square root of its specific gravity, and divide by the square root of the specific gravity of the gas in question.

LEMMA. I am supplying 17,400 feet of gas of a specific gravity of ·400 through a main 1500 yards long, with a pressure of one inch; how much will the main deliver under the same conditions of length and pressure, the gas having a specific gravity of ·600?

Here the square root of 400 is 6325 × 17,400 = 11,055,000 ÷ 7746 square root and 600 = 14,208 feet.

For the application of the preceding rules the following Tables of the square roots of specific gravity and pressures will be found useful:

TABLE OF SQUARE ROOT OF THE SPECIFIC GRAVITY OF GAS FROM ·350 TO ·700.

Gravity.	Sq. Root.	Gravity.	Sq. Root.	Gravity.	Sq. Root.	Gravity.	Sq. Root.	Gravity.	Sq. Root.
·350	·5916	·425	·6519	·495	·7035	·565	·7517	·635	·7969
·355	·5958	·430	·6557	·500	·7071	·570	·7549	·640	·8000
·360	·6000	·435	·6595	·505	·7106	·575	·7583	·645	·8031
·365	·6041	·440	·6633	·510	·7141	·580	·7616	·650	·8062
·370	·6083	·445	·6671	·515	·7176	·585	·7648	·655	·8093
·375	·6124	·450	·6708	·520	·7212	·590	·7681	·660	·8124
·380	·6164	·455	·6745	·525	·7246	·595	·7713	·665	·8155
·385	·6205	·460	·6782	·530	·7280	·600	·7746	·670	·8185
·390	·6245	·465	·6819	·535	·7314	·605	·7778	·675	·8216
·395	·6285	·470	·6856	·540	·7348	·610	·7810	·680	·8246
·400	·6325	·475	·6892	·545	·7382	·615	·7842	·685	·8276
·405	·6364	·480	·6928	·550	·7416	·620	·7874	·690	·8306
·410	·6403	·485	·6964	·555	·7449	·625	·7905	·695	·8337
·415	·6442	·490	·7000	·560	·7483	·630	·7937	·700	·8367
·420	·6481								

TABLE OF SQUARE ROOT OF PRESSURES, ADVANCING BY TENTHS OF AN INCH FROM ONE-TENTH TO FOUR INCHES.

Inches and Tenths.	Square Root.	Inches and Tenths.	Square Root.	Inches and Tenths.	Square Root.	Inches and Tenths.	Square Root.
1/10th.	·3162	1·1/10th.	1·0488	2·1/10th.	1·4491	3·1/10th.	1·7606
2/10ths.	·4472	1·2/10ths.	1·0954	2·2/10ths.	1·4832	3·2/10ths.	1·7888
3 ,,	·5477	1·3 ,,	1·1401	2·3 ,,	1·5165	3·3 ,,	1·8165
4 ,,	·6324	1·4 ,,	1·1832	2·4 ,,	1·5491	3·4 ,,	1·8439
5 ,,	·7071	1·5 ,,	1·2251	2·5 ,,	1·5811	3·5 ,,	1·8708
6 ,,	·7745	1·6 ,,	1·2649	2·6 ,,	1·6123	3·6 ,,	1·8973
7 ,,	·8366	1·7 ,,	1·3038	2·7 ,,	1·6431	3·7 ,,	1·9235
8 ,,	·8944	1·8 ,,	1·3416	2·8 ,,	1·6733	3·8 ,,	1·9433
9 ,,	·9487	1·9 ,,	1·3784	2·9 ,,	1·7029	3·9 ,,	1·9748
1 inch.	1·	2 inches.	1·4142	3 inches.	1·7320	4 inches.	2·

SERVICES.

HERE we may observe that, before the general introduction of meters, indeed, from the first establishment of gas on the Continent, a considerable degree of surveillance was kept over the contract consumers, at least so far as regarded the hours of burning the gas. For this purpose, a tap enclosed in a box, and secured with a lock, was affixed on each of the services outside the premises. At sunset the lamplighters before entering on the duty associated with their name, went their rounds, opening all the taps to supply the gas to the consumers' premises, and at the hour contracted for the supply was discontinued by the same means, often putting premises in total darkness, to the great inconvenience of their occupants. The tap, requiring occasional attention, was rendered a source of revenue to the gas company, who charged a monthly sum for the "preservation of the main tap," by which its cost was paid in many cases by the first year's receipts, and, whether gas was consumed or not, that charge was demanded.

In this respect, apart from the excessive monthly charge, continental companies displayed more wisdom than was evinced by those of the United Kingdom, by whom the gas was left entirely at the discretion of the consumers, many of whom did not scruple to use it in the most extravagant manner possible during the night, but often employed it in the daytime for heating their premises, without any consideration of payment; and this misappropriation, which is now severely punished, was then regarded as a venal act, even by persons of position and reputed respectability.

Among the impediments to the progress of gas lighting in former years were the heavy charges made for services, which we believe still exists in some localities, but by all well conducted companies, unless under special circumstances, they are furnished gratuitously, thus encouraging the use of gas; and it certainly appears an absurdity for a company to incur an expense of many thousands of pounds to erect extensive works, and lay many miles of mains, and these to the very door of the premises, and then refuse to supply gas unless the attending expenses of laying the service is paid by the consumer. To demonstrate the fallacy of this system, we may state that a continental company for several years after its establishment charged an excessive price alike for the service and meter, and consequently made little extension to their business, but on reducing the meter to cost price, and supplying the service gratuitously, their business increased year after year, in a most remarkable manner, this being no less than 30 per cent. per annum.

We have stated that gas mains can be and are laid in such a manner as to be perfectly sound; from that source, therefore, with moderate care, no material loss need be apprehended, and, once laid in this perfect condition, there is every reason to believe that they will so remain during many years, subject to accidents or conditions of the soil in which they are embedded already referred to. The other portion of the distributing plant, namely, the services, now requires our consideration.

Wrought iron, unlike cast iron, is rapidly acted upon in most soils, and even when exposed to the ordinary saline atmosphere peculiar to towns in the neighbourhood of the sea; therefore, under these influences, when without proper protection, wrought iron services are speedily destroyed, and to this cause in some localities is mainly to be attributed the loss by leakage, and so serious did this question become, that a few years ago the Chartered Company decided on embedding all new services in asphalt, which system is now continually practised, no service being laid without that protection. For this purpose common rough wooden troughs are prepared, which are placed immediately beneath and extending the whole length of the service, when the trough is filled with hot pitch intermixed with a small portion of sand, thus covering and protecting the pipe, rendering it of indefinite duration practically, which advantage is obtained at a very moderate expenditure, and the practice is strongly to be recommended wherever these services are employed.

THE SERVICE CLEANER.

There are, however, other positions where the adoption of the asphalt is not practicable, as for instance in the pipes of the lantern columns, or in those on the exterior of walls or other similar localities. In such cases, the pipes should be galvanized, or other equally efficacious means applied to preserve them, and, when galvanized, their duration is still further ensured by a coating of thick tar, taking care not to remove this when screwing up the pipe.

On the Continent, generally the services alike to the houses as well as to the public lamps are of lead, which material is capable of resisting against the action of nearly every description of soil; indeed, we have heard of only one town where that metal is affected by the ground. When laying leaden services a wooden batten of about 3 inches wide, and $\frac{1}{2}$ inch thick, is placed immediately beneath, which prevents any irregularities in the pipe, where water might accumulate and obstruct the passage of the gas. For the supply pipes to the lanterns when leaden pipe is employed, a wooden boss, having an orifice there through, is fitted into the top of the column. To the upper side of the boss is screwed a plate resembling a ceiling plate, and to this is affixed the tap for the burner; and to the underpart is screwed a ferrule or nose piece, to which is soldered the leaden pipe. This arrangement is unquestionably superior to the wrought iron stand pipe, as it admits of the burner being placed always at the same height, whereas with the other, as frequently observed, the greatest irregularity exists. The durability is a further recommendation.

Services, alike of consumers' premises as well as those of the public lamps, are sometimes obstructed with naphthaline, and to remove this, the service cleaner is indispensable. This consists of an air-pump attached to an air-tight chamber (provided with a tap), into which, when required for action, air is compressed, and its action is as follows. On communication being made between the stopped service and the cleaner by means of an india-rubber tube, the workman by means of the pump compresses the air to the desired degree, when the tap is suddenly opened, and the compressed air rushes through and drives the obstructive matter before it, and the service is cleared.

Services should be of course always laid with a slight incline to the main, in order that any liquid condensing in the former may flow to the latter, and hence to the syphon; but when irregularities exist in the level of the ground, which do not permit of the necessary incline, under these circumstances one or more bottle syphons are placed. These are of cast iron, and made to contain one pint and upwards, and attached to the service by a corresponding T piece of the same dimensions as the service, provided with a pipe for pumping the liquor when necessary.

For drilling the mains of services, apparatus are manufactured which permits of the operation being effected with the loss of an inappreciable quantity of gas, which apparatus under some conditions, as in confined localities, are absolutely indispensable.

As the pipes employed for services range from $\frac{1}{2}$ inch to 2 inches in diameter, we will therefore avoid any complication of the simple question by merely giving the dimensions of services as practically applied, together with the quantity of gas they will respectively deliver.

Diameter of service pipes for supplying a given number of lights with a pressure of $\frac{7}{10}$ths in the main, each light consuming 5 feet per hour, together with the quantity the service is capable of delivering per hour, supposing this to be open at the end.

	Diameter of Pipe.		Yards.	Quantity per hour.
From 1 to 5 lights,	$\frac{3}{4}$ inch length of service in	..	150	.. 60 cubic feet.
„ 5 „ 10 „	$\frac{3}{4}$ „	„	.. 100	.. 74 „
„ 10 „ 20 „	1 „	„	.. 100	.. 151 „
„ 20 „ 40 „	$1\frac{1}{4}$ „	„	.. 100	.. 264 „
„ 40 „ 70 „	$1\frac{1}{4}$ „	„	.. 25	.. 528 „
„ 70 „ 100 „	$1\frac{1}{2}$ „	„	.. 25	.. 906 „
„ 100 „ 150 „	2 „	„	.. 100	..1033 „

CONSUMERS' METERS.

THE necessity of a means of supplying gas to the consumers by measure must have suggested itself to Samuel Clegg during the earliest years of his labours in connection with gas lighting. By his son we are informed that the first attempts were made with bladders, similar, we may assume, to the primitive means of storing gas as practised before the invention of the gasholder, to which reference has been made.

This not answering the expectations of Clegg, he adopted another kind of instrument for the object, which, as we are informed by Peckstone, consisted of two small gasholders fixed to a frame. To this was attached a pillar carrying a balanced beam, to each end of which was connected one of the gasholders, and by an arrangement of mercurial valves, actuated by the holders in rising and descending, the gas was admitted to and expelled from the vessels alternately, each holder in the act of filling, expelling the gas from the other. This system seems to have been practically put in operation, but the irregularity of the lights supplied thereby caused its indefatigable inventor to attempt other and more practical means to achieve success. We have already described the rotary gasholder as invented by Clegg, and have supposed this to have been in existence prior to his first patent, and that the holder was suggestive of the construction of the meter now to be referred to.

In December of 1815, Clegg obtained a patent for various inventions, amongst these, to use his own words, for "a gauge or rotative gas meter for measuring out and registering the quantity of gas which passes through a pipe or opening, so as to ascertain the quantity consumed by any certain number of lights or burners."

"This gauge consists of a hollow wheel or drum capable of revolving vertically upon bearings, in the manner of a water wheel. The hollow rim of the wheel is made close on all sides to form a circular channel, which is divided by partitions into certain compartments or chambers to contain the gas, which is introduced into the wheel at one end of the axis and carried off from the wheel at the other end. By certain contrivances it is arranged that each of these chambers will be filled with gas and emptied of the same into the exit pipe every time the wheel makes a revolution, by which means the number of turns the wheel makes, when registered by suitable wheel work, becomes a record of the quantity, or the number of chambers full of gas, which has passed through the gauge."

FIG. 85.

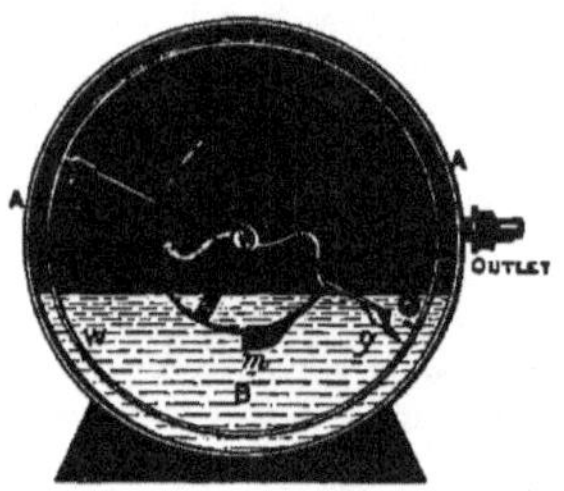

Clegg's Meter.

In the specification there are two kinds of meters described, one of which is exceedingly complicated, the other we will now explain.

With a slight variation for the facility of description, Fig. 85 represents a sectional elevation of the meter as patented by Clegg. A A is the outer case of the instrument, enclosing the wheel or drum, W W, which is a cylindrical vessel, containing a concentric cylinder, both of which are closed at the two ends, thus forming an annular space divided by the partitions into two semicircular chambers, B and C. The drum is secured by the pieces, $k\,k$, to the shaft s, which for the greater portion of its length is hollow, the back part working in a suitable bearing, the front and hollow part revolving in a stuffing box, to which the inlet of the meter is attached. The hollow shaft is connected to the two chambers by the bent pipes, d and e, and in the partitions which separate the chambers are the valves f and g, the lids of which are kept in position by light springs, these lids being opened by the resistance of the water when in the act of entering therein, thus permitting the liquid to flow from one chamber to the other as the wheel revolves in the direction of the arrow. There are also two orifices, h and i, in their respective chambers, through which the gas passes to the outer case for delivery to the burners. Two buckets, m and n, for a purpose hereafter explained, complete the wheel.

The apparatus being supplied with water to the required height indicated, its action is as follows: The gas entering by the inlet situated in the front and at the centre of the meter, passes through the hollow shaft s, and through the bent pipe e, into the chamber B, when by its pressure acting expansively between the surface of the water and the partition r, it causes the wheel to revolve in the direction of the arrow, so expelling the gas contained in the other part of the chamber B, through the orifice h, to the burners. During the revolution, the valve g, on issuing from the water, is shut by the spring, thus closing the communication between the two chambers, and on the valve indicated, f, coming in contact with the water it is opened thereby, when the water flows into chamber C, and expels the gas therefrom by the orifice i. The bucket n, which has been filled with water during its immersion, on the wheel arriving at the desired position, discharges its contents into the bent pipe d, which takes place before the valve g issues from the water. This revolving action continues until chamber B is brought into the position occupied by C, when the operation is repeated until an entire revolution of the drum will be made, the wheel receiving and expelling a quantity of gas equal to its cubical contents. This capacity being known, then by means of suitable wheel-work the number of revolutions of the drum is conveyed to a dial or index, and the quantity of gas passing during any indefinite period will be recorded.

With our present knowledge on the subject, it is obvious that a meter as thus constructed must have been very imperfect in its operation, demanding considerable pressure, and causing serious oscillations in the lights supplied thereby, besides, from the delicate nature and action of the springs it was very liable to disarrangement. These, with other defects, rendered it but ill adapted for the intended object, and although highly ingenious and beautiful in its conception, it was by no means a practical machine.

In 1816 Clegg became associated with Samuel Crosley for the purpose of manufacturing the meters described; little, however, is known of their progress during the brief period that form of instrument was retained.

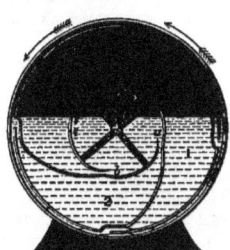
Fig. 86.

Fig. 87.

MALAM'S METER.

The following year John Malam, who was then draughtsman at the Peter Street station of the Chartered, submitted a drawing of a meter to the directors of that company, of which the above is a copy.

Fig. 86 shows a sectional elevation, and Fig. 87 a cross section of the instrument as designed by Malam, and consists of a cylinder or drum enclosed within a case in which it revolves on an axle working in bearings on the bent pipe p, and on the front of the machine. There is also a central compartment x, which conveys the gas from the bent pipe p into the various chambers, as hereafter mentioned. The drum is closed at the two ends and divided into four distinct chambers, 1, 2, 3, and 4, by which, when the meter is charged with water to the desired height, the gas is received and expelled alternately.

Attached to the inlet, as represented in Fig. 87, is a bent pipe p, partly in section to show the passage of the gas, seen also in elevation, which passes through an orifice (without touching the drum) in the centre of the compartment x. This orifice being sealed by the water, the gas is conveyed to the various chambers by their respective inlet passages, a, b, c, and d. As represented, the partitions are placed obliquely for the purpose of offering the minimum amount of resistance to the water during the revolution of the wheel, and each chamber is so formed that at the ordinary water level, on its inlet becoming sealed by the water, the wheel must revolve a short distance before its outlet can become unsealed.

On referring to Fig. 86, the machine will be easily understood, for, as shown, the gas is entering by the bent pipe into compartment x, from there it is conveyed through the orifice d into chamber 1, when by its expansive force the gas presses between the surface of the water and the partition dividing chambers 4 and 1, and acting as in an ordinary gasholder, raises the partition, so causing the wheel to revolve in the direction of the external arrows, at the same time expelling the gas from chambers 3 and 4, by their respective outlet passages on the periphery of the wheel. In due time chamber 2 comes into action, and in filling expels the gas from 4, and brings chamber 1 into that position now occupied by 4, when it neither receives nor delivers gas. Thus the operation is continuous, each chamber receiving its supply of gas and forcing out that from the preceding chambers, identical in action with the meter now generally employed.

The beautiful simplicity of the machine described reflects the highest credit on its inventor, for, in addition to being an instrument to measure gas accurately, it had the further advantage, under certain conditions, of delivering this comparatively free from oscillations.

The instrument in this state of comparative perfection was practically applied in gasworks as early as 1819, and singular enough the two writers of that period, Accum and Peckstone, had each his favourite engineer, and to Clegg the first-named author awarded the most extravagant praise, and to him he gave the sole merit of being the inventor of the meter, Malam's name not even being mentioned therewith. On the other hand, Peckstone, although a stanch supporter of Malam, to whom he was related, did not hesitate to render full justice to Clegg.

Malam's wheel, with the inlet and outlet considerably enlarged, was employed for some years as a station meter; and it was retained among the illustrations of a standard work on gas lighting long after it ceased to be applied.

As thus constructed, the meter required considerable alteration on account of the great pressure necessary to work it, as well as the irregularities of the lights, before it could be generally adopted by consumers.

Its main defect was the small apertures for the admission and emission of the water into and from the various chambers; moreover, for every foot of gas delivered that quantity of water had to pass through the meter, thus producing considerable friction and requiring great pressure to work it. Consequently these difficulties arising, Samuel Crosley and Clegg applied themselves to seek the remedy, and we can imagine that their first thought was to construct a machine by which this excessive amount of labour should be avoided. This would naturally suggest the method of communicating all the chambers at the centre with the water, which effected, the inlet and outlet would be removed to the front and back of the wheel as now constructed, and as already explained when referring to the station meter. But although simple enough to realize after having been accomplished, the task must have been attended with numerous difficulties and disappointments, and could only be carried out by men of genius. Thus by the combined abilities of Clegg, Malam, and Crosley, was this remarkably simple and wonderfully effective machine produced in its present state; and although innumerable patents have been obtained from time to time, with the intention of making other instruments of the kind, or with the view of improving the present form of machine, it is only within the last two years that any positive improvement has been effected therein, and to which we shall have occasion to refer.

Candour compels us to refer to the only weak point in the character of Samuel Clegg, and that was the jealousy evinced by him towards Malam on account of his invention, whom he regarded for years with a strong spirit of rivalry, and only very tardily admitted the share claimed by Malam in the production of the instrument in question.

The exact period when the meter in its present form was first manufactured is not exactly known, but it is believed to have been about 1821; but the incorrodable metal for the wheels was not introduced for ten years after that.

The great difference existing between the meter constructed by Samuel Crosley (for Clegg retired from the firm in 1824), and that patented by the last-named gentleman, induced other firms in various parts of the United Kingdom to manufacture these instruments, which for a period they carried on without interference. But in the course of time Crosley instituted legal proceedings against them, and in all cases recovered heavy damages. At the period great complaints of the injustice of the decision were made; but if we calmly refer to that portion of Clegg's specification already given, we find that he distinctly states that a wheel "divided into compartments," "and which will be filled with gas, and emptied," and that "the number of turns the wheel makes when registered by suitable wheel-work becomes a record of the quantity of gas which has passed through the gauge." After the perusal of this, it is difficult to imagine how any other decision than that favourable to the patent could have been arrived at.

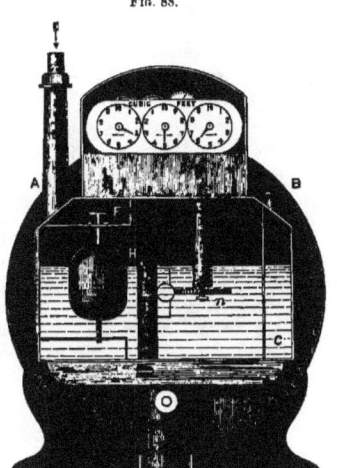

Fig. 88.

Clegg and Crosley's Meter.

The consumers' wet meter, as improved by Clegg and Crosley, differs but slightly from the station meter; this difference consisting principally in the positions of the inlet and outlet pipes, the addition of a box or frame situated in the front, containing the means of protecting alike the interests of the company and the consumer, and the position of the index.

The annexed Fig. 88 represents in elevation the instrument in its simplest form, the front supposed to be glazed in order to facilitate the description. When in operation, the gas passes by the inlet A, through the valve in the box F into the pipe H, hence through the bent pipe or spout into the hollow cover, the action of the wheel being identical with the station meter already described. But as any material deficiency or excess of water would affect the accuracy of the instrument, two simple means are adopted to avoid this. One of these is the float C, which is furnished with a rod at its upper part, and to this is attached a valve; thus in the event of any important deficiency of water, the float descends, closes the valve, and stops the supply of gas, until the proper water level is again restored, by these means securing the interests of the company.

On the other hand, if we suppose an excess of water, which would be prejudicial to the consumer, this will flow down the pipe H into the bottom box E, and eventually fill the pipe and obstruct the passage of the gas, thus assuring an approximative measure alike for the consumer and company; for absolute accuracy other methods are adopted, to be hereafter mentioned. Meters constructed as described, however, are made so as to indicate not more than two per cent. against the company or three per cent. against the consumer, this difference arising from the position of the water line, between the top of the pipe H and the point where the float descends. On the shaft is an endless worm, which gears into the toothed wheel n, so conveying to the index above the number of revolutions of the measuring wheel, or the quantity of gas passed: B is the filling plug; E the box for containing any surplus water; and D the screw for its removal.

One great point to be observed with the wet meter is that it shall be fixed level from back to front in order that it may be effective, and this is particularly in the interest of the company supplying.

We have been unable to ascertain why Malam did not secure the merits of his invention by patent, for although it may be conceded that by Clegg's description in his specification no revolving wheel nor drum could be adopted without infringement, yet the bent pipe or spout, with the means of applying it, was absolutely essential and indispensable to the perfection of the meter, and this secured, at least moderate compensation would have been accorded him. As it was, the only remuneration that Malam ever received for his remarkable ability, was a gold medal from the Society of Arts, for his invention of a gas meter "superior to all others, and likely to be of great benefit to the public."

Sir William Congreve, who had been Government Inspector of Gasworks, in 1824 patented an improved gas meter, in which he proposed to "register the flow of gas through a pipe or cock, when the pressure is uniform, by registering the time the tap was left open."

Fig. 80.

Clegg's Pulse Meter.

For this purpose he applied a small clock movement in connection with the lever controlling the main tap, so that on opening the latter, the clock was set in motion, and was stopped by closing the tap. The number of burners in the establishment, multiplied by the hours the tap had been open, gave the hours to be charged, or the amount of the consumption during the period.

These meters, like Clegg's pulse meter, now to be described, were very extensively applied on the Continent, long before any ordinary meter manufactory was established there; and so recently as 1843, Congreve's "*compteur à l'heure*," or hour meter, was replaced by the "*compteur à volume*," or volume meter. The difficulties attending the hour meter were numerous, and not the least amongst them was the necessity of winding up the apparatus at intervals; or, the fraudulent consumer could with facility stop the action of the timepiece, whilst there was no check on the consumption from each burner.

In 1830 Clegg patented "a meter which shall work without the aid of water," usually termed the pulse meter, which we will describe as briefly as possible, simply in order to make the action of the instrument intelligible.

The annexed figure represents the outer case of the instrument, of about 4 inches in diameter, the actual size of a three-light meter as manufactured; which is surrounded on all sides, except at the bottom part, marked b, with a non-conducting material. Within the case are two glass bulbs, half filled with alcohol, and connected together by the tubes $d\ d$, which are hermetically secured to the frame e. This works freely on pivots at the point p (the front bearing of which is removed), the globes being counterbalanced by the weight w. At the back of the meter is the inlet pipe, made of small proportions, and protrudes into the case at the point marked i, so as to approach close to the upper bulb. The outlet o is at the top of the case.

In the two patents obtained by Clegg for this instrument, he employed different methods of application; but in both he used a lighted jet, which was indispensable to the action of the meter. This we have placed at the point g, immediately under the plate of metal b, at the bottom of the external case, where there is no non-conducting material. The jet, which is very small, being lighted, heats the interior of the meter, where the whole becomes of uniform temperature, and will thus remain so long as gas is not permitted to pass therethrough; and in consequence the liquid within the bulbs remains in the positions indicated.

Now, if we suppose the gas to be passing through the meter, this impinges continually on the upper bulb, so condensing the vapours therein into the liquid state, whilst the lower bulb being in contact with the heated plate, the liquid therein is converted into vapour, which is condensed in the upper bulb, and on this becoming "top heavy" falls into the position of the lower globe, which in its turn becomes subjected to the action of the heat, its vapours ascend to the upper globe, and again are condensed, when the position of the globes is again changed; thus producing a continuous vibration or oscillation of the two globes, at one moment being as represented, and at the next as shown in dotted lines; the velocity of these oscillations depending on the condensing power, or, in other words, the quantity of gas impinging upon the upper bulbs, and consequently the quantity of that passing through the instrument. Then the number of oscillations for a given quantity of gas passing being ascertained by experiment, the index was made accordingly.

In connection with the meter was a method of shutting off the gas when the jet was not lighted. This was effected by the expansion of a metal bar, which, when heated, opened a valve, and on the jet being extinguished the bar by its contraction was intended to close the valve; but in practice it was discovered that this bar, being repeatedly heated, became permanently expanded, and was thus rendered useless, and was one of the practical defects of the instrument. A meter of the kind in our possession is so remarkably sensitive, that by simply breathing on the lower bulb it is set in motion.

This was undoubtedly one of the most favourite of Clegg's inventions, for during eight or ten years he had hopes of its success, and was continually occupied in endeavouring to bring it to perfection; and about 1844 several of these meters were in operation in the metropolis and elsewhere, and, as stated, they were extensively used on the Continent.

It is a remarkable fact that this instrument measured with a very close approximation to accuracy when passing large quantities, but was practically useless as a measurer when working slowly, which was one of the obstacles to its employment; as it was considered that, if not operative under the two conditions, it could not be reliable; and although Clegg claimed several advantages for this meter, it can only be regarded as a beautiful, ingenious, scientific toy. No pretensions have been made to describe the instrument with accuracy, as this would occupy too much space, besides being difficult to understand; we have therefore given only the mode of operation.

Shortly after the production of Malam's invention it was applied as a station meter, and in addition to performing its legitimate duty, it was employed as a motor to drive the wet-lime purifier, as already described; and we can imagine that the augmented leakage at the luting of the mouthpieces, occasioned by the increased pressure thrown on the retorts by the meter as thus used, directed attention to one of the evils of the arrangement, causing it to be abandoned. But the most important defect of high pressure, with high heats, namely, the deposit of carbon within the retorts with all its evils remained unnoticed, and was disregarded for about a quarter of a century.

As already observed, Malam in 1820 patented a method of making a meter to act without the aid of water, as now termed a "dry meter;" but no further steps were taken in the matter until 1833, when an intelligent American, named Bogardus, introduced a system of making these instruments, which was patented in England.

This patent led to the formation of the "Patent Dry Gas-Meter Company," established with a capital of 80,000*l.*; and with the view of carrying on an extensive business, premises of great magnitude for the period were built in the north of the metropolis, adjoining a canal, where the instrument underwent considerable alteration, until eventually it was produced as represented in the accompanying engraving.

This consists of a cylindrical case, the front being glazed in order to explain more clearly the action of the instrument, surmounted on a square box, in the front of which is the index. The upper case is divided in the centre by a loose division of leather, or "bag," or diaphragm, in section, when distended, resembling a hemisphere, thus forming two distinct chambers. Within the square box is placed a three-port valve, with its corresponding slide; two of the ports communicate with their respective compartments in the cylinder, the centre port being the outlet. The slide valve receives its motion from the vertical rod, in connection with the V piece, which is attached to the diaphragm, and acts as a tumbler.

Thus, as the diaphragm receives a quantity of gas in the one side, it expels a corresponding quantity from the other, and each time it is distended the tumbler gives motion to the vertical rod, and so changes the position of the valve cover, and alters the passage of the gas. Therefore, the capacity of each stroke of the diaphragm being ascertained, the index is arranged accordingly.

This was the first dry meter, and our impression is that the machine was never fairly tried, and that the company sadly wanted the commercial element in their enterprise to introduce these meters to the notice of gas companies. As it was, they were made by hundreds, yet in no single instance have we known the meter to be in operation. Eventually the affair failed, bringing with it the ruin of the solicitor who had organized the company, and had invested the whole of his fortune in the undertaking. We are aware that the cause of failure in this meter has been attributed to its want of a maintaining power; which in our opinion is incorrect, as the action of the tumbler supplied sufficient power to move the valve and actuate the index as required. The weakest point in the meter was the speedy destruction of the leather through the wear and tear of the tumbler.

FIG. 90.

DRY METER COMPANY'S METER.

Bogardus in 1836 produced another dry meter, which was patented by Sullivan, and generally bears his name. This consists of a cylindrical vessel, subdivided in the centre, in each division of which is placed a flexible diaphragm, similar to that described. The gas was conveyed to and from the four divisions by means of a rotary valve, identical with those used for the central valves of purifiers, but of course having only four divisions for the ingress and egress to and from the various chambers.

A private company was established for the manufacture of these instruments, which, however, were made in the worst possible manner; but the greatest difficulty in connection therewith was the unsteadiness of the lights supplied by them. Moreover, the directors of this company seem to have created difficulties, for as the experience of many years has since proved, no objection exists against the metal valves used in these instruments, yet the Sullivan Meter Company renounced the use of this for glass.

Again, with the leather forming the diaphragms, although tanned skins as now adopted were employed, and of which the experience of many years has proved the excellence for the purpose; yet as a means of preserving these diaphragms, they were actually coated with gold leaf. The most extraordinary circumstance was that to give this coating the leather had to be varnished, which in drying became hard and cracked whenever motion was given to it; thus the leather was speedily destroyed at those parts, consequently producing the very reverse effect to that intended. This company, like its predecessor, terminated in failure; and a gentleman, well known and highly respected, lost considerably in the affair, and during the rest of his life was the strongest opponent to dry meters.

For the first establishment of the dry meter we are indebted to Defries, who in 1838 patented an instrument with three "bags" or diaphragms, identical in their construction with those of Bogardus and Sullivan.

This meter differs from the others mentioned, in having six measuring compartments formed by three diaphragms, which were worked after the system adopted by Sullivan, and applied in the dry meter now universally adopted, that is by means of perpendicular rods which, receiving their motion from the diaphragms, actuate three slide valves, by which the ingress and egress of the gas to and from the six measuring compartments in succession are controlled. These meters at first were well constructed, the lights supplied by them were perfectly steady, and their measure in some instances was correct, even with a variable supply of gas; but in others the very opposite of this was the case, and from a cause not then understood, namely, the position of the cover of the valve in relation to the chambers it supplied, no certainty in their measure, when delivering different quantities, could be arrived at.

Notwithstanding defects which certainly existed in these instruments, and the two positive failures already referred to, yet Defries, with his indomitable perseverance and energy, succeeded in overcoming the strongest prejudice and opposition, and getting his meters extensively adopted by some of the metropolitan as well as by many of the provincial gas companies.

Eventually, however, it was considered desirable to change the diaphragms from the "bag" form to that shown in the following engraving. This consisted of an outer frame, to which was secured the leather constituting the diaphragms previously blocked to the form of a pyramid. To this the four plates, A, A, A, A, were then secured, as shown in Fig. 91. The flag piece which gave motion to the vertical rod was connected by a kind of universal joint to the four triangular plates. Thus when in action the diaphragm when distended in either direction assumed the form of a cone, as represented in Fig. 92, the space between this and the dotted lines being equal to the volume of gas received and expelled at each stroke. This meter, which was the invention of N. F. Taylor, was certainly one of the most ingenious and beautiful machines ever produced in connection with gas lighting, but, as proved by experience, it was ill adapted for the purpose intended.

Fig. 91.

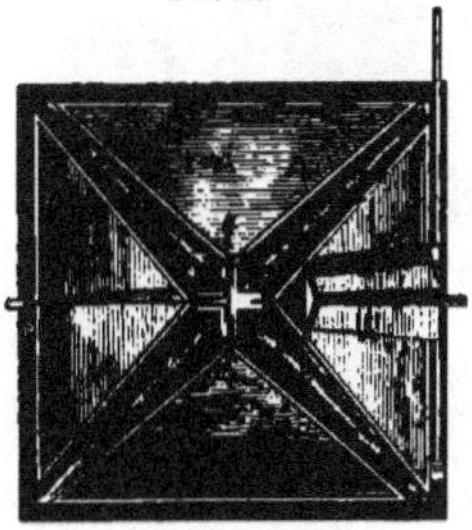

Fig. 92.

It may be now observed, when no longer contending interests exist in the matter, that with Defries' as with Sullivan's meter, one evil was abandoned in order to adopt a greater. For in the "solid" partition, as it was termed, there being but a small margin of leather around it, the wear and tear, particularly at the corners, was so excessive, as speedily to destroy the diaphragm at these parts, and of course rendering the instrument useless. Nevertheless, this meter, for a time, seemed likely to receive general adoption, at least as a dry meter; for in addition to the original manufacturer, it was subsequently made by various other firms in the United Kingdom and America, but at present, so far as we are aware, it is abandoned, and superseded in every sense of the term by the instrument to be described.

In ignorance of what had been accomplished by Sullivan, or rather Bogardus, the author of the present work, in 1843, entertained the idea of making a dry meter, which should be simple and cheap in its construction, unvarying in its capacity for measuring, and of great durability. His first attempts were by no means encouraging, but by persevering he succeeded in inventing the dry meter substantially as now universally manufactured, both at home and abroad; and although others have laid claim to it, the meter was produced and first manufactured on an extensive scale solely by him, of which there is abundance of evidence, even at the present day.

This instrument is represented in elevation by Fig. 93 and in side view by Fig. 94, the outer case being of glass for the purpose of more readily explaining the action. The case A A is divided by a horizontal partition into two compartments, the upper one, E, containing the valves which convey the gas into and from the respective measuring chambers; the lower compartment is subdivided by the vertical partition B, to which are attached two flexible chambers, 1 and 2. Thus there are four distinct chambers, namely, the interior of those numbered 1 and 2, and their exteriors, or rather the spaces the flexible chambers move through in the compartments marked 3 and 4, all of which are in communication with their respective ports of the slide valves hereafter mentioned.

Each of the flexible chambers is formed by the ring a, which is soldered to the partition B, together with the disc C, and a band of thin flexible leather, D, blocked to the requisite form, hermetically secured to the ring and the disc. This disc is guided by the flag piece b, affixed to the vertical rod d, which

FIG. 93.

FIG. 94.

RICHARDS'S METER.

passes through a stuffing box into the upper chamber E; and for maintaining the disc always in the same plane, there is a guide $c\ c$, working freely in its supports, and in the studs attached to the disc.

Fig. 95 represents the plan of the top of the meter, showing the two slide valves which conduct the gas to and from the four measuring chambers, the centre ports of which convey the gas after measurement to the burners. As represented, on the top of each of the vertical rods is affixed a lever ff, carrying a jointed rod, which gives a rotary motion to the "tangent," and so causing the crank to actuate the covers of the valves as desired.

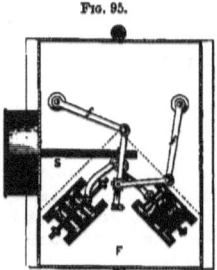

FIG. 95.

Thus the gas, on entering one of the flexible chambers, say No. 2, the disc C will be caused to move through a certain distance, expelling a like quantity of gas from chamber No. 4, equal to that entering chamber 2, and when the diaphragm is distended to the required degree, the position of the cover of the valve is changed. The gas then enters chamber No. 3, expelling that from No. 1, until the disc nearly touches the ring, when the flexible chamber will be collapsed, and the position of the valve cover changed.

The gas enters the chambers in succession, receiving and expelling continuously, so long as any supply from the instrument is required; thus, if we suppose the capacity of each of the flexible chambers to be equal to one-fourth part of a cubic foot, it is obvious that by a complete action of the instrument, the four chambers being filled and emptied, that a cubic foot of gas would be delivered by the company and received by the consumer. Hence, in order to record the quantity of gas passing through the meter, the spindle S is provided with a toothed wheel, and receives its motion from an endless worm on the axle of the valve crank, which with the simplest arrangement of wheel-work completes the meter.

This instrument, as now universally employed, is identical with that invented and manufactured by the author thirty years ago. The main points in the invention were the formation of a meter with two diaphragms and two slide valves; the means of defining the lengths of the levers and the position of the valves, so as to ensure a steady light, which previously had been considered an impossibility with less than three diaphragms; the means of forming large discs and in keeping them always in the same plane when in action; in so distributing and forming the leather band, that any ordinary expansion or contraction would not affect the measurement of the meter; and lastly, that by reason of the small amount of wear and tear of the leather, the durability of that part of the instrument would be assured.

Not only was the meter invented, but the means of manufacturing, as adopted at the present day, were all then established; and although since then upwards of fifty patents have been obtained in connection with dry meters, yet beyond the shape of the external case, no important change nor alteration has ever been made in that under consideration. However, through misplaced confidence, the inventor never derived the slightest advantage by his invention, nor for the labour and time devoted to its production, thus repeating the history of Ami Argand, Reuben Phillips, and others.

Subsequently the meter passed into the hands of the late Mr. Thomas Glover, to whose energy and commercial abilities was mainly due the success it eventually achieved.

Various other descriptions of dry meters have been produced from time to time, some of them having as many as six diaphragms, but gradually the different systems have been abandoned for that described.

It must not, however, be supposed that the meter, whether wet or dry, with all its numerous advantages, was appreciated by those most interested in its employment, but on the contrary; for, singular as it may appear, a rule observed by some metropolitan as well as provincial companies, until comparatively recent years, was to supply by contract wherever practicable.

About 1840 the Chartered Company began to understand the great importance of the instruments in question, and by their directors it was resolved that meters should be introduced to the exclusion of contract burning, or, as it may now be necessary to explain, the system of charging a certain sum per annum for each light upon the premises, as universally adopted before the introduction of the meter. This resolution met with the strongest opposition from consumers; and subsequently, when it was decided that the supply would be discontinued if not furnished by this means, many submitted for a time to that alternative, rather than admit the "mystery box," as it was sometimes called, into their premises.

And it must be admitted when all the circumstances are considered, that there was some excuse for this prejudice, for the action of the instrument itself, measuring constantly and silently, a subtile fluid like gas, was to many no doubt mysterious. Moreover, a most imperfect surveillance had previously been observed with respect to contract consumers, who frequently burned three or four times the quantity of gas they were entitled to, which by the meter was however corrected, when for want of a better means of explanation, the extra charge was attributed to a system of jugglery, of which, by some consumers, gas inspectors were considered adepts.

In France the meter was legalized as a measure in 1846, when the Government made the necessary stipulations as to range or variation from the correct measurement, and other rules to be observed in the construction of these instruments. A few years afterwards most of the Continental countries adopted similar steps.

Some years ago, when the instrument was not so well understood as at present and prior to the Sales of Gas Act, it was by no means uncommon to meet with consumers' meters of the smaller sizes with the spouts of such extravagant height that the meter could be overcharged to such an extent as to register 30 per cent. fast. Singularly enough, there were consumers who always insisted on charging their meters to the highest point, with the belief that by so doing they were obtaining an advantage; and who looked with the greatest suspicion and distrust upon the operations of the company's inspector, who, when taking the index, would withdraw the surplus water. But the moment his back was turned, the consumer would again charge the meter to the full, and, delighted with his cleverness and having the satisfaction that he had benefited by his cunning, would pay his account with pleasure; but still more remarkable, it was utterly impossible to make this class of people understand the truth, that they were by this means really cheating themselves.

2 o

In consequence of some agitation in the metropolis and elsewhere, respecting the accuracy of meters, the attention of Government was directed to the subject, which resulted in the passing of the Sales of Gas Act in 1859, by which gas meters were recognized as legal measures after being duly stamped by the proper authorized inspectors. Amongst other things, it was enacted "that no meter shall be stamped which shall be found by the inspector to register, or be capable of being made, by any contrivance for that purpose, or by increase or decrease of the water in such meter or by any other means practically prevented in good meters, to register quantities varying from the true standard of measure of gas more than two per cent. in favour of the seller, or three per cent. in favour of the consumer; and every meter, whether stamped or unstamped, which shall be found by such inspector to register, or be so capable of being made to register, quantities varying beyond the limits aforesaid, shall be deemed incorrect within the meaning of this Act; and every meter which shall be found by such inspector to measure and register quantities accurately, or not varying beyond the limits aforesaid, and shall be found incapable, by any such means as aforesaid, of being made to register quantities varying beyond the limits aforesaid, shall be considered correct and be stamped as aforesaid in such manner and on such part of the meter as shall be specially directed by the authority appointing him, or in default of such direction, in any part as shall in his opinion be best to prevent fraud."

In this Act, instructions are given as to the means to be employed in testing meters, in ascertaining their freedom from internal leakage, the fees to be paid for testing, and the fines for fixing or using any unstamped meter, and that "any contract, bargain, or sale, made by such" (unstamped) "meter shall be void," and that "every such" (unstamped) "meter so used shall on being discovered by any inspector so appointed as aforesaid be seized, and on conviction of the person knowingly using or possessing the same, shall be forfeited and destroyed."

This law rendered considerable modifications in the construction of wet meters indispensable, particularly in those of the smallest capacity, which was met, however, by increasing their diameter and diminishing their depth; with these alterations no further difficulty presented itself in making them in conformity with all the requirements of the Act referred to.

The general indifference of wet meter manufacturers previous to the passing of the Act mentioned, contrasted forcibly with their subsequent proceedings. As in the first instance, meters were made in such a manner, that although accurate at the water line, in most cases they were susceptible of being overcharged, and consequently their registration affected to an incredible extent in the manner stated. But the extreme limits of variation from the correct standard of two per cent. in favour of the company and three per cent. in favour of the consumer being defined by the Act of Parliament, then various attempts were made to render the instruments still more accurate, by the adoption of means to compensate for the evaporation of the water: to which class the name of "compensating meter" was given.

For this object, various devices were patented, some of them conspicuous for their complication, and by one of these systems, a quantity of water equal to the whole contents of the meter was acted upon by the scoops every three hours that it was in operation, and this simply for supplying the loss by evaporation. The absurdity of this is rendered obvious, when we consider that with a five-light meter placed in an ordinary average temperature, the evaporation will not exceed half a pint in six months, and the abstraction of that quantity would not affect the registration to the extent of one per cent.

Among the various methods of compensating, that of Sanders merits particular mention on account of the simplicity and ingenuity of the invention, by means of which a constant water line and absolutely correct measure are obtained. Numerous other methods are adopted to arrive at the same result, but in the majority of compensating meters, a separate water chamber is attached to the square front or frame, and by means of a small scoop actuated either by a cam or eccentric on the wheel shaft, or the spindle in communication with the index box, a constant supply of water is conveyed to the measuring part of the meter, and thus the instrument is made an accurate measurer.

The Gas Meter Company have recently made an alteration in the meter, by conveying the gas direct to the exterior of the wheel, and so off by the spout, thus reversing the order of working; the advantage of this system consisting in the fact that a more correct measurement under different pressures, with a variable and augmented number of lights, is obtained. A similar system was patented by Pinchbeck in 1861.

The most important improvement made in the wet meter since it left the hands of Clegg and Crosley, is undoubtedly that recently patented by Messrs. Warner and Cowan, by means of which the instrument is rendered practically a correct measurer under the various alterations in the water level, within the extensive range of the float, without the intervention of the mechanical appliances referred to as necessary for compensating meters.

To explain this improvement, let us suppose two meter wheels of the same depth, but one considerably larger in diameter than the other, provided with corresponding spouts. These are placed in an oblong case, side by side and connected together with suitable gearing, in order that they may revolve together, but in opposite directions to each other, for which purpose the partitions of the smaller wheel are placed in the contrary position to those of the larger wheel; the spouts of both vessels being in communication with what we may term the inlet chamber.

This accomplished, on gas being supplied, the larger wheel will deliver the greater portion of its contents to the burners, the other portion being withdrawn by the smaller wheel, and returned to the inlet chamber. Thus by any enlargement of the former, through the lowering of the water line, a corresponding enlargement will take place in the smaller wheel; hence the quantity of gas delivered at each revolution under all the various water lines will be constantly alike. This arrangement explains the principle of the invention, but thus manufactured it would be both inconvenient and costly; in practice therefore, instead of making the small wheel separate, it is placed within the other and affixed to the same axle, as represented in the annexed engravings.

Fig. 96 shows the end view of the meter wheel, which in part is identical with Crosley's, but within it is placed the smaller wheel *b*, represented in side view in Fig. 97, for which object part of the large wheel is cut away. The periphery of the wheel *b* extends a short distance above the highest water line, and in depth the smaller is about half that of the larger wheel; thus occupying a portion of the latter, and preventing the variable influence of the water on the measurement at that point. Hence in lowering the level in the larger wheel, the same depression of water takes place in the smaller; so that any increase of the volume delivered by the larger wheel will be withdrawn by the smaller, and delivered to the interior of the hollow cover, again to be measured. Thus, whatever may be the height of the water line, the same unerring quantity is delivered to the burners at each revolution of the wheel.

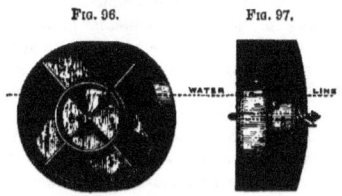

Fig. 96.   Fig. 97.

The extreme simplicity and efficiency of the arrangement is to be much admired, for according to experiments we have made with a three-light meter, on the abstracting of water from the highest to the lowest point in an extensive range of float, no practical difference in the registration existed, and when passing 18 feet per hour through the same, the lights were perfectly steady, and no greater pressure was absorbed than with the meter as ordinarily made.

Hence the gas meter, as now generally constructed with compensating apparatus or as just described, is about the most perfect means of measuring in existence, for with ordinary liquids, with some of the smaller implements employed, it would be an operation of some nicety to arrive at within five per cent. of the true quantity arising from the nature of the liquid and the condition of the measure whether in a dry or wet state; whereas, with the gas meter when properly constructed, a difference of one per cent. will not exist. However, the nature of the fluid measured thereby is another question.

Prior to the Sales of Gas Act, manufacturers generally constructed the wheels of meters of such capacity that the quantity of gas necessary for the maximum number of lights for which they were destined—each light consuming 5 feet—was delivered by 100 revolutions of the wheel per hour. Thus the quantity for ten burners was 50 feet; this divided by 100 gave half a foot as the contents of each revolution of a ten-light meter wheel, which rule was generally applied to all sizes (the three-light excepted), whether for consumers or station meters.

By the Act referred to, the quantity stipulated for the consumption of argand burners is 6 feet per hour; manufacturers however did not increase the capacity of the wheels of meters; the only practical alteration in this respect produced, has been to diminish the capacity of the station-meter wheel, for this, as now constructed, is calculated to work at the rate of 120 revolutions per hour, instead of 100 as formerly. Thus a wheel of 10 cubic feet capacity is considered sufficient to supply 1200 feet per hour. This is the basis of calculation of some of the most eminent makers of these instruments, to which there can be no objection so long as the working pressure is not increased.

Whenever a meter is made to supply a greater quantity of gas than it is calculated for, the pressure required to work it is augmented in proportion. Thus a five-light meter will deliver 30 feet per hour, with a loss of pressure between the inlet and outlet of about 1·5, or 2 tenths; but if 50 feet be passed through the same meter, the loss of pressure will be increased to 3 tenths or 4 tenths, but from that point any augmentation in the quantity of gas passing, a very considerable loss of pressure occurs.

In several towns and parishes, both in the United Kingdom and elsewhere, the method of supplying the public lamps by meter, either entirely or by the average system, is adopted. For this purpose both the dry and wet meter are employed, and fixed either within the lamp columns or enclosed in cast-iron cases, and placed beneath the paving. The first plan requires an account to be opened for every lamp in the district, the indices are taken once a month or quarter, from which the total account is made. By the average system a meter is fixed on every twelfth lamp, or any other number according to arrangement, and the total amount of consumption of the whole of the lamps is determined by the indication of the average meters. We are of opinion that with our modern resources of regulators and rheometers, hereafter referred to, the utility of meters to public lamps may be regarded as somewhat doubtful, for here is required a given quantity of gas to be supplied to each lamp, which within a certain percentage can be assured by the instruments named; the question therefore arises whether the expense of the meters and boxes, together with the trouble of inspection and keeping the accounts, is compensated by any corresponding advantage.

In a parish in the metropolis some economy appears to be effected by the use of meters, but this may arise, as has been suggested, from the diminished hours of lighting and by the use of smaller burners, and could be carried out by arrangement with the company equally as well without as with the meter. It must be stated, however, that gas companies, like municipal bodies, are not always infallible; and in some cases, although rarely, the meter system may be desirable.

Opinions as to the respective merits of wet and dry meters are very conflicting; for whilst we find some companies directly opposed to dry meters, and who will only permit of their employment under extraordinary circumstances; on the other hand there are other companies who strongly advocate their usage; and dry meters are further supported by fitters generally, on account of the facility with which they are fixed; and in consequence of the absence of the liability of condensed liquor collecting in the pipes, greater facilities are afforded for placing the fittings throughout a building.

That the dry meter possesses several advantages there can be no question, for being complete in itself and not demanding the aid of water to render it effective, the inconvenience and annoyance of this freezing in cold climates and stopping the supply of gas is prevented; hence for certain localities they are indispensable. For the same reason the dry meter does not require the periodical attention of the wet meter, and by the absence of water the evils of evaporation and condensation in the fittings, the oscillation and total extinction of the lights are avoided; nor does it present the facilities to fraud which the other certainly offers to the initiated. These are, undoubtedly, points of superiority possessed by the dry meter; but as a measurer under certain conditions it is not considered so reliable as the wet meter, which is in consequence always employed for experimental purposes.

The systems adopted by different gas companies concerning meters both at home and abroad are variable, as we find that some companies reserve to themselves the right of purchasing all meters, which they supply to the consumers at prices often extravagantly high, by which the progress of their business is retarded ; at other times they are sold at cost price, the company undertaking to repair the meters, which is frequently done gratuitously. There are companies who supply meters at a fixed rental, averaging about 10 per cent. of their cost price, as authorized by recent Acts of Parliament, the consumers having the privilege of purchasing their own meters if they think proper. Again, there are other companies who supply their customers with meters gratuitously, but giving the option of purchasing them should the consumers desire it.

But these different systems are not so surprising as the choice of meters, as there are companies who employ only the dry meter to the total exclusion of the other; whilst other companies insist on supplying wet meters only, and if the consumer requires a dry meter, he must purchase it, and in some instances even this is not permitted. Again, there are companies who use wet meters, and these with cast-iron cases only; and, on the contrary, there are others by whom none but tinned plate cases are permitted. The arguments sometimes employed in favour of the latter being that, in the event of an explosion, the cast iron might be splintered and have the effect of a shell, which, however, may be considered about the weakest of all reasons. Hence, with all these different systems and opinions expressed by the actions of the various companies, any personal opinion must carry but little weight.

If, however, we may express an opinion as to the durability of meters, we believe that the cast-iron case wet meter is to be preferred to all others ; as within our knowledge some of these instruments have been in operation upwards of twenty years, and are still doing good service, without having required the least repair. All that has been done has been to wash them out occasionally every three years or so; and for works abroad not possessing the ordinary means of repairing, these are particularly well adapted, as we have found from a lengthened experience.

One of the most serious sources of loss to a gas company is the imperfect condition of a portion of the meters, which permits of a small quantity of gas passing without being registered, occasioned either by decay, or wear and tear, or imperfect construction; and through want of attention to this point, the loss sustained by a company some years ago materially affected their dividends. To avoid this a periodical investigation of the state of the whole of the meters should be made, which can be accomplished without the necessity of removing them. For this object caps will be required corresponding with all the various sized meters, each having a small orifice therein capable of delivering with the ordinary day pressure one-twentieth part of the hourly supply. Thus with a 50-light meter the orifice in the cap will be required to deliver 15 feet per hour, or about the eighth part of an inch in diameter, and if desirable a flexible tube can be attached for the purpose of conveying the gas into the atmosphere or to burn it. The outlet of the meter being detached, one of the marks of the drum of the index being brought previously opposite to the pointer, the cap is then placed on the outlet, the gas is turned on, and an observation of a quarter or half an hour will be sufficient to ascertain the soundness of the meter.

In smaller instruments the same method can be observed, when for a three-light the orifice would be made to deliver nine-tenths of a foot per hour; for a five-light, 1½ foot; and so forth. With an inspection of this kind, which in some of our largest companies would require a staff continually employed for that purpose, unaccounted-for gas would be considerably reduced. Undoubtedly a more accurate method would be to attach a test meter to that under investigation, but in all cases meters should be inspected every three years, and would well repay the expenses incurred in the operation.

We must not leave this part of our subject without referring to an application of the meter to which only a passing allusion has been made, that is, the means of increasing the pressure of the gas at any desired point or premises, by its employment as an exhauster, or, as generally termed, "motive-power meter."

The necessity for this may arise from various causes, such as manufacturing or other premises, or a number of public lights being situated considerably below the level of the works, when in order to supply these, a greater pressure will have to be given to the whole of the town, which may be objected to by the company supplying on account of the increased leakage; or a manufacturer may require large quantities of gas during the day at irregular periods, which the ordinary pressure furnished will not meet. Under these and similar circumstances the "motive-power meter" may be advantageously employed.

This instrument is remarkable for its simplicity, and when of small dimensions consists of an ordinary meter, with the shaft of the wheel protruding a short distance through a stuffing box at the back of the case. On top of the case is a horizontal tube or drum working in suitable bearings, which for a five-light meter is about 2 inches in diameter and 6 inches long, having a toothed wheel at its end. To this drum a cord is secured, which passes over a pulley placed at a convenient height, and to the end of the cord is attached a weight of about 28 lbs., which just clears the ground when the cord is unwound. To the end of the shaft of the wheel is fixed a toothed wheel or pinion, gearing by intervening wheels with the toothed wheel of the drum; and when required for action, the drum is turned by a key similar to winding up a clock, and the weight raised to its full height, and this in descending gives motion to the meter, exhausting the gas from the main and delivering it with the pressure desired into the premises.

In large meters the gear work will require to be in a separate frame similar to a clock frame with a weight corresponding with the size of the machine and the intervals of winding up. To give a vague idea of this, a weight of 4 cwt., with a fall of 25 feet arranged with proper gear work, will be sufficient for a 50-light meter during four hours, or for 400 revolutions, and will deliver the gas with a pressure of about five-tenths of an inch above that existing in the main, or, in other words, it will exhaust 2000 feet of atmospheric air and deliver it at the pressure stated.

For the purpose of showing the efficiency of the apparatus under consideration, we may observe that some years ago, when engaged at a Continental town, a motive-power meter was placed by us on the premises of a manufacturer, a large consumer, who required the gas principally during the daytime for his business; and previous to commencing some extensive alterations in the mains, which occupied about a month, due notice was given of the inability to continue the supply in the daytime during that period, which was met without demur or complaint. But in the course of time the alterations being completed, the manufacturer was informed that he could renew his operations, which intimation was received with a smile and the acknowledgment that he had never ceased working a single hour. On inquiry, this statement was ascertained to be true, and was confirmed by the meter supplying the premises, and although during the alterations the main was of course occasionally opened, the consumer had no difficulty in obtaining an abundant supply of gas. In explanation of this apparent enigma, it should be stated that there was a considerable distance between the premises supplied and the locality where the mains were being laid; thus the gas was exhausted from the pipes in the immediate neighbourhood of the consumer's premises, and as a natural result for every foot of gas exhausted, a foot of air entered the main. However, this at once accounted for the annoyance nightly experienced by the quantity of air intermixed with the gas, which had hitherto been such a mystery.

Under various conditions, the motive-power meter is of the greatest assistance to the gas manager, but its use has never been thoroughly appreciated.

GAS BURNERS.

PRIOR to 1784 the only method of burning oil in lamps was by the simple wick dipping into the oil, which yielded but a limited degree of light, often accompanied with a considerable amount of smoke and numerous other annoyances. In the year mentioned Ami Argand, a French gentleman, obtained a patent for his admirable invention, which he describes as "a lamp that is constructed to produce neither smoke nor smell, and to give more light than any lamp hitherto known, which is effected in several ways. First, by causing a current of air to pass through the flame. Second, by increasing the current of air and causing another current of air on the outside of the flame by means of a chimney enclosing the flame, so that the fresh air which enters continually into the chimney replaces that which is decomposed or rarefied by the flame." Hence the origin of the argand burner, the vast importance of which was speedily recognized, and its adoption to lamps became general; but it is to be regretted that whilst others were enriched, the great inventor derived no benefit by his extraordinary abilities, thus realizing the proverb that "one soweth and another reapeth."

Although Murdoch is stated to have employed the argand burner in his early experiments, still in practice this was not applied for many years afterwards, as the only methods of burning gas represented in the early editions of Accum's 'Treatise' were by the single jet and the cockspur burners, the latter being formed by three single jets placed at angles to each other. It can be readily imagined that the argand oil-lamp suggested the means of employing gas in a similar manner, and argand burners for the purpose are described both by Accum and Peckstone in their works published in 1819.

For the account of the invention of the fishtail burner we are indebted to Mr. Johnson, of Glasgow, who states that this burner was the joint invention of Neilson, the inventor of the hot-blast, and James Milne, of Edinburgh, the founder of the well-known firm of the same name. These gentlemen on the first introduction of gas into Glasgow, devoted much attention to all matters in connection therewith, and in the course of experiments Mr. Neilson found that by placing two single jets of equal size, at an angle to each other, the flame was flattened at right angles to the flowing gas, and that increased pressure only further flattened the flame and increased the light instead of causing it to smoke and go to waste, as was the case with the single jets. A socket to hold the two jets was then made, which was, however, both clumsy and ineffective, on account of the method of making, and the difficulty experienced in getting the jets at the proper angle. Mr. Milne then hit upon the idea of drilling the two holes in the burner diagonally to each other, thus forming the union of the two jets, hence the name of the "union jet," or "Scotch burner," as it was commonly called some years ago. We regret our inability to give the name of the inventor of the batwing burner, which was in use some years before the fishtail, and has done good service to the advancement of gas lighting.

About the period of the invention of the union jet, Drs. Christison and Turner entered into some interesting experiments in connection with gas burners, the results of which were published in the Transactions of the Philosophical Society of Edinburgh, in 1826. According to the investigations of these gentlemen, the light of an argand burner increases in a much greater ratio than the consumption of gas. That for all burners there is a certain point where the maximum amount of light is obtained. That the sizes of the holes in the burners, and their distance from each other, as well as the central aperture, influence most materially the quantity of light yielded by the gas. That the greatest light is obtained when the flame is on the point of smoking, and that the glass chimney should correspond with the size of the burner and the quantity of gas consumed.

All these important facts relating to the proper method of burning gas and constructing burners which are now well understood and appreciated, like the discoveries of Van Helmont, Boyle, and

Dr. Clayton, were disregarded and ultimately lost sight of, and for many years argand burners were commonly made with six or eight holes, each jet being detached, which clearly proves the ignorance then existing on the subject. Subsequently, about twenty-five years ago, in the 'Chemistry of Gas Lighting,' Mr. Lewis Thompson entered with great minuteness into the various considerations concerning the consumption of gas, in which we meet with the assertion that a "burner may be known to be acting with its maximum effect when the flame issuing is on the verge of smoking," which statement, as far as regards the argand, is undoubtedly correct. This was followed some time afterwards by the assertion of Alexander Wright, that "the best result is obtained when the burner gives a regularly formed flagging flame, and the worst condition is when the flame is wire-drawn and irregular; both of which opinions only confirm the conclusions arrived at by Drs. Christison and Turner many years previously.

Attempts at improving the argand were made about thirty years ago, which resulted in the production of burners under the name of "the Sun," "the Eclipse," "the Universal," and other names equally pretentious, without, however, making any positive progress, as no law or theory was followed. This state of affairs continued until the passing of the Metropolitan Gas Act of 1860, by which it was defined that the gas should be tested with "an argand burner having fifteen holes, consuming 5 feet per hour, with a glass chimney 7 inches long." The extreme vagueness of such a description is now well understood, for, as will be hereafter shown, the light from a given quantity of any gas can be reduced to any extent by simply diminishing the size of the orifices of the burners; and according to the late Dr. Letheby, there was a difference of at least 35 per cent. between the burners submitted to him at that period for the purpose of testing the quality of gas in accordance with the Act.

Practical men, probably animated by considerations in the interests of the gas companies, then directed their attention to the subject, and by them important improvements were effected, and at the present time the method of constructing burners leaves little to be desired, although each description possesses inherent disadvantages inseparable from it.

Whatever may be the illuminating power of any particular gas, the quantity of light yielded therefrom is dependent on various circumstances, such as the pressure of the gas at the point of issue; the size of the orifices of the burners; the temperature of the surrounding atmosphere; the quantity of air combining with the gas; the quantity of gas consumed; the form and dimensions of the glasses enclosing the flame, and other minor circumstances hereafter referred to.

Firstly, as regards the pressure, to take an extreme view, if an argand burner be made with holes of such dimensions that $\frac{4}{3}$ths pressure be required for the emission of 5 feet per hour, the gas will give simply a blue flame, and be practically devoid of light. The same result is obtained with a fishtail or batwing burner, which, under ordinary conditions, consumes 5 feet per hour, and yields the light of twelve candles; but on increasing the pressure to a certain extent, then, instead of the illuminating power of the gas being increased, it is diminished, and by further augmenting the pressure a roaring flame is produced, and with a consumption of about 22 feet a peculiar coloured flame is obtained, but yielding no appreciable light.

These experiments may, however, be regarded as so extravagant as to deviate from the question, and it may be argued that such burners and pressure are never practically applied. It must, however, be borne in mind that the evil effects arising from imperfect burners, or excessive pressure, exist between those points where no light is yielded by the gas, to the condition where the maximum degree of illumination is evolved. In illustration of this we have to direct attention to the following experiments in which four argand burners, arranged especially for the purpose, were employed, having their respective orifices drilled in such a manner that the pressures stated were requisite to deliver 5 feet per hour from each of the burners:

Experiment No. 1. Gas issuing from the burner at a pressure of $\frac{1}{10}$th gave the light of 12 candles.
,, 2. ,, ,, ,, ,, $\frac{4}{10}$ths ,, ,, 6 ,,
,, 3. ,, ,, ,, ,, $\frac{10}{10}$ths ,, ,, 2 ,,
,, 4. ,, ,, ,, ,, $\frac{40}{10}$ths gave no appreciable light.

Thus, from the preceding statements, resulting from our personal investigation, and of the accuracy of which we vouch, we learn the important facts that when gas is burned under excessive pressure, or with the orifices of the burners of restricted dimensions, it gives no light; and in proportion as that pressure is diminished, so is the illuminating power of the gas increased, until arriving at that point where it is consumed to the greatest advantage.

Within our knowledge, so imperfectly was the gas burner understood, that about thirty years ago, as a check on the extravagant consumption of contract consumers, the holes of the burners were drilled of such dimensions as not to permit more than 5 feet per hour to pass at a pressure of five-tenths of an inch, when, under these circumstances, as demonstrated in the above table, only one half of the available light was derived from the gas. At that time, moreover, when iron burners were exclusively used, in consequence of their corrosion and being cleaned repeatedly, their orifices were eventually enlarged to that degree, by which, according to our present knowledge, the best results are obtained, and the gas gives its full amount of light. But the gas inspectors, totally ignorant of the advantage derived by the consumer, without prejudice to the company (chimneys being employed with the burners), realized only what the burner was capable of passing, without reflecting on the quantity of gas actually burned, and thus the consumer was compelled to purchase new burners, by which the gas produced only half the amount of light for which he paid.

To demonstrate that the illuminating power of gas is affected by the state of the surrounding atmosphere, and that under other conditions gas gives practically no light, we have only to direct attention to the gas flame from a stand pipe, as employed during the alteration of gas mains or in other public works, which at one moment yields the light of thirty or forty candles, and at the next instant by a gust of wind all that is perceptible is a blue flame devoid of light. The same effects are observed in all exposed lights in windy weather, also in candles when moved rapidly through the atmosphere; which is stated to be due to the flame being cooled, and until a more powerful reason is given for the phenomenon this must be accepted as an explanation, although, perhaps, far from satisfactory.

With respect to our next proposition, the principal conditions for consuming gas to the best advantage would appear to consist in the flame having just a sufficient supply of air, and that any excess of this tends to diminish the amount of light. This effect is strikingly illustrated with the argand burner, which, at the point of smoking, with just a sufficiency of air, yields the maximum of illumination, but on retaining the same consumption, and by substituting a longer chimney, thus causing a greater draught, the luminosity of the flame is at once reduced in a marked manner. Again, if the flame be reduced to a consumption of 2 feet, its light will be equal to a candle only, but on raising the chimney about half an inch from the gallery the light is greatly increased. Or, with the same flame, by simply placing a disc on top of the chimney, according to an experiment of the Referees hereafter mentioned, the light is increased ninefold, both of which results arise from a diminished current of air. Lastly, with the ordinary glass moon so universal in private dwellings with the fishtail burner, as usually employed, in consequence of the strong current passing, whether 4 feet or 6 feet of gas are consumed per hour, little difference exists in the degree of light given; but if the top of the glass be enclosed with a talc cover having an orifice in the centre of about one inch in diameter, then the conditions of the burner are completely changed, and the economical advantages of the argand are approached, for under these circumstances, whenever there is an excess of gas there is a tendency to smoke, to avoid which the supply is diminished, and consequently the gas is burned with economy. However, a practical difficulty existing against this system is, that in the event of the gas not being lighted at the moment it issues from the burner, an explosive mixture will be formed within the glass which detonates on igniting. Hence it is evident that the maximum of light from burners is also dependent on the flame having a sufficiency, but with no excess of air, which indeed embraces all the points of excessive pressure, small orifices, and the quantity of gas consumed.

That the illuminating power of gas is influenced by the quantity consumed is demonstrated by the following experiment, made with a Sugg's London argand burner, consuming the quantities indicated.

Quantity consumed.	Height of flame.	Illuminating power as indicated by photometer.	Percentage of light, 5 feet being 100.	Foot of gas equal to standard candles.
5·0 feet	3·0 inches	equal to 15 candles	100	3·00
3·5 ,,	1·6 ,,	,, 7 ,,	66	2·00
2·5 ,,	1·1 ,,	,, 3 ,,	40	1·20
2·0 ,,	·9 ,,	,, 1·1 ,,	15	·45

These results are corroborated by the experiments of the late Mr. Bannister in the following table.

Quantity consumed.	Illuminating power as indicated by photometer.	Reduced to 5 feet per hour.	Value of one foot of gas in standard candles.
5·0 feet	14·0 candles	14·0 candles	2·800
4·0 ,,	9·5 ,,	11·9 ,,	2·375
3·0 ,,	4·3 ,,	7·2 ,,	1·433
2·5 ,,	2·8 ,,	5·5 ,,	1·120
2·0 ,,	1·0 ,,	2·5 ,,	0·500

Hence we learn that the quantity of light derived by the combustion of gas from the argand burner diminishes very rapidly with reduced consumption, for, as shown in the preceding tables, when only consuming 2 feet per hour, 85 per cent. of the light of the gas is lost, and with 2½ feet the loss is 60 per cent., and with 3½ feet it is 34 per cent. of that derived from the gas, when burning the full quantity for which the burner is constructed.

The conditions which control the efficiency of the argand burner are entirely different to those of the fishtail and batwing; for in the argand, the size and number of the orifices for the emission of the gas; the area of the central as well as the annular space for the admission of air to the burner; the height of the chimney to give the necessary, but no excess of draught, are all essential points in its formation. Nor is this all; for, as Dr. Bowditch has demonstrated, the form of flame is influenced by any extraordinary width of the ring whence the gas issues, and by the position of the holes, from which we may conclude that this part of the burner should be no wider than is absolutely necessary to contain the "adamas" or "steatite" forming the orifices. The consideration of all these points is essential for the proper construction of argand burners, and that their importance is now well understood, will be made evident when we state that there are manufacturers who make no less than eight different sizes of these burners, consuming respectively from 2½ feet of cannel (25-candle gas), or 3 feet of (15-candle) common gas, to the largest size burning 5 feet of cannel or 7 feet of common gas; and in all these burners when consuming the maximum quantity for which they are made, the light of three candles is obtained from each foot of common gas, and an average of 4½ candles per foot of cannel gas, which degree of perfection, for reasons hereafter stated, can never be realized with the fishtail or batwing burner. But when more gas is consumed than that for which the burner is constructed, it has a tendency to smoke, and when consuming less than its quantity a material loss of light is sustained as represented in the preceding tables, which is an inherent defect in the argand burner.

Argand burners are manufactured to consume from 3 feet to 7 feet per hour, but seldom more than this, the number and dimensions of the holes depending on the judgment of the maker; but in all cases they are of such dimensions or number as to permit the full quantity of gas to pass for which the burner is constructed at a pressure of about $\frac{1}{10}$ths. The argand burner most generally in use consumes 5 feet of common gas per hour, and when this is of good quality, and the burner well made, usually gives the light of fifteen sperm candles; but, as hereafter shown, with a badly formed burner about two-thirds of the available light from gas may be lost. For cannel gas the burners have frequently the holes of smaller dimensions, but in greater numbers, than those used for common gas; thus the richer the gas, the thinner is the stream required for its proper combustion, a condition equally applicable to the fishtail and batwing burner.

The flame of the batwing is formed by the pressure of the gas issuing through the slot, whilst that of the fishtail is produced by two jets of gas impinging against each other, so forming a "sheet" of gas, if we may use the term, at right angles to the jets, and in both cases if the slot or orifices be too fine, an excess of air comes in contact with the gas and its illuminating power diminishes. When the slots or orifices are very fine, the flame of the gas becomes blue, and they may be diminished to an extent that, with only moderate pressure, no light will be evolved by the gas. With batwing and fishtail burners the principal controlling influences are the pressure at the point of issue, together with the size of the orifices, and the quality of the gas; for the richer the gas, the smaller will the orifices be required.

That the amount of light evolved by fishtail and batwing burners is materially influenced by the degree of pressure at the point of issue is proved by the following extract from a table of experiments made by Mr. Kirkham, in which the first, the Sugg and Letheby burner, consuming 5 feet per hour, and giving the light of 14·32 candles, was adopted as a standard of comparison.

No. of Experiment.	Illuminating power in candles.	Decrease in candles.	Percentage of decrease of light.	Pressure at point of issue.
2. Sugg Letheby argand	14·32			inch. 0·85
BATWING BURNERS.				
10. Flat flame	13·27	1·05	7·4	0·175
11. „	13·02	1·30	9·09	0·250
19. No. 5 iron	5·41	8·91	62·26	1·150
FISHTAIL BURNERS.				
24. No. 4	6·37	7·97	55·55	1·162
26. „	11·11	3·21	22·48	0·52
31. „	9·36	4·96	34·64	0·565
33. No. 3	7·02	7·3	50·98	1·
34. „	4·67	9·65	67·42	1·575
35. „	4·31	10·01	69·96	1·245
36. „	5·56	8·76	61·24	1·13

Thus we learn from the above the prejudicial effects of high pressure, by which the light of a batwing when the gas issued at a pressure of 1·15 inch, was upwards of 50 per cent. less than when issuing with ·25 inch pressure; whilst with the fishtail burners, in two instances, the loss of light is still greater, this being nearly 70 per cent., arising from the same defect. In experiment No. 24, the percentage of decrease is 55·55, but with the same burner, by simply reducing the pressure to ·52 inch, the degree of light is nearly doubled. In fact, this evil effect of heavy pressure at the point of issue is evident throughout the whole of the table, of which we have extracted the most prominent points, but at the same time it may be observed that most of the burners were too small for the proper combustion of the quantities passed through them.

To further demonstrate the effects of excessive pressure, or small orifices for the issue of the gas; if an ordinary fishtail, consuming, say 2 feet per hour under a pressure of 1 inch, be surmounted during ignition by a cap having a burner nib of much larger dimensions, the light evolved will in many cases be doubled, and is always considerably increased, the result arising simply by the reduction of the pressure at the point of issue, and from which cause an incalculable loss of light, or gas, or money, daily occurs.

The remedy for this is to have the orifices of the burners of sufficient capacity, and the batwings sufficiently wide in order to permit the quantity of gas required to be consumed to pass at the necessary low pressure, and to adjust this by means of a regulator or rheometer to counteract the varying pressure in the company's main. A similar object is attained in Brönner's burners, by making the orifice for the admission of the gas into the burner considerably smaller than that at the point of issue. Other manufacturers place a packing of wool or cotton within the fishtail or batwing burner, which not only checks the gas at a moderate pressure, but as the packing becomes more compact under heavy pressure, the passage of the gas is controlled accordingly. But by whatever means it may be effected, we think that sufficient has been stated to prove that low pressure at the point of issue, within certain limits, is indispensable to economy in burning gas.

There is, however, a degree of pressure below which the flame of the batwing or fishtail is not properly formed, when the gas is not economically burned, and it is only when the flame is full and "flagging" that it is properly consumed; but whenever a flame shows any irregularity on its surface, arising from the pressure, the gas is being disadvantageously employed. The wider the slots or orifices of burners, the greater is the degree of pressure necessary to form the flame; with common gas, this will be from $\frac{2}{10}$ths to $\frac{4}{10}$ths, and with cannel gas a very slight increase of pressure may be necessary. In the argand, the flame is mainly formed by the passing current of air, therefore in this, as shown, a pressure of less than a tenth is sufficient to consume gas under the best conditions, and in few instances is more than $\frac{2}{10}$ths requisite. The effect of high pressure at the point of issue in the argand is also made evident in the Referees' experiments hereafter mentioned, but with that burner this is only one of the several controlling influences by which it is affected.

With batwing or fishtail burners, but particularly the latter, when of small dimensions, the light evolved, no matter the number of burners, has always a blue dull tinge, which is very perceptible in premises so lighted, arising from an excess of air combining with the gas, or the gas containing an insufficiency of carbon. To obviate this evil, some years ago the "twin" burner was introduced; which consisted in having two fishtail burners of limited consumption, connected together by a joint, so that they could either be used separately, or their flames made to combine, when in the latter case the light was greatly increased, arising no doubt from the increased thickness of the stream of gas; and to this combination of flames is due the excellence of the argand burner, for if the jets constructing it were burned separately, the light evolved would be considerably less.

Eventually, however, it was accepted as a fact, that a single burner having the orifices sufficiently large gave the same results as the twin burner, and this was abandoned. But more recently a series of experiments have been made by Mr. A. H. Wood, according to which by the combination of two batwing burners in the manner stated, the light evolved for a given quantity of gas is on the average increased 25 per cent. above that obtained when the burners are separated, which experiments are well worthy the attention of practical men.

That the light evolved by the batwing burners is increased in proportion to the augmented dimensions is further confirmed by the following table of the light derived from batwing burners of different sizes, all burning under the same favourable conditions, namely, yielding the maximum of illumination according to the consumption of each burner.

Quantity of gas consumed.	Illuminating power by photometer.	Percentage of light, 2 feet taken as a standard.
2·0 feet	2·5 candles	100
2·5 ,,	3·5 ,,	112
3·5 ,,	6·0 ,,	137
4·5 ,,	10·0 ,,	177
6·0 ,,	15·5 ,,	206
7·5 ,,	22·0 ,,	234

From this we find that the smallest batwing burners are for the quantity of light obtained, nearly two and a half times more costly than the largest size, which variation is even greater with the fishtail burners. Therefore it may be accepted as an axiom that for the quantity of gas consumed, the light of the fishtail and batwing burners is increased by augmenting the size of the burners. This, although contrary to the opinion of the Referees hereafter mentioned, is, notwithstanding, correct, and indeed is confirmed by their experiment No. 6, where the largest consumption at moderate pressure gave the greatest amount of light.

As observed by the Gas Referees, every description of batwing and fishtail evolves the maximum amount of light at one particular point of consumption, which is illustrated in the following table of experiments on burners by Mr. Blackburn.

No. 2 Fishtail.

Consuming feet of gas per hour	1·5	2·0	2·5	3·0	3·5
Light in sperm candles	2·40	3·87	4·65	5·50	8·3
Light per foot of gas in candles	1·60	1·93	1·86	1·83	1·80

No. 3 Fishtail.

Consuming feet of gas per hour	2·0	2·5	3·0	3·5	4·0
Light in sperm candles	4·31	5·42	6·60	7·45	7·92
Light per foot of gas in candles	2·15	2·17	2·20	2·13	1·98

No. 4 Fishtail.

Consuming feet of gas per hour	3·0	3·5	4·0	4·5	5·0
Light in sperm candles	7·30	8·62	10·10	10·89	11·27
Light per foot of gas in candles	2·43	2·46	2·52	2·42	2·25

Thus with the No. 2 fishtail, the maximum of light is obtained with a consumption of 2 feet per hour, with the No. 3 of 3 feet, and the No. 4 of 4 feet per hour; the coincidence of the numbers is accidental; but on increasing the consumption beyond the points stated, the loss of light augments considerably, until reaching that point where, as observed, no light is yielded by the gas. In the preceding tables the statement that the light evolved from a given quantity is increased by burning the gas in large burners is also confirmed; thus, with the No. 2 burner, the greatest degree of light is evolved when consuming 2 feet per hour, the light of which quantity is considerably augmented when burned in the No. 3; whilst the light yielded by 3 feet of gas when burned in the No. 2 is increased 33 per cent. when consumed in the No. 4 burner. In the larger burners, both batwing and fishtail consume the gas economically, and it is to be regretted that these tables were not carried to higher consumption, in order to show their advantages.

The want of care observed by some engineers and managers in the selection of proper burners for their consumers, or at least in permitting the use of bad burners, is made strikingly palpable by Mr. Henry Woodall, who, when reporting on the quality of the gas of a provincial company, stated that a large portion of the burners used by the consumers, called "the tulip," when consuming 5 feet of gas per hour, only gave the light of 5·25 candles, whereas a like quantity of the same gas when burnt with a London argand burner produced the light of eighteen candles, and consequently a loss of 71 per cent. of the light to be derived from the gas was sustained by the use of these burners; or, in other words, for the amount of light obtained, the consumer paid upwards of three times more than would be necessary if proper and good burners had been used. Defects like these are by no means uncommon; indeed, in one shape or the other they exist in every street.

However, it is not only by the proper pressure and form of the burner that the maximum light is evolved, but for this object the form of the globe enclosing the fishtail burner must also be considered; for these, as usually made with orifices of 2½ inches at the bottom, and of 4½ inches at the top, cause a current sufficiently strong to pass, which prevents the flame smoking, no matter the quantity consumed, at least with common gas, consequently a considerable portion of this passes insensibly away without giving light; but if the opening at the bottom is made of ample diameter and corresponding with that at the top, the current will be then diminished, and these evil effects in a great measure counteracted. The best form of globe and size of orifice are well worthy the attention of manufacturers.

Another difficulty is the kind of glass employed, as we find by the experiments of Mr. King, of Liverpool, and Mr. Wood, of Hastings, the amount of light lost from these sources to be as follows:

Light obstructed by a clear glass globe..		about	12 per cent.
,,	,,	clear globe engraved with flowers	,, 24 ,,
,,	,,	globe of ordinary pattern	,, 35 ,,
,,	,,	globe obscured all over	,, 40 ,,
,,	,,	an opal globe	,, 60 ,,
,,	,,	painted opal globes	,, 64 ,,

It must, however, be observed that, under some conditions, economy is not a consideration; hence with the opal globes the light is of that peculiar soft agreeable nature, that, as proved by their very general application, cost of light is often not regarded. But where economy is desired and globes indispensable, the half ground is the most suitable.

In concluding this part of our subject we may observe that the argand, when consuming its full quantity, and in a locality free from draughts or currents of air, is the best and most economical burner, and the advantage of the argand is increased by its flame smoking with any excess of gas, which causes attention to be directed to it, and thus prevents loss. But, on the other hand, in proportion as the flame is decreased so is the amount of light diminished, until burning a quantity which in a fishtail or batwing would give a good light, but which with the argand is practically useless; and when consuming 2 feet per hour the light of one candle only is obtained, whereas with the other burners more than double that degree of illumination would be evolved from that quantity of gas. Therefore, when consuming small quantities the batwing and fishtail burners are preferable.

We have shown the disadvantages of high pressure at the point of issue, and the loss of light arising from the use of small batwing and fishtail burners; also the effect of the current in the ordinary glass moon, and suggested the means of obviating this, as well as to render the consumption of the fishtail enclosed in the moon glasses both economical and more agreeable. With the argand burner all the various circumstances which control it can be arranged in the burner and chimney, but with the other burners no such means exist, as the light produced will constantly vary according to the thickness and surface of the stream or sheet of gas; the temperature of the surrounding atmosphere; as well as the quality of the gas; and it may be probably influenced by any alteration in the particular angle at which the holes of the fishtail are formed. With common gas, alike in the argand as in the batwing or fishtail burner, a thick stream will be necessary to do justice to the gas, and this can never be obtained with small burners, for which reason they must always be extravagant, and in proportion as the burners increase in size so will the stream be augmented in thickness and the light increased.

A theory established by Mr. Farmer and accepted in America, is that the light evolved by different burners is as the square of the quantity consumed, thus, if 2 feet give the light of four candles, 3 feet will give the light of nine candles, respecting which we will offer a few remarks. Firstly, we have shown that argand burners when properly made, and consuming their maximum according to the various sizes, whether these be large or small, produce relatively the same amount of light per foot of gas consumed. Secondly, we may observe that the variable illuminating power of an argand from one to fourteen or fifteen candles, as compared with the consumption of gas, to which we have referred, corresponds with no mathematical law.

Moreover, according to experiments instituted in conjunction with Mr. Hartley, this variation changes with the quality of the gas, as with poor gas, for the quantity consumed, a remarkably uniform progressive proportion exists to produce the light of each candle as indicated by the photometer, from one to eighteen candles. Again, this variation is influenced in an extraordinary manner according to the construction of the argand burner, as demonstrated in the experiments of the Referees in page 298.

Lastly, an argand burner consuming 2 feet per hour will yield about three times more light when the glass chimney is removed. Therefore, with these discrepancies arising from the quantity and quality of the gas consumed, the form of burners, the use of the chimney, and other points already mentioned, it is obvious that no general law can be applied concerning the light produced from an argand burner, under a variable consumption of gas.

With reference to the fishtail and batwing, in these we have shown that increased light accompanies augmented consumption, or, in other words, small burners with very limited orifices approach that condition when no light is yielded by the gas, whilst burners with large orifices permit a sufficiently thick stream of gas, or carbon vapours, to combine with the atmosphere, by which the maximum of light is evolved; but if from this point the pressure is increased, then the amount of light for the quantity of gas consumed gradually decreases, until eventually, as explained, no light is yielded. Under all these various conditions, we venture to express the opinion that there is no foundation for the Farmer theorem.

So far we have endeavoured to illustrate the various theories recognized in relation to the consumption of gas, and little remains to be said on the best method of constructing burners. For these the parts for the emission of the gas were formerly made either of steel or iron, and consequently, by their rapid oxidation, the orifices were obstructed, thus diminishing materially the illuminating power of the gas, and in some cases entirely preventing its issue. This defect was remedied by the introduction of a material called "steatite," and subsequently an artificial silicious compound called "adamas," by Mr. Leoni, of which the parts for the emission of the gas, properly termed the burner, whether for the argand, batwing, or fishtail, are made. With these two excellent substances, not susceptible of oxidation, practically everlasting, iron burners of all denominations should be abandoned, and a further reason for this recommendation is, that the steatite and adamas burners are made with a degree of care, according to the required consumption and quality of the gas, not attained with the ordinary iron burners.

The most important and comprehensive publication bearing on our subject, is undoubtedly the report to the Board of Trade by the Gas Referees, respecting the construction of gas burners, in reference to gas illumination, and from this we make some abstracts which cannot fail to be of interest.

According to that report:

"Every improvement in the construction of gas burners is equivalent in its economical effects to the discovery of a method of cheapening the manufacture and supply of gas; for it enables the public to obtain more light from the gas which they consume and pay for. By using good burners instead of bad ones, consumers may obtain from 30 to 50 per cent. more light, while their gas bills remain the same.

"The improvement of burners is also important as a measure of sanitary reform, for, as by this means the required amount of light is obtainable from a smaller quantity of gas, the atmosphere of rooms and workshops is less vitiated. Not only is an unnecessary amount of heat avoided, but in consequence of less gas being burnt, the pernicious products of combustion discharged into the air (viz. the carbonic acid gas and sulphur impurities) are equally diminished; so that the condition of the occupants of private dwellings, and still more of the workpeople employed in factories and other large establishments, is rendered more comfortable and healthy than it could otherwise be.

"The economy to the public arising from the use of good gas burners instead of bad ones, is so obvious as hardly to need remark. The gas rental of London amounts annually to more than *two millions sterling*.* Taking a very moderate estimate, upwards of *one-fourth* of this sum (500,000*l*. per annum) might be saved by the use of good burners. This is the saving which might be made in London alone; how much vaster the sum thus economized if good gas burners were to come into general use throughout England! In truth, the economy arising to the public from the use of improved burners is as large as can be produced for many years to come from any improvements in the manufacture of gas. The question of burners, indeed, although hitherto so little considered or investigated, meets one at every turn in matters connected with gas light, whether these be regarded as problems of science, or in the more widely useful and practical form as a means of economy for the gas-consuming public.

---

* This report was made in 1871. The actual yearly gas rental of the London companies is upwards of two million six hundred thousand pounds sterling.

"As a scientific question the illuminating power of gas has given rise to much discussion, and there are several points of this, which, owing to their practical bearing, must be determined at the outset; the first of these questions being: Is the illuminating power affected by the quantity in which gas is burnt? In other words, is it more economical to use small burners or large ones?

"In a case like this, the first and most natural suggestion is, to inquire whether the observed variations in the illuminating power are not due, wholly or in part, to the mechanical apparatus employed for developing it. If two tons of identically the same coal do not, when burnt separately, give out the same amount of heat, is not the explanation to be sought in some difference in the mode of combustion? If two gallons of the same water, when weighed separately, do not show the same weight, must there not be some difference in the balances or in the details of weighing? In like manner, if 4 feet of gas do not give exactly two-thirds the amount of light which 6 feet of the same gas gives, is not the difference first to be looked for in the nature of the burners employed in the experiments? Not to take into careful account the influence of the burners, when testing the illuminating power of gas, is as great an oversight as if, in weighing, one were to make no examination of the balances; or as if an engineer were to take no account of the boilers he employed, and then, finding that a ton of coal in some circumstances raised a greater proportion of steam than when half a ton was used, were to jump to the conclusion that the heat-giving power of coal became greater, relatively to the quantity consumed, when a ton was used than when half that quantity was employed.

"What a boiler is to coal and the generation of steam, so is a burner to gas and the development of light. One ton of coal in a locomotive of the present day generates as much force as six tons did forty years ago, simply owing to the superior construction of the locomotive. In like manner, as regards the illuminating power of gas, there are good burners and bad ones. Moreover, as every scientifically constructed boiler is devised especially for a given amount of coal, by the consumption of which the boiler develops its maximum of power relative to the quantity of fuel used; so every well-constructed burner is devised to consume a fixed quantity of gas. Indeed, for every burner, whether good or bad, there is a certain rate of consumption at which the burner does more justice to the illuminating power of the gas than at other rates, whether greater or less. To disregard these considerations is to render experiments wholly useless and misleading.

"First, as to good burners and bad ones. Take two burners, each of which gives its maximum of light, i. e. does most justice to the gas, at the same rate of consumption; nevertheless the light emitted by one of the burners may be much greater or much less than the other. Secondly, as to the misuse of burners. Take a burner which does most justice to the gas when the rate of consumption is 5 feet per hour; then, if the rate of consumption be either increased or diminished from that point, the gas will of course give out less light than before in proportion to the quantity consumed.

"The proper regulation of the supply of air to the flame of an argand burner is the chief secret of developing a maximum amount of light from gas. The greater the quantity, or the richer the quality of the gas, the more air is required, and hence the better will the flame bear contact with the atmosphere; for the greater is the quantity of matter to be oxidized or consumed by combustion. And it is important to observe that the greater the velocity with which gas issues through a burner, the greater is the supply of air to the flame and the more air is the flame brought in contact with. The stream of burning gas from the burner rising through the (we shall say) quiescent atmosphere of the room draws in the air upon itself, just as a rapid stream passing through a pool or lake disturbs the stillness of the pool, and draws in upon itself in eddies the surrounding water; and the more rapid the upward stream of gas, the greater the quantity of air thus drawn in upon the flame.

"The important bearing which the above statements have upon any question connected with gas illumination is manifest at a glance. They illustrate the chief conditions which affect the illuminating power of gas; they show how great may be the variations of that illuminating power in which the gas is consumed, even in the same burner. The subjoined statistics exemplify both of these points.

"The experiments here given were made at intervals during the last two years, and the earlier experiments of the series were made simply in order to ascertain correctly and fairly the capacity of burners to develop the illuminating power or light-giving property of gas. The gas with which the experiments were made was common gas, having an average illuminating power of fifteen sperm candles, when consumed at the rate of 5 feet per hour, in Sugg's London Argand No. 1. This Argand, burning at the above-mentioned rate, was taken as the standard in the experiments, and its light reckoned as 100. The other burners, namely, those experimented with, were made to consume gas at various rates, from 1 to 6 feet per hour, and the mode of computing their light with reference to the standard burner was the ordinary and simple one, as follows: Suppose the test burner to give a light of 40 per cent. (compared with the standard burner) when burning at the rate of 4 feet per hour, then instead of 40, its light is stated in the fourth column as 50, because the standard burner was consuming 5 feet per hour against the 4 feet consumed by the tested burner. In like manner, if a tested burner gave a light of 60 per cent. (compared to the standard) when burning 6 feet per hour, then, instead of 60 its light is stated in the fourth column as 50—as it was consuming one-sixth more gas than the standard burner. On the other hand, if the tested burner consumed exactly 5 feet per hour, then the figures in the fourth column would be the same as those in the third; for, in this case, both the standard and the tested burner were consuming the same volume of gas."

First, let us give the results of the experiments with fishtail and batwing burners:

FISHTAIL BURNERS.

Pressure of gas as delivered to burner, in tenths of an inch.	Consumption of gas in feet per hour.	Actual illuminating power,— Sugg's London No. 1 at 5 feet being taken as 100.	Illuminating power,— calculated to 5 feet,— Sugg's London as 100.	Pressure of gas as delivered to burner, in tenths of an inch.	Consumption of gas in feet per hour.	Actual illuminating power,— Sugg's London No. 1 at 5 feet being taken as 100.	Illuminating power,— calculated to 5 feet,— Sugg's London as 100.
\multicolumn{4}{c}{I.}	\multicolumn{4}{c}{IV.}
·1	1·1	6·8	31·2	·05	1·0	6·9	34·1
·2	1·7	12·2	36·0*	·17	2·0	18·8	47·0*
·4	2·5	17·6	35·3	·26	3·0	27·6	46·1
·6	3·0	20·2	33·7	·61	4·0	30·9	38·6
·8	3·7	22·0	29·8	·95	5·0	31·5	31·5
1·0	4·2	23·0	27·4				
\multicolumn{4}{c}{II.}	\multicolumn{4}{c}{V.}
				·3	1·4	4·8	17·4
·1	1·7	19·1	54·8	·5	1·8	6·8	19·0*
·2	2·6	30·7	59·0	·65	2·0	7·2	18·2
·3	3·1	38·4	62·0*	·8	2·4	7·4	15·6
·4	3·6	44·3	61·6	\multicolumn{4}{c}{VI.}
·5	4·1	48·7	58·7	·05	1·1	8·0	36·6
·6	4·6	57·5	56·0	·1	1·5	16·6	53·7
·73	5·2	55·9	53·2	·2	2·2	28·1	63·8
				·3	2·8	39·2	68·7
\multicolumn{4}{c}{III.}	·4	3·3	46·2	69·0
·35	3·0	29·7	49·6	·6	4·2	62·0	73·0*
·45	3·3	34·3	52·0*	·75	4·7	67·8	72·2
·57	3·8	38·1	50·2	·85	5·1	71·9	70·5
·67	4·2	40·9	48·7	·95	5·4	76·3	70·0
·77	4·6	43·5	47·3	1·0	5·6	77·7	68·7
·87	5·0	45·3	45·3	1·1	5·9	79·5	67·4
\multicolumn{8}{c}{BATWING BURNERS.}
\multicolumn{4}{c}{VII.}	\multicolumn{4}{c}{VIII.}
·05	1·3	13·2	53·0	·1	2·0	28·7	70·7
·1	2·2	16·4	74·8	·2	3·4	52·4	77·0
·2	3·6	62·0	85·0	·3	4·6	75·6	82·2*
·3	5·0	86·5	86·4*	·4	5·7	87·2	76·5
·4	6·2	106·0	85·4	—	—	—	—
·5	7·2	111·2	79·4	—	—	—	—
·6	8·1	127·4	78·6	—	—	—	—

"These experiments show at a glance what a difference the burner makes on the light emitted by gas. The quantity of gas was in each experiment the same, yet how serious the difference in the amount of light given by the several burners, one of them (No. V.) giving, at its best, barely one-fifth of the light obtainable from the gas."

* These lines show the points of consumption at which each of the burners gives the greatest proportion of light from the gas.

Here we may remark, although it appears to have escaped the notice of the Referees, that the cause of the diminished light arises from the circumstance with which we commenced, namely, the limited orifices of the burners or, what is the same thing, the excessive pressure, together with the small consumption of gas, as in the experiment No. V. $\frac{7}{20}$ths are requisite to expel 1·4 foot per hour, whereas in the following experiment, which is the most satisfactory of any, a greater quantity of gas is delivered with $\frac{1}{10}$th, and yielding nearly four times the light of the former. Our impression is that all the fish-tail burners examined were too small for the object intended.

The Referees, after referring to the Farmer theorem already mentioned, state:

"Instead of the gas giving more light as the rate of consumption is increased, it will be seen that *in every case there is a point beyond which the light* DECREASES *relative to the proportion of gas consumed.* In every case too, this point lies below, and in most cases *far below*, the maximum of ordinary gas consumption, observing the turning points in the case of different burners. In No. VII. the maximum of light given out is when the gas is burning at the rate of 5 feet per hour, beyond which point the more gas burned the less is the proportion of light which it gives. In No. VIII. the maximum of light is at 4·6 feet per hour; in No. VI. it is at 4·2 feet; in Nos. II. and III. it is about 3 feet; in No. IV. it is at 2 feet, and in Nos. I. and V. at only 1¾ foot."

"As already said, the chief means of obtaining the maximum of illuminating power from gas, is to ensure an adequate supply of air to the gas flame, and with argands this point is easily found, for it immediately precedes the state of combustion at which the flame smokes, i. e. when the air supply becomes deficient and a portion of the gas is not thoroughly consumed. Indeed we may state as an absolute rule, that *every burner gives its* own *maximum of light* (relative to the quantity of gas consumed) when its flame is just on the point of smoking. But with argands (owing to the glass chimney which encloses them and regulates the supply of air), it is always possible to increase the consumption of gas to such a point as will make the flame smoke, and hence every burner of this kind can be used in a manner which will give the full illuminating power of the gas, *so far as that is dependent upon an adequate air supply*. Now, as the common fault of argands is, that the gas issues under too great a pressure, i. e. with too great a velocity, thereby bringing the gas into contact with too much air—it follows that the worse the argand the better will it become when a large quantity of gas is burnt in it; for the air supply as regulated by the chimney being nearly a fixed quantity, any excess in the air supply can be neutralized by increasing the quantity of gas consumed. But with all argands, whether good or bad, the larger the quantity of gas consumed in them (short of smoking), the greater will be the proportion of light which they give from the gas.

Pressure of gas as delivered to burner.	Consumption of gas per hour.	Actual illuminating power,— Sugg's London at 5 feet being taken as 100.	Illuminating power calculated to 5 feet,— Sugg's London as 100.	Pressure of gas as delivered to burner.	Consumption of gas per hour.	Actual illuminating power,— Sugg's London at 5 feet being taken at 100.	Illuminating power calculated to 5 feet,— Sugg's London as 100.
I.				III.			
inch.	feet.			inch.	feet.		
·05	2·1	5·4	12·7	·2	2·6	10·3	19·7
·07	2·8	19·5	34·2	·3	3·3	22·4	33·8
·10	3·3	34·1	51·6	·4	3·9	36·4	46·7
·14	4·0	60·5	75·0	·5	4·4	49·2	55·7
·17	4·4	77·0	86·1	·6	4·9	64·0	65·3
·218	5·0	100·0	100·0	·7	5·4	75·6	69·4
				·78	5·8	90·6	77·4
II.							
·02	1·6	2·0	6·2	IV.			
·04	2·2	7·7	17·1	·1	1·8	1·6	4·5
·07	2·9	19·7	33·1	·2	2·7	7·8	14·3
·10	3·5	32·2	45·4	·3	3·4	15·0	21·8
·15	4·5	58·9	64·8	·4	4·2	21·9	26·0
·20	5·1	74·4	72·7	·5	4·6	29·3	31·5
·22	5·6	89·1	78·8	·6	5·2	34·7	34·3

"Here are the results of experiments made with four kinds of the argand burner; the first (Sugg's London burner, No. 1), one of the best that has ever been constructed; the second and third are ordinary good argands; and the fourth one of the worst argands we have met with. The experiments with each burner were carried up to the smoking point; beyond which point experiments are useless, as there is a manifest waste of gas by imperfect combustion.

"Here it is shown that the larger the quantity of gas consumed in argands, short of smoking, the higher the proportion of light which they give from the gas. But it is to be noticed, that even at this most favourable point of consumption there is a vast difference in the amount of light given by the different burners (a most important fact, the causes of which will be fully considered in the sequel). In short, then, these experiments with argands, like the previous ones with batwings and fishtails, show the paramount influence which the burner has upon the amount of light obtained from the gas. For while No. I. gives (what we may for the present call) the full illuminating power of the gas, which is taken as 100, Nos. II. and III. give a light only equal to 78, and No. IV. a light only equal to 34. In other words, the last of these burners, *taken at its best*, gives only one-third of the light which may be obtained from the gas by a really good burner.

"Every burner is fitted, and every scientifically constructed burner is expressly devised, for a certain rate of consumption; and to use a 6-foot burner with 3 feet of gas is as absurd as to use a 3-foot burner with 6 feet of gas.

"There is one point more connected with the question of illuminating power, relative to the quantity of gas consumed, which remains to be noticed. It will be seen, in the case of every one of the burners used in the preceding experiments, there is a point in the rate of consumption *below* which the proportion of light given by the gas falls short of the maximum of light which such of the burners respectively give. Indeed, in every case we have purposely commenced our experiments with each burner at an unduly low (sometimes extremely low) rate of consumption, in order that this fact may be exhibited. And arguing from this fact, the most cautious and moderate upholder of the doctrine that the illuminating power of gas is greatest when the gas is burnt in large quantities, maintains that whatever be the rate of consumption, a fixed portion of the illuminating power of the gas must always be low, and that therefore it is necessary to burn gas in large quantities.

"No burners have as yet been devised for the consumption of gas in very small quantities, nor are they much needed.* But lacking such burners, let us make use of a rude apparatus for diminishing the air supply, viz. a metal disc placed above the upper end of the chimney of an argand, and observe the result. The burner to which it was applied was Sugg's London argand, and which gives its maximum of light when burning at the rate of 5 feet per hour, the gas being of 15-candle power. The following table shows the extraordinary change made by the apparatus upon the light of this burner when consuming gas in small quantities:

Pressure of the gas as delivered to the burner.	Consumption of gas per hour.	Illuminating power *without disc.* Calculated to 5 feet per hour. Compared with Sugg's London No. 1, burning 5 feet an hour, as 100.	Illuminating power *with disc.*
inch.	feet.		
·064	1·65	7·36	63·5
·075	2·05	15·2	90·1
·1	2·4	30·7	93·3
·2	3·9	88·3	101·0
·25	4·5	103·9	Smokes.
·27	5·0	109·6	Smokes.

"Here it is seen that in the initial experiment the application of the disc instantaneously increased the light from the gas *ninefold!* And in the second experiment the application even of this rude apparatus sufficed to develop from only *two* feet of gas an amount of light proportionably greater than is obtainable from ordinarily good argands (or any other kind of burner) when burning at the usually higher rates of gas consumption. So that the loss of light when gas is burned in small quantities, so far from being a fixed and constant quality, is shown to be simply owing to the burner, not to any variation in the light-giving power of the gas, but to the mechanical apparatus employed for developing that power. Just as (as best shown in batwings and fishtails) a similar and equally great loss of light takes place when the consumption of gas is raised *above* a certain point, which varies with each burner."

* The Referees appear to ignore the single jet.

Let us now summarize the results of the preceding experiments:

"I. In the case of the batwing and fishtail burners there is a point of consumption *above* which every increase in the rate of consumption produces a *decrease* of light relative to the quantity of gas consumed.

"II. The point of consumption at which each of those burners gives its maximum of light, relative to the quantity of gas consumed, varies enormously; two burners (Nos. I. and V.) giving most light from the gas when the rate of consumption is only 1¾ *foot per hour*.

"III. With argands, on the other hand, the light from the gas steadily increases in a higher ratio than the consumption. In other words, the larger the quantity of gas consumed in argands (up to the smoking point), the greater the amount of light obtained relatively to the quantity of gas consumed.

"IV. Alike with argands, batwings, and fishtails, whatever be the rate of consumption at which the maximum of light is obtained (in other words, taking each of the burners at its best), there is nevertheless a striking difference in the degree of light obtained from the same quantity of gas, some burners giving a light equal only to 20, while others give a light equal to 60, 80, and 100.

"V. The best kinds of argands give a nearly equal amount of light, relatively to the quantities of gas consumed; the experiments with them tending to show that, within the ordinary range of consumption, the illuminating power of gas remains the same.

"VI. Finally, even as regards very low rates of consumption (rates, indeed, at which gas is never burnt for illuminating purposes), the application merely of a rude apparatus for regulating the air supply suffices to make only *two* feet of gas give a light equal in proportion to the greatest amount of light obtainable from the gas when consumed at any higher rate in a really good burner.

"The establishment of the true facts of the case, simple and natural as they are, not only serves to remove a stumbling-block and perplexity from the path of science, but, as will be evident in the sequel, is of importance in the practical question as to what is the best mode of consuming gas with a view to obtain efficiently and economically its highest amount of illuminating power."

The report of the Gas Referees, from which the foregoing is extracted, must be regarded as a most valuable contribution to gas lighting; yet it is to be regretted that these gentlemen did not extend their researches to burners consuming larger quantities, and that those with which they experimented were not more suitable for doing justice to the gas, as by employing proper burners by which the gas could issue at a low pressure, the results of their experiments, and their conclusions, on some points, would have been very different.

The following are the dimensions of the Standard London Argand Burners, as recently fixed by the Gas Referees for testing the gas of companies under their control:

Number of holes	24
Diameter of holes	0·045 inch.
Internal diameter of steatite top	0·480 ,,
External	0·840 ,,
Height from the upper surface of the gallery on which the glass stands, to the level of the holes or upper surface of the steatite top	0·75 ,,
Cone, internal diameter at top	1·08 ,,
,, ,, bottom	1·50 ,,
,, height of, externally	0·75 ,,
Chimney for 16 candle gas, 6 inches diameter by 1⅞ inch high.	
,, ,, 14 ,, ,, 6 ,, ,, by 1½ ,, ,,	

The diameter of the chimneys is *inside*.

When burners are employed for heating alone, the conditions are entirely changed, as in this case air is intermixed with the gas before it issues from the burner, by which means smoke is prevented. For this purpose the Bunsen burner is very generally adopted, and in its simplest form consists of an open tube connected to and forming part of the gas burner, by which tube air enters and mixes with the gas. These are made by manufacturers to suit all ordinary requirements.

REGULATORS.

As already explained, the quantity of light derived from a given quantity of gas is in a great measure dependent on the pressure of the gas at the point of issue when being consumed, therefore to reduce and control this pressure on the consumer's premises, and obtain the full advantages of the gas in an economical point of view, the regulator is indispensable. This may be applied either to control all the burners on the premises, or a portion of them, or separately on each burner, which last method is beyond all question the most advantageous.

Regulators, like the governors already explained, are either wet or dry; the former, as generally applied to consumers' premises, is represented in section in the annexed Fig. 98, and differs but slightly from the Station Governor, except that the whole is enclosed within its case. The bell is guided at the bottom and top by rollers, and carries the cone plug which is suspended within the inlet pipe similar to those already described, weights being placed on the top to give the desired pressure. The cover of the apparatus opens with a hinge, thus permitting the pressure being adjusted as required.

FIG. 98.

The dry regulator bears so strong a resemblance to that invented and patented by S. Crosley, and described in page 228, as to render further explanation here unnecessary, but when speaking of its application to street lighting a drawing of the instrument as applied for that purpose is given. Both kinds of instruments are manufactured to supply from five lights and upwards for the use of private dwellings, manufactories, and other buildings. We are informed that a very extensive and successful application of the dry regulator for consumers' premises has been effected by Messrs. Peebles and Co., of Edinburgh, and undoubtedly the very general application of gas in Scotland, together with the tenements being composed principally of "flats" or floors, afford great facilities for the introduction of these apparatus.

When a regulator is fixed to control the pressure of any premises, it is only strictly effective with those lights on a level with the instrument; for as gas in consequence of its low specific gravity increases in pressure, as already stated, about one-tenth of an inch for each 10 feet of augmented elevation, it follows that in a building of which the highest floors are situated 50 feet above the regulator, there will be a difference of $\frac{5}{10}$ths between the pressure of the two localities. For this reason in buildings or manufactories having several floors, to ensure proper economy a separate regulator is required for each floor, and by this means the various burners can be controlled to such a degree that the maximum amount of light is derived from the gas, without the chance of waste either through the negligence of the workpeople or irregularities of pressure from the company's main.

Regulators, whether wet or dry, for many years after their first manufacture, were not regarded with favour by either gas companies or consumers, and indeed at one period there was a strong prejudice against them. This arose partly from errors in their construction, but more particularly through the objections raised by gasfitters to their use, for wherever a defect existed, either by a stoppage in the fittings or through the pipes being too small, in the event of a regulator being used on the premises, the fault was always attributed to that, when its removal was considered necessary, and of course a better supply was the result, at least whenever the pressure in the main was greater than that to which the regulator was adjusted. In addition to this it may be stated that the benefits to be derived from the instruments in question were not then understood, and although in this respect our knowledge is tolerably complete, yet the general application of regulators leaves much to be desired, as hundreds of thousands of pounds could be saved annually by consumers on the more general adoption of these instruments, without in any manner diminishing the quantity of light obtained.

The most important application of the dry regulator is to control the supply of gas to the public lamps, which we believe was introduced by Mr. Paddon about sixteen years ago, since which period the system has become almost universally adopted with the greatest advantage to gas companies, assuring at the same time, when proper burners are placed, the maximum amount of light for the quantity of gas consumed. The lamp regulator, represented in section by Fig. 99, is always supplied with the burner and socket complete. The flexible diaphragm in the centre acting on the cone controls the pressure, at the same time the quantity of gas passing per hour; and, as represented by the arrows, the gas enters beneath the diaphragm and then passes to the burner. These instruments are made to deliver any stated quantity of gas per hour according to the contracts of the gas company with the parochial and other authorities. When first introduced great doubt existed as to their durability, which the experience of several years has entirely removed; and it must be admitted that they have done good service, as by their use the constant varying pressure in the main is completely controlled, and the pressure of each lamp is adjusted to that degree where the maximum amount of light is evolved by the gas. Of these apparatus Messrs. Wright and Co., of Westminster, are accredited manufacturers.

Fig. 99.

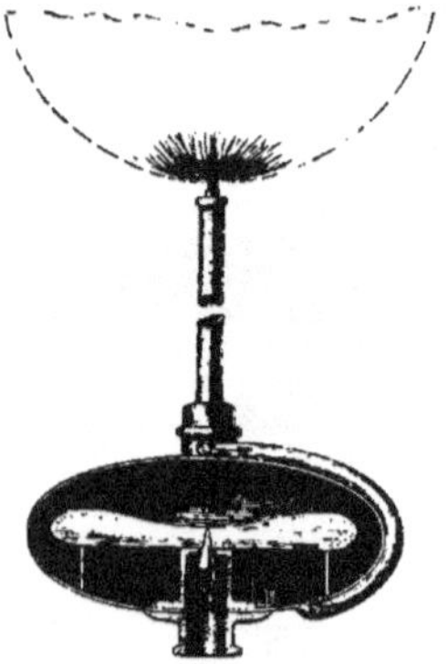

The instrument connected with our subject which merits particular attention, is Giroud's Rheometer, or "Flow Measurer," which is represented in Fig. 100 full size, as applied to public lamps, or when attached to the burners of private dwellings. Within the outer cone A A, which we may call an annular tank, is enclosed the bell C C, working in a luting of glycerine. On top of the bell is a cone D, corresponding with the orifice above, the two forming a valve similar to those of other regulators already referred to, and when in action, the gas passes the orifice O on the top of the bell, hence through the valve to the burner. With the view of ascertaining the merits of this instrument, and being provided with the usual appliances of experimental meter and pressure gauge, we entered into a series of experiments for that purpose, choosing one at random, which was stated to be regulated to deliver 5 feet per hour.

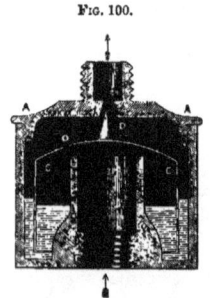

Fig. 100.

The first trial was with a pressure of $\tfrac{7}{10}$ths, when the instrument was ineffective and the flame of the burner imperfectly defined, the consumption being 3 feet per hour. The pressure was then increased to $\tfrac{4}{10}$ths, when the batwing burner assumed the proper form and the consumption was $4\tfrac{1}{10}$ feet per hour, instead of 5 feet for which it was adjusted. By other experiments at 1 inch pressure the same consumption was rigidly maintained, and when the pressure was increased to 2 inches, the meter indicated $4\tfrac{1}{10}$ feet. On this the burner was removed and the gas consumed from the open socket, the pressure at the same time being increased to $2\tfrac{1}{4}$ inches, all that could be had, when again the consumption was $4\tfrac{1}{10}$ feet per hour. These trials were made repeatedly and always with the same results.

Thus we find that the rheometer is inoperative when a quantity of gas passing is less than that for which it is calculated, a condition inherent with all regulators. That under the other various circumstances of pressure and orifice for the issue of the gas it is practically correct, as there was only a variation of two per cent., a range which is permitted in the gas meter. We may state that although we had heard previously the most favourable opinions expressed concerning the instruments in question, we were not prepared for such remarkable results; therefore, with this degree of excellence, the vexed question of the proper and uniform supply of public lighting no longer exists, for all that is requisite to ensure regularity in this respect is to maintain the pressure at the lowest, or the most remote parts of the district, at a minimum of $\frac{4}{10}$ths; to have the rheometers properly tested as to their delivery previously to fixing; and to have proper burners and sufficiently large, by which means the quantity of gas contracted for will be supplied, and from which the greatest amount of light will be obtained.

This instrument, as may be conceived, is of the greatest simplicity, is made of white metal, the glycerine requisite for each being about the tenth part of an ounce, and, as is well known, this is not susceptible to freeze, thicken, or evaporate at any temperature our variable climate is exposed to. In reply to our inquiries we are assured by engineers who have had these instruments in active operation during upwards of sixteen months, that they have given the most entire satisfaction, and on being re-examined after some months' service they have been found not to have varied to any material extent. The importance of this invention, according to our opinion, in the interests of gas companies and consumers has induced us to extend unusual space to its description. Another instrument for a similar object, invented by the same gentleman, M. Giroud, also merits attention. This consists simply of a disc of metal, within the outer case, which is less than three-quarters of an inch in diameter, and attached to the disc is the cone of the valve; thus by the simple action of the pressure the disc is elevated and the supply controlled.

All regulators are very desirable alike for public lamps, manufactories, or private dwellings, and particularly for the latter where often 50 per cent. of the gas consumed passes off without yielding any light whatever, which is a remarkably common occurrence with the ordinary fishtail burners enclosed within the glass moons, and to which reference has been made in the chapter on burners. However, to obtain the full benefits to be derived from the instruments under consideration, the most suitable burners and glasses for the proper and economical consumption of the gas are indispensable.

GAS FITTINGS.

During many years a strong prejudice existed among a certain class of *proprietaires* against the use of gas, and who persistently refused its introduction into premises owned by them; and although fire insurance companies insured premises thus lighted at more favourable rates than when illuminated by other means, yet, at one period, the opposition to gas was incredible, and from this cause a popular metropolitan theatre continued to be lighted by candles, twenty years after every other similar place of amusement in London had adopted the use of gas.

However, since then, a generation has passed away, and this prejudice now no longer exists, but, on the contrary, as we find in the majority of new dwellings in the metropolis, of an annual value of 40*l*. and upwards, that during the operation of building, the pipes are placed to supply gas to the principal parts of the house, whilst in superior dwellings the fittings are carried throughout. Hence with this general adoption, it may be predicted, with tolerable certainty, that fifteen years hence the gas consumption of London will be double that of the present day, at which period its cost may be estimated at upwards of twenty shillings per head of the population per annum.

Some years ago as a means of preventing accidents by the escape of gas, it was ordered in France that wherever practicable all pipes for conveying gas should be exposed to view, and thus the ceilings of the finest cafés were often peculiarly ornamented for the purpose of disguising the unsightly gas pipes. This system had, however, a directly opposite tendency, as by the exposed position, the pipes were often mischievously or maliciously punctured, frequently occasioning the very evil intended to be averted. Unquestionably the best description of pipes for all confined localities, as beneath floorings, or other inaccessible places, is the wrought-iron gun-barrel, which has contributed in an extraordinary manner to the development of gas lighting. This material, under some circumstances, as when placed in dry localities, may be pronounced practically everlasting, and possesses the further advantage of avoiding those recesses which occur in pipes of soft metal, hence, once the gun-barrel properly laid, there is no likelihood of those obstructions by condensation so frequently experienced with lead or similar pipes. For factories of every denomination all the fittings so far as practicable should be of wrought iron, and although at times some inconvenience may be experienced at the end of several years by the accumulation of the oxide of iron therein, with this evil in view, and those complained of by a writer, where he described how the "walls and paper-hangings of dwellings were pulled to pieces in order to clear out this obstruction," we believe that whenever ordinary intelligence is displayed in fitting up premises, on any such accumulation occurring, it can be readily removed without occasioning the difficulties mentioned.

For this purpose all that is necessary is to carry one or more vertical pipes direct from the bottom to the top of the building, which pipe is provided with a T piece and plug at the bottom where it may be exposed to view; hence, any accumulation in the pipe will fall into the T piece. From these vertical pipes, branches are taken to the respective floors, always avoiding as much as possible the introduction of bends, which invariably harbour the oxide; and in the event of any obstruction by the removal of the plug at the lower part of the building, and the application of the service cleaner to the part obstructed, the pipes will speedily be cleared. But by the adoption of galvanized pipes all these evils are averted, or at least practically so; and although the cost is about 25 per cent. more than the common or black iron pipes, all superior work, in our opinion, should be executed by the former.

Soft metal pipes should not be laid in confined places, as Dr. Bowditch mentions an instance where some lead pipes were fitted beneath the flooring to supply gas to a gentleman's mansion, but before the gasfitters had terminated their work the pipes were gnawed through by rats; and similar circumstances are mentioned by other writers. Hence from a simple neglect of this kind, and from such unexpected cause, serious accidents might arise.

Whatever it is to be attributed to we cannot pretend to say, but we never heard of the accumulation of naphthaline in the fittings of dwellings, and its absence may be probably due to the moderate temperature of the locality; but in the event of this occurring, it would of course be readily removed by the apparatus mentioned. Here we may observe, that naphthaline is deposited under peculiar circumstances, whenever the gas is poor, but generally in the autumn and spring, often occasioning stoppages in the services and sometimes in the mains. It is readily dissolved by naphtha, but its accumulation is prevented by the admixture of a small percentage of cannel, with the caking coal carbonized.

The following tables, copied from 'Molesworth's Pocket Book,' give the maximum quantity of gas supplied by pipes of various dimensions under different pressures, which differ slightly from the tables by Barlow already given, but the variation is not susceptible of practical appreciation:

SUPPLY OF GAS IN CUBIC FEET PER HOUR.—LENGTH OF PIPES 10 YARDS.
PRESSURE OF GAS IN TENTHS OF INCH.

Diameter of Pipe in Inches and Parts.	1	2	3	4	5	6	7	8	9	10
⅜	13	18	22	26	29	31	34	36	38	41
½	26	37	46	53	59	64	70	74	79	83
¾	73	103	126	145	162	187	192	205	218	230
1	149	211	258	298	333	365	394	422	447	471
1¼	260	368	451	521	582	638	689	737	781	823
1½	411	581	711	821	918	1000	1082	1162	1232	1299
2	843	1192	1460	1686	1886	2066	2231	2381	2530	2667

SUPPLY OF GAS IN CUBIC FEET PER HOUR.—LENGTH OF PIPES 100 YARDS.
PRESSURE OF GAS IN TENTHS OF INCH.

Diameter of Pipe in Inches and Parts.	1	2	3	4	5	7·5	10	12·5	15	20
½	8	12	14	17	19	23	26	29	32	36
¾	23	32	42	46	51	63	73	81	89	103
1	47	67	82	94	105	129	149	167	183	211
1¼	82	116	143	165	184	225	260	291	319	368
1½	130	184	225	260	290	356	411	459	503	581
2	267	377	462	533	596	730	843	943	1038	1193

From the preceding, the quantity of gas that any pipe on the list will deliver is readily ascertained; the pipe being open at the end, and laid in a straight line; elbows and bends, and particularly the former, diminish very materially the flow of gas; besides, in practice, it is always advisable to fit larger pipes than those indicated by theory to be sufficient, which is, moreover, rendered necessary on account of the variable pressure given by the Gas Company, and the requirements of the premises, and particularly in anticipation of any augmentation of the number of lights.

With the general adoption of gas lighting, the apparatus for its use in dwellings have acquired a degree of elegance, rendering them in many cases the most prominent decorations; but with this degree of ornamentation, there is a weak point which certainly merits particular attention; we allude to the hydraulic joint, to be met with in almost every house where gas is employed, and from which more accidents have arisen than from all other causes combined. Nor can this be surprising, when we reflect that in the majority of cases a house is fitted up with gas, and the hydraulic joints are supplied with water by the fitter and left to chance. The inhabitants enter their dwelling in utter ignorance of any necessary precaution to be observed, and in the course of time a small quantity of water evaporates, when the disagreeable odour arising from the escaping gas directs attention to the prevention of this, at the same time avoiding a disastrous accident; and, as already stated, if gas had not this repulsive odour, its employment would be attended with hourly danger.

For accidents arising from the cause in question, the manufacturers of the apparatus are in some measure responsible, in consequence of the exceedingly limited size of the cup at the top of the joint, which is often not capable of containing more than an ounce of water; and by the evaporation of that small quantity on drawing the chandelier down to the lowest point, with the ordinary night pressure of the works, the gas escapes, and if not attended to, results in an accident. Moreover, the facilities for the evaporation of the limited supply of water mentioned is increased by the comparative large area of its surface, and the temperature to which it is exposed when the gas is lighted. The remedy for this evil is simple, and consists in increasing the dimensions of the cup by making it three or four times the depth usually adopted, when sufficient space will be allowed to permit of a small quantity of oil being poured on the surface of the water, in order to prevent evaporation, and thus diminish the chances of accidents from this source to the minimum. But with the present method of making these cups, the oil, when placed as described, is liable with the least jerk to fall on the furniture beneath; hence the annoyance from this cause prevents the adoption of the ordinary means of avoiding the accidents in question.

All public places of resort, as theatres, lecture halls, assembly rooms, and similar places, should be supplied by two meters at least, in order that in the event of one of these failing in action, the supply would be continued with the other. In churches the pulpit is invariably lighted with candles, so that, under these circumstances, a total extinction of the gas would not put the locality in total darkness, as in other places where the illumination is entirely by gas.

Every class of merchandize when displayed in shop windows, is always seen to advantage when the gas lights are above the line of sight, and still better when the eye is protected entirely from their glare. For public halls and churches, the best means of illumination is by the sun-light suspended from the ceiling. These are commonly used in banking houses and other large offices of the metropolis; and in addition to lighting, they serve the purpose of ventilation.

The subject of gas-fitting is very comprehensive, and in itself would occupy a volume, and, as our space becomes limited, we have therefore referred to the most important points.

RESIDUAL PRODUCTS.

In the process of gas manufacture, when caking coal is carbonized, the most important of the residuals is the coke; and in some localities such is the value of this, that after supplying the necessary fuel for heating the retorts, the remaining portion realizes the total cost of the coal carbonized. The variable quantity of ash, as already observed, existing in the coke derived from the numerous descriptions of cannel, renders it impossible to estimate the value of these residuals, unless each class were taken separately, which would occupy considerable space, without possessing general interest, and is moreover foreign to our purpose. Our remarks will therefore be confined to the coke obtained from caking coal, of which the remarkable uniformity in composition has already been observed.

According to the balance sheets and returns of various gas companies, the quantity of coke produced as well as the percentage used for fuel for carbonizing, is very variable. For, on referring to the returns of the metropolitan Gas Companies for the year 1875, we find the average yield of this to be 34 bushels per ton of coal, of which 31 per cent. was used as fuel, consequently the average sales were 24·9 bushels of coke for each ton of coal carbonized. At the works of one company, the coke employed as fuel was 20 per cent., whilst at another it amounted to no less than 37 per cent. of the production, the yield in the latter case being but 33 bushels per ton.

On the other hand, from whatever cause it may arise we will not pretend to offer an opinion, some provincial works produce as much as 40 bushels, and of this 30 bushels of coke are sold for each ton of coal carbonized. With these great discrepancies, with the fact that one works sells within 3 per cent. as much coke as the total production of another, we are led to the conclusion either that the operation of gas manufacture has not attained that degree of perfection which is generally supposed, or, if that exists, it is not practically carried out. Nor are these remarkable variations confined to the coke alone; as, in the metropolitan returns already mentioned, we find, according to the operations of the respective companies, a difference of no less than 11 per cent. in the yield of gas, one company producing 10,334 feet, whilst another obtains only 9332 feet per ton, the average being 9892 feet per ton of coal carbonized. But these differences in the production of gas might be supposed to arise from the varied percentage of cannel employed, which, however, is not confirmed by the returns, inasmuch as the company using the smallest quantity of this, is second on the list for high production, whilst the greatest percentage of cannel rates as the third in the quantity of gas per ton. Therefore, with these examples before us, considerable latitude may be allowed for any estimates in connection with gas engineering.

Referring to the gas, it can be readily understood that the higher the heats employed, the greater would be the yield, which would necessarily demand a larger percentage of cannel in order to bring the gas to the desired standard; but in this again there is no kind of relation between the operations of the respective companies, except only in one instance where the large yield exists, the standard of illuminating power is lower than that of other companies.

As regards the large amount of fuel used for carbonizing at the metropolitan works, as compared with some provincial companies, perhaps this may be attributed to the system there generally adopted of having the arch immediately over the furnace extending the whole length of the oven, by means of which the passage of the caloric is undoubtedly obstructed. This opinion is further favoured by the fact, that in works where the settings are built so as to allow the heat to pass to the retorts with all facility, there, as we are prepared to show, as in the provincial works already alluded to, and many others, the fuel accounts are the lowest. Hence we find, in addition to the extraordinary variation in the yield of coke, and that consumed for heating the retorts, that the "percentage of fuel," without stating the quantity produced, means nothing; therefore to understand the operations of a company in this respect, the quantity of coke sold is the only guide.

The next consideration is the price obtained for the residual in question, which is governed by innumerable circumstances, but in large towns principally according to the neighbourhood in which the works are established. This fact is made very palpable in the returns of the metropolitan companies, from which we learn that the price realized per chaldron by the Gas Light Company is the lowest on the list, arising undoubtedly from the isolated position of their Beckton works; whilst the other companies procure from 5 to 40 per cent. more per chaldron than the Company mentioned, the highest prices generally being obtained in the most populous neighbourhoods; therefore the sale of coke is materially influenced by the locality where the works are situated. Among other circumstances which affect the value of the residual in question, is the class of manufactories established in the neighbourhood of the works, to some of which coke is almost indispensable, and as a matter of course it commands a high price. Some years ago we remember, at a small town in Belgium, by the combined circumstances of the low price of coal and the extraordinary value the coke possessed, the gas was regarded as a secondary product, the principal object being to produce coke.

Where coke is not sold readily, but is stacked, in the course of a short time a portion of it is resolved into breeze, and, as a natural consequence, the coke is reduced in size and value. This deterioration is considerable, depending on various circumstances, as the quality or hardness of the coke, the height of the stack, the kind of weather it is exposed to, and the time it remains on hand.

The importance of a rapid sale of coke—that is, sold as produced—is hardly susceptible of being over-rated. The amount of loss sustained by stacking, for reasons stated, can only be matter of conjecture; but we believe there is no exaggeration in saying, that it is more profitable, taking labour and other things into consideration, to dispose of coke as produced at 12s., than to stack it for six months and then sell it for 16s. per chaldron. It is better to sell under the conditions and at the price stated, rather than stack it for twelve months, and then obtain 18s. per chaldron for it; unless, indeed, breeze in the locality may have a most favourable market, and can be sold at an extraordinary price, which would alter the conditions.

Let us now consider the difficulties attending the sale of coke, which in some localities are no doubt numerous, particularly on the first establishment of a gasworks, and especially in warm climates, where the domestic fire is unknown, when it becomes part of the gas manager's duty to seek the means of creating a sale for this article, which otherwise is lost, or, what is the same thing, ordinary care not being observed in the retort house, it is all consumed there. To promote a sale, the coke must be adapted to suit the various industrial purposes of the locality; for steam boilers, and even locomotives, should a railway exist in the neighbourhood; in kitchens, as a substitute for charcoal; in hotels, cafés, hospitals, and public buildings. For all these applications it is admirably suited, simply requiring some slight alteration in the construction of the furnace or stove, and, like gas, when once adopted and properly applied, the use of coke is rarely abandoned.

There are many continental towns and cities where, at the commencement of the operations of the gas company, the residual under consideration was a drug of little or no value, for the simple reason that there existed no means of employing it; but after protracted and continuous losses, it became necessary to find a remedy for the evil, and for this object, in some instances the company interested constructed various kinds of kitcheners, sometimes combining these with stoves for warming apartments adapted to the wants of all classes. These were produced at a cheap rate, and sold regardless of profit, the coke being sometimes given gratuitously in the first instance for trial. By these experiments coke was found to be more economical than other fuel; and, in a short time, a market was established for all that could be produced. Afterwards, on coke becoming a direct necessity, its price was augmented, and the former drug constituted the principal source of excellent dividends.

We believe also that some companies at home may profit by the foregoing remarks, and so remove the mountains of coke sometimes to be seen at their works. For what can be more unsuitable and wasteful than the means of cooking and warming as applied in the dwellings of our working classes? And might not the stoves and kitcheners referred to be applied to their humble dwellings, alike to the advantage of the occupants and the interests of gas companies? Let there be a good, cheap, and suitable stove made, and, this appreciated, there would cease to be any large stock of coke at gasworks, at the same time the price of that commodity would be necessarily augmented.

In our opinion, it should become part of the gas manager's education, when going abroad, to understand the various applications of gas coke. From experience, we know that a ready and remunerative sale can always be effected in any town sufficiently civilized to employ gas lighting, and at a price corresponding with the value of coal. It may not be a manufacturing town, requiring steam boilers; there may be no railways, and, in consequence, no locomotives to supply; but there will surely be establishments, public or private, where coke can be applied alike to the interest of the purchaser and seller, and only requiring knowledge and energy to cause it to be used.

When burning coke, a point often lost sight of is the necessity of regulating the size of the pieces in proportion with the dimensions of the stove where it is consumed. Thus, if we suppose a cylindrical stove of 6 inches in diameter and 4 inches deep, in which large pieces of ignited coke are placed, the fire would be speedily extinguished; but, if the pieces of coke do not exceed a chestnut in size, then the combustion will be sustained, and as the dimensions of the stove are increased, so is the size of the coke proportionably augmented.

The one great test of good working and economy in fuel is the money test, viz. the amount received for coke per ton of coal carbonized. Necessarily the returns from this source must always vary according to local circumstances, but a great deal depends on the management. We knew of a foreign works, held by an English company, where the returns from the sales of coke did not amount to 2 per cent. on the value of the coals distilled. This, however, arose from a peculiar kind of mismanagement, of which, let us hope, there are but few specimens. But in many instances, however, where there is no reason for doubt, the receipts for the residual in question do not amount to 20 per cent. of the cost of coal carbonized. Under such circumstances, in a works of medium capacity, employing Newcastle coal, or with a small percentage of cannel, and supposing there to be no extraordinary extenuating circumstances, a gasworks producing only this amount for coke may be pronounced badly managed, and the fault will probably consist in the settings of the retorts, or the want of proper care to control the orifices of the dampers, or a deficiency of draught; but from whatever cause it may arise, there can be no excuse, as, from experience, wherever a gasworks is established, the coke will find a ready sale for some purpose or another, and wherever coal is high in price, coke will as a rule be proportionately increased in value.

Before the establishment of railways in France, fuel of every denomination was exceedingly dear in Paris, wood and charcoal being that generally employed for culinary and other domestic purposes. This excessive price gave rise to the manufacture of an artificial charcoal, by mixing sawdust, chips of wood, and such like material, with a small portion of tar, which mixture was compressed into cylindrical blocks of about 6 inches long and 2 inches diameter. These blocks were afterwards baked, so as to expel all the volatile matter, when a substance was produced similar in appearance and nature with charcoal, and was called "Charbon de Paris," and made the fortune of the first manufacturers.

The large accumulation of breeze at the French gasworks opened another source for developing this industry. In this case the breeze was treated in a similar manner to the sawdust, but manufactured in large blocks of the size of ordinary lumps of coal, to which the fuel bore some resemblance; and this became, and is now, an important and economical fuel for the poorer classes of the community.

This application of breeze seems to have had but little attention in England; and, probably, in localities where it has no sale, there are hundreds of tons of it thrown away annually; and by this misapplication of so much that might be rendered useful and valuable, a great wrong is committed; for not only may this breeze—even the smallest kind—be converted into fuel for sale, or, by preparing it as hereafter indicated, it may be used for heating the furnaces of gasworks, thereby economizing nearly its weight of coke.

We are only aware of this method of applying breeze to heat retorts being practically carried out in England at the Dover gasworks, under the direction of Mr. Anderson; but it is hoped its application is more general than supposed.

Fig. 100 represents a drawing of the apparatus for compressing the breeze, which is described in the following words by Mr. Anderson:

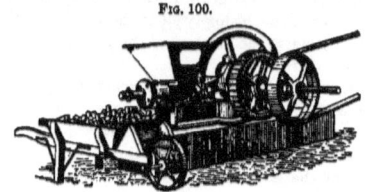

Fig. 100.

"The object of this machine is to convert the breeze, or small coke, that has hitherto been sold to brick-makers and lime-burners, at from 1s. 6d. to 3s. per ton, into a fuel capable of heating the retorts, and by which its value is increased to the extent hereafter stated.

"The breeze, in a dry state, is thoroughly saturated with gas tar, and afterwards passed through an ordinary pug-mill, such as is used for the pugging of clay and mortar.

"This mill has one or two passages in the bottom, communicating with corresponding cylinders and pistons, according to the construction of the machine, whether single or double.

"The pistons are actuated by an excentric, placed immediately under the hopper, which compresses and forces the fuel out at the open ends of the cylinders, as shown in the drawing.

"As it issues from the machine, the fuel breaks off in pieces convenient to be employed in the furnaces, therefore no means have been taken to cut it of uniform size; but if this were requisite, a 'cutter, actuated by the machine, could be easily applied.'"

The following is Mr. Anderson's account of the cost of converting a chaldron of gas breeze, formerly sold at 2s., into the patent fuel:

	Ton.	Cwt.	Qr.	Lb.		s.	d.
1 chaldron of breeze	1	1	2	7	at 2s.	2	0
54 gallons of tar		5	3	14	at 1d. per gallon	4	6
Total weight	1	7	1	21			
Labour, working the machine				1s. 8d.			
Coke for boilers				8d.		2	4
Thus 27 cwt. costs 8s. 10d., or at the rate of 4d. per cwt.						8	10

This fuel is supplied to the furnaces as it comes from the machine, without either breaking or drying, but is not used exclusively, the means of employing it being as follows. After clinkering, the bottom of the fireplace is made up with coke, on which the breeze fuel is placed so as to fill the furnace, the door of which is so arranged as to admit air for the proper combustion of the fuel when first charged, and thus avoid the loss of heat and the issue of smoke from the chimney. And on all occasions on introducing fresh coke, a proportionate quantity of breeze fuel is supplied.

To produce six tons of this fuel in twelve hours, an engine of about four horse-power is required; and it is needless to say, that in localities where breeze possesses little or no value, it may either be converted into fuel, as stated, for supplying the furnaces, or for sale; but for the latter purpose it would be necessary to bake it in an oven having a temperature of about 300°, by which the fuel becomes hard, and is capable of being transported with facility without breakage. The vapour arising from this process should be condensed, in order to retain the naphtha and other volatile liquids.

Manufacture of the Sulphate of Ammonia.

In most gasworks where the ammonia is manufactured on the premises, on account of the simplicity of the operations, the general demand for the product, together with its remunerative price, this is usually converted into the sulphate of ammonia; and for the following description of the manufacture of that compound, and tar, as practised in gasworks, we are indebted to Mr. Whimster, engineer of the Perth Gas Commissioners, whose experience in such matters is well known.

"The process of utilizing this secondary product is simple. In the operation, the condensed liquid, as well as that derived from the scrubber or washer, is pumped into a large separating reservoir, from the bottom of which the tar is run into its corresponding boiler or still, and from the upper part the ammoniacal liquor is conveyed to the boiler intended for its distillation.

The boiler for the ammoniacal liquor is fitted in the ordinary way with a swan neck at the top, from which a pipe passes into the precipitating pan, which is situated immediately at the back of the boiler. This pan is formed by preference of ½-inch sheet lead, set in brickwork, and covered in with 1¼-inch planking, as shown in section. A leaden pipe in connection with the boiler passes through and projects about a foot above the centre of the cover, and terminates by an inverted cup or "cracker," also of lead, and perforated so as to break up the ammonia vapours into small streams, in order that they may be brought into the most intimate contact with the acid. On commencing operations, a quantity of "mother liquor" from the previous distillation is left in the pan, to which is added as much sulphuric acid as to cover the cracker to a depth of about 4 inches; and in the course of distillation, when this has become saturated or nearly so, and crystals or salts are forming rapidly, more acid is supplied, and so on till the end of the distillation, when the ammonia has been expelled from the liquor in the boiler, and, this effected, the boiler is then emptied ready for another operation.

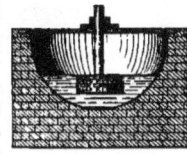

When the liquor in the precipitating pan begins to increase rapidly, the distillation should be stopped, at which point the mother liquor should be nearly neutral, or just sufficient to redden blue litmus paper. The distillation being stopped, the liquor cools, and the crystals of sulphate of ammonia are precipitated, when they are withdrawn and placed on a drainer, and after being freed from liquid are conveyed to the stores.

By the process as thus conducted, from 18 lbs. to 22 lbs. of sulphate of ammonia are obtained from every ton of Scotch cannel carbonized in the manufacture of gas, the average produce of ammoniacal liquor of 4½° Twaddle being about 30 gallons per ton of cannel. Of this liquor, one ton, or 200 gallons, will yield about 1 cwt. of sulphate of ammonia, and will require nearly that weight of acid at 144° Twaddle, the actual proportion being in ordinary working about 92 per cent. With good sulphuric acid made from sulphur, not from pyrites, a good light-grey commercial salt is produced, containing 97 per cent. of pure sulphate of ammonia, or 25 per cent of pure ammonia.

When the process is carried on near a town, certain precautions must be observed, to prevent any nuisance arising from the gases issuing from the precipitating pan during the combination of the ammonia and the vitriol, and for this object a flue is carried from the top of the pan through a condenser to the retort-house stack, where the gases are brought in contact with red-hot gases from the benches, and all nuisance prevented.

Cost and Returns by this Process.

One ton of cannel yields on the average 20 lbs. of sulphate of ammonia, which at 17*l.* 7*s.* 6*d.* per ton, the present net price at the works, is equal to 37·5*d.*; from which, if we deduct 9·7*d.* as the cost of acid, fuel, and labour, then there remains 27·8*d.* net return from this source for every ton of coal carbonized. But from this, interest on plant, together with wear and tear, will have to be deducted, which will vary according to the style of the works.

In some large works it is found more convenient to distil by steam, instead of fixing each boiler separately, in which case the interior of each of the boilers is fitted with coils of malleable iron pipe, through which the steam passes, a trap being provided at the outlet of the worm to receive the condensation without wasting steam. In these boilers, for the purpose of preventing loss of heat by radiation, the outside is covered with a thick coating of non-conducting material. The precipitating vessels in such cases are made with a curtain or division plate, formed by planking lined with lead, which dips some depth into the liquid, so that the vessel is divided into two distinct chambers, one of which is closed by the cover provided with the tube from the boiler and the flue for conveying away the gases, whilst the other is left open for the purpose of removing the crystals of sulphate of ammonia during the process of distillation, by which means the operation is continuous. In other works a cylindrical boiler is used, and provided with a stirrer actuated by gear work, similar to the wet-lime process. In this a quantity of slaked lime is placed, when the boiler is charged with the liquor and the man-hole cover secured, the advantage of this method being that the whole of the sulphuretted hydrogen is retained and the sulphate improved. However, Mr. Whimster does not consider a small portion of sulphuretted hydrogen prejudicial to the sulphate for ordinary purposes.

As observed elsewhere, the yield of sulphate from common caking coal is less than that from cannel, the former averaging about 16 lbs. per ton of coal carbonized.

With reference to the method of ascertaining the value of ammoniacal liquids described in page 133, Mr. H. L. Greville observes, in a communication to the 'Journal of Gas Lighting,' that "by the method of evaluing gas liquors by means of sulphuric acid, any ammonia present in the free state is neutralized, whilst the carbonate and sulphide of ammonium are decomposed with expulsion of carbonic acid and formation of sulphuretted hydrogen."

There are, however, present in the gas liquor, compounds of ammonia which are not affected, and therefore not indicated by the acid, and which (assuming the liquor to be sold to the manufacturer) "are thus altogether lost to the gas companies," "which fixed ammonia is combined with sulphuric, cyanic, sulphocyanic, and hyposulphurous acids, in various relative proportions." And according to experiments instituted by Mr. Greville, in nine different samples of liquors, he ascertained that the quantity of ammonia not indicated by the acid process, was an average of 20 per cent. That gentleman also says, "It must be borne in mind, that what is the company's loss is the buyer's profit, as the process to which the liquor is submitted in the manufacture of the ammoniacal salts, is such, that practically speaking, the whole of the ammonia is obtained."

APPLICATIONS OF TAR.

When this residual is manufactured on the works, the crude tar is run from the separating cistern into the boiler provided for its distillation, which is either effected by a direct fire or by steam, as described in the ammonia process. From 3 to 6 inches of water are run on top of the tar, to assist the operation of distillation.

The products of distillation are carried off by a swan-neck pipe affixed to the boiler, leading to a worm or other form of condenser, of cast or wrought iron, set in a tank filled with water, by which they are condensed. This water is kept cool by a constant supply, regulated according to necessity, running in at the bottom, causing the warm water to pass off at the top in a corresponding quantity. When the distillation is effected by means of steam after the lighter products are expelled, the remaining tar is conveyed to other stills for conversion into pitch, in which operation much heavier oils are obtained than in the preceding. When direct firing is employed, the distillation is continued in the same boiler after the naphtha and light oils come over, which are generally collected separately from the heavy pitch or "dead" oil.

The first products are crude naphtha and water, which, in consequence of their relative densities, at once separate, and during all the various stages of the distillation, the products are recognized according to their specific gravities. Thus the crude naphtha is considered all off when the liquid has reached the specific gravity of ·930, the average specific gravity being about ·885 to ·890; the light oil when it has reached ·970, its average being about ·950; and the heavy or dead oil which comes over when the tar is converted into pitch of average medium hardness, is of a specific gravity of 1·03 or 6° Twaddle. It is obvious that the whole of the water is expelled from the tar long before the heavy oil comes over. When it is desirable to procure anthracene, the distillation is continued until the oil attains a specific gravity of 1·08 or 16° Twaddle, by which the pitch is rendered so hard as to be commercially valueless, but it can be brought back to a proper consistency by being mixed with the naphthaline crystals which are deposited from pitch oil in cold weather.

The commercial test for crude naphtha is the percentage of distillate it yields at 120° centigrade (248° Fahr.), of which from 25 to 30 per cent. may be taken as the yield from the average of Scotch cannels.

Scotch cannel in the manufacture of gas yields on the average 20 gallons of crude tar (caking yields but the half of that quantity), and this proportion of tar from cannel yields about 10 per cent. crude naphtha, 3 per cent. of light oil, 25 per cent. pitch oil, and 38 per cent. of pitch. The quantity of pure anthracene in coal tar is very small, not exceeding ·75 to 1 per cent.

Products from the Tar of a Ton of Cannel.

		s.	d.
2 gallons of naphtha at 1s. 10d., present price	.. ..	3	8
0·6 gallon of light oil at 3½d. "	.. ..	0	2·1
5 gallons of pitch oil at 1½d. "	.. ..	0	7·5
89·6 lbs. of pitch at 0·16d. "	.. ..	1	2·3
		5	7·9

The cost of fuel and labour averages about 8·4d., making the net return per ton of coal carbonized, exclusive of capital, rent, and wear and tear, about 4s. 11½d. Were the process carried only so far as the manufacture of the crude naphtha, as in many works, having a favourable sale for all the rectified tar at about 2d. per gallon, the return per ton of cannel carbonized would be about 5s. 10d., and the cost of fuel and labour about 2d., making the net return in this case 5s. 8d. per ton of cannel carbonized, excluding, as before, capital, rent, and wear and tear.

Formerly the tar was supplied to the boiler without the water, when the fire was applied very gradually for some time, until the noise within the boiler, which always accompanied the operation, ceased; and on this occurring the fire could be increased to any desired extent. But on neglecting this precaution, on the tar acquiring only a moderate temperature, it issued in a frothy state, until the boiler was emptied.

However, with an inferior class of men, the manufacture of tar is attended with some degree of risk; and whenever the works are situated near a town, in consequence of the alarm created by its accidental ignition, which at times occurs, it is better to employ it in other ways, either by disposing of it to the manufacturers of these products, or consuming it for heating the retorts or other purposes. An application of this residual, which can be adopted in many towns both at home and abroad, is for paving the footways of streets, as practically carried out on an extensive scale in London, or for the flooring of warehouses, stables, or similar places. For this purpose gravel is required; but in the absence of this, broken stones or bricks, the largest pieces of which are about the size of a chestnut, intermixed with smaller material so as to fill the interstices, will answer the purpose. When preparing this, the gravel or broken stones is passed through a sieve, the meshes of which do not exceed $\frac{1}{4}$th of an inch in width; by these means a portion of the small material is separated from the larger.

The ground to be paved is levelled at about $1\frac{1}{2}$ inch beneath the intended surface of the pavement; and should it be a footway, the edge or edges will require to be supported either by wood or stone, so as to prevent the material from breaking away; but when required for flooring, this precaution will not be necessary, as it will be supported by the walls. The larger stones are then saturated with tar, the most volatile portions of which have been evaporated, and placed over the whole area and levelled, and on this is placed a portion of the screenings, also saturated with tar, so as to fill all the interstices. This effected, a layer of about $\frac{1}{4}$ of an inch of dry fine screenings is placed over the whole, when it is well levelled with a heavy garden roller. By this means a very good and substantial pavement can be made at a moderate cost, and although for a few days there will be an odour of tar in the neighbourhood, this is but a slight inconvenience attending the process.

During a number of years various experiments were made with the view of manufacturing gas from tar; indeed, in the early stages of gas lighting this was one of the most favourite pursuits of engineers, but which invariably ended in failure; and, however simple and easy the operation may appear, it is attended with innumerable difficulties; in short, in a commercial point of view, is impracticable. Therefore the best and most economical method is to avoid any excessive formation of tar, by maintaining the retorts at a proper heat, and avoiding excessive charges, as stated under the head of carbonization.

Foul Lime.

In many localities the foul dry lime may be rendered a source of revenue, although but small, for which purpose the following abstract of a communication, by Professor Voelcker, to the 'Journal of Gas Lighting,' on its employment for agricultural purposes, is given:

"When judiciously applied, all samples of gas lime derived from the purifiers can be employed with advantage, for increasing the production of land suitable for its reception. But this is not a universal manure, like farmyard dung, which benefits, more or less, every kind of soil or description of crop. Nor is it a concentrated fertilizer acting like guano or bone manures. Briefly, gas lime exercises a most beneficial effect on some soils, but is entirely useless in others. Its successful application, therefore, mainly depends upon the proper selection of land where it is intended to be distributed.

"The principal advantages of gas lime are, that it exercises a beneficial mechanical effect upon heavy, stiff, clayey soils, by rendering them more friable and porous, and consequently better adapted for cultivation, and by consolidating, on the other hand, light sandy soils."

Gas Lime supplies Food to Plants.

All our cultivated plants on burning furnish ashes, containing a good deal of lime, which is essential to the healthy development of all vegetable produce. As plants have not the power of generating lime, it is clear either the soil upon which they are grown, or the manure which is put upon it, must contain a sufficient amount of this constituent, so necessary for the very existence of all plants. Gas lime not only supplies lime to plants, but also sulphuric acid, a combination not present in any quantity in quicklime. For leguminous crops, such as peas or beans, for clover, and other crops specially benefited by sulphate of lime or gypsum, gas lime, when obtainable, as is generally the case, at a trifling expense, is certainly preferable to quicklime as a manure.

The crops which are particularly benefited by gas lime are clover, sainfoin, lucerne, peas, vetches, and turnips. It is also a most useful fertilizer for permanent pasture, especially if the land is naturally deficient in lime. On natural grasses the best farmyard manure often produces little improvement, until a dressing of lime, marl, or gas lime, has been applied. The latter more particularly destroys the coarser grasses, and favours the growth of a sweeter and more nutritious herbage.

By virtue of its alkaline properties, gas lime exercises a beneficial effect upon the organic matters in the soil, by destroying the remains of previous crops and converting them into plant food.

"Gas lime also kills moss, heath, feather grass, and other plants characteristic of peaty land, and is therefore a valuable means for improving peaty or mossy meadows.

"For improvement of peat land, the liberal application of gas lime cannot be too highly recommended. On such land it is best to use gas lime in the form of a compost, which should be kept in a heap for a period of ten or twelve months, and turned once or twice before spreading.

"On land naturally deficient in lime, turnips often refuse to grow at all, produce but a scanty crop, which is moreover very liable to be attacked by a disease known to practical farmers as 'fingers and toes.' A large dose of gas lime applied to the stubble land in the autumn, before it has been turned up by the plough, in many instances is an effectual cure for this disease.

"With regard to the quantity of gas lime which ought to be put on the land, no general rule can be laid down, for the quantity should be regulated by the relative deficiency in calcareous constituents which different soils exhibit. Speaking generally, however, 2 tons per acre may be used with safety, and, in many instances, a heavier dressing will not be amiss.

"The proper time for application is autumn, or during the winter months, when vegetation is at a standstill.

"On arable land, gas lime should be applied to the stubble, spread out evenly, and left exposed to the air, before ploughing up, for three or four weeks. On grass land it should be spread during the months of December or January, or, at any rate, before vegetation is making a fresh start.

"It should be remembered that gas lime acts beneficially as a fertilizer mainly in virtue of its calcareous constituents, and therefore is most usefully applied to land naturally deficient in lime.

"On land abounding in this substance it has little or no effect. Though by no means a substitute for farmyard manure, guano, and other concentrated artificial manures, gas lime, judiciously used, unquestionably is a valuable auxiliary manuring agent, which frequently can be used with greater economy than quicklime or marl."

APPENDIX.

THE COMPOSITION OF COAL GAS.

By LEWIS THOMPSON, Esq., M.R.C.S.

WHEN we consider the great variety that exists in the kind of coals used for making gas, we might easily be led to suppose that an equal variety will be found in the different gases thus obtained; but it is not so, for whatever may be the kind of coal used, the gases produced from it are almost identical in their nature in every instance, and differ chiefly in their relative proportions towards each other. This general resemblance is the result of the process to which they are subjected, and which in chemical language is known by the term "destructive distillation," an expression which must always be understood to indicate the destruction of vitality, or, as we might say, the complete death of the substance operated upon.

It may perhaps seem singular that we shall speak of the vitality of a thing like coal, which has been buried for thousands of years in the earth; but the evidences of its remaining vital forces are easily demonstrated by a comparison of its chemical qualities with those of a substance altogether destitute of the peculiar powers conferred by life. Thus we say in common language that a tree is dead when it is cut down and made into sawdust, the composition of which is in fact merely charcoal and water; but then it is charcoal and water united by means of vitality, and this makes all the difference, for we can convert that sawdust into sugar, and that sugar into whisky or alcohol, although we can do nothing of the kind with charcoal and water. Similarly, dead bones contain carbon, hydrogen, nitrogen, and oxygen, which we can convert into gelatine; but we cannot make gelatine from carbon, hydrogen, nitrogen, and oxygen, because in the so-called dead bones there is still an impress and power of vitality remaining, which is not to be found in the elementary bodies we have named. Pursuing this argument a little further, we find that coal can be converted into humus and humic acid, and a part of the gas made from it can be turned into alcohol or whisky, all of which things come under the general designation of organic bodies, or substances whose elements have been united by organic action, and therefore they possess the characteristics by which all organized or vital compounds are distinguished from the inorganic, consequently coal still retains a portion of that vitality which originally united its elements.

The celebrated mineralogist Hauy was of opinion that "crystals might be looked upon as the flowers of minerals," and if nothing more than a poetical metaphor was intended, the simile may be allowed; for undoubtedly crystals display occasionally powers that approach very closely to those of the lowest order of organized substances; as, for instance, when we allow a drop of a concentrated solution of the hydrochlorate of ammonia to evaporate on a piece of glass whilst we regard it through a good microscope; the salt then crystallizes, and in so doing takes on a fern-leaf appearance that cannot fail to produce an impression of resemblance to life. Then again, if we dissolve two parts of nitre and three of sulphate of soda, in five parts of hot water, and having divided the solution into two portions, we put a crystal of nitre into one, and a crystal of sulphate of soda into the other, we find that as the liquid becomes cold, the two halves give crystals similar only to the crystal put into them; so that one yields nothing but nitre, and the other nothing but sulphate of soda; as if the crystals put in had a kind of vital power to make a selection in each case, although the effect is merely the result of electrical induction.

We must not therefore allow ourselves to be led astray by this semblance to vitality in substances of an inorganic nature, because in point of fact it has no kind of analogy to the condition of flowers, for these do not originate flowers, but seeds or fruit, out of which new plants capable of continued propagation can be formed, and so the process of life is preserved; nor is it entirely lost or destroyed until the elements which have been united by vitality are completely separated and resolved into their simplest condition. This fact is of considerable importance to the gas maker, since it enables him to arrange the products obtained by the destructive distillation of coal into two classes, the dead and the living: the first class may be looked upon as hopelessly lost to any other use than that of producing heat and light, and some of the substances that belong to it are utterly useless, or even worse than useless, since they require to be removed from the others, and thus create expenditure of labour and money; of this class we may mention hydrogen, carbonic oxide, carbonic acid, sulphuretted hydrogen, nitrogen, bisulphuret of carbon and steam, or the vapour of water as the most prominent. The second class furnishes us, however, with a great number of substances endowed with the characteristics of organic life, and it is in this class that we may expect to find the materials for increased discoveries and improvements. It has been stated already that from the gas alcohol may be made, and the substance which enables the gas to yield alcohol is the bicarburetted hydrogen or olefiant gas, the organic or vital quality of which has escaped destruction in the retort, so that it has been merely distilled in the ordinary sense of that word, just as alcohol or whisky is distilled without undergoing any change in the affinities of its elements.

In addition to alcohol, there is, however, another substance having organic qualities, and which indeed so strongly resembles tallow, that it is difficult by mere inspection to distinguish the one from the other. This substance may be produced from crude coal-gas, by merely passing the gas a great many times backwards and forwards through a porcelain tube heated to dull redness; or we may use for the same purpose a mixture of ten parts of olefiant gas, twenty parts of light carburetted hydrogen, and one of carbonic acid, the result of which process is that the two cold ends of the tube become gradually lined with a fatty matter having the appearance of suet or tallow, but possessing a somewhat crystalline aspect with a peculiar empyreumatic odour. Nevertheless, this artificial tallow may be made into candles, which burn like common candles, and clearly enough prove that the disposition to enter into arrangements of an organic nature still exists in both the kinds of carburetted hydrogen given off from coal, so that a large field of discovery is open to the gas manufacturer in this direction, and more especially so, when we come to consider the great number of other semi-organic bodies contained in coal gas, and by which the living class of products from coal is completed. The most important are perhaps those which contain nitrogen, as, for instance, ammonia, hydrocyanic acid, aniline, quinoline, pyrrole, picoline and petinine, or butyriac as it has been called, and (though not yet separated from gas) yet probably there is also amyliac amongst the nitrogenized products; and with these we have a corresponding list of hydrocarbons possessing different degrees of volatility, and therefore liable to condense in diffcrent parts of the street mains and service pipes; these are naphthaline, naphtha or benzoline, propyline, butyline and carbolic acid, or hydrate of phenile, as some of the French chemists call this substance. It ought, however, to be borne in mind, that although the names here given to bodies that are contained in coal gas have been given in consequence of the supposed identity of those bodies with the real substances to which the names belong, it is not absolutely certain that they are identical in every case; for it is now a well-established fact in chemistry, that two substances may have exactly the same composition, the same boiling or fusing point, and the same specific gravity, and yet be very different substances in many other respects.

On this account we think it necessary, at this stage of our labour, to give a minute description of the real substances regarding the gaseous or coal namesakes, of which any doubt exists in respect to their complete similarity; and first we will begin with aniline, although the identity of this fluid as derived from coal, and that from indigo seems quite established.

This substance appears to be produced under a great variety of circumstances, and from a great variety of azotized matters; but the best mode of obtaining it is to distil a quantity of indigo, mixed with an excess of fused potash, by which we procure an amount of pure aniline equal to about one-fifth of the indigo employed. This aniline is colourless, and gives a greasy stain to paper, which, however, gradually disappears by evaporation at a heat not exceeding the boiling point of water; but the boiling point of the aniline is about 378° Fahr. Its specific gravity is 1·028, and it dissolves readily in alcohol and ether, although but little soluble in water; it has a hot unpleasant taste, is poisonous, and when poured upon chloride of lime, or mixed with sulphuric acid and bichromate of potash, it becomes blue; or when a thin strip of fir wood is dipped into a mixture of it with hydrochloric acid, the wood takes on a bright yellow tint that even chlorine will not remove; so that we have several very satisfactory means by which to identify aniline.

The next of this class is quinoline or leukol, as it was called by its discoverer, M. Runge, in consequence of its non-production of any coloured compound, in which respect chiefly it differed from aniline; but as the same substance was afterwards obtained in a more pure form by distilling quinine with fused potash, the name was changed to quinoline. Like aniline, it is a colourless oily fluid, which has a repugnant odour, and a bitter burning taste, and is most probably poisonous. Its specific gravity is $1 \cdot 081$. It is combustible, and boils at about 460° Fahr., but does not differ from aniline in regard to its conduct with alcohol, ether, water, and the fixed and volatile oils. It produces, however, no colour with the agents which colour aniline, and its only peculiarity consists in a singular power of resisting the action of heat, for it may be passed unchanged through a red iron tube which would totally destroy aniline.

Our next compound is pyrrole, the nature of which is so volatile that we may almost call it a gas, and hence but little is known of its true character, except that it is obtained in the largest quantity by distilling bones, and that it forms a compound with hydrochloric acid, which will not crystallize, but is neutral, and may be decomposed or set free unchanged by lime or potash, when it affords an odour like that of turnips; and if a piece of fir wood wetted with hydrochloric acid is exposed to its vapour, the wood becomes tinted of a deep purple colour, which chlorine will not destroy. We may here be allowed, perhaps, to remark, that we have on more than one occasion met with gas which gave a purple tint to fir wood so prepared, and therefore seemed to indicate the presence of pyrrole, although none could be obtained by passing the gas through a mixture of hydrochloric acid and water.

We now come to picoline, a liquid which not only resembles aniline in appearance, but is identical with it in composition, and has the same combining number or atom. It differs, however, in its specific gravity, which is $0 \cdot 955$, and also in its boiling point, which is 271° Fahr.; in addition to which it is more soluble in water than aniline, and does not become blue by the action of the chloride of lime: moreover, it has never yet been obtained except in gas and gas tar.

Petinine, like pyrrole, is a very volatile compound, which, however, unlike picoline, has been obtained from several sources, as well as from gas and gas tar; thus it may be formed by acting on cyanic ether with fused potash, or by distilling at a dull red heat any kind of animal matter; or in a still purer condition than by the above methods, it may be produced from the cyanuric ether of butylic alcohol, one of the constituents of what is called potato oil, a kind of fusil oil produced by those distillers who make whisky from potatoes; and this name butylic alcohol has been given in consequence of the fact that potato oil is converted into butyric acid by the action of caustic soda upon it at a low red heat; and hence also the change of name from petinine to butyriac, a change that has been before alluded to, and may serve to illustrate the gradual manner in which knowledge is sometimes developed; for petinine was first discovered in coal tar, but has been traced back, step by step, until we now find it as a product from butter, a substance recognized as a vegetable fat that in all human probability existed in the vegetable matters from which coal was formed, thus giving us a strange proof of the tenacity of organic vitality. Petinine or butyriac is a light etherial fluid, possessing decidedly alkaline qualities, and having an ammoniacal odour, with a hot pungent taste, which remains long on the palate and is very unpleasant. Butyriac boils at a temperature of 176° Fahr., and dissolves readily in water, alcohol, and ether, forming solutions which precipitate metallic salts in general, but which form clear blue solutions with the salts of copper, exactly like those formed by ammonia; and like ammonia, the vapour of butyriac renders turmeric paper brown, and gives a cloud with hydrochloric acid, so that these indications are not a positive proof of the existence of ammonia in coal gas. As we have suggested that amyliac may occasionally be present in gas, we think it proper to give a description of that substance. It may be obtained by distilling the iodic ether of potato oil with potash, or by acting upon the iodic ether of potato oil with ammonia, in both of which cases we obtain amyliac in the form of a colourless etherial fluid of a strong ammoniacal odour and a hot taste, bearing, in fact, much resemblance to butyriac. The specific gravity of amyliac is $0 \cdot 750$, and it boils at 203° Fahr.; its vapour, like that of butyriac, renders turmeric paper brown, and gives a white cloud with hydrochloric acid, so that here again we find a possible source of error for those persons who rely upon such indications as a proof of the presence of ammonia in gas.

Having thus briefly described the azotized compounds, we will now notice those generally known as hydrocarbons; omitting, however, naphthaline or benzoline, because these are too well known to require description here.

When the fluid called fusil oil is distilled with any substance having a strong attraction for water, the oil is converted into amyline, a compound which is said to exist in some kinds of gas, and more especially in the poor kinds of gas, though we have never yet met with decisive evidence of its presence in any instance. It is a colourless mobile liquid, having a peculiar but not disagreeable odour, and it boils at about 100° Fahr., evolving a highly inflammable vapour which burns with a white flame but no smoke: this vapour is rapidly absorbed by olive oil, from which the amyline may be recovered by distillation in a water bath, and that appears to be the only test we have of its presence in gas; a test which we venture to think is very inconclusive, because it equally applies to the vapour of the next two substances, propyline and butyline. Propyline may be obtained by passing the vapour of potato oil through a red-hot tube and condensing the products by cold in the usual way. It is a light, fragrant, oily fluid, which boils at a temperature not much exceeding 96° Fahr., though the exact point has not yet been sufficiently determined; like amyline, its vapour burns with a bright flame and little or no smoke, and it is rapidly absorbed by olive oil and also by concentrated sulphuric acid, from the latter of which it may be recovered in the shape of fusil oil, by mixing the acid compound with water and distilling the mixture. As in the case of amyline, so in that of propyline, we have never met with satisfactory evidence of its existence in any sample of gas which has come under our observation; but with regard to the next compound, butyline, we constantly find it.

This compound was first obtained in a separate or pure state by Mr. Faraday, from the fluid produced by the Portable Gas Company, in consequence of the pressure employed in filling the vessels used by that Company for transporting the gas, which was made from oil, and contained therefore a large amount of hydrocarbon compounds. This substance was at one time called ditretyle, because it was found to contain four volumes of hydrogen and two of carbon condensed into one volume; but as it has since been obtained from various sources, and particularly from potato oil or butylic alcohol, the name butyline has been given to it. At ordinary temperatures this substance is gaseous, and it becomes a liquid only under pressure or below the freezing point of water; it is perhaps the lightest fluid yet discovered, for its specific gravity is only $0\cdot 625$. Its vapour is rapidly absorbed by olive oil and also by sulphuric acid, from the latter of which it cannot, however, be separated (as happens with propyline) by distillation with water; and in this respect it also differs from olefiant gas, because when sulphuric acid has absorbed olefiant gas, and we distil the mixture with water, alcohol is produced; but it may be as well for us here to remark, that, in common with olefiant gas, all the above hydrocarbons are condensed by chlorine and bromine, so that the amount of this condensation, taken alone, is no index of the value of a gas for lighting purposes.

The vapour of carbolic acid no doubt exists in gas; but the quantity must be trifling, for at no very low temperature this substance becomes a crystalline solid, and it requires a heat of 370° Fahr. to distil it, a circumstance which we may regard as providential, for carbolic acid is very poisonous, and therefore any great amount of its vapour in coal gas would prove a serious obstacle to gas lighting. When carbolic acid is saturated with ammoniacal gas, and passed through a tube heated to about the boiling point of mercury, we form a considerable amount of aniline; and when gas or even atmospheric air is charged with the vapour of carbolic acid and passed through water, the water acquires the power of coagulating albumen, and this is pprhaps our best test of the existence of carbolic acid in gas.

We regret that Mr. Thompson was prevented by unforeseen domestic circumstances from concluding this article, which contribution, like all his valuable labours in connection with gas lighting, has been of the most disinterested nature, and to whom our sincere thanks are due.

CARBURATING GAS, AIR GAS, AND WATER GAS.

VARIOUS have been the propositions and patents obtained from time to time, since the first establishment of gas lighting, with the view of superseding the ordinary process of manufacturing coal gas. Most conspicuous among these may be enumerated the processes of intermixing hydrocarbon vapours with hydrogen, or atmospheric air, to produce an illuminating compound; or by saturating rich gas with hydrogen; or enriching coal gas by an admixture of hydrocarbon vapours.

The first patent recorded which bears upon our subject, wherein hydrogen is mentioned for the purpose of illumination, is that of Ibbetson, obtained in 1824. In his specification he proposes to admit steam, intermixed with tar or oil, into a decomposing chamber, containing ignited coal or coke; but from the general description, the impression is conveyed that the patentee possessed only the most vague ideas on the subject.

Six years afterwards a patent was granted to Donovan, for "an improved method of lighting places with gas." He states, "This invention relates to the enriching those gases, which afford little light when burning, by substances which will impart to them a higher illuminating power; and may consist in bringing the gas (by preference), produced by the action of steam on coke, &c., heated to redness, into contact with the liquid, or vapour of spirit of turpentine, spirit of vegetable tar, coal naphtha, or similar substances, heated or otherwise, and employing the carbonized mixture as the illuminating agent." In this specification, therefore, we have the first description of intermixing hydrocarbon vapours with hydrogen, or other gas, in order to render it an illuminating agent; and it may be regarded as the basis of all the various processes now to be referred to.

In 1832 George Lowe patented "the means of increasing the illuminating power of coal gas as usually produced in gasworks," by "impregnating such gas with naphtha, commonly called spirit of coal tar, or with any other hydro-carbonaceous liquid by any convenient method." Here it will be observed, that the difference in the two processes consists simply in the description of gas to be operated upon; for while Donovan specified gases which afforded "little light when burning," and by preference hydrogen, Lowe proposed to enrich, or, as now termed, carburate, ordinary "coal gas."

At the epoch of their production both patents created considerable excitement, the first in particular, as it threatened to effect an entire revolution in the manufacture of gas; whilst by the second, the interests of companies appeared to be menaced through the anticipated economy to consumers to be realized by the process.

In carrying out Lowe's system on a small scale, an urn containing a sponge saturated with naphtha was connected with the supply pipe, and thus the gas in its passage became enriched by the vapour; but when applied on an extensive scale, an oblong metal box, called the "naphthalizing box," was provided. In this were a series of shallow trays, on each of which was placed a layer of wool or cotton, for the purpose of obtaining augmented surface, and so arranged, that when supplied with naphtha, the uppermost tray was first filled; and on this overflowing, the second tray was charged; hence to the others, and so charging all the vessels.

The gas entered at the bottom of the box, and passing in a zigzag direction over the whole of the area of the trays, in its transit absorbed the vapours of the naphtha, by which the gas was enriched in proportion to the degree of saturation. The process, singular as it may appear, was much favoured by the directors of the Chartered Company, and was tried on an extensive scale, among other places, at Millbank Penitentiary; but, for reasons hereafter stated, it did not meet with success, and was eventually abandoned. Since then the same system has been revived repeatedly, with all pretensions to originality, but in no case has it been successful.

The names of Mollerat, Val Marino, Cruckshank, Manby, and various others, appear also as patentees at different periods, for processes very similar to, if not identical with, those of Donovan and Lowe, neither of whom, however, made any progressive steps in the matter.

In 1847, Stephen White, who afterwards distinguished himself by his energy and ability in endeavouring to establish his process, termed the "hydrocarbon process," patented the means of "producing gas adapted for the purpose of illumination through the decomposition of water, by bringing it into contact with charcoal or coke, or small pieces of thin plates of iron at a very high temperature; and thereby producing from such combination of materials, hydrogen gas and oxide of carbon, and afterwards, in combining such compound gas with carburated hydrogen gas, produced by the destructive distillation of oil, or fat, or tar, so as to obtain a compound gas fitted for the purpose of illumination." This invention differs from the preceding, inasmuch as the object is to increase the volume of the richer gases derived from fat, tar, and oil, of course at the same time diminishing their illuminating power.

Closely following White's patent was that of Mansfield, who, if we may judge from his specification, possessed a most advanced knowledge of the subject on which he treated. In his specification he says, "This invention consists in the manufacture from bituminous matters, by acting upon them at suitable temperatures, of spirituous substances, which are so volatile that a current of atmospheric air at ordinary temperatures passed through them may, when ignited, continue to burn with a luminous flame till all or nearly all such substances are consumed." Further, "in the manner of passing a current of atmospheric air, or of other more inflammable gases with the vapour of spirituous substances, which are so volatile and of such nature that a current of such air at the ordinary temperature of the atmosphere may burn with a luminous flame at a distance from the reservoir."

This patent, therefore, describes two processes: one, the manufacture of highly volatile hydrocarbons, which does not come within our province; and the other, which merits particular attention, being the means of combining the vapours of hydrocarbons with a current of air, so producing a compound capable of being consumed and yielding light in the same manner as ordinary gas.

The practical application of this system is of great simplicity, and for the purpose an ordinary holder of ten feet, more or less, will be sufficient. A tin vessel, called the "carburator," is then provided, which may be a cylinder of about eight inches in diameter and three inches deep, having its inlet and outlet at opposite sides and near the top; but before the cover is soldered, the vessel is filled with wool or cotton in a loose state, for the purpose of presenting a large surface for evaporation. When finished, the carburator is charged with about a pint of the spirit of petroleum, or any other very light hydrocarbon, when, the inlet being connected to the holder filled with air, and the outlet in communication with the burner, the apparatus is ready for use. The air in the holder, having an ordinary pressure, is then passed through the carburator, and the "air gas" ignited, and if the spirit be moderately good, a very rich light will be obtained, and for the purpose an argand burner is the best. But by continuing the operation for some hours, gradually, as the lighter spirit evaporates, the flame diminishes materially, and eventually when the vapour ceases to issue, the light is extinguished, leaving a considerable portion of the spirit within the vessel, which does not evaporate at the ordinary temperature of the atmosphere. In practice, the holder is replaced with a motive power meter, as hereafter explained.

With a carburator of half the capacity, or even with the vessel described, by attaching an argand burner to the outlet and a flexible tube to the inlet, on a person blowing into the vessel the "air gas" may be ignited; and as it is capable of being carried from place to place, it may be regarded as the simplest of all portable gas apparatus, although at the same time rather inconvenient in its practical application. The contrivance is mentioned on account of its novelty, for utility it has none.

The discovery of Boghead Cannel, about 1849, in consequence of the extreme richness of the gas derived therefrom, afforded to Stephen White an excellent opportunity of carrying his invention into practical application, and for this object he secured the valuable assistance of Dr. Frankland and Samuel Clegg, who entered into extensive experiments in connection with the process, in order to be enabled to appreciate its merits.

According to Dr. Frankland's experiments with the hydrocarbon process, the yield of gas from Boghead, Lesmahago, and other cannels was increased about threefold; whilst the illuminating power in some cases was not materially diminished. In Clegg's report he stated, that whilst only 13,500 feet of gas was obtained from a ton of Boghead by the ordinary method, 52,000 feet of twenty-candle gas, or 75,000 feet of twelve-candle gas, was derived therefrom by the hydrocarbon process. He further estimated that the cost of gas by the new system varied from $9\frac{1}{4}d.$ to $11\frac{1}{2}d.$ per 1000 feet, according to the quality of the coal carbonized, the gas derived from the richer coals being the cheapest.

With these favourable reports from gentlemen of such eminence and unquestionable probity, it is not surprising that the innovation created the greatest excitement; and in consequence several small works and many mills abandoned the old system of making gas, by changing the retorts and other apparatus necessary in order to apply the hydrocarbon process; and among the towns where it was adopted were Ruthin, Southport, and Dunkeld. The system was also tried on an extensive scale at the gasworks of Manchester, and the South Metropolitan Gasworks in London; and the late Mr. Livesey, at one period during the investigation, expressed to the author a very favourable opinion concerning it; but gradually practice demonstrated the fallacy of the process and the difficulties attending it. By experience it was ascertained that no uniformity in the illuminating power of the gas could be maintained; for at one period, through an excess of hydrogen, the gas was deteriorated in such a remarkable manner as on some occasions hardly to yield any light whatever, whilst at other periods, through an excess of the hydrocarbon vapours, the gas passed off as smoke, with all the accompanying evils. These irregularities were fatal to the process, and gradually in those establishments where the hydrocarbon process had been applied, after sustaining severe loss, the old method was again adopted.

Twelve years after the date of Mansfield's patent, after his process had been put into practical operation, its defects ascertained, the system condemned, and, in a manner of speaking, forgotten, it was again revived by Mongruel, who patented a means of carburating air, differing only in the form of the apparatus, but in every other respect identical with Mansfield's process. The patented invention was exhibited in London under the name of "Photogenic Gas;" and eventually a company was established, with a capital of 40,000l., for the purpose of carrying it into operation. Of this capital, one half was paid to the fortunate patentee for his "invention," the remaining half dwindled away in useless experiments; and in less than three years from the formation of the "Photogenic Gas Company," it ceased to exist, with the loss of the whole of the capital.

This was followed a few years afterwards by the issue of the prospectus of the "Air Gas Light Company," with a proposed capital of 200,000l., for the purpose of "presenting to the world the manifold advantages of the system." Although it was well known that the process was not original, together with the failure of the company referred to, yet more than half the capital stipulated was raised; and during several years great efforts were made by the parties most interested to attain success. For this purpose a manufactory was established for the construction of the necessary apparatus, such as motive-power meters and carburators, and the system was practically adopted at several gentlemen's mansions; but in the course of time its defects were recognized, and the "Air Gas Light Company" shared the fate of its predecessor. It must, however, be stated that a degree of honesty was displayed by the patentee which is seldom met with under similar circumstances, as he actually invested 60,000l., the amount received for his patent, in the shares of the company.

Within the last three years an attempt has been made to add to this class of enterprises, by the establishment of the "Eupion Gas Company," with a capital of 250,000l.; the patent under which they intended to carry on their operations differing in no respect from the others. This company, however, was never formed, and the peculiar circumstances in connection with its promoters must be fresh in the memory of the majority of readers.

Hence we find that the three enterprises mentioned—the "Photogenic," the "Air," and the "Eupion" gas companies—were all based on the invention of Mansfield; and that beyond the form of the apparatus there was no difference whatever between the one and the other. They all signally failed, without having the slightest prospects of success; yet at one period, such was the mania for this process, that seldom a month passed without the patent register containing a provisional specification for "air gas," "means of carburating air," or something analogous.

Some years ago, with the view of effecting considerable economy, Lowe's process of carburating was strongly recommended by an eminent chemist, and supported by an equally eminent engineer, to be applied to the public lamps within the city of London, which was accordingly adopted. However, the experience of a few days only was sufficient to show the errors of the system, which were due to several causes, the principal of these being the great irregularity in the evaporation of the spirit, which, when first supplied, would give off its vapours in such abundance as to blacken the lantern glasses in the course of a few hours; and in a short time, as the volatile portions evaporated, the gas, in the absence of the vapours, would be consumed in its ordinary state; and as the burners were much reduced in size in order to effect the anticipated economy, under these conditions, according to the well-known law, the light of the gas was reduced in a remarkable manner. Hence, by the combined circumstances of the smoke, the irregularity of the light from the different burners, the unsightly appearance of the naphtha boxes in the lanterns, and the shadows cast by them, and the want of the promised economy, the system was abandoned.

More recently, Dr. Bowditch applied the spirit in a vessel immediately over, but situated some distance from the burner, which method possessed the advantage, that in consequence of the high temperature the greater portion of the spirit was available; but there still existed the irregularity of supply of hydrocarbon vapours.

The main causes of failure in the various systems, whether water gas, carburating ordinary gas, the hydrocarbon process, or air gas, were the irregularity of the evaporation of the spirit; the danger and inconvenience of removing the old and supplying fresh spirit necessary for the process, and lastly its expense. It may be observed, that in addition to these defects, the expense of obtaining hydrogen in a moderately pure state precluded its use in competition with coal gas.

Lowe's process was practically applied when gas was 9s. per thousand feet, and naphtha 1s. 9d. per gallon; and although every facility existed for its development, yet it was never successful, and had there been any economy in the system, it would have been employed at that period. The hydrocarbon process commenced under the most promising auspices, but failed completely, whilst the want of success with air gas was mainly due to the irregularity of evaporation mentioned, and the fact that only a portion of the spirit evaporated at the ordinary temperature of the atmosphere. Finally, in all the various methods, the expense has prevented their practical adoption in England.

It must be stated, however, that the process of carburating air and rendering this available for illumination, is employed in America with a degree of success; but it must be remembered that the price of gas in that country is very high, and the cost of the spirit of petroleum comparatively low, which circumstances admit of the application of the process. For this purpose, a vessel similar to Lowe's naphthalizing box is sometimes employed; in other instances, the spirit is placed within the vessel, which exhausts the air and gives the desired pressure. This, in its most complete form, consists of a motive-power meter, and according to the quantity of air required (which corresponds approximately with the supply of gas for a given sized burner), so will be the capacity of the wheel, the gear work, the actuating weight, and the height of fall, as already mentioned in the chapter on consumer's meters.

CONSIDERATIONS ON THE ESTABLISHMENT OF GASWORKS.

WHEN defining the magnitude of a gasworks, or that of the various apparatus employed therein, or in estimating the probable returns for any establishment of this nature, but particularly when situated abroad, there are numerous circumstances which have to be taken into consideration before any approximation to the magnitude of the plant required, or to the true state of affairs in a commercial point of view, can be arrived at.

These circumstances principally consist in the geographical position of the town to be lighted; in the number and nature of its population; the resources of the locality, either in point of industry or commerce, together with the nature of the contract made for supplying the public lights, as well as the cost price of the coal required for the production of gas; for it is assumed that the price of gas is known.

To enter, first, into the geographical position, we may observe that at or near the equator the days and nights are within a few minutes of like duration throughout the year; but as we proceed north and south, they vary considerably, according to the respective latitudes, until reaching the poles, where there exists continuous darkness or light during nearly six months of the two periods of the year; all towns, therefore, in the different latitudes at any one particular period of the year have the nights and days of varied length. Hence the table of the hours of public lighting in each month during the year, which has been reproduced from time to time in several books on gas, is only applicable to certain localities in one defined latitude; and though it may be accurate for London, it would be incorrect for Edinburgh, and still more so for Aberdeen, to say nothing of more distant places.

It must be observed, however, that in all parts where gasworks are established—whether in the tropics, where the sun rises and sets every morning and evening at six o'clock approximatively throughout the year; or at the north, where the longest night is of twenty and the shortest of four hours' duration—there is the same average of twelve hours of day and night per diem.

With these points in view, it follows that a works situated in the tropics, where the consumption of gas is nearly uniform each month, so far as regards the capacity of the plant required to supply a given quantity of gas, and the consequent capital invested therein, is placed in a much more favourable position than a works established in a more northern or southerly latitude.

In the first instance, in consequence of the uniformity of consumption, a given number of retorts are employed, with little variation, throughout the year; and these, once in operation, remain so until rendered useless by wear and tear. The holders and other apparatus are also for the same reason limited in capacity, whilst the mains, not being required to deliver a greater quantity of gas at one period of the year than another, are likewise proportionably small; therefore under all these favourable circumstances the plant and consequently the capital requisite for a works so situated are limited. On the other hand, in all the gasworks in the north, a considerable portion of the plant—at least one half of the whole of the works—is lying idle during six months of the year, and in many localities, in the height of summer, the consumption of gas per diem is not equal to the eighth part of that required in the depth of winter. Besides, in such localities an extensive portion of the plant is only brought into requisition during a very brief period, while the exigencies of the climate, the lengthened hours of the night, the use of gas as a source of warmth in dwellings, as well as for cooking purposes during the day, with the occasional uncertainty of daylight arising from fogs; all tend, at times, to try the capacity of the most complete works. With these differences duly considered, the geographical position of the town to be supplied with gas becomes one of great consideration.

When designing gasworks for a foreign city or town, the numerical population has frequently been the guide of the engineer; and this without having the opportunity of visiting the locality, or possessing any reliable information concerning the commerce or industry of the inhabitants, and in some instances a foreign population has been compared with one of a like number of the United Kingdom. This, however, as experience has proved, is very erroneous; for, with the exception of a few European cities and towns, and some of those of the United States, no fair comparison can be made between the one and the other for the purpose of gas lighting. Nor can this be surprising, when we consider the comparatively low price of gas at home, the industrial character and the comparative wealth of our countrymen, and the luxurious establishments existing in many cities of the Continent and United States.

A populous town or city in some foreign localities may be largely composed of straw or mud huts, inhabited by the most unproductive, and consequently the poorest, classes, whose wants and necessities are restricted within the most narrow limits. These people, although entering into the number of the population, do little or nothing towards enriching the town, and a single manufactory, such as may be found by dozens in the vicinity of Manchester, would be more productive for a gas company than a town containing thousands of those idlers which constitute the population of many parts of the world.

Numerical population, therefore, goes for little without a personal knowledge of the town to be lighted, the number and class of shops, public buildings, manufactories, and private dwellings suitable for the adoption of gas. As a proof of this assertion, we may state that a foreign town once came under our notice, reputed to have a population of 16,000 inhabitants, with a quay for discharging vessels of 300 tons; and although it was the capital of the province and the port for receiving the products of all the neighbouring country, and was of necessity the mart to supply their wants, yet in the whole of the place there was not a manufactory of any description, and it was proved, upon close investigation, that the small number of 200 private lights alone could not be calculated upon.

The next consideration is the nature and conditions of the contract for supplying the public lights of a town, which are much more variable than is generally known. In most cities and large towns the contract is made for lighting the public lamps from sunset to sunrise every night throughout the year, averaging twelve hours per diem, or about 4000 hours per annum, which is the most favourable condition for a gas company, as their mains are thus employed continuously. In other places, however, the public lighting is suspended on the nights of full moon, and two or three nights before and after, according to arrangement, the rest of the month the town being illuminated during the whole of the night. By another method, in addition to suspending the lighting on moonlight nights, the lamps are extinguished at a certain hour, say twelve o'clock in summer and eleven o'clock in winter, and in some cases the lighting is entirely suspended during the summer months.

Finally, another class of contract for public lighting, although more profitable than the last, is certainly more onerous for the gas company. By this arrangement advantage is taken of every hour of moonlight, or at least when the moon should appear, and the lighting is either continued to a certain hour, or throughout the night. In the latter case, on the night of full moon and the two preceding and following, the street lamps are not lighted at all; but on the third night, in the event of that luminary appearing an hour after sunset, the public lamps are lighted for that period and extinguished on the rising of the moon. The fourth night the lighting is extended about three-quarters of an hour, and so on, progressing each night until the moon has disappeared, when the town is illuminated the whole of the night during five nights. Advantage is then taken of the new moon, and the lights are extinguished as it rises, advancing about three-quarters of an hour each night.

By this system of contract the hours of lighting are reduced to about an average of six hours per night, or 2000 hours per annum. But against this supposed economy, the alleged moonlight nights are often of the darkest, for even in the tropics the moon is no more to be depended upon at certain periods of the year than the sun in other less favoured climates; hence, under this arrangement, a town is often left in complete darkness for nights together. Besides, the system naturally gives rise to great inconvenience by the irregularity of the nightly consumption and the difficulty of enforcing the proper fulfilment of the lamplighters' duties, it moreover renders a large gas storage indispensable in order to continue the operations of manufacture uniformly during the period that the public lighting is not required, as well as to provide for the extraordinary demand for supplying the successive nights of illumination. The system last referred to is adopted in many towns abroad.

The other considerations mentioned, viz. the price of coal and the cost of transport of all material entering into the erection of gasworks, as well as the customs duties levied on these commodities, become also of serious importance. There are many inland towns having roads which are quite impracticable for the transit of heavy goods, and in which therefore the cost of coal is greatly enhanced. Again, there are towns in Europe where the duty on coal is equal to the cost of the material itself delivered at works in London; and in these places gas can only be supplied at a price at which the modern paraffin lamp comes into formidable competition with it.

Another important controlling influence is the nature of the contract entered into by the *concessionnaire* when obtaining the exclusive privilege for lighting any foreign town, and it is to be regretted that some of these engagements are of the most onerous character for a gas company. In some instances within our knowledge, where the undertaking is carried out with British capital, the most tyrannical terms exist, such as exacting a fine of double the nightly value of the gas, that is, the price paid per burner per night, for every public lamp found extinguished, no matter from whatever cause it may arise; also a flue of the full value of the gas is exacted for every "dim light." Moreover, the right of deciding these questions is invested in a government official, whose interest it is to obtain the gas at the lowest possible rate, and against whose decision there is no appeal, and often an arrogant employé of this description will boast of the economy effected, or in other words, the revenue derived from the gas company by those means. Under these circumstances, the only course to adopt in order to avoid serious losses, is to empower the manager to arrange in a "friendly spirit" with the government representative; when by such an arrangement a marked and surprising change will be experienced in the public lighting, and fines will become almost unknown. This is our experience, which can be confirmed by many managers abroad, and should teach a lesson of prudence in entering into contracts of this nature, and with such authorities.

Other sources of loss to a company consist in the low price at which the public lighting is taken, or the conditions which are sometimes accepted; such as, that the public buildings of the town shall be lighted gratuitously; or that on certain nights of the year, as those of general holidays, the supply to the whole of the public lights shall be gratuitous. But still worse than this, as it often happens, perhaps, no proper guarantee exists for payment of the public lighting when due; and in the absence of which, the authorities of the town, after being indebted to the gas company for one or two years' supply, on this being discontinued in consequence of non-payment, a second company is invited to oppose the former, which in due time is treated in the same fraudulent manner. This unfortunately is too often repeated, to the great prejudice of the capital of this country; and we can name a city where there are three contending companies, each of which in its turn has been subjected to this system of fraud as practised by local governments, and for which there is no remedy.

In concluding these remarks, we express our belief that sufficient has been adduced to prove that no defined rule can be observed in the construction of the various parts of a gasworks, without a thorough knowledge of all the details connected therewith to which reference has been made, and that more care should be observed when contracting with the authorities of any foreign town; and above all, that proper and substantial guarantees for payment should be obtained.

From the preceding observations, it is evident that the capacity of a works, or the various parts of which it is composed, can only be decided upon after due consideration of all the several questions referred to. Thus, for cities and towns where the public lighting is continued nightly throughout the year, the gasholders or storage will be required of less capacity than when the public lighting is irregular.

On the other hand, where the public lighting is irregular, and particularly when every hour of moonlight is taken advantage of as described, the storage or holders will be required equal to about three nights' consumption, in order to meet the demands of the five consecutive nights of illumination; as also to provide for the storage of the gas during the moonlight nights, and permit of the operations of manufacture being continuous. Therefore, only on understanding the various contingencies connected with the town to be lighted, can a proper estimate of the capacity of the various apparatus be formed, and the knowledge acquired of the necessary capital to be invested.

In the works of the metropolis as well as other large establishments in the United Kingdom, the greatest number of retorts in use and the storage provided are about equal to the demand for gas during the longest night and shortest day, that is, for twenty-four hours' maximum consumption. But it sometimes happens that the prevalence of foggy weather during several hours taxes the resources of these works to the utmost.

The capacity of the coal stores of a gasworks will be entirely dependent on its position and the facilities of obtaining that material. In the event of the works being situated in the neighbourhood of the mines, the stores may be limited to those in the retort house, as hereafter described, so long as immunity from strikes can be assured; but as the difficulties of obtaining a regular supply increase, so must the stores be augmented in proportion. In distant localities, where the transport of coals occupies weeks or months, a continuous and regular supply should be maintained; for, as already shown, any slight advantage that may be derived from the occasional reduction of freight is in some cases more than counterbalanced by the deterioration of the coal. Another evil attending the endeavours to obtain low freights is, that at times the manager is left without that material, when he is driven to the last extreme, having often to purchase at a fabulous price whatever he can meet with, in order to continue the demand for gas.

In some contracts for lighting foreign cities and towns, which have come under our notice, the company is bound to keep a supply of coal sufficient for their requirements for a lengthened period, in some cases of six months; the obligation is very onerous, but is seldom complied with; and if it were, would certainly be highly prejudicial, on account of the deterioration of the coal; and the best system is to maintain a regular and uniform supply as recommended.

GASWORKS OF VARIOUS MAGNITUDES.

THE smallest of works, in a practical sense, are the experimental apparatus in use at most establishments of large and medium capacity, by which the engineer or manager is enabled to ascertain in a very short time the quantity and quality of the gas, as well as the coke and other residuals to be obtained from any particular coal or material, and thus be in the position to determine its commercial value. This apparatus is employed at some large works to test every cargo of coal delivered, thus preventing the possibility of an inferior quality being substituted without detection. Although desirable in all establishments, the experimental apparatus is absolutely indispensable in many cases, as in gasworks situated abroad, where it may be necessary for the manager to make a hasty trial, in order to be enabled to form a decision respecting the value of coals, perhaps forming the cargo of a passing vessel lying outside the port or harbour, and probably arriving very opportunely when the stock is nearly exhausted. Under these circumstances, when a decision has to be formed in the course of two or three hours, with the apparatus in question a sample of the coal can be carbonized, and the quantity and quality of the gas and coke promptly ascertained; and if the price is acceptable, the purchase may be effected without the slightest fear of incurring risk or loss.

It also often happens that the gas manager abroad has specimens of minerals, as bituminous schist obtained in the locality, brought under his notice, when, in the interests of the company he represents, with the view that the mineral might be advantageously employed for the production of gas, it becomes his duty to investigate its nature, and for this purpose the experimental apparatus is indispensable. A trial made with this will make known at once the true value of the mineral, and so put an end to the extravagant recommendations which sometimes accompany these samples.

The apparatus in question consists of all the usual parts essential for an ordinary gasworks, namely, retort and setting, ascension pipe, condenser, dip pipe, purifier, meter, and holder, all of which should adjoin the photometer room, in which may be the appliances for ascertaining the quality of the gas produced.

The retort used for the purpose is always of cast iron, and for the sake of simplicity of calculation, is made to carbonize a quantity of coal equal to the multiple of a ton, as 2·24 pounds the thousandth part, or 22·4 pounds the hundredth part of a ton; thus, by adding the corresponding zeros to the quantity of gas produced, the yield per ton is at once ascertained. The retort is usually made cylindrical, the smaller kind being about 16 inches long and 5 inches in diameter, having two lugs cast on the end, to which the arms are attached, the lid and cross bar and screw being similar to those used in larger apparatus.

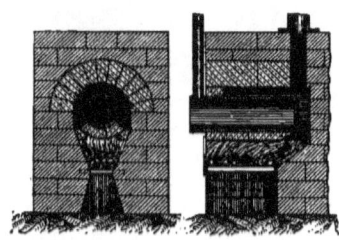

The accompanying diagram represents the retort as usually employed, set in an oven, similar to Murdoch's first trials. Beneath it is protected by a guard tile, and a piece of flange pipe constitutes the chimney to the oven. A wrought-iron pipe of 1¼ inch diameter forms the ascension pipe, which at about three feet above the retort is bent at right angles, and becomes part of the condenser. This is made of four or five 1-inch pipes, each about six feet long, with corresponding semicircular bends, forming a zig-zag, all deviating slightly from the horizontal position, in order to allow the products of condensation to flow into the receiver, which is formed by a short piece of barrel with a tap at the lower end; and as one of the objects is to bring the retort speedily to the required temperature, the furnace is made comparatively of large dimensions. A point of the utmost importance with this apparatus is, that the retort shall be properly heated, otherwise, for the reasons stated, the full quantity of gas will not be derived from the coal; it is also essential that a small portion of various pieces should be chosen for the purpose of having a fair sample of the coal under investigation; at the same time it should be observed that the quantity of gas obtained by experiment is invariably greater than that produced in practice. By preference the dry-lime purifier is employed, and consists of a tin-plate or sheet-iron box, similar to the ordinary apparatus; this is about 18 inches square, and having two or three tiers of pierced sheet-iron grids.

The gasholder has a capacity of about 20 cubic feet, and not being required of the same precision as the instruments for testing meters, this with its tank can be manufactured by any local tinman. The holder is suspended by a cord passing over suitable rollers affixed to a beam and counterbalance weights. For the meter, a five-light wet meter, with large dial, or other means to indicate the hundredth parts of a foot, will be required. Or in the absence of this the holder can with ordinary intelligence be graduated. However, all these apparatus as usually manufactured are much more suitable for the purpose.

With this apparatus the gas manager is enabled to assure himself of the quality of any kind of coal, and of the variable effects of different temperatures below that of cherry red in the production of gas. When employing it a portion of the coal or other mineral should be first distilled, and its gas allowed to pass through the apparatus and blown off from the holder, so as to get rid of all the atmospheric air, the presence of which, as already stated, would materially affect the illuminating power of the gas produced, and thus mislead the operator.

For larger works the retort should be capable of containing $22 \cdot 4$ lbs. of coal, and be about 4 feet long and 9 inches diameter, all the rest of the apparatus corresponding. In setting iron retorts, even in these small apparatus, it is necessary to protect the lower part from the direct action of the fire, by placing a fireclay slab immediately underneath, as shown in the preceding diagram; it is also absolutely imperative to leave a space, depending on the length of retort, between the end of that retort and the wall, to allow of the expansion of the iron; and it must be remembered that, when expanded by a high degree of heat, iron never contracts to its former length in cooling, so that at each time of heating and cooling the retort is permanently expanded in a slight degree.

Next in the order of magnitude are the gasworks employed for lighting schools, gentlemen's houses, or establishments of similar nature or magnitude, when situated beyond the district of any public gasworks. The plant and operations of works of this kind differ materially from larger establishments, where the stokers are employed continuously for the production of the necessary supply, and the various apparatus are constructed proportionately. For localities like these under consideration, in which we may suppose the consumption to be about 300 or 400 cubic feet, supplying twenty or thirty burners an average of four hours per night, it is undoubtedly more convenient and economical to manufacture the gas at intervals of eight or ten days; and for this purpose the holder would be required of corresponding capacity, and the rest of the apparatus in proportion, and naturally the capital invested will be large for so small a consumption. Nevertheless, there can be no doubt that this is the proper manner of supplying small establishments, and the economy effected by the saving of fuel and labour will be sufficient to pay a good interest on the capital invested, to say nothing about avoiding the inconvenience of a small apparatus by which gas has to be manufactured every second or third day, as often practised, and the increased chances of being without a proper supply of that indispensable article. The only objection that can be raised against this system is the possible deterioration of the gas by storage, which, in our opinion, for so short a period, is not important.

Should economy not be a consideration, these kinds of apparatus may be rendered ornamental to suit the premises wherein they are placed; or an ornamental structure, corresponding with the style of the adjacent buildings, may enclose the whole of the different parts, as the retort house, purifying and lime shed, and coal store; so that all can be hidden from the dwelling if necessary. In most localities the gasholder will be regarded as ornamental, and may be exposed to the general view; but should this be objectionable, it then becomes simply a question of slightly augmenting the expense, by enclosing it within the building referred to.

SMALL GASWORKS.

We now pass on to gasworks for supplying manufactories, where the operations are carried on with a degree of regularity similar to public gasworks, although, in most instances, only during a portion of the year. However, we may observe that there are many manufactories, and even private business establishments in the metropolis, where the yearly consumption of gas is equal to that of some towns of 10,000, or even 15,000 inhabitants; hence, so far as regards magnitude, no difference exists between these and the works of many public companies.

For supplying a large manufactory with gas, the principal part of the capital is invested in the apparatus, little being required for mains within the premises; and it would be supposed from these favourable circumstances, that in such establishments the gas could be procured at a cheaper rate than when supplied by the company, but such is rarely the case. This is easily explained, for it is not to be supposed that the various points of excellence in a company's works, with the engineer's experience, can be attained by the attendant in charge of a small place, who is often engaged, in the first instance, without the slightest knowledge on the subject. An example of this kind recently existed at a large manufactory in the eastern suburbs of London, where the production of tar was at least three times greater than would be necessary with ordinary good working; and, of course, the loss accruing can be readily understood.

Some of our extensive manufacturing establishments are constructed with that degree of taste and elegance as to require the gasworks to be designed accordingly, when they become a prominent point of attraction; while in other instances the opposite of this is the case, and the gas apparatus is placed in any available corner in the most remote part of the establishment in order that it may be out of sight. Therefore, with these two diametrically opposite views, it is obvious that only an intimate knowledge of all the various circumstances can enable the engineer to form a correct judgment as to the requirements of any locality.

In small gas apparatus, where the process of manufacture is intermittent, and sometimes at intervals of several days, or weeks, or months, in consequence of the capacity of iron to withstand the effects of varied temperature without injury, retorts of that material are always adopted. We have found that by coating their exterior to a thickness of $\frac{1}{4}$ inch with fireclay in a semi-fluid or plastic state, which is afterwards allowed to dry before the retort is set, the formation of the carburet of iron on the external surface is very much impeded, which system can be recommended wherever iron retorts are used. When treating on iron retorts, other observations respecting their usage have been already made.

When retorts are charged at intervals of two or three days, a good system to be observed in order to preserve the heat is to "bank up the fire," or well charge the furnace with coke, to shut the damper and the entrance to the ash pan, and to leave the incandescent coke in the retort; by these means the heat will be retained a considerable time, mainly depending on the methods adopted for the prevention of radiation from the oven. Works for large manufactories so closely resemble those of public companies that any separate reference to them is unnecessary.

Such has been the progress of gas lighting in the United Kingdom, that there are small towns and villages, with populations ranging from 300 to 400 persons, which are supplied with gas from works having mains especially provided for the purpose. The convenience which this affords to the inhabitants cannot be questioned; but unless the most favourable conditions exist for procuring coal, these enterprises are not often commercially successful. An illustration of this fact occurs at the present moment in the case of a small English inland market-town, with a population of about a thousand inhabitants, having its own gasworks and mains, with a capital of 1500*l*., and a gross gas rental of 260*l*. per annum, the price of gas being 6s. per 1000 feet. The company during two or three years carried on their operations at a loss, when it became a question of stopping the works, and to prevent this a gentleman of the locality, having other occupations, took them on lease for a period; but, with all his business capacity and energy, they could not be made to pay. Nor can this be surprising, when we learn that the coal for carbonizing cost more than half the yearly income, and that no coke was obtained for sale, and that out of the other part of the revenue (125*l*.), the wear and tear had to be provided for;

rent and taxes and wages had to be paid; the collector and secretary (for these offices existed, although in the same person) had to be recompensed, and the various petty expenses to be met. Under these circumstances, it cannot be wondered that an establishment on such a limited scale should be unproductive.

The price of coal in the locality referred to is 24s. per ton; but if this were obtained at 10s. or 12s., at which it is sold in many towns in England, then the affair would pay a very moderate dividend. Hence works of the most limited capacity are mainly controlled by the cost of the first material.

In an establishment of this description the whole of the laborious duties devolve upon one man, who, perhaps assisted by his wife, or a boy, to keep up the fire in his absence, does the duty of stoker both night and day, and besides fulfilling the other offices attached to the works, he attends to all complaints, executes all repairs, lays all services, lights and extinguishes, as well as cleans, the glasses of the few public lamps supplied, and perhaps is the only gasfitter in the locality. Such are the duties of the gas manager of the most unpretending class, and under these varied circumstances his position is not to be envied, the more particularly as his remuneration is generally in the inverse proportion to his duties.

But as gasworks increase in magnitude under the average conditions of price of coal and gas, so do the prospects of deriving profit therefrom improve. In all cases, however, where a manager can render a works of a production of a million and a half or two million feet per annum satisfactory to the shareholders, he may be congratulated on his success. With coals at a moderate price, and an average amount of capital invested in the undertaking, and a fair charge for gas, a works producing three million feet per annum conducted with ordinary care and intelligence should always yield a moderate profit.

Confining our observations to the gasworks of England, the average yearly consumption in small towns is about 1500 cubic feet per head of the population. In manufacturing towns this is materially exceeded, and in the metropolis it increases to nearly 4000 feet for each inhabitant; the production of the Metropolitan Gasworks in 1875 being nearly fifteen million of thousands of cubic feet of gas. The mains in small towns average one mile in length per thousand inhabitants, but in London the mileage is considerably less than this, which difference is compensated by their large diameter.

Hence, from the preceding observations, we learn that a works of three million feet per annum will be ample for an ordinary town having a population of 2000 inhabitants; and as in practice small works do not obtain so large a yield of gas per ton as larger establishments, for such an establishment about 330 tons of coal will be required per annum, or producing little more than 9000 feet of gas per ton, which comparative small yield arises from the circumstance, that by reason of the limited demand for coal the company is not always placed in the position to choose the best quality; another cause is, that in consequence of the protracted storage, the coal, as already observed, becomes deteriorated. The maximum consumption in a town like that in question will be about 12,000 feet, and the minimum 5000 feet per diem; but this as well as the gasholder storage will depend mainly upon the nature of the contract for supplying the public lights. In small works, however, large holders offer great facilities for working the retorts, more particularly when passing from one period of the year to another, as when entering on the "heavy lighting," or the reverse.

Here we may observe that in designing gasworks there are some general rules to be considered which should be strictly adhered to—as, for example, in order to avoid unnecessary labour in barrowing, the coal stores should immediately adjoin the retort house, and they should also be so situated as to give the greatest facility for cartage or other means of bringing the coal into the works. The coke-spreading floor in small works is usually in the yard near the retort house, and sometimes in the coal store; and, like the retort house, both these places should be paved, either with firebricks or good hard stock bricks. The purifiers and tar well, as a matter of precaution, are placed at a considerable distance from the retort house, and free ventilation should always be provided for the former.

The station meter, governor, and exhauster, when the works are of sufficient capacity for this apparatus, should all be readily accessible, without exposing the man in charge unnecessarily to the influence of the variable weather.

The roof of the retort house should be made with iron principals and covered with slates. In small works, a very common error is to cover this with sheet iron, and sometimes corrugated for the purpose of diminishing the thickness of the sheets, which on account of the various gases and steam existing in a retort house is very speedily destroyed, and ought not to be used, more particularly as the price of slates is very moderate, these being, when fixed in the neighbourhood of London, about 40s. per square, or 100 square feet.

The gasholders may be placed at a distance from the rest of the apparatus, depending on the conformation of the ground.

When treating on gasholder tanks (page 172), the method of estimating the quantity of brickwork was given, which is also applicable to such buildings as retort houses, coal stores, and other edifices in a works. In the measurement of these structures, the whole area of the walls, including the doors and windows, is estimated, unless otherwise specified; it being considered that the extra labour occasioned by these, together with the cost of centres, and other expenses, counterbalance the material and labour saved by the openings. In these cases, however, the style of building and the height and thickness of the walls are considered when calculating the price per rod, which will vary from 12*l*. to 16*l*., according to circumstances, particularly the cost of building materials.

In all places, no matter where, the building material generally employed in the locality should be adopted, whether it be brick, rubble, sandstone, or other kinds of stone; and although the bricks made in the neighbourhood may vary considerably in size and quality with those the engineer has been accustomed to, it would be unwise to endeavour to make any change, as such a course would be attended with considerable loss and vexation. Where rubble is employed, the resident workmen are, as a rule, skilful in building with it; and although soft sandstone does not appear very suitable for the buildings of a gasworks, the engineer will at times find this the only material to be procured for the purpose, and he must, therefore, employ it.

The drawing in Plate 20 represents the elevation and plan of a gasworks suitable for the production of from 3,000,000 to 4,000,000 cubic feet per annum, allowing for considerable extension of business by the addition of an extra gasholder. The retort house has four settings, of which there will always be retorts in reserve, in the event of an accident with those in use.

The coal store is of large capacity, for the double purpose of employing a portion of it for a spreading floor where the coke is quenched and afterwards stacked; also for allowing for the extension of the retort house at any future period, when by extended business it may become desirable; and by having the spreading floor immediately adjoining the retort house, unnecessary labour in transporting the red-hot coke is avoided. An important feature in the plan is that the whole of the night duty can be conducted by the man in charge, without him being exposed to the weather. Thus the coal store, the spreading floor, the meter, the governor, and even the purifiers, are under cover, and accessible without the necessity of going out of doors.

Should it be considered desirable to employ an exhauster, which will be controlled entirely by the amount of pressure existing on the works, the locality occupied by the meter will be well suited for the boiler and other apparatus, whether a mechanical or steam-jet exhauster be employed; under such conditions the meter and governor may be placed in the situation designated for general stores.

According to established rules, the grid surface of the purifiers in action, for the maximum production of the work under consideration, would be equal to 80 square feet; this, however, in the plan, for various reasons, is considerably increased. Firstly, because, in a small work, where all the duties have to be performed by one person, labour, so far as practicable, should be reduced to the minimum consistent with a moderate expenditure of capital, and for this purpose the purifiers should be of a large capacity, thus enabling the vessels to pass several days without changing the purifying material. Secondly, by having purifiers and connections of ample capacity, the back pressure is reduced, and the deposit of carbon in the retorts diminished accordingly. And, lastly, with large purifiers the gas is more readily purified, and the expenses of alterations occasioned by augmented business will be deferred.

The general store is for bulky or heavy goods, as lanterns or brackets. The lime store is indicated at the point most suitable for the purpose; the space in the yard opposite being the spreading floor for the oxide, should that material be used. The tar well is in the most isolated place, in order that any nuisance arising therefrom may be confined to the locality.

The cottage for the foreman is situated at the entrance, in order that all material entering or leaving the works shall come under his immediate notice. Adjoining this is the office, which may be further employed as a testing room, for ascertaining the illuminating power or purity of the gas. The store room adjacent is for articles of extra value, as lead and iron pipes, fittings, taps, brass, and similar articles.

The gasholder is 36 feet in diameter and 15 feet deep, capable of containing 15,000 feet of gas, and the site for a second holder of the same dimensions is provided whenever that may be desirable, which can be constructed without interfering with the other buildings. The pipes of communication between the various apparatus are represented in dotted lines.

The following may be considered an average result from a works producing 4,000,000 feet per annum, making 9250 feet per ton, coal at 22s. per ton, the price of gas being 5s. 6d. per 1000 feet, and loss by leakage 12 per cent. The capital estimated at 4500l.

Expenditure.

	£	s.	d.
432 tons of coal at 22s.	475	4	0
Foreman	75	0	0
Stoker	54	12	0
Wear and tear 2d. per 1000 feet	33	6	8
Rent of ground, and taxes	24	0	0
Laying services and repairs, fixing meters, lanterns, and repairs to tools	16	0	0
Purifying materials	12	9	9
Sundry expenses, stationery, &c.	6	10	0
Collector and secretary	25	0	0
Bad debts 1d. per 1000	15	0	0
Carried to reserve fund	64	12	7
Interest 5 per cent. on capital	225	0	0
Total	1026	15	0

Revenue.

	£	s.	d.
3,520,000 feet of gas sold at 5s. 6d.	968	0	0
50 chaldrons of coke at 16s.	40	0	0
3000 gallons of tar at 1½d.	18	15	0
Total	1026	15	0

Beyond all question there is no country in the world where, considering the number of inhabitants, the use of gas is so extensively applied as in Scotland, for there are to be found several gas companies for supplying localities possessing less than four hundred persons; and one place, having a population of one hundred and fifty individuals, can boast of having a limited company to supply it with gas, the annual consumption of the locality being about 334,000 feet, or about 2000 feet of gas per head per annum, which appears to be the average of towns throughout Scotland; but as the illuminating power of the gas is nearly double that supplied by the Metropolitan Works, for the light obtained, the requirements of that country are, for the population, equal to those of London, and far in excess of any English provincial town within our knowledge.

As may be imagined, on account of the isolated position and limited description of some of the works, together with the high price of coal and other disadvantages, the price of gas in Scotland is very variable, ranging from 4s. 2d. to a general average of 5s. 10d. per 1000 feet. There are other localities where a higher price is charged, and in a very few instances exceeding 11s. per 1000 feet, the average illuminating power of the gas being twenty-five candles. The contracts of some of these companies are somewhat remarkable, for in certain places the whole or a portion of the public lighting is supplied by the gas company gratuitously; and in one instance, such is the extent of the exaction, that the gas company is bound to light the town clock without payment. At one period, when the privilege of establishing a gasworks in a town was considered equivalent to the discovery of an *El Dorado*, the imposition and acceptance of such terms were not surprising; but at the present time, with the stern reality of a gasworks before us, and with our knowledge of the difficulty experienced in many small establishments of earning a modest dividend, it is astonishing that such conditions should be retained, and more particularly in Scotland, where the inhabitants are such thoroughly practical people.

In consequence of the diminished percentage of fuel for carbonizing, the reduction of labour in the production of gas, with proportionate amount of wear and tear, and other advantages possessed by augmented operations, works become more profitable, and permit of gas being sold at a more moderate price. Thus, when based on the grounds already assumed in English works producing 10,000,000 feet yearly, the average price of gas is 5s. per thousand feet; and in other works, producing 25,000,000 feet per annum, it is reduced to 4s. 6d. per thousand. But of course these circumstances are controlled principally by the extent of the capital invested, for there are instances where, by extraordinary circumstances, the capital is so extravagantly high, as to be out of proportion to the capacity of the works, when, necessarily, a corresponding high price is required to be charged in order to realize dividends for the shareholders.

When works reach the capacity of a production of 50,000,000 or 60,000,000 feet per annum, or a maximum make of 250,000 or 300,000 feet per diem, they possess the same facilities for economical working, so far as regards carbonization and purification, as larger establishments. The plan and elevation of a works approaching this magnitude is represented in Plate 21, as erected to supply a Continental town, the principal defect consisting in the limited site, not permitting of further extension, which arose from the continued growth of the company's business, until eventually the works, after repeated alterations, assumed their present form, and may be regarded as a good specimen of works of this magnitude. The premises are situated at some distance from water communication, so that all the coals have to be carted.

These works consist of a retort house 80 feet long and 55 feet wide, having sixteen benches of single retorts, with a passage in the centre communicating with both sides of the building, which may be objected to by some engineers on account of the loss of space that might be occupied by the retort beds; however, in this case it was rendered imperative by the form and position of the ground. The coal store is capable of containing about 600 tons. The condenser is formed by seventy-two 12-inch pipes, with top and bottom boxes, which at the present time could be advantageously substituted by a Graham's condenser. The scrubber is, according to the most modern ideas, too small, being 30 feet high and 6 feet in diameter. The purifiers are four in number, each 12 feet square internal dimensions, with five tiers of wood grids; the connections are 12-inch, controlled by central slide valve. The shaft is 60 feet high, with a clear opening of 10 square feet throughout.

The boilers, engines, exhausters, and pumps are in duplicate; thus, in the event of any accident occurring to one, another is in reserve. The liquor and tar tanks are underground. The lime shed and workshop are represented, the space opposite being employed for the revivification of the oxide. As in many Continental works, the meter and governor are placed in a separate building. The gasholders are each 80 feet in diameter and 24 feet deep, containing 120,000 feet.

The tanks are of stone, this being the building material of the locality. The offices of the works, the manager's dwelling, laboratory, general store-room, and porter's lodge, are all represented in the plan.

Having at our command an estimate furnished by a firm of respectability for the material composing the apparatus for a works of this magnitude, the holders having cast-iron tanks; a summary of it is annexed:

Estimate.

Two gasholders, each capable of containing 120,000 cubic feet, constructed in such a manner that they may be hereafter rendered telescopic, with cast-iron tanks, columns, girders, counterbalance weights, chains, chairs, connections, and valves, all complete, at 4600*l.* each	£9200 0 0
Eighty clay retorts, with their corresponding settings, mouthpieces, lids, and fittings, ascension and dip pipes, hydraulic main, furnace doors, bars, buckstaves, all complete	1400 0 0
Condenser of seventy-two 12-inch pipes, with top and bottom boxes, complete	295 0 0
Four dry-lime purifiers, each 12 feet square, with connections and lifting apparatus	620 0 0
Scrubber, 30 feet high and 6 feet in diameter	230 0 0
Two engines and exhaustors	250 0 0
Two boilers	110 0 0
Station meter	150 0 0
Governor	160 0 0
Iron roof for retort house	610 0 0
Slates for same	88 0 0
Pipes, valves, and connections on works	187 0 0
Weigh-bridge	70 0 0
Barrows, tools, rakes, &c.	45 0 0
Total	£13,415 0 0

The whole delivered free on board, or at a railway station, and labour and materials being provided for setting the retorts and boilers in the United Kingdom, for the price stated.

We find in the best designed of modern gas establishments, that in many instances the coal is conveyed by rail direct from the mines into the retort house, which is constructed of such ample capacity as to serve the purpose of a store. The coal is shot from the railway trucks, and deposited in the retort house just opposite the retort beds, ready for use; by this means, after leaving the mine, in some instances, as at Newcastle, the coal passes only through the hands of the stoker to the retort, and the expenses of labour are reduced to a minimum. A further advantage of this system of operation is, that, by the temperature of the locality, a portion of the humidity existing in the coals is expelled, rendering them more suitable for carbonizing.

An excellent specimen of this method of construction is represented in Plate 23 and the three following at the end of the volume, which are descriptive of the retort house at the Redheugh Gasworks, at Newcastle-on-Tyne, designed and erected by Mr. Wyatt, the constructing engineer of the Beckton Works, already mentioned. Plate 23 is the elevation of the retort house, remarkable for the excellence of design, which in execution was attended with but little increase of expenditure over the old-fashioned unsightly buildings formerly adopted for gasworks. This will be understood when it is stated that the piers, which have a prominent appearance, are the only parts possessing an extra degree of thickness, whilst the large surface of intervening space, and panelling, is considerably thinner than the walls of an ordinary structure. These points, together with the judicious use of terra-cotta and coloured bricks, produce an elegant building at a cost little superior to the meanest class of a corresponding capacity. Plate 25 is the end elevation of the retort house on a smaller scale.

As represented, there is a central shaft, with two ventilating shafts; by these and the windows perfect ventilation is assured. The entrances to the firing stage are reached by the two inclines, with railings, beneath which are entrances to the coke-hole; in addition there are two other entrances, as represented. The end elevation on a smaller scale is shown in Plate 25, having several entrances to the coke-hole, an incline and two entrances to the firing stage, and the openings for the two railways by which the coal is conveyed to the retorts.

The interior of the retort house is represented in longitudinal elevation, with part of the retort stack in section, in Plate 24, which conveys an excellent idea of the general arrangement. It will be observed that the passage of communication between the two sides of the building is immediately beneath the stack enclosing the inner shaft, which is in communication with the retort settings; the space outside being for the purpose of ventilation. By the part of retort benches in section the mode of building is clearly explained.

On referring to the cross sections of retort house, Plate 26, the method of forming the firing stage, railways, and roof, is readily understood. For these purposes there are twelve cast-iron columns, each cast in one piece, fixed on both sides of the building, which, with the small intermediate pillars, support the girders of the firing stage, this being composed of cast-iron plates. These columns, with the assistance of the outer walls, also carry the railways; in addition to which, a large portion of the weight of the roof is supported by them, and as its span is in consequence diminished by about 22 feet, the roof is constructed proportionately light. The lower lattice-girders support the railway, whilst the upper ones carry the intervening principals of the roof in Plate 26, of which the main ones are trussed, as represented. In order to show the importance of the space in front of the retorts as a coal store, it may be stated that it is capable of containing about 50 tons of coal in front of each bed, sufficient for upwards of a fortnight's operations. With these remarkable facilities it can be easily imagined that in certain localities no other storage for coals than this is desirable, hence by these means effecting considerable economy in avoiding the expenses of loading and unloading, carriage, trimming, and barrowing the coals, all of which must speedily repay the capital invested.

The whole of the structure is worthy of admiration, alike for its simplicity and excellence of design, and reflects the highest credit on the engineer.

From works of medium capacity we pass to those of the first magnitude, of which there are but few in number, and the most important of these on every account are the works of the Gas Light Company at Beckton; and whether these are considered for their vast magnitude, the excellence of their combined arrangements, the extent of their operations, or the engineering difficulties which presented themselves in the course of construction, on each and all of these accounts they possess the greatest interest to all connected with the gas profession.

The circumstances which gave rise to their construction were the following. The business of the Chartered Company having extended far beyond the capacity of their three stations, and after having purchased land for extension from time to time at fabulous prices, the company resolved to remove their works to more extensive premises. For this object the necessary Act of Parliament was obtained, and after encountering considerable difficulties in the choice of site, eventually that on which the works are erected was decided upon. This consists of about 150 acres, adjoining the Thames, situated about a mile from North Woolwich, and nearly opposite Woolwich, and about ten miles from London. In addition to this extent of land for the works, about 100 acres more were purchased, partly with a view to its future value as building ground, through which a branch railway to a station of the Great Eastern Railway, also a main road, extending a length of $3\frac{1}{2}$ miles, to Canning Town, on the road to London, were constructed.

As already stated, the site was the worst that could have possibly been chosen for constructing works, but in other respects on several accounts it was indispensable, situated, as it is, adjoining the river, at such a distance from any town as to preclude the possibility of complaints arising on sanitary grounds, and possessing extraordinary power of extension. These and other advantages of the site caused the expense of construction to be a secondary consideration. The plans for the various edifices were prepared by the company's engineer, Mr. F. J. Evans, assisted by Mr. V. Wyatt, and the whole carried out under the able direction of the former, and with what success we will attempt to describe.

The plan of the works in question is presented in Plate 29, having a river frontage of 1500 feet, of which a width of about 300 feet was reclaimed from the river, the wall being built about 5 feet above high-water mark. The main pier, shown in plan, is upwards of 800 feet long, extending a distance of 400 feet into the Thames; its width is about 44 feet, and is 28 feet above high-water mark. From the main pier there are several curves, formed on viaducts of a width of 25 feet, which continue to the shore. The pier as well as the viaducts are massive ornamental structures; the former is constructed of cast-iron cylinders, each being 6 feet in diameter, and sunk a depth of 18 or 20 feet below the bed of the river, and filled with Portland cement concrete. The superstructure is composed of wrought-iron longitudinal girders, of an average length of 62 feet and 4 feet in depth. The transverse girders correspond with the width of the pier and viaduct respectively, and on these are laid the timbers for the rails for conveying the coal, which is effected by locomotives. At each end of the pier there are six steam cranes, for the purpose of discharging the coals from the steamers, each crane being provided with a length of chain and two iron buckets, each of which is capable of containing about 13 cwt. of coal; one of these being filled in the hold of the vessel whilst the other is raised and delivering its contents into the waggon on the pier. By these means, with such appliances at command, 10,000 tons of coal can be discharged from the vessels and placed in the retort houses and other stores in the twenty-four hours. The vessels for unloading lay either within or outside the pier. The engines for working the cranes are situated at the end of the pier beneath the platform.

From the shore and extending the whole length, and midway between the blocks of the retort houses, is formed a viaduct 22 feet high and 12 feet wide, which consists of cast-iron columns, 20 feet long and from 15 to 12 inches in diameter, placed 25 feet apart longitudinally, and 12 feet apart transversely from centre to centre, and on the transverse girders, which are formed of wrought iron, the timbers are laid to carry the rails of the viaduct, which is represented in elevation in the plan, also in the view from the Thames, Plate 27. In consequence of the nature of the ground, each of these columns is supported on and bolted to a cast-iron pile upwards of 30 feet long, and 15 inches in diameter, driven at a depth of more than 30 feet below high-water mark. The superstructure is composed of wrought-iron girders, 25 feet long and 2 feet deep.

There are also lines of rails which extend throughout the length on both sides in the interior of all the retort houses, and communicating with the pier. Thus a vessel is discharged and the coal placed in any retort house where desired ready for charging.

Railways also exist throughout the works as indicated, communications being made between the viaduct or "high level," and the ordinary ground or "low level," by means of the incline as shown. Thus goods can be conveyed from any part of the works to the pier, or *vice versâ*. Rails run also through the coke-holes, so that on drawing the coke it is quenched, and immediately carried away in waggons. The chalk required for burning, necessary for the lime for the purifiers, or for building purposes; the bricks, retorts, iron, and other materials are discharged at the eastern and western jetties; which also serve for the purpose of loading the coke into the barges, a portion of which coke is conveyed by railway. The tar and ammoniacal liquor are pumped by steam from their respective reservoirs into tank barges constructed for the purpose, and conveyed to the premises of the manufacturers of these compounds.

For the transport of the coal, coke, chalk, building and other materials upon the works, thirty horses and nine locomotives are necessary.

From the works there are two mains, each 48 inches in diameter, which convey the gas to the works at Bow Common, and the distributing stations of Blackfriars, Westminster, Goswell Street, and Whitechapel, telegraphic communication existing between all these points and the works. The mains are laid in the road formed by the company, and from there over the river Lea, where a handsome bridge was built especially for their passage to London.

2 x

THE BECKTON WORKS.

The works are situated about two miles from the station of the Great Eastern Railway, and between these places a substantial railway has been constructed by the gas company, which runs trains periodically in correspondence with those of the railway company, and appears to have been principally established for the convenience of the workmen, of which there are on the average about seventeen hundred employed on the works.

There are ten retort houses, and two others designed for construction, of which nine are in active operation. Eight of these retort houses are each 360 feet long and 90 feet wide, containing 240 double retorts or 480 mouthpieces capable of producing from two million five hundred thousand to three million cubic feet per diem. The two others are of larger dimensions, each being 460 feet long and 100 feet wide, containing 640 mouthpieces.

The retorts are set eight in a bench, as shown in Plate 1. The hydraulic main is rectangular, and of wrought iron, divided and separated in lengths corresponding to every two benches, a main being carried from each division to an 18-inch main in the coke vault on both sides of the retort house. These join a 24-inch main underground, and continue to the condenser. The condensers are of the form of Graham's, already described, and consist each of 2600 feet of 12-inch main, or 3 feet of surface per 1000 feet per diem, maximum production.

Each retort house has its separate set of purifiers, condenser and station meter, the magnitude of which is in keeping with the rest of the establishment, and has two chimney shafts and three ventilating shafts which render the temperature within the building, unless when near to the settings, little above that of the external atmosphere. In the galleries connecting the retort houses are kitchens, dining rooms, lavatories, and baths, these being among the points of consideration for the comfort of the workmen.

At Beckton the operation of purification is effected, to a great extent, by means of the condensers and scrubbers, which deprive the gas of a great deal of its impurity; it is then passed into clean lime purifiers, called "carbonators," where the carbonic acid is eliminated; from there the gas enters the purifiers containing the sulphide of calcium, or as generally termed *foul lime*, that is, lime impregnated with sulphuretted hydrogen; which absorbs the bisulphide of carbon, and the sulphur compounds "other than sulphuretted hydrogen," the purification being completed by means of oxide of iron. Thus, as practised at these works, the ordinary method of purification is in part reversed, which is rendered necessary as described in page 154, in order that the sulphide of calcium may be effective. Lastly, there is a check or "catch" purifier, by which any casual trace of impurity is removed. The first scrubbers erected at these works were so limited in height as to require the liquor to be pumped through the vessels a second or third time, in order that it might acquire the requisite strength. Recently ten other scrubbers have been erected, each of them being 60 feet high, similar to those represented in Plate 10. Pure water entering at the top absorbs all the ammonia during its progress downward, until at the point of issue it obtains the strength of 20-ounce liquor, as already described.

The system of purification is so complete as to leave nothing to be desired, and the sulphur compounds, which for years contributed so much to the annoyance of our engineers; as we are informed, are now removed within the limit prescribed by the Referees under the company's Act of Parliament.

In the laboratory the gas is tested as produced, both for its illuminating power and purity, every two hours.

At these magnificent works there are at present six gasholders, four of which are single-lift, and of the respective capacity of one million feet; the other two are telescopic, each of the capacity of 1,500,000 feet, making a total storage of seven million cubic feet, which is estimated at only one-fourth of the maximum day's make. Two additional gasholders, marked 7 and 8, of a capacity of two million feet each, are now in the course of construction.

The maximum daily make during the past winter was about twenty-one million feet, produced from about 2100 tons of coal, yielding 2100 chaldrons of coke, of which 1470 chaldrons and about 200 chaldrons of breeze were delivered daily from the works. In addition to this, there were produced about

250 butts of ammoniacal liquor and 600 barrels of tar, carried away in tank barges into which the liquid was pumped from the reservoirs by steam power. These products, together with the quantity of coal received, must, at times, tax even the extraordinary power employed in conveyance. Here we may observe that the weight of the maximum daily production of gas will be approximately 320 tons, and its illuminating power, burning, say for six hours, equal to ten million candles lighted during the same period.

The isolated position of these works requires that they should be, as far as possible, self-dependent; and for the purpose of repairing their engines and machinery, as well as to produce and repair implements of daily necessity, there are some excellent workshops which will bear favourable comparison with many small engineering establishments. Here are forges actuated by steam blowers, lathes, planing machines, drilling machines, as well as machines for slotting and shaping, and even a steam hammer. Thus in the event of disarrangement in any parts of the machinery, the appliances exist by which it can be speedily repaired.

In the neighbourhood of the workshops are four lime-kilns constructed on the continuous system, the chalk, and breeze (which is used for fuel) being supplied in layers at the top, and when the former is properly calcined, it is withdrawn from below, thus securing a constant supply of fresh lime, either for the purpose of purification or building operations, as may be required.

As shown in the plan, there are several cottages erected for the workmen, all of which are inhabited, and their cleanly external appearance conveys a favourable impression of the comfort within. Near to these cottages is the "canteen," which the company have provided, and conduct for the comfort of their men, where refreshments of various kinds are furnished by a disinterested manager, and all irregularity or excess is strictly prohibited. Meals are provided at regular hours for the men, at the most moderate charges; and we were assured by the attendant that the company, desirous of supplying at as reasonable a price as possible, actually sustain a slight loss on every dinner furnished. A capacious and lofty hall, capable of accommodating about two hundred men, serves as a dining room; in addition to this there are the other conveniences already mentioned.

The number of persons employed on the works in the depth of winter is about two thousand, and to their great credit, as well as to that of the various officials controlling them, a single policeman is sufficient to maintain order; and indeed his services are but rarely called into requisition. A large number of the men live at Canning Town; and they are conveyed to and from their work at a mere nominal charge, sufficient, we suppose, to remunerate the railway company. Certain trains leave London daily direct for the Beckton Works, and perhaps this is the only instance of a railway of ten miles in length communicating with and issuing tickets for passengers to a gasworks.

To convey a further idea of the magnitude of these works, we may state that the combined *length* of the charging floors of all the retort houses, when completed, as represented in the plan, will be about 3000 yards, or one mile and three-quarters. Of course, the coke-holes, the viaducts in the retort houses, the hydraulic mains, all are of a corresponding length. There are 5120 mouthpieces, and nearly twelve miles of retorts placed in their benches. The ovens, or arches of the benches, are upwards of a mile in length; and, roughly estimated, about ten miles of railway, including the high and low level, exist on the company's works.

A view of the Beckton Works, taken from the river and copied from a photograph, is given in Plate 27, in which no part of the pier is shown, but the continuation of the viaduct from that point to the first retort house is represented. In the centre is an ornamental clock-tower, the utility of which for the whole of the works is obvious, and the following Plate 28 represents a view of the works, as photographed from the clock tower. All the parts can be readily understood by referring to the plan Plate 29. The buildings in front in the centre (Plate 28) and adjoining are the offices, that to the left is the meter house, and attached thereto is the laboratory. The range of buildings to the right are the retort houses, which extend to the extreme end of the ground.

Behind the principal central edifices are two blocks of buildings, containing purifiers, and between

2 x 2

them are ten scrubbers. A similar arrangement of buildings, purifiers, and scrubbers will be observed in the plan on the other side of the retort houses. The positions of the lime-kilns, workshops, and locomotive house are all represented; the blocks to the north of the two large gasholders being for the purifying house, scrubbers, meter, boiler, and exhauster for the two large retort houses. To the right will be observed a quantity of retorts, covering from two to three acres of ground, which gives some idea of the magnitude of the operations of the establishment. To the left are represented the check purifiers, the covers of which are raised by the hydraulic ram as already mentioned, and beyond these are the buildings for the exhausters, boilers, engines, and station meters, as shown in plan. The elegance of the design of the various edifices cannot be questioned, and at first sight conveys the impression of great expenditure, but on examination we find this is not the case, the great secret being in the design, and in the application of terra-cotta to all the light parts, and to the framing of the windows, and having occasional pilasters, which give a bold appearance to the whole; and when on the spot and analyzing the various buildings, it becomes a matter of surprise to find the extraordinary effect of so little ornamentation, and had the buildings been simply plain walls, instead of having their present appearance, the economy effected would have been merely nominal.

We have thus endeavoured to describe this colossal establishment, and unhesitatingly assert that it would be utterly impossible to design a better arrangement than that applied for carrying out the gigantic operations of the company. For, by the adoption of the viaducts and the "high" and the "low level" railways, all confusion is avoided; to use a technicality, the former may be regarded as the *inlet* and the latter as the *outlet* of the works; for by the high level all the coals for carbonization are admitted into the retort houses, whilst the low level conveys away the immense quantities of coke and other residuals daily produced. This, therefore, with the 20 miles of 48-inch main, may be termed the outlet, and by the high and low level, the operations of the company are conducted with the same facility as they were formerly at the small stations of Westminster or Brick Lane.

As already stated, to Mr. F. J. Evans is due the great merit of having designed this remarkable establishment, and carrying it to a successful issue, and which will remain a lasting monument of his engineering skill.

Within a short period of the first operations of the Beckton-Works, the facilities afforded by them for the economical production of gas, led to the amalgamation of the City of London, the Great Central and Equitable Companies, with the Chartered; subsequently the Western Company, and more recently the Imperial and Independent Companies entered into the amalgamation. Thus the seven companies which formerly supplied the greater portion of the metropolis are now combined, under the name of "The Gas Light and Coke Company," and, as a short experience has proved, this combination has been alike highly beneficial to the general public as well as to the shareholders of the various enterprises.

It should be stated that the name of Beckton was given in honour of Adam Beck, Esq., who had been a director of the Chartered Company for upwards of a quarter of a century, and during the greater portion of that period had filled the office of governor, and who in that capacity, aided by the secretary, Mr. Phillips, and Mr. Evans, has greatly contributed towards the present prosperous condition of that company, by raising it from the secondary place it formerly had among the gas companies of the metropolis to the high position it now holds.

The Imperial Company's new works are situated at Bromley-le-Bow, about four miles from Beckton, for which a site of 130 acres was purchased. The buildings at these works are of a more extensive character than those described, each of the retort houses being constructed for a maximum production of 5,000,000 cubic feet per diem, and is provided with separate purifying apparatus, all of which are worthy of admiration. There is, however, wanting the river-side communication which is so advantageously employed at Beckton, and the amalgamation of the Imperial Company will undoubtedly cause a considerable change in the arrangements of the two establishments in question, in order to profit by the great facilities for discharging coal at Beckton. The combined capacities of the two works of Beckton and Bromley are equal to a daily production of 40,000,000 cubic feet. In addition to these, the

Gas Light Company has other works at Bow Common, Fulham, Kensal Green, St. Pancras, Pimlico, Haggerstone, and Hackney, making a total capacity bordering on 60,000,000 cubic feet per diem, and to produce that quantity of gas no less than 6000 tons of coal will be required.

We conclude this part of our subject by giving the following particulars respecting the operations of the various London gas companies for the year 1875, obtained from Mr. Field's 'Analysis of the Metropolitan Gas Companies' Accounts':

	£
Total capital of the London gas companies	12,516,009
Capital employed	11,005,589
Capital unraised	1,510,420
Total gas rental	2,606,818
Cost of coals	1,455,407
Receipts for coke and breeze	492,927
Receipts for tar	162,151
Receipts for ammonia	111,951

	Thousands of feet.
The total quantity of gas produced	14,888,133
Gas sold	13,622,639
Unaccounted-for gas 7·32 per cent. of the make	1,090,081

	Tons.
Coal carbonized, of which 4 per cent. was cannel	1,505,000

	Chaldrons.
Coke produced at the rate of 34 bushels per ton	1,417,654
Of which 31 per cent. was used for fuel or	440,685
Coke sold	976,969

	Cubic feet per ton.
The average production of gas was	9,892

	Per 1000 feet.	
	s.	d.
The average selling price of gas	3	9·19
„ net profit	1	2·58
„ bad debts	0	0·25
„ cost of lighting and repairs to public lights	0	0·74

	Per ton.	
	s.	d.
The average net profits	10	11·31
„ total working expenses	12	3·47

Thus from the small works existing in Peter Street in 1814, with its gasholder of 14,000 cubic feet, has sprung up the present colossal establishments, and, according to the rate that gas lighting progresses, we may fairly assume that twenty years hence the consumption of gas in the metropolis will be doubled.

CHIMNEY SHAFTS OR STACKS.

It is an accepted axiom, that the success of a gasworks is mainly dependent on the operations of the retort house, and particularly the proper settings of retorts, and these, in like manner, are eminently dependent on the chimney shaft, aided by the dampers.

The high heat or vivid combustion so indispensable for working retorts, and the proper carbonization of coal, can only be obtained by having the fuel concentrated within certain restricted limits, and by a forced supply of atmospheric air, or pure oxygen. The expense of the production of the latter precludes its use; therefore, in practice, it is derived from the atmosphere as in the blast furnace, the forge, the blow-pipe, &c., and the chimney stack of a gasworks fulfils a similar purpose to the appliances mentioned.

The effect of a good draught is to cause an ample supply of atmospheric air to intermix with the coke or coal in the furnace, thus producing the vivid combustion referred to. At the same time, the orifice of the damper, or the point of separation between the interior of the oven and the atmosphere, is proportionately reduced, in order to prevent unnecessary loss of caloric at that point.

But it is to be regretted that the importance of the part of a work now in question is often sadly overlooked by engineers. In confirmation of this assertion we may state that a few years ago, when visiting one of the best managed metropolitan gas establishments, we found that there existed an orifice, similar to but larger than a sight-hole, in the front of the bed, situated just beneath the arch of every one of the settings, from which flames from the furnaces issued, and we were informed that only by these means could the proper heats be obtained. This was one of many similar instances of the kind which have come within our observation.

Under such circumstances, however, we have not to seek far in order to find the cause of the evil, as this arises either from the dampers being too much closed, or the flues choked; or, as in the case mentioned, where all the beds were affected alike, the shaft being too small, or the main flue obstructed at the entrance to the shaft. Either of these contingencies occurring, the flames must issue from the sight-holes instead of ascending to the shaft; and, although by making an orifice in the front of the bed as described, the draught or current would be slightly increased, yet it would still be very far from realizing the effects of a well-constructed shaft, with unobstructed flues.

The theory of the action of the chimney stack may be described as follows: Air, when heated, expands in proportion to the temperature it acquires, and, by this expansion becoming lighter than the surrounding atmosphere, it possesses an ascending power, which, in a chimney, produces what is generally termed a draught or current. The force or velocity of the current will depend mainly on the difference of the weight of a given volume within the stack, and a like volume of air at the ordinary temperature of the atmosphere.

To explain this, we may observe that, according to the best authorities, 1000 measures of air at ordinary temperatures, when heated to something like 550° Fahr. (which may be considered the average temperature of the interior of retort stacks when in active operation), are augmented to 2000 measures; or, in other words, by the additional heat mentioned, the volume of the air is doubled. The weight of 1000 cubic feet of air, at ordinary temperatures and barometrical pressure, is approximatively 80 lbs. With these data, we shall be enabled to explain the *modus operandi* of the structure under consideration.

For this purpose, let us suppose a chimney stack of an area of 5 square feet (about 2 feet 3 inches square), and 20 feet high above the main flue, having a capacity of 100 cubic feet. If the air within the chimney be heated to the degree already stated (viz. 550° Fahr.) it would weigh but 4 lbs., whereas the weight of a like volume of the surrounding atmosphere would be 8 lbs. The heated air, therefore, would ascend with a force equal to the difference of the two weights—viz. 4 lbs.—exerted over the area of the chimney, which force or draught would be maintained so long as the temperature remained as supposed, and the supply of air within the stack was continued.

But, if the height of the stack be 40 feet, we find the draught doubled in force; and if increased to 80 feet, all the other conditions being as represented, then the draught will be equal to 16 lbs. over the area. Thus the ascending force of the 80-feet chimney would be four times greater than that of 20 feet high.

In the preceding remarks, for the facility of illustrations, we have confined ourselves to the effects of heated air only; but, as the products of combustion issuing from the shaft are widely different from these, being composed of several compound gases, of which some are much heavier than atmospheric air at ordinary temperatures, the results are considerably less favourable than those described. The accepted rule being that the draught of a chimney is in proportion to the square root of its height, which draught is subject to the influence of every change of atmospheric pressure.

We find in a standard work on gas lighting, the question of chimney stacks passed over with a few simple observations, to the effect that a high chimney is essential for the purpose of carrying off the smoke, which, if not allowed to spread (we assume at a great elevation), would become a nuisance to the neighbourhood. It is further stated that the draught given by an ordinary coke oven is all that is absolutely required for the combustion of the fuel beneath the retorts, and it is further recommended, in order to avoid an excess of draught, that a valvular opening in communication with the atmosphere should be made at the bottom of the shaft. Beyond these remarks, little is mentioned as to the object of this most important part of a gasworks.

These statements, however, are erroneous, because in well-conducted establishments smoke is not permitted to issue in any quantity from the chimney. It is produced principally when discharging or charging the retorts, and for the escape of this, other provisions, as louvres or air shafts, are always made; secondly, the shaft has a more important duty to fulfil than that assigned to it; thirdly, a good draught is indispensable for economical carbonization, and the heat of a coke oven (about 1800° Fahr.) in the furnace would be insufficient for the retorts; and, finally, to make a valvular communication between the chimney and atmosphere, would be equivalent to diminishing the capacity and efficiency of the former, and would necessarily increase the consumption of fuel for carbonization.

The utility of a stack consists not only in its height, but in its area, and should be of such proportions that the heated air or products of combustion arising from the furnaces in combination therewith may pass off with all freedom. A very common error committed is to contract, to a considerable extent, the stack at its summit, by which means it is supposed that the draught is increased. This, under ordinary circumstances, is just as absurd as to suppose that, by contracting a main from 10 inches diameter to 4 inches, the pressure required for the delivery of a given quantity of gas would be diminished. That, under peculiar circumstances, it may be desirable to reduce its area at the top, there can be no question; as we once observed in a very small works, when the wind blew in a certain direction, or, as the stoker said, "chops over that hill," it was impossible to obtain any draught; in fact, at times the wind blew down the chimney, so that it was utterly impossible to heat the settings. In this case the top of the stack was materially diminished, and protected from the hill-side, which cured the defect. In dwelling-houses we frequently witness a similar evil in a chimney abutting against the wall of a building. The wind blowing against the wall, rebounds and descends the chimney, producing that most disagreeable of domestic annoyances—a smoky chimney.

The chimney stacks of gasworks of ordinary or large magnitude, however, differ entirely from that mentioned; and should never be contracted, but built as near as possible, consistent with economy of construction, of uniform area from the bottom to the top.

As in many other subjects in connection with gas apparatus, the opinion of engineers is divided as to the best means of giving the necessary draught to the furnaces. By some the high shaft is considered the most effective, whilst others maintain that the dwarf chimneys, of a height of from 20 to 30 feet above the level of the beds are the most suitable, and it should be stated that in establishments which have come under our notice, where the two systems are adopted, preference is given to the shorter description.

The dwarf chimneys are invariably square, and built of firebricks, of 9 or 14 inches thick, of an internal width of about 12 or 18 inches, and of the height stated, one of which serves for every six or eight double settings of retorts, four or more chimneys being placed in the retort house. An advantage claimed for dwarf chimneys is economy in construction, and as they are necessarily situated at short intervals along the settings, the draught throughout the retort benches is more uniform, and consequently there is the absence of the "cutting heat" which occurs in beds immediately exposed to the strong draught of the high shaft. It is true that the latter objections may be obviated by proper attention to the dampers, but as facility in all the operations of a gasworks becomes a consideration, these may be accepted as forcible reasons in favour of the dwarf chimneys, and as they are capable of producing a draught of four or five tenths exhaust, which is more than is actually requisite for furnaces, they meet all requirements.

Dwarf chimneys are always built upon the benches, and being limited in height and weight, no extraordinary precautions or skill are necessary in their construction, but with high stacks the opposite of this is the case.

At the period of the publication of the first edition of Clegg's 'Treatise,' considerable importance was attached to the construction of chimney stacks; but since then they have been erected in such numbers, for various mills and manufactories, that the manner of building them is now well understood, so that any good bricklayer, on being furnished with drawings and instructions, will undertake to build them when of moderate dimensions. Although in this, as in most other matters, it is generally wiser to employ men who are accustomed to the work, so long as their charges are moderate, which, however, is frequently not the case.

In the work referred to, mention is made of a chimney built with good concrete foundations on a quicksand, which afterwards settled bodily a depth of $16\frac{1}{2}$ inches, and from this it is asserted "that the only disadvantage attending a bad natural foundation is the expense of making an artificial one." With these statements we do not agree, as it is absolutely impossible to estimate, with such a foundation, the extent of settlement; nor can the vertical position of the shaft be assured under such conditions.

The first essential for a chimney shaft is, good firm ground for its foundation, and where this does not exist, and proper solidity cannot be obtained, either by piling or otherwise, it becomes too hazardous to risk any settlement like that mentioned.

The foundations of shafts are generally square and of a superficial area proportionable to their height. Some writers give the side of the foundation as one-eighth of the height of stack above the level of the ground, which rule is observed in the Edinburgh chimney, hereafter mentioned, and may be sufficient, but as any slight increase of cost of this part of the structure will be insignificant as compared with its safety, it would in most cases be preferable to err on the safe side by adopting the sixth part of the height as the length of the side of the foundation. Thus, for a stack 36 feet high, the foundation would be 6 feet square, or for another 60 feet high, 10 feet square.

Shafts are usually built either square, octagonal, or circular, the first-mentioned form being the cheapest is usually employed for small establishments. For a works producing four or five million feet per annum, a chimney 35 feet high, with an opening of 14 inches square, will be sufficient, and on an average will cost about 35*l*. For a works of from ten to twenty million, a stack 45 feet high, with opening throughout of 20 inches square, is ample. The stacks of small works should always be built outside and detached from the retort house, and the main flue connected thereto in such a manner, that by any expansion of the brickwork by the heat, it shall slide into and not abut against the chimney.

In building a square shaft of 45 feet high, the excavation being made of 7 feet 6 inches square to the solid ground, the concrete foundation of 2 feet thick is then laid. On this is commenced the brickwork of three courses of 6 feet 6 inches square, with two offsets of 9 inches high, reducing it to 5 feet 6 inches square, which is continued to the surface of the ground, all carried up in solid brickwork.

OCTAGONAL SHAFTS.

At this point commences the pedestal of 6½ bricks (4 feet 10½ inches) wide, 1½ brick (13½ inches) thick, carried up plumb inside and outside; and at a suitable height is left the opening for communicating with the main flue of the retort stacks. From the level of the main flue the chimney is sometimes lined with firebricks, 4½ inches thick, for about 12 feet in height; but this is a precaution which in a small works or stack, when rigid economy is necessary, may be dispensed with. At the height of 13¼ feet is formed a cornice of two courses, and the whole levelled, and on this is built the stack, which is 1½ brick thick for one-half the height, where there is a set off; the rest of the shaft being continued one brick thick, the top part being rendered slightly ornamental by one or two projecting courses.

A chimney of this description contains about 10,000 bricks, and costs about 65*l.*, when lined with firebricks as described.

For works of medium capacity or large magnitude, the shafts are generally made either circular or octagonal, for both of which the bricks are required of a particular form, and considerable more skill is necessary than in building a square shaft.

Octagonal and circular shafts are often built of very neat design, at very moderate cost; the former description is extensively adopted at the various manufactories in the neighbourhood of Manchester. For octagonal chimneys the bricks for the angles are specially made, and they are so laid as to break joint at the sides, but the angles from the top to the bottom are without joints. For building a chimney of this description, of 80 feet high, with an opening of 4 feet diameter at the narrowest part, the concrete foundation will be required 13 feet 6 inches square and 2 feet 6 inches thick. On this commences the footings in solid brickwork of 12 feet square, diminishing by offsets for a height of 3 feet, and is thus continued to the level of the ground, where it is 10 feet 8 inches square. At this point is commenced the pedestal of 10 feet 4 inches square, which is carried up four bricks in thickness to the height of 3 feet, where an offset of 2 inches wide is formed by reducing the pedestal to 10 feet square and 3½ bricks thick to a height of 12 feet, leaving the opening for communication with the main flue at the desired point. A cornice one foot wide is then built, at the same time the interior is so formed as to receive the octagon or circular stack.

The exterior diameter of the stack at the base is 9 feet between the flat sides of the octagon, and is built 2½ bricks thick for a height of 16 feet; it is then reduced by an offset in the interior of half a brick, and so continued 2 bricks thick to 32 feet high, where there is a second offset, again reducing the shaft to 1½ brick thick to a height of 48 feet, the last 16 feet of the stack, except at the capital, being one brick thick. For a height of 32 feet, as well as the interior of the pedestal, from the lower part of the entrance for the main flue, it should be lined with firebrick 4½ inches thick, to be counted in the thickness of the base and the stack. The stack is surmounted by a capital, according to the taste of the engineer, and these in octagon shafts may be rendered very ornamental.

Some engineers build hoop-iron bonding every 3 or 4 feet; others build a portion of the courses at intermediate distances in cement, which is considerably more secure than the hoop iron. In many large chimneys there are two shafts, one within the other, having an intervening space between them, as represented in the description of the Redheugh Works, Plate 24; by this arrangement the excessive heat is confined to the internal part; thus the exterior is not liable to crack, as frequently happens when the shaft is composed of one wall only. However, when a shaft is well built, and has ample time to dry before being brought into action, the cracks or fissures are not likely to occur. As represented in the Plates, in modern works of the first magnitude, two or more high shafts are built within the retort house, ventilating shafts also being provided.

In rare instances it becomes imperative to construct the stack at a distance from the retort house, and to erect a subterranean flue from one to the other. With this class of structure a difficulty in obtaining a current is generally experienced at the commencement of operations, arising from the humidity, combined with the materials forming the flue, as well as that of the surrounding soil. Under such circumstances, a temporary fire is placed within the stack, which, deriving its air through the lateral flue, conveys away the moisture, and, on the flue becoming warm, no further difficulty is

2 Y

experienced. It is, however, most desirable, in such a structure, to prevent the humidity from penetrating to the flue; for this purpose it should be rendered on the exterior with Portland cement, and other precautions observed to avoid the loss of heat by transmission to the soil.

Although we are aware that trials have been made, on a limited scale, of applying a forced current to the furnaces by means of fans, and that the system was patented some years ago by Messrs. Lowe and J. Kirkham, we have never heard of it being practically adopted; and would venture to suggest that the system, during the depth of winter, when fresh benches are brought into action, and sometimes only for a period of three or four weeks, might be advantageously adopted. To explain our views, let us assume that a certain amount of caloric is requisite to carbonize a given quantity of coal, say in six hours. Now, if by an additional supply of air to the furnaces by the means proposed, the caloric could be increased so as to carbonize the coal in four hours, then by the application of the system to a few benches, the expense and inconvenience of lighting other beds, and this for a very limited period, would be avoided. All that would be required to carry out this system would be a fan or blower, with pipes in connection with the ash-pans of the furnaces, an arrangement which, with our modern appliances, is capable of being readily carried into operation.

The highest chimney existing at any gasworks is that of the Edinburgh Company, and was designed about thirty years ago by Mr. Mark Taylor, then their engineer. The base of the foundation of this shaft is 40 feet square and 6 feet 6 inches thick; on this commences the stone foundation, carried up to the level of the ground. At this point is a pedestal of dressed stone built to a height of 65 feet, being at its base 30 feet square, and having a slight batter, is 27 feet 9 inches square at the top. From the top of the pedestal the stack is 264 feet high, its outer diameter at the bottom being 26 feet 3 inches, and at the top 13 feet 6 inches. Within this is a second shaft or firebrick lining 90 feet in height, and of a diameter of 18 feet 10 inches at the bottom, and 16 feet 4 inches at the top; thus the smallest area of the interior of the shaft is upwards of 200 square feet.

To convey to the mind the magnitude of this chimney, it may be stated that its pedestal of 65 feet is equal in height to an ordinary shaft for works producing 40 or 50 million feet per annum. Its height, 329 feet, may be compared with the Monument of London, which is only 202 feet high; whilst the base of its foundation is almost equal to the width of some of the old-fashioned retort houses containing double settings, and the area of the base of the stack is equal to the area of many gasholders employed for supplying very small towns and villages. The cost of this shaft was upwards of 4000l., and the draught produced by it is equal to about 2½ inches exhaust.

The necessity for this shaft arose from the circumstance that the Edinburgh Gasworks are situated in a deep valley which intersects the city, and divides the "Old" from the "New Town," the slopes on each side as well as the heights being thickly populated. In Scotland formerly, and probably the same system exists at the present day, in consequence of the inferiority of the coke obtained from the cannel, which is solely used for the manufacture of gas, another description of coal was employed for heating the retorts, and it may be assumed that the smoke arising from this becoming offensive to the neighbourhood, the colossal structure in question was erected.

We believe that no definite rule exists to determine the area of a chimney; but according to the estimate of Mr. Evans, each square foot of an ordinary high shaft is sufficient to carbonize five tons of coal in the twenty-four hours, and according to this datum the area of the Edinburgh chimney is ample for one thousand tons of coal per diem; and when we consider that the draught of this is three times greater than any ordinary shaft, it may be assumed that it is capable of carbonizing a much greater quantity than stated.

From the preceding we may conclude that for a small works one detached chimney is sufficient. In larger establishments a series of dwarf chimneys may be applied, or if high stacks are built, two or more of these should be placed within the retort house, so that each shaft should control the benches in its immediate vicinity, and in some localities high shafts may be indispensable in order to avoid complaints on sanitary grounds.

It has been proposed to place dampers on the main flue instead of applying them to the respective settings. This, after what has been stated, cannot be too strongly deprecated, inasmuch as the stack answers the purpose of the blast, the bellows, and the blow-pipe, and having such an excellent appliance, without cost or trouble, beyond its first construction, the folly of checking it by a damper must be palpable. The proper system of working is to have a good draught throughout the whole of the settings, and to check each bed carefully by its damper; and in many works it would pay handsomely to have an intelligent man, whose duty might be confined to this alone, and thus prevent the loss of an incalculable quantity of fuel, which often passes uselessly to the atmosphere, without any indication whatever of the loss.

By properly adjusting the dampers of the settings, the influence of excessive draught is avoided, and economy of fuel effected, inasmuch, as before stated, the damper is the point of separation between the oven and the atmosphere, and the smaller the opening, the more complete must be that separation. Thus, if we suppose, with an insufficient draught, the opening of the damper to be, say, 30 square inches, and with a good draught to be only 6 square inches, it is evident that the heat in the latter case will be retained in the oven, whilst in the former, on account of the large opening and consequently the imperfect separation of the interior of the oven from the atmosphere, great facilities will be afforded for the caloric to pass uselessly away.

COMPETITION OF GAS COMPANIES.

ALTHOUGH the effects of competition between gas companies in the United Kingdom are now happily almost at an end, a reference to the past may be productive of some good practical lessons for companies established in other parts of the world, and for this purpose we will confine ourselves to the description of the competition which existed for many years in the metropolis.

For a considerable period after the establishment of the Chartered Company, on account of the danger that was supposed to be inherent with gas, it was difficult to induce the ordinary class of tradespeople to employ it in their shops. To counteract this prejudice, persons were engaged to lecture in the metropolis and provinces on the advantages of the use of gas, and in which a considerable amount of money was expended, without, however, achieving any degree of success; and for some time gas met with but little encouragement, as we find that the whole of the storage in 1814, on the occasion of the Committee of the Royal Society visiting the works at Peter Street (the only station then existing), was but 14,000 cubic feet. But in the course of time public opinion changed in favour of the new light, when the demands for gas were so great that the company was unable to comply with them, which led to considerable dissatisfaction.

During many years gas was only employed in shops, but it was eventually admitted into private dwellings; and first, as a matter of precaution, it was employed to light the entrance hall; from there it made its way to the kitchen, and subsequently to all the other parts of the house.

On the first establishment of rival gas companies in opposition to the Chartered, there were neither boards of works nor district local boards. The only authorities of the kind in the metropolis beyond the City boundaries were the ordinary parish vestries, in whom all local powers were vested; and it was supposed that those vestries had power to authorize the opening and occupation of the streets, and unde this belief new companies were established at different parts of London; in fact, all the metropolitan gas companies, except the Chartered, commenced operations in this manner. Moreover, at that time there were no means of forming a joints-tock company other than by Act of Parliament, and every company not incorporated was simply an ordinary partnership, in which every individual member was liable for the debts of the whole undertaking. This, with other disadvantages, rendered it almost impossible for companies to continue their operations without being empowered by Act of Parliament, and, as stated, this by the first companies was gradually effected.

The two or three first Acts passed authorized different gas companies to supply the same districts throughout the metropolis, and for a time opposition existed between the Chartered, the Imperial, and City of London Companies in a portion of their districts. The attention of Parliament was, however, directed at a very early period to the evils arising from this competition; and Sir William Congreve, who was then Inspector of Gasworks, on reporting to the House on the matter, expressed the strongest disapprobation of it, and recommended that no future Act should be passed without limiting the district of each company; consequently in all Acts passed during several years each of the companies was limited to a certain district, under a heavy penalty for every light supplied beyond it.

Other companies were, however, formed without statutory powers for supplying the metropolis, and in 1842 the principle which was established in 1821 for preventing competition among gas companies was abandoned, and an Act was passed authorizing a comparatively new enterprise to supply districts already lighted by several other companies, and closely following upon that four or five other Acts were passed upon the same principle, thus strengthening the competition which had existed for years in various parts of the metropolis, and brought on a state of affairs that cannot be too strongly condemned.

COMPETITION OF GAS COMPANIES. 349

The competing companies laid their mains in all the leading thoroughfares where there was a large consumption; hence in some streets there were as many as six mains, and Oxford Street was supplied at one time from no less than six different works. This state of affairs led to the greatest disorder, and in some instances, whether accidentally or by design, the main of one company was connected with that of another; and by the frequent change of supply the wrong service would be at times connected, so that one company supplied the gas whilst another received payment; and it was not an uncommon occurrence for a consumer, on entering a house where there were two or three services, to state to the inspector of each company in turn that he had arranged to take the supply from the *other* company, and so kept out of the books of all the companies, and had his gas for nothing, which deception was favoured through the secrecy observed by the various competing enterprises.

This competition of companies made them reckless of their respective interests. Canvassers were employed not only to get new customers, but to induce customers to change their supply to the company represented by the canvasser; and most extravagant offers were made which consumers did not fail to take advantage of. Sometimes the company in the case of a large consumer would refit the house and keep the fittings in repair. In other cases two or more burners were allowed on the premises without passing through the meter, and of course were not paid for; or in the event of the nominal standard price being charged according to mutual arrangement with the companies, a large discount was allowed on payment in order to disguise the bargain.

Many consumers having driven a hard bargain with the company in the first instance, would report that more favourable conditions had been offered by a rival company, and give notice to change; when that supplying, rather than lose the consumer, would consent to the same terms; and thus by the secrecy referred to, observed by the various enterprises, were they imposed on in every way by the unscrupulous portion of the consumers. Moreover, at one time no uniformity whatever existed in the charges, and premises were frequently lighted at one-sixth part of the value of the gas they consumed; the rule was, " take what you can get, and get what you can; but don't lose the customer."

Again, with the loss by leakage which frequently occurred in consequence of the imperfect manner in which the mains were laid, and the repeated changes of services, whenever the mains of two or three companies existed in the same street, instead of the companies' officers co-operating with the view of detecting and remedying the escape, it was disregarded, when often the parochial authorities had to interfere and insist on the nuisance being abated; hence it was not surprising that accidents at times occurred. It must, however, be stated that many of these evils never were brought before the directors, but were entirely owing to the rivalry existing between the officers of the respective companies, who, in the interests of those they represented, often resorted to tricks that they would have been ashamed to have come to the knowledge of their directors. In one instance within our memory, the inspector of a company, being informed that one of their consumers was supplied by the opposition company, had the service secretly changed to what he believed was the proper main, whereas it was actually changed from the right to the wrong main; and a few weeks afterwards on the second company ascertaining that the premises were supplied by them, they claimed the whole amount that had been received from the consumer during several years, which was paid; as it was hardly likely that the inspector would admit his share in the affair. Here we may observe that when the mains of two or more companies are in the same street, the company supplying each house is readily ascertained by the action of pumping a neighbouring syphon at night, by which all the lights on the main to which it is attached are caused to jump or oscillate.

Although a few consumers profited by this discreditable state of affairs, it was by no means advantageous to the general public, and was ruinous to the gas companies, two of which at least during several years never paid a shilling of dividend to their shareholders, whilst the profits of the old companies were most materially reduced. From these statements a faint idea of the evils attending competition among gas companies may be formed, but it did not cease here, as by the multiplicity of mains a large amount of capital was literally thrown away, and upon which the public are now paying dividends. In addition to this, there was the frequent breaking up of the streets and stoppage of the thoroughfares for laying or altering the mains and services, or transferring the supply from one company to another, which at one time caused a universal outcry throughout the metropolis; and on account of these proceedings the public mind was for a period kept in a constant state of irritation by the press, all of which proceeded from the encouragement that had been given to competition by the public.

The result of all these difficulties was that the companies agreed among themselves to cease all competition; to confine their operations to separate districts without interfering with each other; to discontinue all special arrangements, and charge one uniform price over all the parts of their respective districts. These proceedings caused in some cases an increased charge of from 25 to 50 per cent., and gave rise to considerable agitation in the metropolis; public meetings were held, and resolutions were passed to petition Parliament for an inquiry, which was eventually granted. At that inquiry the public and the companies were both represented, and after a careful investigation of all the circumstances, the districting arrangements were confirmed, and the companies secured in their respective districts, and all other companies or persons prohibited from supplying gas without first obtaining the sanction of Parliament; and a public Act was passed embodying these provisions, although at the time the companies considered they were subjected to some very arbitrary regulations.

As already observed, in the early years of gas lighting it was supposed that the local authorities possessed the power to authorize the opening of the streets, therefore the main object in applying for an Act was to incorporate a company.

At that time, when the cost of obtaining an Act of Parliament was, if unopposed, seldom less than 600*l.*, but, if opposed, would often cost several thousands, the shareholders of the different companies preferred carrying on their operations as a private partnership, and taking all the risks in preference to incurring the expenses of being empowered by Parliament. Since then a public Act has been passed authorizing the formation of joint-stock companies, by a very simple and economical process, with the same powers, rights, and privileges as an Act of Incorporation, and nearly every gas company not subject to statutory regulations have since availed themselves of that Act and become incorporated.

Concurrent with these proceedings it has been conclusively established in several instances before the Courts, that local authorities cannot give gas companies the powers they require to open and occupy the streets; consequently any company supplying gas without an Act, or its equivalent, is liable to have its operations stopped in a very summary way. The object, therefore, of applying to Parliament now is to procure powers to open the streets, not to incorporate a company as heretofore, and it is almost needless to say that the necessity for procuring these powers is far greater than the other.

As many gas companies are but very small undertakings, to whom the expenses of an Act of Parliament, although now not more than half its former cost, would be notwithstanding a burdensome expense to the companies in question; to meet these circumstances Parliament, in the year 1870, passed an Act to authorize the Board of Trade to confer upon gas companies, by means of what is professionally termed a "Provisional Order," all the powers they could acquire by a special Act. The expense of procuring one of these Provisional Orders, if unopposed, is estimated at about 150*l.*; but if opposed it will be very little less than an ordinary Bill.

The powers conferred upon the Board of Trade for granting Provisional Orders are limited to companies; but recently a similar power has been conferred upon the Local Government Board with respect to local authorities.

Every gas company, therefore, that is supplying gas without Parliamentary powers is in peril of having a Provisional Order granted to a local authority over its head. The effect of this would be, that as it is illegal for any company or person to open and permanently occupy the streets without express Parliamentary sanction, any local authority having obtained a Provisional Order, granting such a power for its own district, could stop a company from any further operations as to the opening of the ground in that district, and take the supply into its own hands.

When any existing company applies to the Board of Trade for a Provisional Order to authorize the maintenance of works in their present site, it is required to deposit with the Board a map of the site and surrounding neighbourhood, for which purpose almost any published map is sufficient; but when such a company, or any new company, applies for powers to construct works on any new site, it is required to deposit with the Board a plan showing the site, and how it is to be laid out. It is also required to deposit with the Board the conveyance of the land to parties representing the company, or an agreement for the conveyance of the land, as the Board will not grant an order on the mere chance of the company procuring the land after the order is granted. Plate 22 is an illustration of the plan required by the Board for a new site.

We have been favoured with this contribution by a gentleman of the most extensive experience in gas matters, and whose opinion on the legal points raised is entitled to the highest consideration.

INFLUENCE OF BAROMETRIC PRESSURE AND TEMPERATURE.

The weight of the atmosphere pressing upon the earth as well as upon all living beings is equal to about 15 lbs. on each square inch of surface.

The existence of this pressure is demonstrated in a very simple manner, for which purpose a glass tube of about 33 inches long and exactly one-fourth of a square inch in sectional area, having one of its ends hermetically closed, is requisite. This is filled with mercury, and the top secured by the finger when the tube is reversed, and being held in a vertical position, its end is immersed in some mercury contained in a cup. This accomplished, on removing the finger, the top of the column of mercury will descend to a height of about 30 inches from the surface of the mercury in the cup. Hence the pressure of the atmosphere is sufficient to support a column of mercury of the height indicated, the weight of which is 3·75 lbs., and being of an area of one-fourth of a square inch, it corresponds with 15 lbs. on the square inch. This experiment is identical with the action of the barometer, by which the varying pressure of the atmosphere is constantly indicated, the column of mercury rising as the pressure increases and descending as it diminishes.

If our atmosphere possessed the same density throughout, its height would be about five miles, and its influence at all the various elevations would be alike, but the reverse is the case, inasmuch as its lower portions which press upon the earth bear the weight of those next above them, and these again are pressed upon by other portions still higher, so that the amount of pressure, and consequently the density of the air, decreases continually as we ascend above the level of the sea, whether in mountainous districts or other places of great elevation, or in a balloon. The following table by Professor Graham gives the volume of air corresponding with the height, together with the barometrical pressure, at the respective elevations.

	Volume of Air.	Height of Barometer.
At the level of the sea	1	30 inches.
2·705 miles above the sea	2	15 ,,
5·410 ,, ,,	4	7·5 ,,
8·115 ,, ,,	8	3·75 ,,
10·820 ,, ,,	16	1·875 ,,

As gas and air are influenced alike by atmospheric pressure, we gather from the preceding that a cubic foot of gas, under ordinary conditions, if carried to a height of little more than eight miles, would be increased in volume to eight cubic feet; or a balloon containing 1000 feet of gas, supposing the vessel of sufficient capacity to hold it, on reaching a height of 2·705 miles (which is perfectly practicable), the gas, in consequence of being relieved of one-half the pressure, would expand to 2000 feet or double its former volume. Or, if we suppose a town to exist at the elevation stated, and a gasworks to be there established, the yield of gas would be doubled, and a production of about 20,000 feet would be derived from each ton of Newcastle coal.

But in this augmented volume, as all the atoms of the gas and hydrocarbon vapours are separated from each other proportionably, the smaller volume contains precisely the same amount of illuminating agents as the larger, the increased volume being accompanied by a corresponding reduction in the quality of the gas; on the other hand, by increased atmospheric pressure the density of the gas and the illuminating power of any given quantity are augmented. It therefore follows that according to the height of a town above the level of the sea, so will the yield of gas per ton of coal be proportionably increased, and the quality of the gas reduced.

INFLUENCE OF TEMPERATURE.

In addition to this varying influence of barometrical pressure according to the elevation above the level of the sea, in all places this pressure changes daily and hourly, the extreme limits of which in any particular locality do not exceed 2½ or 3 inches throughout the year, but it frequently happens that a variation of one inch takes place within the twenty-four hours, the result of which is, that the gas in the holders is increased or decreased in volume to the extent of 3·3 per cent. by such change. Or a difference of 3 inches increased, or diminished, atmospheric pressure causes a difference of 10 per cent. in the volume of the gas, which of course occupies the greatest volume when the pressure is least. The standard for the barometer usually adopted in experiments on gas, and retained by the Gas Referees, is 30 inches; therefore to reduce a quantity of gas at any given pressure to the standard, the following rule is applied. As the standard pressure is to the volume, so is the given pressure to the quantity required. Thus 1000 feet of gas, at 31 inches pressure, what would be its volume at 30 inches?—Then as 30 : 1000 :: 31 : 1033 cubic feet. Hence it follows that 1000 feet of gas with the higher atmospheric pressure, is augmented to 1033 feet by the pressure being diminished one inch, which variation, as observed, sometimes occurs within twenty-four hours. By reversing the rule we find 1000 feet of gas at 30 inches increased to 1033 feet at a pressure of 29 inches, or 1000 feet at the lower pressure contains but 967 feet when at the standard pressure, as represented in the table of corrections.

Barometrical pressure has also considerable influence on the draught of the furnaces of a gas-works, the current increasing with diminished atmospheric pressure.

The volume of gas is also controlled to a considerable extent by the temperature it is submitted to. To illustrate this in a familiar manner, if an ordinary bladder be about two-thirds filled with gas, and its neck firmly secured, on holding it near to a fire, the gas will speedily expand and completely fill the bladder, but on again exposing this to the former temperature, the gas contracts to its original volume. This rate of expansion is $\frac{1}{480}$th part for each degree of Fahrenheit, or nearly 2 per cent. increase or diminution of volume for 10° of augmented or diminished temperature. Consequently 1000 feet of gas at 60° would, on attaining a temperature of 80°, expand to 1039 feet, or a volume of 1000 feet at that temperature would be reduced to 961 feet at 60° Fahr.

Hence, when rigid accuracy is required, these variations demand the necessary corrections to reduce the quantity to a given standard of pressure and temperature, which, as already stated, for the barometer is 30 inches, that for the thermometer being 60° Fahr. to which the tables, page 246, referring to the method of ascertaining the illuminating power, as well as the purity of the gas, according to the instructions of the Gas Referees, are calculated.

In page 66 we referred to the errors incidental to the method of determining the value of gas as an illuminating agent, by means of its specific gravity, arising from the nature or composition of the gas under examination, and for the same reasons the bromine and chlorine tests are not to be relied on. We have therefore confined our observations to the most modern methods of ascertaining the value of gas, according to its degree of purity and its illuminating power, as indicated by the photometer, and that these are the only positive and correct methods is made evident by the operations and instructions of the Gas Referees. And when we state that these gentlemen have been nominated by the Government, in consequence of their scientific knowledge, and which they have brought to bear on the subject, we may feel satisfied that the course they have adopted, in addition to simplifying the operations, must fulfil all possible requirements.

RETORT SETTING.

Of all the apparatus of a gasworks, unquestionably the most important are the settings of retorts, since on these depend the most effective means of carbonizing the coal and obtaining therefrom the full quantity of gas, together with the minimum consumption of fuel, and consequently the maximum production of coke for sale, which residual, with caking coal, constitutes the second source of profit to a company; and whether hundreds or thousands of tons of coke are allowed to be wasted and dissipated in the atmosphere, or the value of this retained as profits to the shareholders, depends entirely on the construction of that part of a gasworks now in question.

Yet such is the apathy often displayed concerning this subject that retorts are frequently set by any local bricklayer, without either system or rule, and as a natural consequence the consumption of fuel is excessive. This fact is the more surprising on account of the facilities now existing for the proper erection of this description of work under consideration, as there are engineers who furnish drawings, and firms established in various parts of the kingdom especially engaged in retort setting, and some of whom guarantee that the furnaces shall be heated with a given percentage of fuel. Therefore, under these conditions, any company which is not provided with regular retort setters, or good working drawings to guide them in the erection of this part of their works, must in the majority of cases sustain a serious and continuous loss, which a comparatively insignificant expenditure would avert.

We may here observe that it is not our intention to advocate any particular system, and in venturing to express opinions in opposition to the plans adopted by some of our first engineers, it is simply to ventilate facts, and we are encouraged in this by the undoubted excellent results obtained by a large majority of engineers, who have deviated from the old-fashioned system of setting retorts.

Plate 1 represents various views of settings of retorts for a work of the first magnitude, having a firing stage and coke vault, the details of which are given in Plate 24.

The setting is shown in elevation; in cross section through the line B B; and in longitudinal sections through A A and C D. As seen in cross section, the retorts are D shaped; but as these are provided with Morton's air-tight lids already mentioned, it was considered necessary in order to avoid the irregularity in the expansion by the heat that any other form might occasion, to make the lids and the front of the mouthpieces circular, the flange ends of these being suited to the retorts. Recently we have been informed that these lids are made alike for all forms of mouthpieces, whether D, oval, or round.

In the longitudinal section through A A above the furnace is an arch extending the greater portion of the length of the oven, having six nostrils on each side, two of which are shown in section through B B. The heat is conducted directly and indirectly through the arch and nostrils, and after acting on the retorts, returns to the front, and enters the flues beneath the retorts. It then passes to the vertical flue, and hence to the main flue, where there is a damper to check the draught. This system of arch over the furnace has been retained from the iron settings described in page 89, when that metal was employed for retorts and required protection from the action of the fire, but is now only adopted in works of magnitude, where, as we have had occasion to observe, the percentage of fuel for carbonizing is often considerably higher than in works of medium capacity, and which we have ventured to assume may arise from the adoption of the arch over the furnace, presenting a serious impediment to the conduction of the heat. The gas and tar are taken off at the end and side of the hydraulic main as represented. At the Beckton Works, as a precautionary measure, the hydraulic main is constructed in such a manner that any two lengths of the hydraulic can be thrown out of action, and for this object a main passes the length of the retort house parallel with the hydraulic, each section having its valve as represented in the drawing.

Plate 2, Figs. 1, 2, and 3 represent respectively an elevation, longitudinal section, and cross section of a setting of one iron retort; in this the heat acts by direct conduction beneath the retort, whilst the current passes along the length of the oven, and returns to the orifice in front, thence into the small flue, where it is checked by a damper on entering the main flue.

Figs. 4, 5, and 6 represent the elevation, longitudinal section, and cross section of a setting of retorts as constructed by Mr. Anderson for localities where the retort house space is limited, and consists of one brick and two clay retorts. The brick retort is built of Stourbridge firebricks especially made for the object, having rabbeted joints to hold the fireclay cement. The arch which forms the bottom of this retort, in consequence of the heat being continually carried off by the gas in the process of carbonization, does not attain that excessive temperature incidental to arches which are employed especially as shields or guards, the durability of the brick retort is therefore as great as that of the clay retorts.

As seen in the longitudinal section, the oven is divided by two walls into three separate compartments, other walls being formed to support the retorts. The heat as generated is conducted directly to all the parts of a setting, whilst the current ascends the four passages, as shown by the arrows, over the top of the uppermost retort, descending in a zigzag direction through the other three passages, and when reaching the bottom it meets with a supply of air at the flue a which materially assists the combustion, thence the heated air passes through an orifice not shown to the main flue. Thus in a setting of this description the current has to travel some twenty feet before it reaches the main flue.

The advantages claimed for this class of setting by Mr. Anderson, after an experience of eight or nine years, are economy of space, low percentage of fuel for carbonizing, and cheapness in erection, all of which may readily be expected, by the confined limits of the setting, the direct action of the caloric, and the circuitous current; on all these accounts the setting appears admirably adapted for small works.

We may here observe that according to the opinion of Mr. Anderson, whose extensive experience is well known, the percentage of fuel in large works should not exceed 20 per cent. of the coke produced.

Plate 3 represents two settings of retorts with which we have been favoured by Mr. Skoines; the one being a setting of two, the other of three retorts in a bench, both of which are well suited to meet the requirements of small works, and have the recommendations of simplicity of construction, and requiring but a small quantity of brickwork. In the setting of two, the retorts are placed side by side, with the view of keeping the stack of uniform height, when a single retort is set in an adjoining arch. The furnace is large, in order to hold a good quantity of fuel, and thus diminish the necessary attention for firing, a very desirable consideration where the stoker has numerous other duties; but in such cases too much care cannot be devoted to the proper adjustment of the damper in order to avoid waste of fuel.

The retorts are set on the front wall and four piers; thus the lower parts of the retorts are subjected to the direct heat of the furnace. The current of heated air or gases passes along the channels $a\ a$, at the sides of the retorts, to the end of the oven, where it enters the flue formed by the retorts and the block, thus returning to the front of the bed. The current then passes off by the top flue to the main flue.

The setting of three retorts resembles the other so far as regards the furnace and the passage of the heat through the flues $a\ a$ to the end of the bed. The current then passes through an orifice at the end of each of the blocks $b\ b$, returning to the front along the sides of the top retort, and then enters the flue c beneath the top retort, hence to the main flue, either by an orifice in the crown of the arch or at the end. The whole is clearly represented on the drawings with the letters of reference.

Mr. Skoines informs us that he has adopted these settings for very small works, and that they have been found desirable and economical, whilst there is the absence of the "cutting heat" which exists wherever the furnace is covered by an arch having small nostrils.

Plate 4 shows a method of setting five retorts in a bed, as frequently adopted whenever the quality of the retorts can be relied on, and is built without transverse walls or any other brickwork in the interior of the oven beyond the piers, the slabs, the saddle pieces, and the few bricks employed between the two side retorts and the wall to direct the passage of the caloric. Beneath the two lower retorts are two flues, communicating with the ascending flues marked by dotted lines, which unite and continue in a vertical direction (and should be also in dotted lines) to the main flue ; a space of 6 inches separating the end of the flues from the front wall. Each of the upper retorts is supported on the lower by three saddle pieces, the intervening spaces being filled from end to end with brickwork in order to prevent the passage of the current of caloric at these points. For a like purpose a stop is built on the side of each of the upper retorts which extends from the front wall to within 15 inches of the end wall; hence the current of caloric must pass from the furnace to the back of the bed, when it returns, as shown in Fig. 4, along the sides of the lower retorts, enters the flue in front of the setting, and hence to the vertical and main flues. The centre retort is supported by the slabs placed on the three piers.

A decided benefit to be derived from this kind of setting consists in having the retorts entirely detached from the arch, except so far as regards the stops, which can have no injurious effect ; hence any settlement or expansion of the brickwork, which must of necessity crack all retorts embedded in transverse walls, cannot be prejudicial in the setting described, and we repeat that we have taken down retorts after a lengthened active service and in one entire piece, which would be utterly impossible with retorts embedded in brickwork. Transverse walls are never used with brick retorts, and if the scurfing bar were dispensed with (and its use ought never to be permitted), then no fear would exist concerning the stability of clay retorts when properly made.

Plate 5 represents a setting very similar to that last mentioned, as adopted by Mr. Hutchinson, of Barnsley. In this the only brick material within the oven consists of the saddle, the block bricks and the slabs, and in consequence it may be regarded as the simplest of all settings. The advantages of this, like the preceding setting, consist in the absence of transverse walls, consequently there is no obstruction to the passage of the heat ; all the lower retorts are in direct contact with the fuel in the furnace ; and the current is conducted in such a manner as to economize the fuel.

These settings have been constructed by Mr. Hutchinson during several years, giving the greatest satisfaction, and a further recommendation to them is, that the balance sheet of the company represented by that gentleman, for the returns on the value of the coal, is perhaps second to none in the kingdom.

Plate 6 represents the different views of a setting of seven clay retorts as constructed by Mr. Cathels, which can be either applied to the single or double setting, that in question being a double setting broken off at about the centre. According to this system, each single oven is divided by the walls a and b (see longitudinal section) into three chambers, in addition to which there are the chambers beneath the lower retorts, shown in section through F G, seen also in sectional elevation through B C. Immediately over the furnace is a guard arch, seen in plan at G H, having very large orifices for the passage of the caloric, and as shown in longitudinal section the current passes in the direction of the arrows, through the orifice in the wall a, and descends beneath the lower retorts, when, after making the circuit indicated by the arrows, represented in section through F G, it passes through the openings shown in sectional elevation through D E at the bottom, and ascends to the main flue, where the passage of the caloric is controlled by a damper.

One great advantage peculiar to this description of setting is the large orifices in the arch for the passage of the heat, thus rendering practically no obstacle to its conduction, and on which we believe mainly rests the high reputation these settings possess ; for although the circuitous course of the caloric must be conducive to economy in fuel, yet this could never be obtained without the free communication between the furnace and the retorts. We are assured by Mr. Newbigging, who has had considerable experience with this description of settings, and on whose word we have the utmost reliance, that the fuel for carbonizing varies from 18 to 23 per cent. by weight of the production of coke, that the average yield is $13\frac{1}{2}$ cwt. of coke per ton of coal, and the duration of the setting, when built of good bricks and properly used, is two years and a half, but during which period occasional repairs are required to the furnace when it becomes enlarged.

RETORT SETTING.

A plan adopted by Mr. Cathels, although not shown in the drawing, is to have a seal pipe at the end of the hydraulic main where the gas issues, which reaches to a short distance above the level of the tar, but is open at the bottom, so that the tar always flows away from the lower part, thus avoiding the possibility of accumulation of thick tar in the main.

Plate 7 represents several views of a setting of retorts as adopted by Mr. Skoines, who has made this branch of gas engineering his special study, and whose extensive experience is entitled to some consideration. In this setting the current of caloric is caused to pass four times along the length of the oven. The retorts are each 23 in. × 15, the largest usually employed; their mouthpieces being slightly tapered in order to make the lids of diminished size, and thus reduce the labour in lifting them. As shown in the longitudinal section, the current of caloric descends into a chamber beneath the lower retorts, as represented in plan in section through G H; it then passes to the front underneath the bottom retorts, and rises freely through the flues half sunk in the piers, and passes along the outer sides of the four lower retorts. The current then returns to the front by the flues at the upper sides, as shown in plan in section through A B, and then enters the flue formed by the bricks on the top of the upper retorts, and so off to the main flue, where there is a damper to control the current.

It appears that Mr. Skoines has erected settings of this description in various works of the United Kingdom, applied to 5, 7, 8 and 10 retorts in a bed, with the best results on all the considerations of uniformity of heat throughout the bench, economy of fuel, and durability of settings; and we are informed that the average consumption of fuel used by them in carbonizing is under 20 per cent. of the coke produced. A bed of retorts like that described should carbonize, in the ordinary way of working at six hours' charges, 3 tons of coal per diem, producing nearly 60,000 feet of gas, and 2 tons of coke, of which 8 cwt. would be employed as fuel.

The furnace bars, as represented in the plan, are continued to the level of the front wall, which system offers great facilities for occasionally removing the clinkers without opening the furnace door. This plan will be found far superior to the old method of placing the ends of the bars 9 or 10 inches back from the wall.

The point of excellence in this setting undoubtedly arises from the uninterrupted passage of the heat from the furnace to the retorts, in which case the direction of the current is only a secondary consideration, as the inner sides of all the retorts are exposed to the direct heat of the furnace.

Plate 8 represents a setting of 10 retorts in a bed, as employed by the Imperial Gas Company during many years, and more recently introduced into other works with a slight modification wherever ground is very valuable and the capacity of the retort house restricted; but in order to draw and charge the four upper retorts, a travelling stage which runs on rails placed in the retort house is indispensable.

Fig. 1 is the elevation. Fig. 2 the longitudinal sections through A B and C D. Fig. 3 is a cross section through I K. Fig 4 is a plan through E F. Fig. 5 a plan through G H, and Fig. 6 the same through L M.

The furnace is 4 feet long and 30 inches deep, having only one fire-bar, and as shown there is a free passage for the heat, as the three short arches in the oven on each side are intended simply to support the upper retorts. The heat ascends in the centre of the setting, and the current, after passing over the top retorts, descends on the outside by the walls, when it enters the flues beneath the retorts, and passes to the vertical flue in the centre of the bed. This setting is sometimes constructed with the retorts on each side placed exactly perpendicular to each other, the ascension pipes being in this case all straight. A peculiarity possessed by the system is that the cross bars and screws can be secured by hinges to their respective mouthpieces, the other end of the cross bar dropping on to the ear, where it is held by a lug; and it is desirable that this should have a more general application, in consequence of the saving of labour to the men in taking down and lifting the cross bars. Some years ago we were informed by Mr. Methven that the fuel used for carbonizing by these settings was from 17 to 20 per cent. of the coke produced.

Hence we learn from the various authorities mentioned, that the percentage of fuel for heating the retorts, when the arch over the furnace is dispensed with, is about 20 per cent. of the coke produced, whereas, as shown elsewhere, in some large works this is exceeded by 50 per cent., thus clearly proving the loss sustained by that serious impediment to the passage of the caloric—the arch. And with the knowledge that this is retained in many works, and some of those of the largest magnitude, we may be charged with presumption in expressing views in opposition to the practice of some of our first engineers. In answer to this we leave the statement of facts to the judgment of the reader.

As the use of coal for heating the furnaces is of comparatively limited application, and as each class of coal requires a method of treatment according to the amount of volatile matter in combination therewith, no definite law or rule can be applied for the construction of the furnace. But for the object in question, usually the fire-bar surface is of large area, in order to permit the necessary supply of air to intermix with the volatile constituents of the coal; which, however, presents the difficulty of permitting an excessive quantity of air afterwards to pass when these are given off, which has an injurious effect. A very simple and effectual remedy for this evil has been applied by Mr. Tindall, of Walsall, who contracts the furnace-bar surface to the most limited extent, using only one fire-bar, and on each side of the furnace there is a passage for the admission of air, which is capable of being controlled, by very simple means. Thus the air in its transit becomes highly heated, and enters the furnace on the top of the coal, combining directly with its most volatile constituents, preventing the formation of smoke, and employing the fuel to the best advantage. By this system, according to the results of four years' working, Mr. Tindall has effected a very considerable economy, and wherever coal is used as fuel we believe the plan is worthy of consideration, particularly when the supply of air is under control.

INDEX.

	PAGE
ACCIDENT to Clegg	23
Accum's evidence	18
—— settings of retorts	83
—— treatise on gas lighting	83
Ackerman's gasworks	21
Air discovered to be ponderable	1
—— carburization of	320
—— intermixed with gas, effects of	125
—— tight retort lids	77
Alchemists, their pursuits	1
Alkalimeter	133
Ammonia, effect of, in gas	126
—— tests for	152
—— to ascertain strength of liquor of	133
—— manner to make sulphate of	311
Analyses of coal	60, 64
—— of cannel	59, 63
Anderson's scrubber	131
—— exhauster	121
—— breeze machine	310
—— retort settings	354
Annular tanks	195
Audouin's (Pélouze and) condenser	115
—— scrubber	132
BANISTER's experiments on burners	290
—— opinion on the formation of naphthaline	72
Barlow's table of the discharge of gas	263
Beale's exhauster	120
Becher's theory of combustion	5
Beckton Works	336
Bench of retorts	89
Black first produced carbonic acid gas	8
Blackburn's experiments on burners	293
Board of Trade, some Metropolitan companies under the control of the	146
Bogardus's dry meter	278
Bowditch on the argand	290
—— his experiment	116
—— his process of carburating	323
Boyle's experiments	3
Braddock's governor	230
Broadmeadow invented the exhauster	118
Brothers, H., traveller	144
—— specification of tank	177
—— —— of gasholder	211
Bude light	32
Bunsen's photometer	240
—— burner	300
Burners, Argand's patent	287
—— Murdoch's first experiment with	287
—— fishtail, invented	287
—— investigations of Drs. Christison and Turner on	287
—— variable lights yielded by	288
—— attempts at improving	288
—— gas gives no light when burned under certain conditions	288
—— light from gas affected by the atmosphere	289
—— —— and by mechanical appliances	289
—— —— evolved from argand	290
—— argand burners as usually made	290

	PAGE
Burners, batwing and fishtail	291
—— Kirkham's experiments with	291
—— effects of heavy pressure on	291
—— remedy for	292
—— degree of pressure necessary for	292
—— Wood's experiments on	292
—— light of gas augmented with large	292
—— Blackburn's experiments on	293
—— H. Woodall's report on	293
—— effects of glasses	293
—— the Farmer theorem	294
—— materials employed for	294
—— the Gas Referees' report on	295
—— tables of experiments	297, 299
—— dimensions of burner ordered by Referees as standard	300
CAPACITIES of station meters	224
Capital of the metropolitan companies	340
Carbonaceous deposit in retorts	71
Carbonic acid, test for	153
Carbonization, superiority of coal for producing gas	69
—— importance of proper heat	69
—— varied degree of temperature	69
—— effects of low temperature	70
—— —— moderate temperature	70
—— —— high heat and pressure	70
—— incrustation of carbon	71
—— method of removing the incrustation	71
—— Edge's process of scurfing	71
—— method of working the exhauster	72
—— the term exhaust explained	72
—— evil effects of using wet coal	72
—— the action of coal when undergoing carbonization	73
—— the average yield of coke per ton of coking coal	73
—— fuel for carbonizing estimated	73
—— necessity for good draught	74
—— the importance of the damper	74
—— different coals vary in the time for delivering their gas	74
—— tar used as fuel	75
—— luting of retort lids	76
—— air-tight lids, Morton's	77
—— mechanical stoking	77
—— Foulis's machine	78
—— West's machine	79
—— cost of carbonizing	80
Carburization of air and gas	320
Cast-iron mains, prices of	260
—— cost of laying	258
—— weights of	257
—— tanks, specification of	191
—— —— weights of	193
—— —— prices of	192, 193
Cathels's washer	127
—— district governor	233
—— retort settings	356
Cavendish's discovery	9
Chartered Company established	22
—— first meeting connected with the	16
—— first premises of the	23

INDEX.

	PAGE
Chemical evidence	68
Chemistry of gas manufacture	39
Chimney shafts, theory of	343
—— —— various kinds of	345
—— —— octagonal	345
—— —— at Edinburgh	346
Church and Mann's photometer	243
Clay retorts, first applied	92
—— —— patented by Grafton	91
Clayton's, Dr., experiments	5
Clegg, first works erected by	20
—— received silver medal	20
—— erected works at Ackerman's	21
—— engaged by the Chartered Company	23
—— accident to	23
—— first attempts to supply by meter	25
—— retired from the Chartered	25
—— invented his meter	220
—— his collapsing gasholder	86
—— —— his rotary holder	161
—— his retorts	85
—— the hydraulic main	106
—— his web retort	87
—— his pulse meter	276
—— his opinion on the hydrocarbon process	322
Clegg, jun., his Treatise on Gasworks	88
Cleland's exhauster	122
Coal, its origin	53
—— derivation of name	54
—— earliest mention of	55
—— Romans acquainted with	55
—— first worked	55
—— became a commercial product	55
—— patents in connection with	55
—— Lord Dundonald's patent	56
—— various descriptions of	56
—— method of appreciating coal	57
—— analyses of cannels	59
—— —— coals	60
—— value of coal for gas making	62
—— analyses of Scotch cannels	63
—— —— of gas coals	64
—— variable ash in coal	65
—— specific gravity no indication of quality of	66
—— schists and shales	66
—— lignite	66
—— pitch	67
—— chemical evidence	67
—— importance of employing fresh coal	68
—— advantage of coal for producing gas	60
—— deterioration of coal by storage	68
—— liability to spontaneous combustion	60
—— wet coal produces naphthaline	72
—— non-conducting quality of	72
—— varied facilities for delivering gas	75
Coke, production of	307
—— used as fuel	307
—— sales of	308
Combustion, Hooke's theory of	1
—— phlogistic theory of	5
—— spontaneous in coal	68
Compensating meters	282
Competition of gas companies	348
Composite tanks	183
Composition of gas	316
Compound tanks	198
Concrete	170
Concrete tanks	185
Condenser, object of the	108
—— earliest method of employing the	108
—— John Malam's	108
—— Perks's	108
—— the air	109
—— effects of cold on gas	110
—— Wright's	111
—— Kirkham's	112
—— Warner's	112
—— observation on	112

	PAGE
Condenser, Graham's	113
—— the South Metropolitan	114
—— the Beckton Works	114
—— Pélouze and Audouin's	115
—— Dr. Bowditch's experiment with	116
—— various influences on the action of the	117
—— the breeze box to assist the	117
Congreve's meter	276
Connections of purifiers	141
Considerations on establishing gasworks	324
Consumers' meters	272
Contrivances for lifting covers of purifiers	143
Cowan and Warner's meter	283
Creighton's account of Murdoch's operations	135
Croll patented the use of oxide of iron for purification	148
—— first employed clay retorts in London on an extensive scale	32
Crookes's radiometer	248
Crosley, S., invented the pressure register	295
—— became partner with Clegg	273
—— invented the dry governor	228
—— in conjunction with Clegg perfected the meter	274
DAVY, Sir H., essay on flame by	7
—— his experiment	248
Defries' dry meter	278
Dempster's lifting apparatus	143
Details of retort setting	102
District governor	233
Donovan's patent	310
Douglas's concrete tank	185
Dry-lime purification	147
Dry meters	276
EDGE's system of scurfing retorts	71
Edinburgh chimney shaft	346
Effects of excessive exhaustion in retorts	125
Evans discovered the revivification of oxide of iron	149
—— his photometer	244
—— engineer of the Gas Light Company and constructor of Beckton Works	337
Exhauster, object of	118
—— invented by Broadmeadow	118
—— Grafton's	119
—— Beale's	120
—— Jones's	120
—— Anderson's	121
—— Cleland's	122
—— Körting's	122
—— Wright's regulator for	124
—— excessive exhaustion, effects of	125
FARMER theorem	294
Fish revived oxide by steam	144
Fittings	305
Foulis's stoking machine	78
Gas produced by Paracelsus	2
—— discovered and named by Van Helmont	2
—— produced and stored by Boyle	3
—— ignited by Shirley	4
—— accidental ignition of	4
—— obtained by distilling coal by Hales	5
—— " " " Dr. Clayton	5
—— " " " Dr. Watson	7
—— carbonic acid discovered by Black	8
—— hydrogen discovered by Cavendish	9
—— oxygen discovered by Priestley	9
—— from coal, Murdoch's first experiments with	11
—— Soho Works lighted with	12
—— Lebon's patent for manufacturing	12
—— employed for lighting a house in Paris	13
—— Winsor's first experiments with	13
—— Winsor's first exhibition in London of	14
—— early experimental apparatus for making	13
—— Winsor's efforts to establish a company for lighting by means of	15
—— petition to the King by projected company	16

INDEX. 361

	PAGE
Gas Committee of House of Commons on	17
—— Accum's evidence	18
—— evidence of James Watt, jun.	19
—— works established for lighting with	20
—— the first works erected by Clegg	20
—— the first company established	22
—— Westminster Bridge lighted by	23
—— accident with	23
—— first parish lighted with	24
—— Winsor lighted the Passage des Panoramas	25
—— Luxembourg Theatre lighted with	26
—— Paris lighted with	26
Gas fittings	304
Gasholder, primitive bladder	158
—— invented by Lavoisier	159
—— described	159
—— first practical application of the	160
—— Clegg's rotary	161
—— —— collapsing	86
—— Ackerman's	21
—— conditions which affect the	200
—— untrussed roofs for	201
—— table of the weight of sheets for	202
—— application of table	202
—— method of calculating table	202
—— for ascertaining pressure and weights of	204
—— specification of a	205
—— engraving of a	206
—— cost of gasholders	208
—— telescopic holders	209
—— drawing of	210
—— specification of	211
—— specification of Redheugh holder	214
—— weight of	217
—— large holder at Fulham	218
Gasholders at Beckton	218
Gasholder tanks, various sites for	163
—— site at Beckton for	164
—— general construction of	166
—— materials employed in building	167
—— Portland cement	168
—— puzzolana	169
—— concrete	170
—— caissons	171
—— price of material for brick	172
—— specification of a brick tank	172
—— conditions of contract for building	174
—— Methven's specification for a	175
—— H. Brothers' specification for a	177
—— specification of small	179
—— prices of tanks	180, 188
—— stone tank, specification	181
—— composite tank, specification, Wyatt's	183
—— concrete tank, Douglas's	185
—— —— G. Livesey's	187
—— cast-iron tank, specification	191
—— weight of cast-iron tanks	193
—— prices of	192, 194
—— annular tanks formerly at the City of London Gasworks	196
—— comparison of dimensions of	197
—— at Rotherhithe	197
—— price of	197
—— compound tanks	198
—— —— constructed by G. W. Stevenson	198
—— —— the method of construction	198
—— —— comparison of cost with iron tank	199
Gasworks of various magnitudes	327
—— for private establishments	329
—— for manufacturers	329
—— for varied populations	330
—— in Scotland	332
—— of three million feet	332
—— estimate of operation of	335
—— of sixty million feet	334
—— at Redheugh	336
—— at Beckton	337
Giroud's rheometer	302

	PAGE
Governor, the object of the	228
—— as first made by Clegg	228
—— made to counteract the irregularity of the supply of the first meters	228
—— Crosley invented a dry	228
—— Parkinson and Co.'s	229
—— accident occasioned by a	230
—— Braddock's	230
—— Hartley's	231
—— Peebles's	231
—— Hulett's	232
—— the rule for calculating the	232
—— by pass of the	232
—— Stevenson's, for districts	233
—— Cathels's, for districts, how applied	233
Grafton patented clay retorts	91
—— his experiments	118
—— his exhauster	119
Graham's condenser	114
—— opinion on ammoniacal liquor	115
HALES', Dr., experiments on gas	5
Handbook for gas engineers	142
Hartley, improvements in governors	231
—— on photometry	243
Heard, compressed gas	25
—— patented oxide of iron	148
Henry's, Dr., experiments	22
Hill's patent	179
Hislop's analyses of coals	63
Holman's hydraulic ram	144
Hooke's micrographia	2
Hulett's regulator	232
Hutchinson's retort settings	356
Hydraulic main invented	106
—— proposed as purifier	106
—— formation of pitch in	106
Hydrocarbon process	320
INFLUENCE of barometric pressure and temperature	352
KING, A., adopted the Bunsen photometer	240
—— invented the turned and bored joint	252
King's, W., experiments on globes	294
Kirkham, his experiments on burners	291
—— constructor of the largest holders	218
—— his condenser	112
Körting's blower	146
—— exhauster	123
LAMING's patent	149
Lamps, number of public, in London	261
—— supplied by each company	261
Lavoisier, discoveries by	7
—— invented the gasholder	153
Lebon, his patent for gas making	12
—— additions thereto	13
—— lighted a house in Paris with gas	13
Leoni's adamas	295
Livesey, G., scrubber invented by	127
—— composite tank erected by	183
—— concrete tank erected by	187
—— his scrubber	130
London Portable Gas Company	25
—— and Westminster Oil Gas Company	27
Longworth's form of contract	174
Lowes' scrubber	128
—— evidence on the oil gas	29
—— jet photometer	250
—— carburating process	320
MAINS of wood formerly employed for water	251
—— first used for gas	251
—— various materials employed for	251
—— the Chameroy pipe	252
—— the leaded jointed	252
—— the turned and bored jointed	252
—— effects of expansion and contraction in	252
—— durability of	253

3 A

INDEX.

	PAGE
Mains, action of various soils on	253
—— the "life" of	253
—— necessity for preserving	253
—— cause of leakage from	253
—— supposed loss of light by gas passing through	253
—— leakage from	254
—— unaccounted-for gas	254
—— diameters of	254
—— obstructions by naphthaline and other causes in	255
—— method of detecting stoppages in	255
—— necessity for sound	256
—— excess of syphons to be avoided in	256
—— necessity for ascertaining the leakage in	256
—— means of ascertaining leakage from	256
—— table of the weight of	257
—— present system of laying	257
—— Newbigging's table of prices of laying	258
—— effects of restricting prices in laying	258
—— a substitute for the clay belt	258
—— temporary valve	259
—— accident from the use of bladder valve	259
—— effects of inhaling gas	259
—— effects of small	259
—— Newbigging's table of the cost of mains per yard at various prices per ton	260
—— length of the sewers of London	261
—— for water in London	261
—— length of the streets of London	261
—— estimated length of the Companies', of London	261
—— mileage of, compared with population	262
—— estimate of, embedded in London	262
—— unanimity of opinion concerning the capacity of mains for distribution	262
—— Barlow's table of the discharge of gas through, of various diameters	263–267
—— application of rules, with table	267
—— formulæ for the distribution of gas	268
—— table of square roots of pressures	269
—— " " specific gravities	269
Malam's improvements in gas apparatus	24
—— dry meter	26
—— arrangement of purifiers	27
—— treble purifier	136
—— first retort settings	85
—— gas meter	273
Mann's photometer	243
—— scrubber	129
Mansfield's patent	320
Mechanical stoker, Foulis's	78
Meter, Clegg and Crosley's	275
—— station, necessity for	220
—— action described	221
—— capacities and prices of	224
—— tell-tale for	234
—— influences of variation of water line in	225
—— Parkinson's	225
—— dimensions and weight	225
—— consumers', Clegg's first attempts to produce a	272
—— Clegg's patent rotative	272
—— Malam's improved	273
—— Malam's dry	277
—— Clegg and Crosley's improvement in the	275
—— Congreve's	276
—— Clegg's pulse	276
—— Bogardus's dry	277
—— Sullivan's	278
—— Defries' and Taylor's	278
—— Richards'	279
—— compensating	282
—— legalization of	281
—— Warner and Cowan's	283
—— opinions concerning meters	284
—— motive-power	285
Methven's retort settings	356
—— specification of tank	175
Milne and Neilson invented the fishtail burner	287
Mixture of air and gas	125
Morton's air-tight lids	77

	PAGE
Murdoch, his first application of gas lighting	11
—— description of the same	11
—— the first method of purification adopted by him	135
—— the first burners employed by	287
—— the first retorts used by	81
Newbigging's handbook	142
—— table of cost of mains	260
—— —— cost of laying	258
—— opinion on fuel	356
Oil gas applied for lighting	25
—— works erected at Bow	26
—— various towns lighted with	27
—— London and Westminster Company proposed	28
—— alleged advantages of	28
—— evidence concerning	28
—— failure of	28
—— Portable Gas Company	25
Oxygen discovered by Priestley	9
Paddon's scrubber	132
Pall Mall first lighted by gas	16
Paracelsus first produced gas	2
Paris, Winsor's attempts at	25
—— gas established at	26
Parkinson's governor	229
—— station meter	225
Paterson's, J, description of gas coals	57
—— analyses of cannels	59
—— analyses of coal	60
Patterson's, R. H., discovery	154
Pélouze and Audouin's condenser	115
—— scrubber	132
Perks's setting of retorts	83
Phillips invented purification by hydrate of lime	25
Phillips's W. R., system of purification	50
Phœnix Company established	24
Photogenic gas	320
Photometer, theory of the	238
—— Count Rumford's	239
—— Ritchie's	239
—— Wheatstone's	239
—— Bunsen's	240
—— first introduced by A. King	240
—— Wright's	241
—— rules for making	242
—— Letheby's	243
—— Church and Mann's	243
—— Evans's	244
—— instructions of Referees	245
—— table of	245
—— Davy's experiment	245
—— Crookes's radiometer	248
—— Lowes' jet photometer	248
Portable Gas Company	25
Portland cement	168
Pressure gauge, objects of	234
—— ordinary kind	234
—— King's gauge	234
—— register, Crosley's	234
—— Wright's	235
Priestley, Dr., discovered oxygen	9
—— his experiments on air	9
—— his pneumatic inventions	158
Progress of setting retorts	81
Purification, considerations on	146
—— as defined by the Board of Trade	146
—— the best system of	45, 146
—— wet-lime purification	147
—— quantity of lime required	147
—— dry-lime process of	148
—— oxide of iron patented by Heard	148
—— " " " Croll	148
—— all oxides not suitable	149
—— method of employing	148
—— oxide of iron, mode of using for	149
—— to ascertain the value of oxide for	150

INDEX.

	PAGE
Purification to ascertain the quantity of sulphur in oxide..	151
—— sulphate of iron for	152
—— mode of preparing..	152
—— various methods of	153
—— to prepare turmeric paper	152
—— to prepare blue litmus paper	153
—— to prepare red „	153
—— to prepare acetate of lead	153
—— to prepare test for carbonic acid..	153
—— the constant test	153
—— discovery of Mr. Patterson	154
—— instruction of Gas Referees..	154
—— the delicacy of modern tests	154
—— Referees' instructions for testing	154
Purifiers, Murdoch's first	135
—— Clegg employed lime in	135
—— Heard's patents for purification ..	135
—— Winsor's patents	136
—— Malam's treble purifiers	137
—— Phillips' hydrate of lime for	137
—— Malam's arrangement of	138
—— Cookey's rotary valve	140
—— Walker's rotary valve	141
—— Newbigging's 'Handbook' on	142
—— dimensions of	142
—— dimensions of connections for	143
—— object of several layers of material	142
—— depth of seals of	143
—— wooden grids for	143
—— contrivances for lifting covers	143
—— H. Brothers' traveller	166
—— Messrs. Cockey's apparatus ..	144
—— hydraulic ram	144
—— Messrs. R. Dempster and Sons	144
—— Wyatt's	144
—— Holman's	144
—— Körting's reviving apparatus	145
—— dimensions of purifiers	145
Puzzolana..	169
RADIOMETER, Crookes's	249
Redheugh Gasworks	310
Referee, Metropolitan, instructions for testing the purity of gas	154
—— instructions for testing the illuminating power of gas	245
—— report on burners..	295
Regulator, object of	301
—— Crosley's dry	228
—— Peebles'	301
—— consumers' wet	301
—— lamp	302
—— Wright's lamp	302
Residual products, coke the most important	307
—— coke produced per ton of coal	307
—— prices of coke realized	308
—— difficulties attending the sale of coke	308
—— means of advancing the sale of coke ..	309
—— breeze used for fuel	309
—— Anderson's compressing machine	310
—— sulphate of ammonia, the manufacture of..	311
—— cost of	311
—— defects of the methods of evaluing ammoniacal liquor	312
—— the applications of	313
—— foul lime, application of	314
Retort setting ..	354
Retorts first employed by Murdoch	81
—— setting described by Accum	83
—— Perks's settings	83
—— Rackhouse's setting	84
—— Clegg's rotary	85
—— Clegg's web retort	87
—— setting described by Clegg, jun...	89
—— bench of retort	89
—— cast-iron, the defects of	90
—— clay retorts	91
—— Grafton's patent for	92
—— first made and used	92

	PAGE
Retorts adopted at Brick Lane station	93
—— adopted by Mr. T. Livesey	93
—— Grafton's brick oven	92
—— arguments for and against clay retorts	93
—— various forms of clay retorts	94
—— influence of form of „	94
—— advantages of brick	94
—— the importance of good quality of	95
—— theory of retort setting..	96
—— considerations on the most advantageous manner of applying heat	96
—— capacity of furnace	96
—— description of fuel..	96
—— fuel assisted by the steam from the ash pans	97
—— best means of conducting heat	97
—— obstruction of non-conducting material	97
—— prevention of loss of heat by radiation	98
—— practice of setting retorts	99
—— single and through retorts	99
—— advantages of through retorts	99
Rey discovered air to be ponderable	1
Rheometer, Giroud's	302
Richards' dry meter	279
Rotary holder, Clegg's	161
—— retorts, Clegg's	85
SERVICES, wrought-iron pipe	270
—— means of preserving	270
—— leaden pipe	270
—— method of clearing obstacles in	271
Shirley, description of the ignition of gas by	4
Skoines's retorts	356
Specific gravity of coal no indication of quality	65
—— of gas not reliable	66
Station meters	220
Stone tanks	181
Sugg's burner	299
Sullivan's dry meter	277
TABLES of specific gravity of liquids	134
—— of weights of iron	201—203
—— of discharge of gas through mains	263—267
—— of square roots of specific gravities	269
—— of square roots of pressures	269
Tar employed as fuel for furnaces	77
—— manufacture of	313
—— applications of	314
Tests, various	152
Thompson, Lewis, chemistry of gas manufacture	39
—— analyses of coal	64
—— purity of gas	146
—— on oxide of iron	149
—— composition of coal gas..	316
Twaddle, hydrometer	133
VAN HELMONT, first recognized gas	2
—— its name given by..	2
—— his experiments with gas	3
—— conclusions drawn by	3
WALKER's scrubber	129
—— valve for purifiers..	141
Warner and Cowan's meter	283
Washer and scrubber	126
—— object of	126
—— estimated value of ammonia	126
—— value of ammonia disregarded	126
—— patent obtained by Wilson	127
—— affinity of water for ammonia	127
—— Livesey's washer	127
—— Cathels's washer	127
—— Lowe's scrubber	128
—— dilute sulphuric acid used as purifying agent	129
—— defects of the system	129
—— chemical manufactories for tar established	129
—— value of ammonia for agricultural purposes understood	129
—— Mann's scrubber	129

3 A 2

INDEX.

	PAGE
Washer and scrubber	126
—— Walker's scrubber	130
—— means of determining the capacity of scrubbers	130
—— yield of ammonia from coal	130
—— Anderson's brush scrubber	131
—— Paddon's scrubber	132
—— Cleland's steam scrubber	132
—— Pélouze and Audouin's spray scrubber	132
—— Whimster's exhauster and scrubber	133
—— the value of ammoniacal liquor	133
—— the means of ascertaining, by hydrometer	133
—— " " by sulphuric acid	133
—— defect of acid test "	312
—— table of the specific gravity of weights and liquids	134
—— the value of ammonia	134
Watson's, Dr., experiments	7
Watt, his letter on the kinds of air	6
Watt, jun., his evidence	10
West's drawing and charging apparatus	79
Whimster's exhauster and washer	152
—— description of the manufacture of ammonia	311

	PAGE
Whimster's description of the manufacture of tar	313
Whitehouse's patent for gun-barrel	30
Winsor, first acquainted with gas	13
—— first experiments	13
—— first exhibition of gas in London	14
—— his prospectus	15
—— introduced gas into Paris	25
—— his patent for purification	136
Wood's process of ascertaining the quantity of sulphur in oxide	150
—— experiments on burners	292
—— on globe glasses	294
Woodall's, C., application of steam blower	145
Woodall, H., report on burners	203
Wright's condenser	111
—— pressure register	237
—— photometer	241
—— lamp regulator	302
Wyatt, V., specification of gasholder by	214
—— specification of tank by	183
—— drawings of Redheugh Works by	335

LONDON: PRINTED BY WILLIAM CLOWES AND SONS, STAMFORD STREET AND CHARING CROSS.

RETORT SETTINGS OF

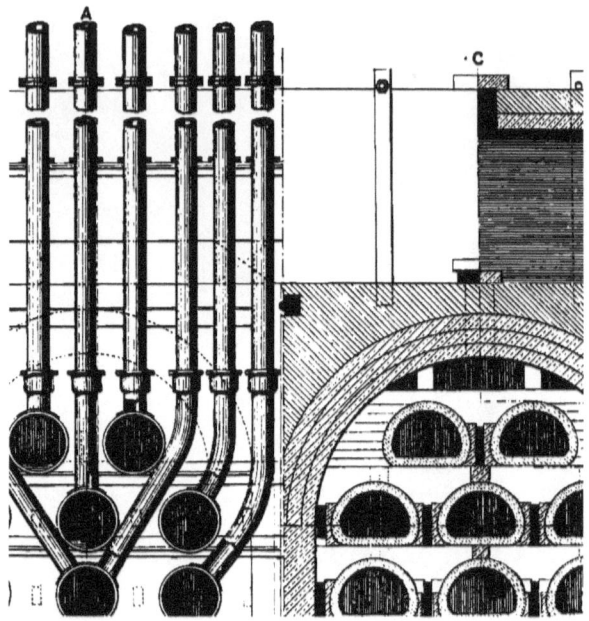

THE FIRST MAGNITUDE.

SECTION THROUGH A.A.   SECTION THROUGH C.D.

Fig. 1.

Fig. 4.

Scale, ½ Inch = 1 Foot.

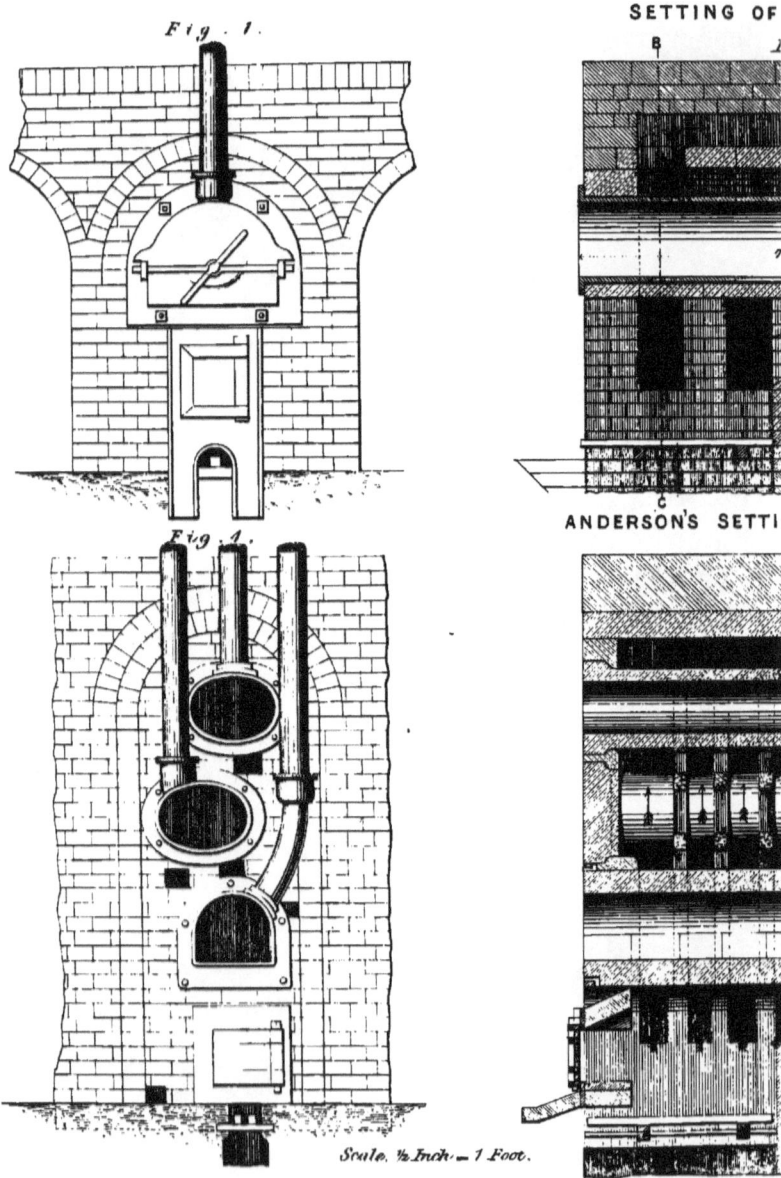

SETTING OF

ANDERSON'S SETTI

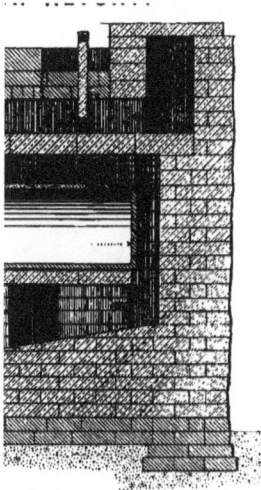

Fig. 3.

REE CLAY RETORTS.

Fig. 6.

SETTING OF TWO CLAY RETORTS.

Scale of Feet.

SECTION THRO R.S.

SECTION

SECTION THROUGH G.H.   SECTION THROUGH E.F.

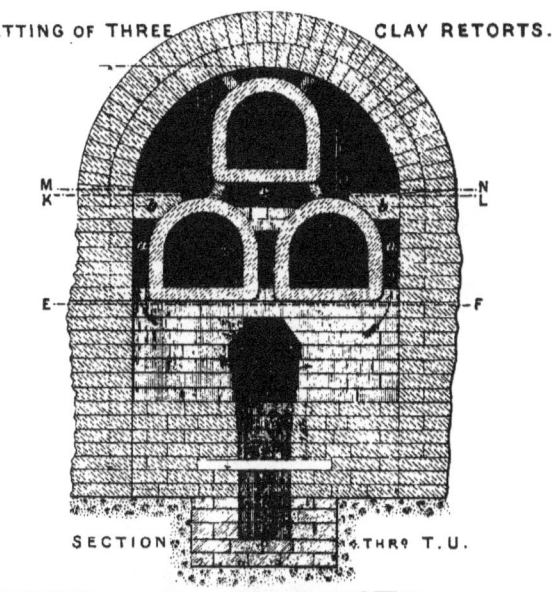

SETTING OF THREE CLAY RETORTS.

SECTION THRO T.U.

SECTION THROUGH M.N.   SECTION THROUGH K.L.

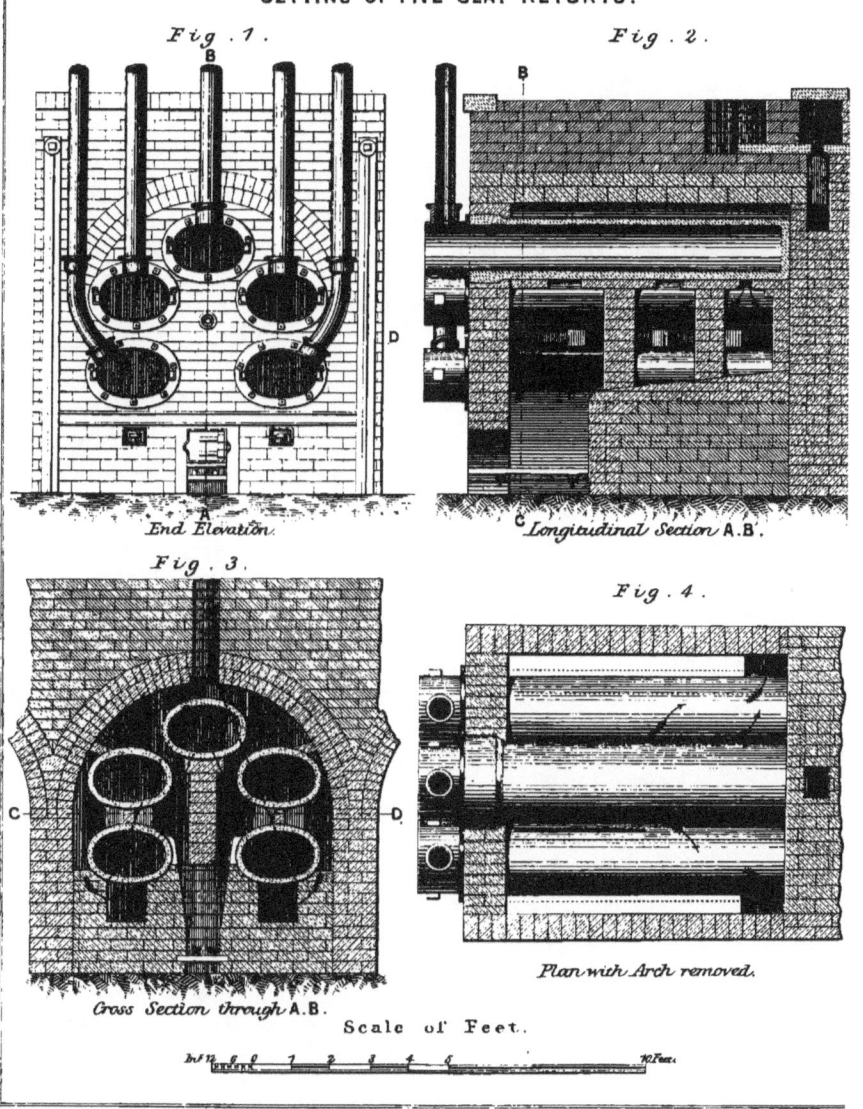

HUTCHINSON'S SETTING OF FIVE RETORTS.

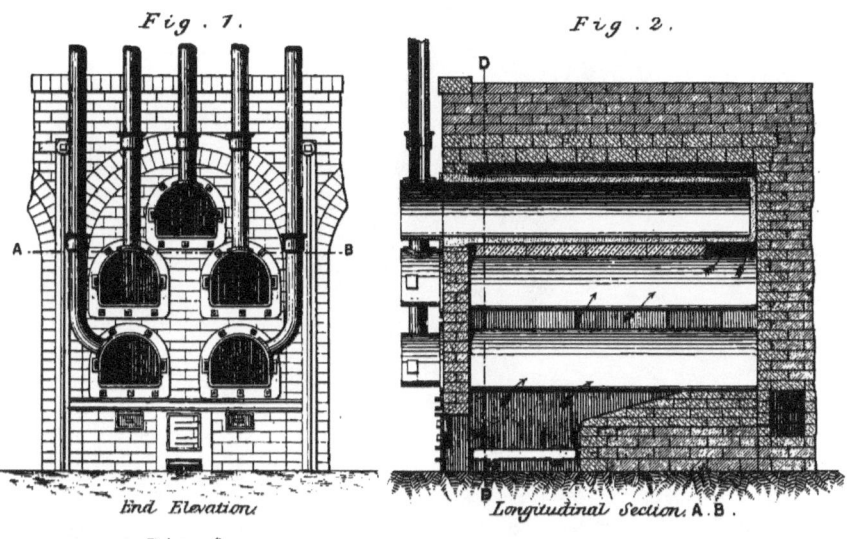

Fig. 1. End Elevation.
Fig. 2. Longitudinal Section, A.B.
Fig. 3. Cross Section through D.D.
Fig. 4. Plan with Arch removed.

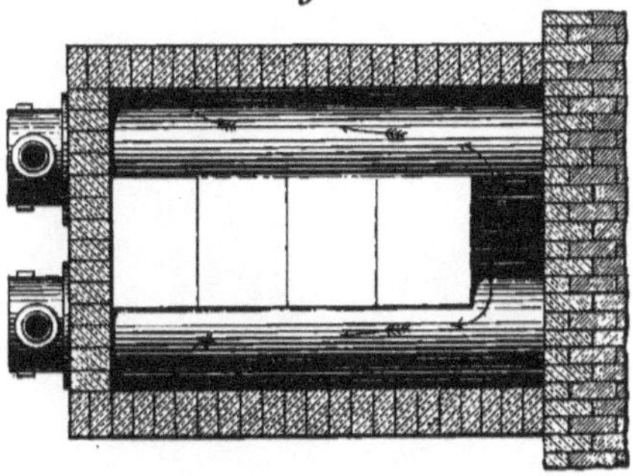

Scale of Feet.

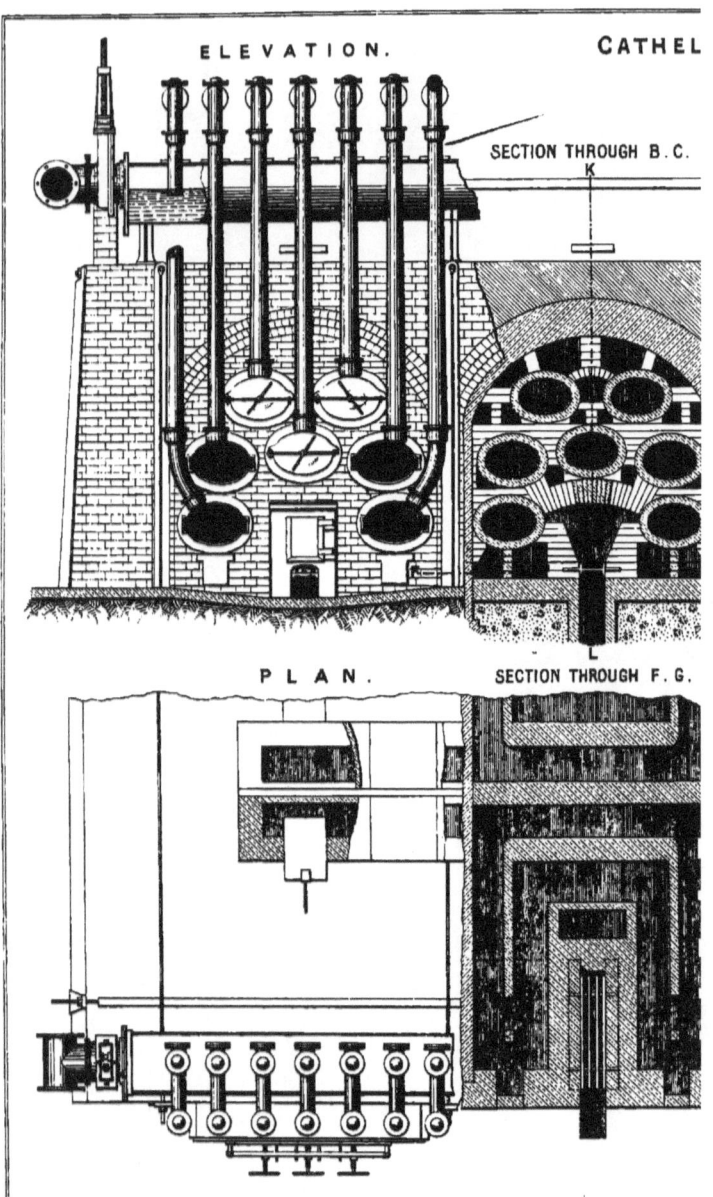

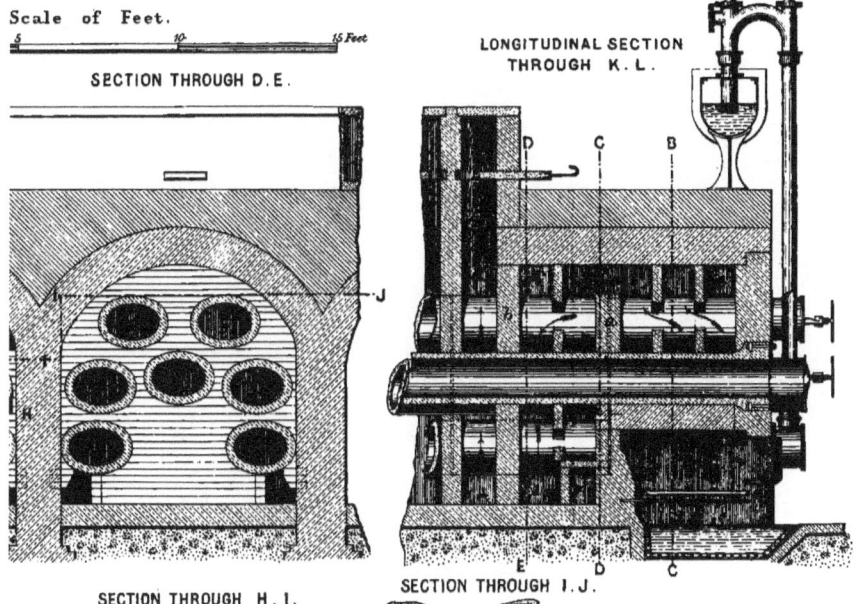

RED SETTING OF SEVEN CLAY RETORTS.

SECTION THROUGH D.E.

LONGITUDINAL SECTION THROUGH K.L.

SECTION THROUGH H.I.

SECTION THROUGH I.J.

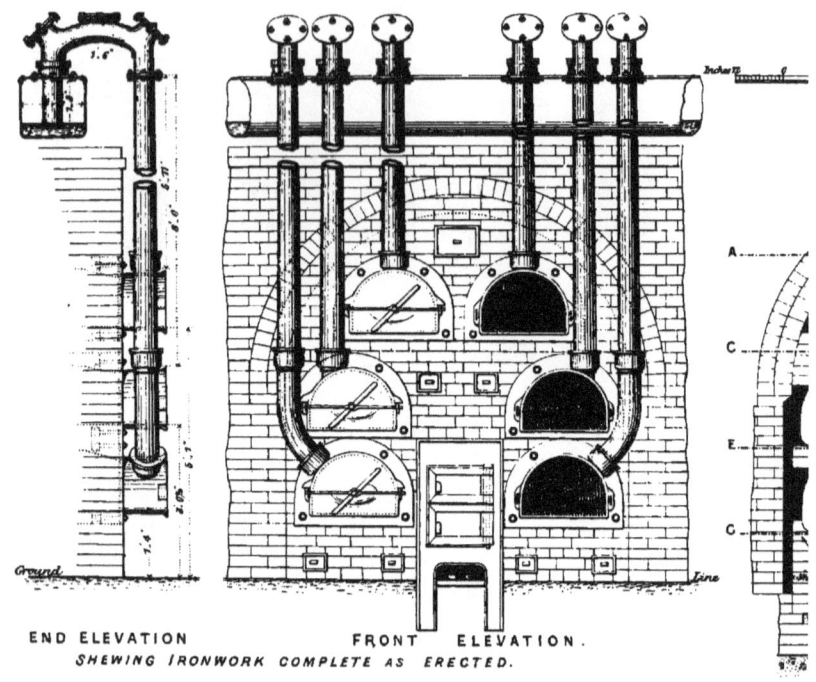

END ELEVATION FRONT ELEVATION.
SHEWING IRONWORK COMPLETE AS ERECTED.
SECTION THROUGH A.B. SECTION THROUGH C.D.

SKOINES' SETTING OF SIX CLAY RETORTS.

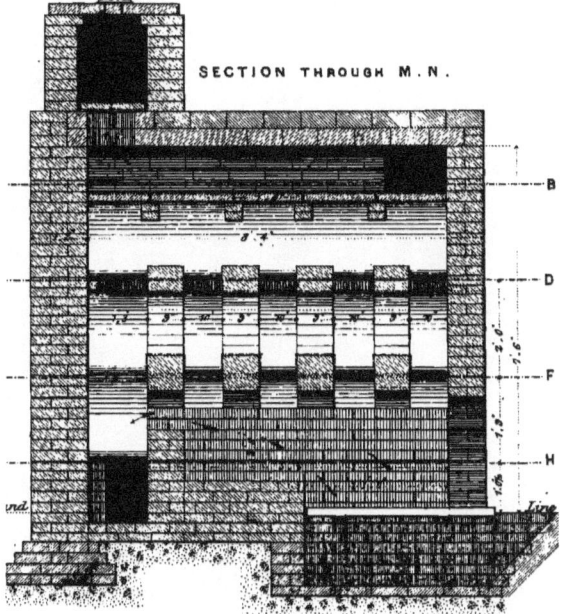

SECTION THROUGH M.N.

SECTION THROUGH G.H.

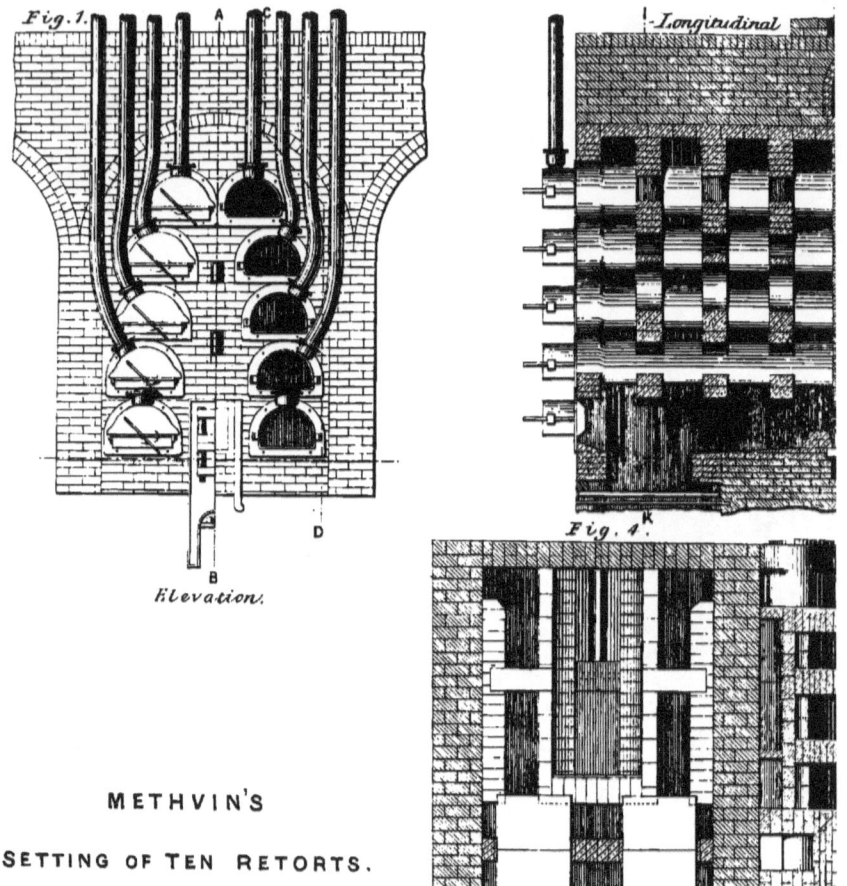

METHVIN'S SETTING OF TEN RETORTS.

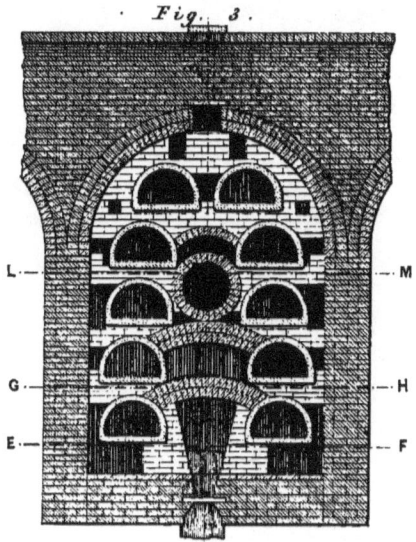

Fig. 3.

Section through I. K.

Scale of Feet.

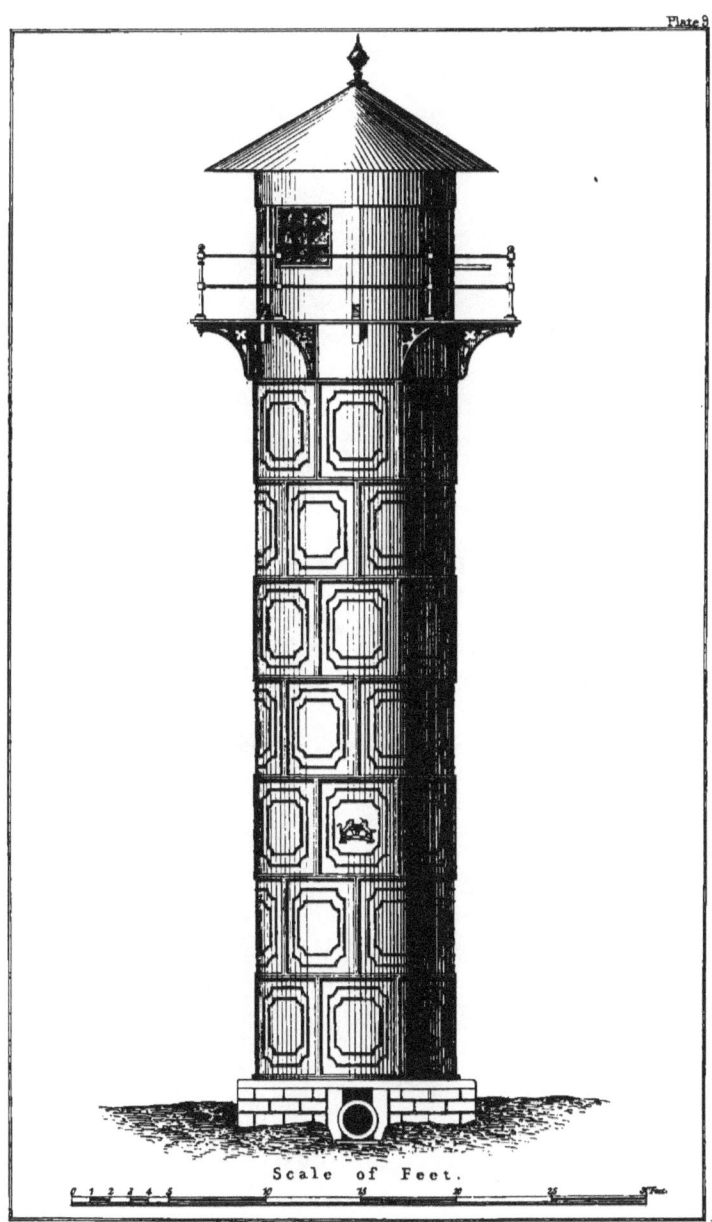

Scale of Feet.

E. & F. N. Spon, London & New York.

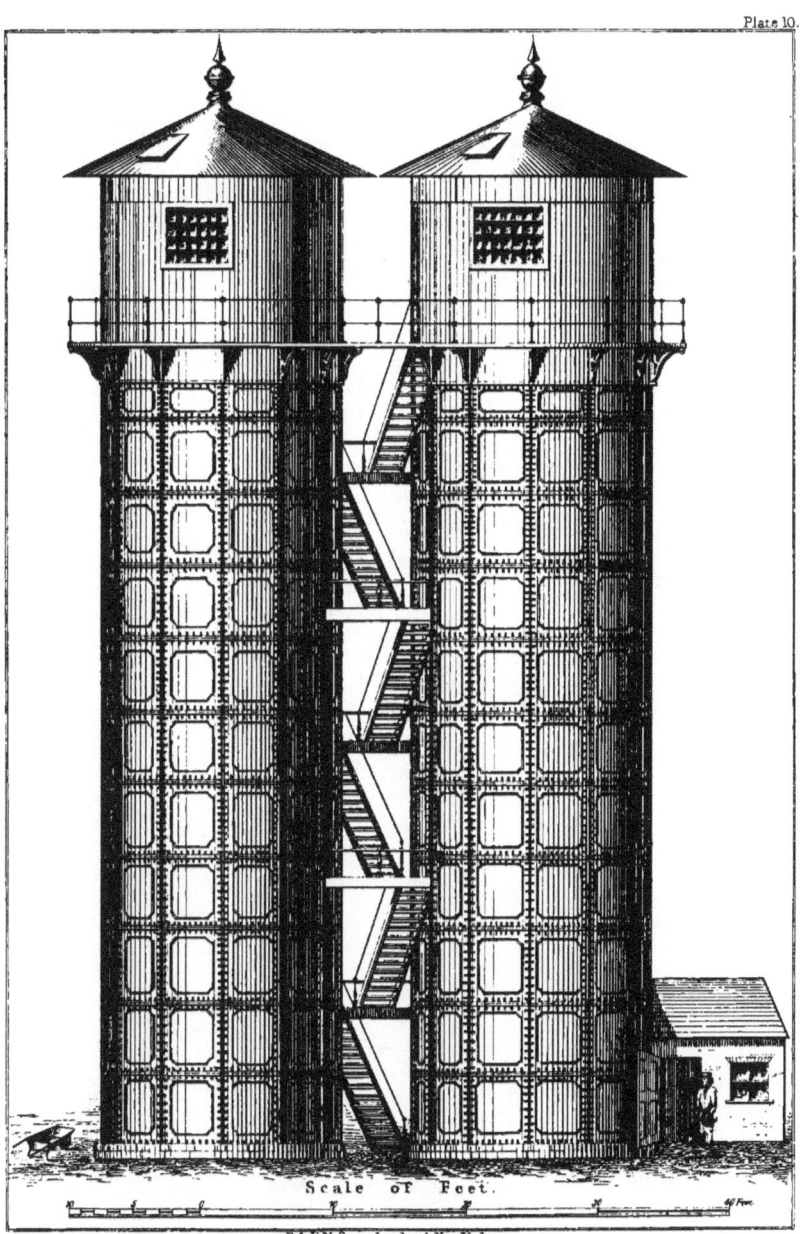

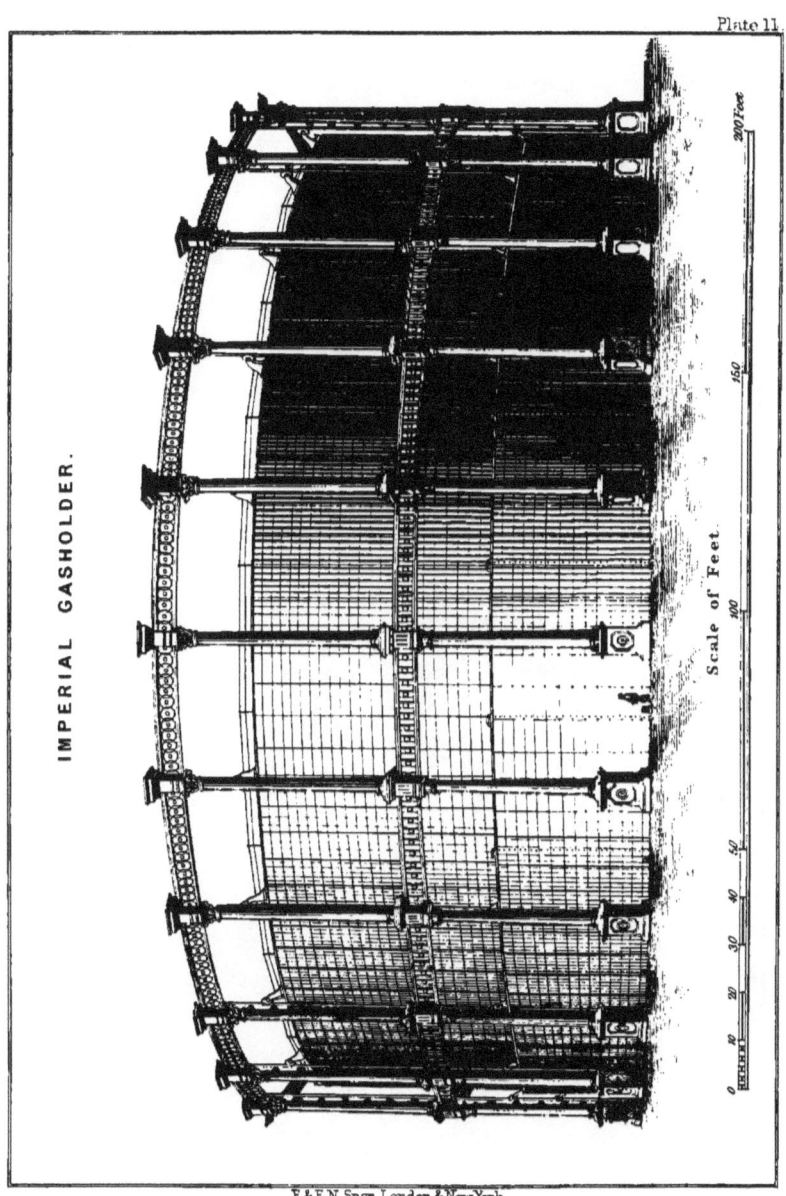

GASHOLDER IN IRON TANK.

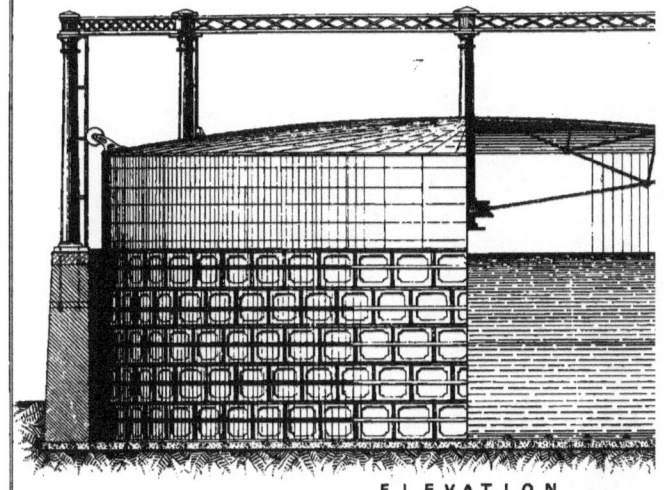

ELEVATION.

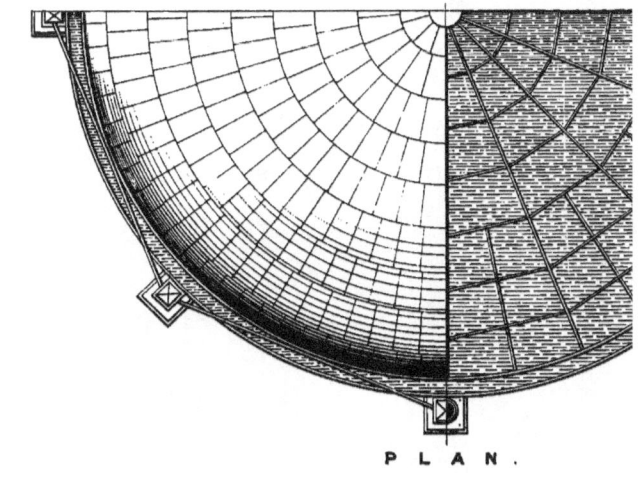

PLAN.

E. & F. N. Spon, London & New York.

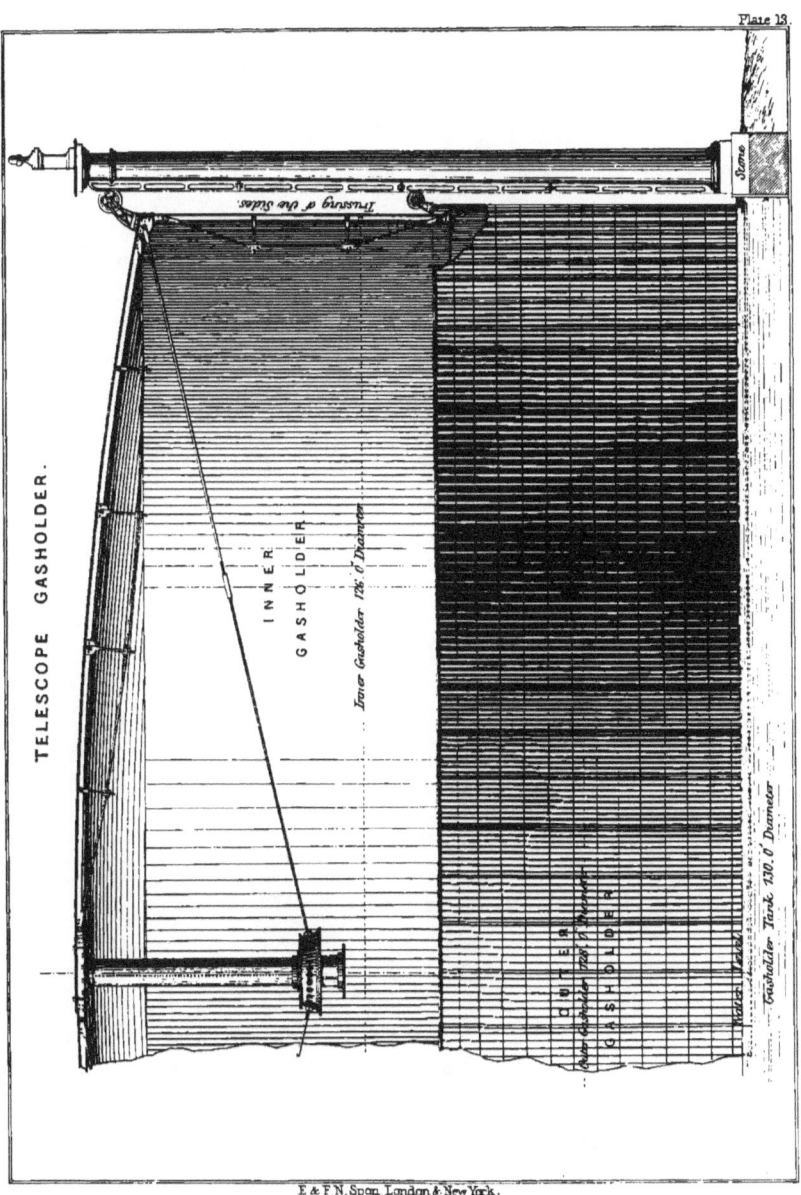

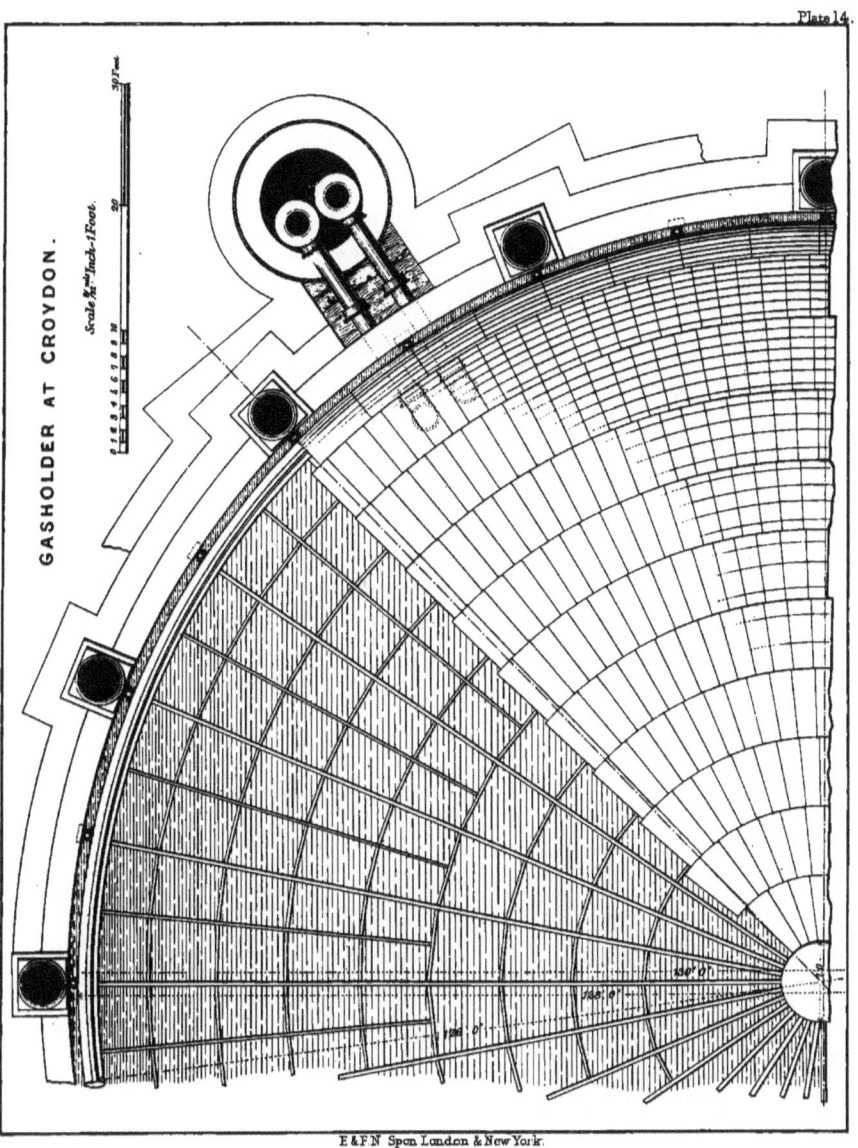

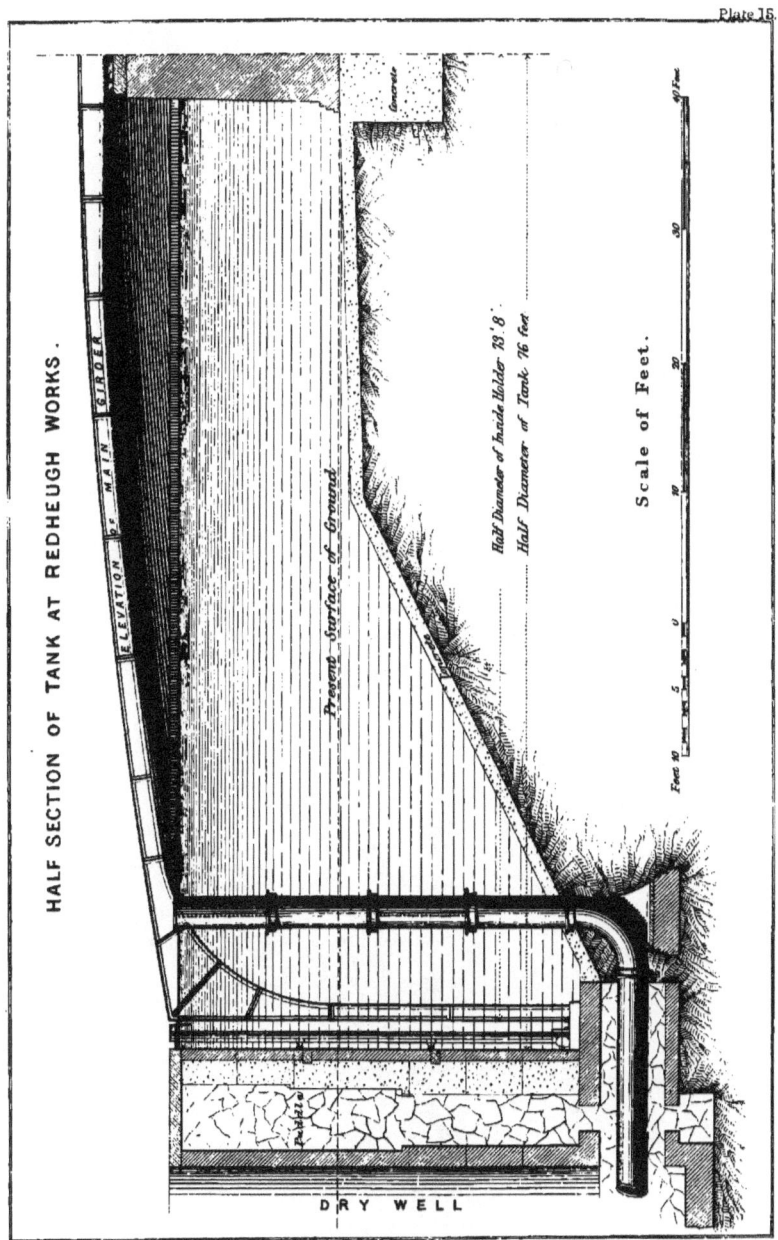

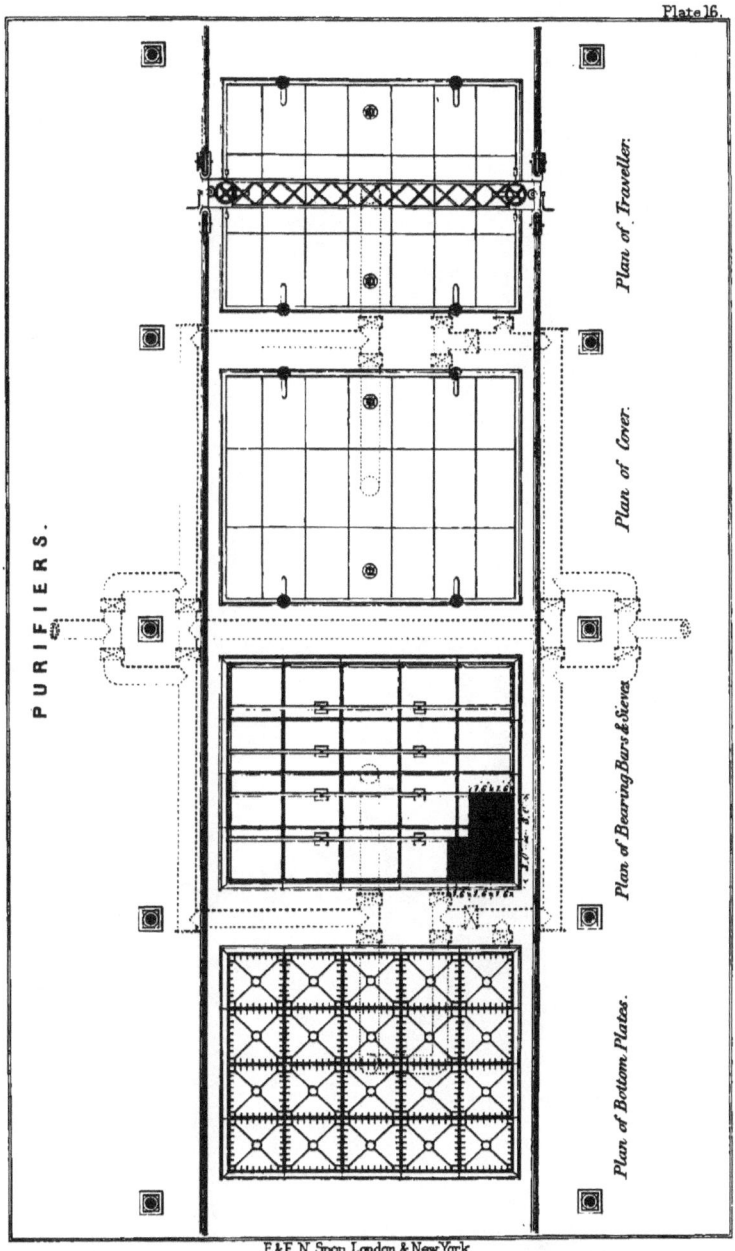

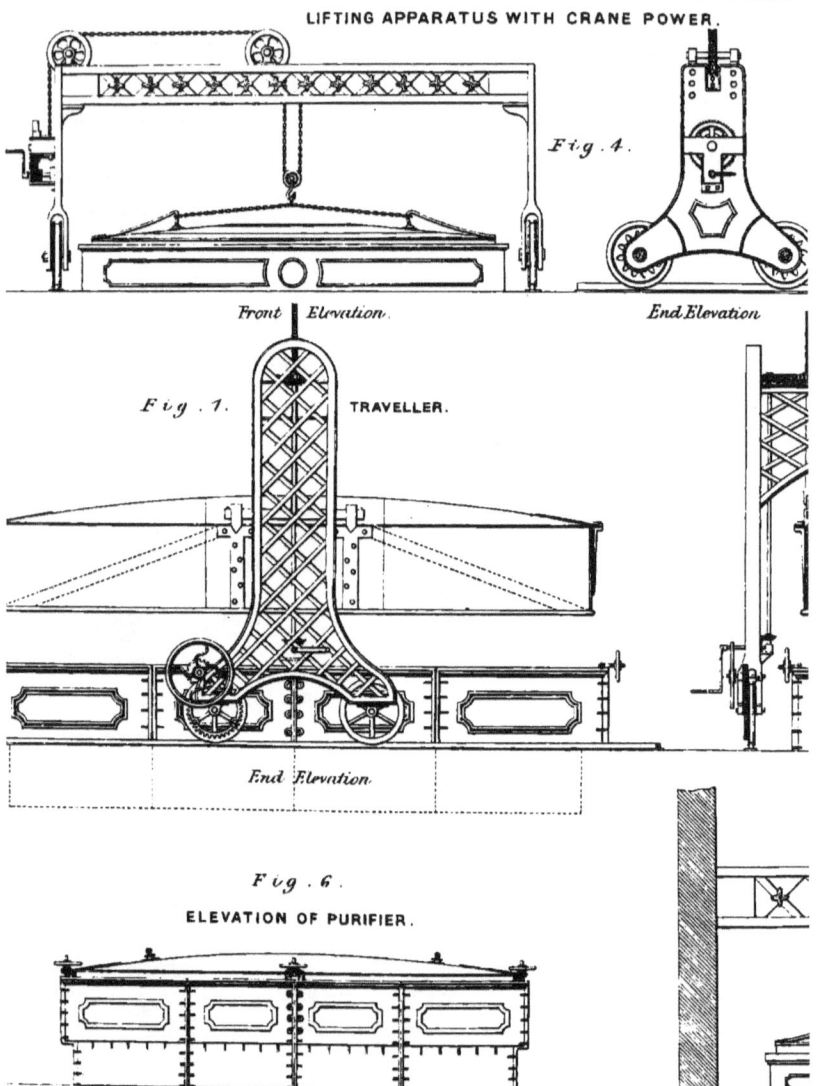

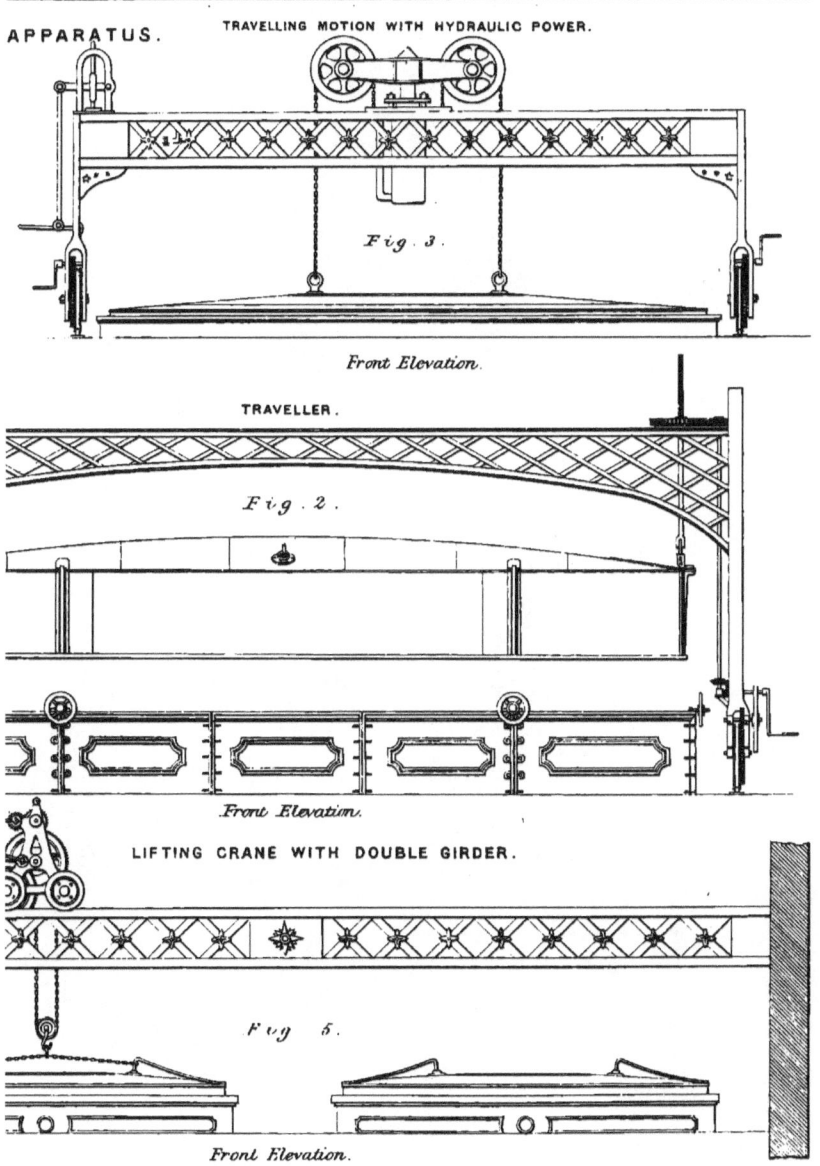

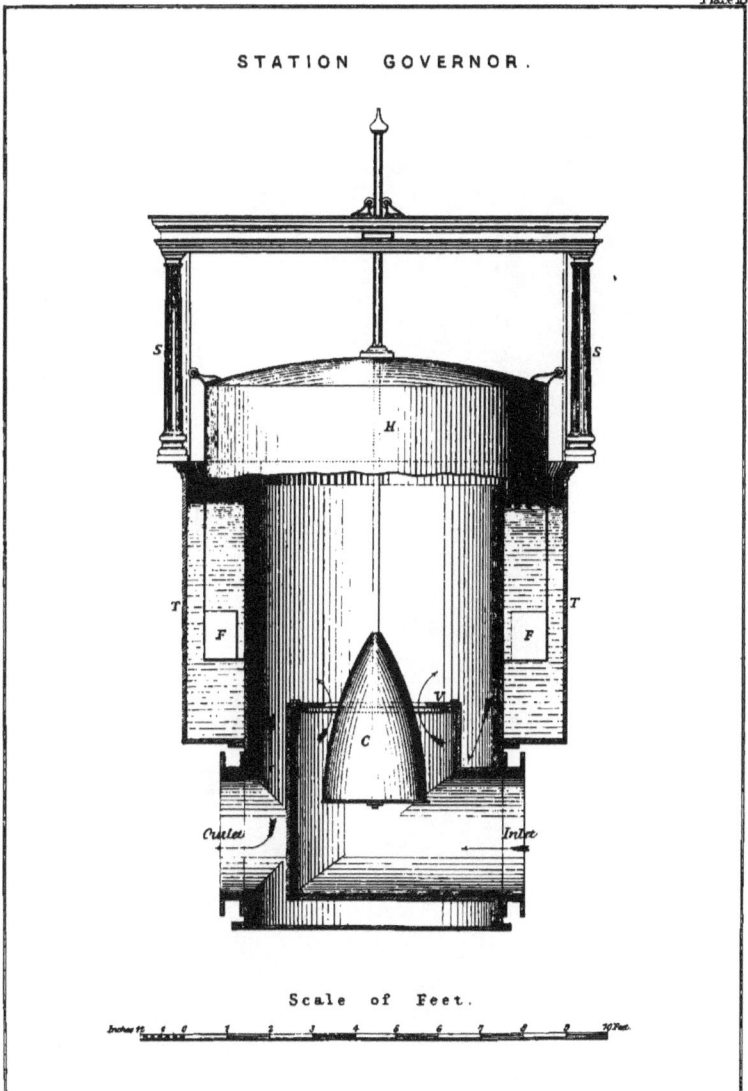

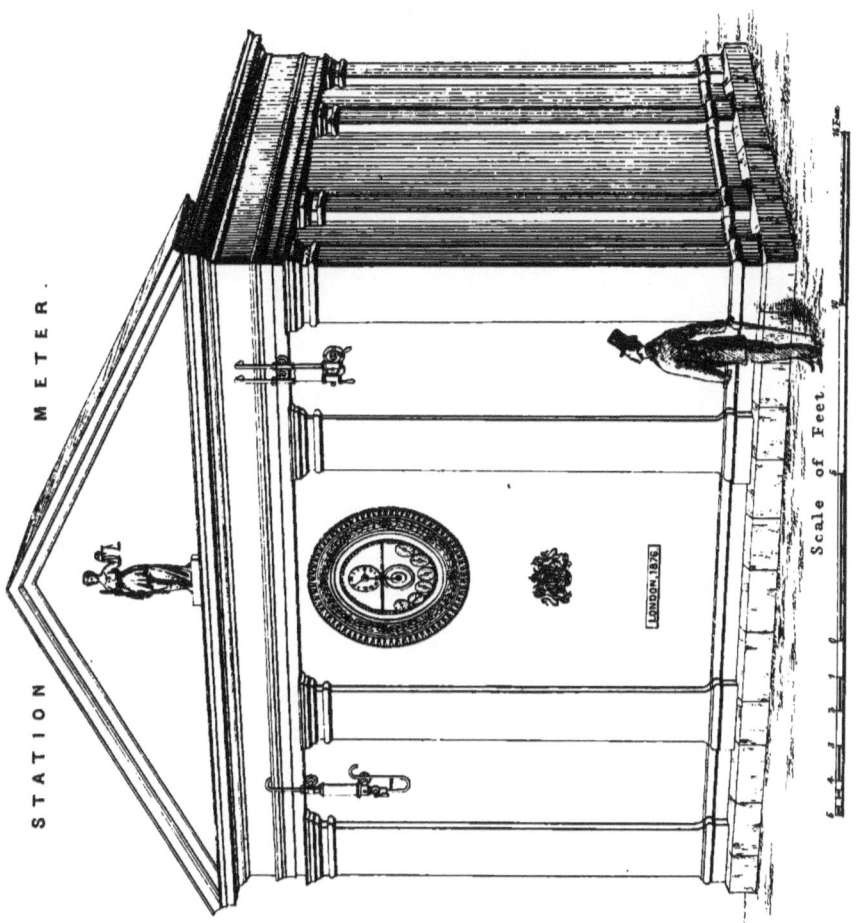

ELEVATION.

PLAN.

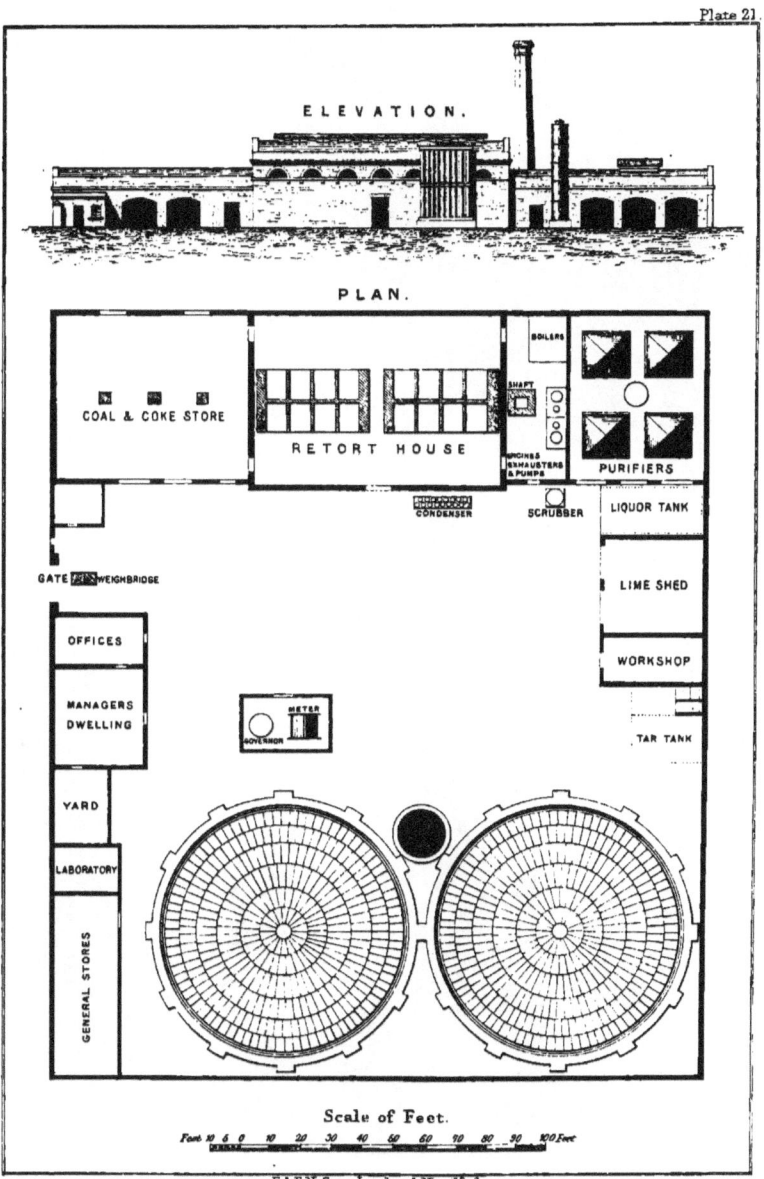

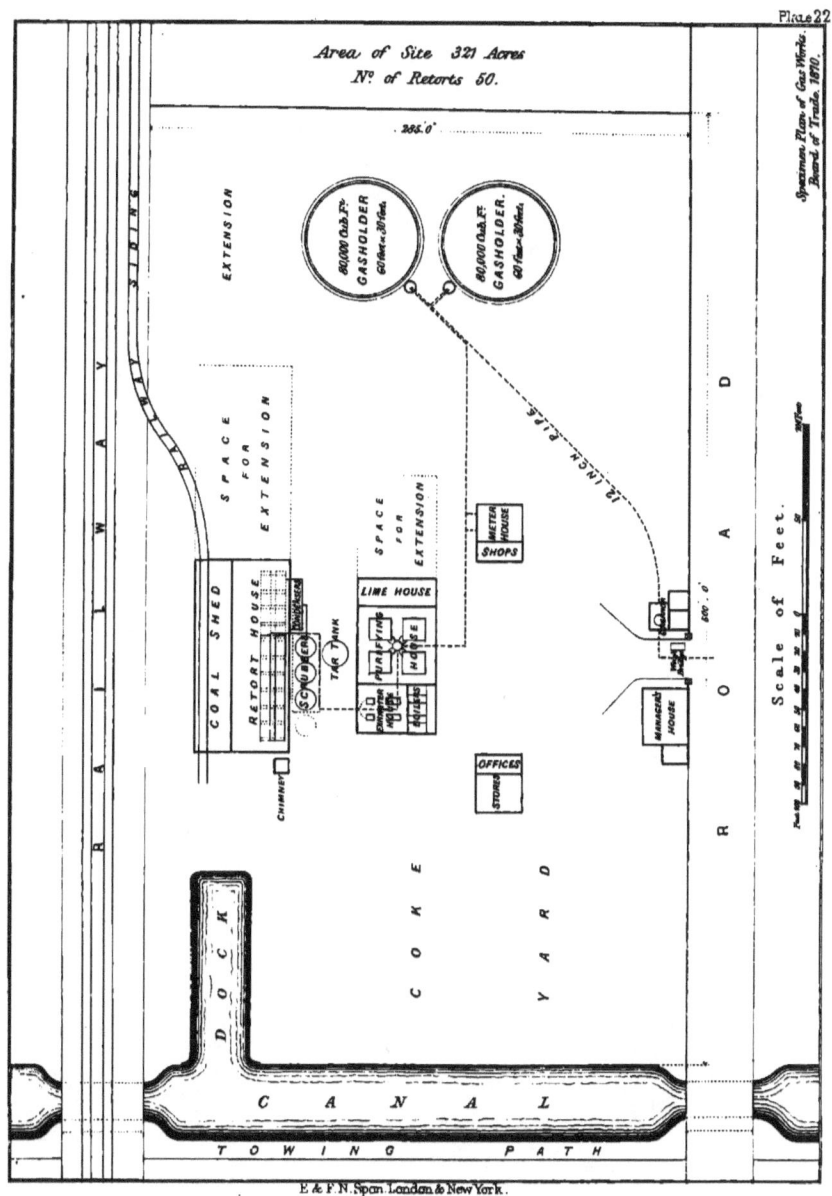

NEWCASTLE & GA
REDHEU

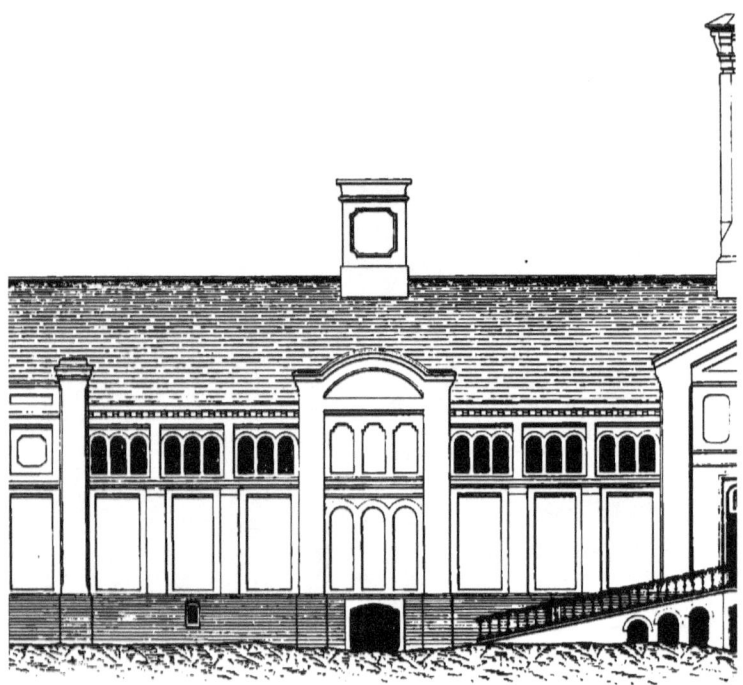

E L E V

Scale

HEAD GAS COMPANY.
WORKS.

REDHEUGH.

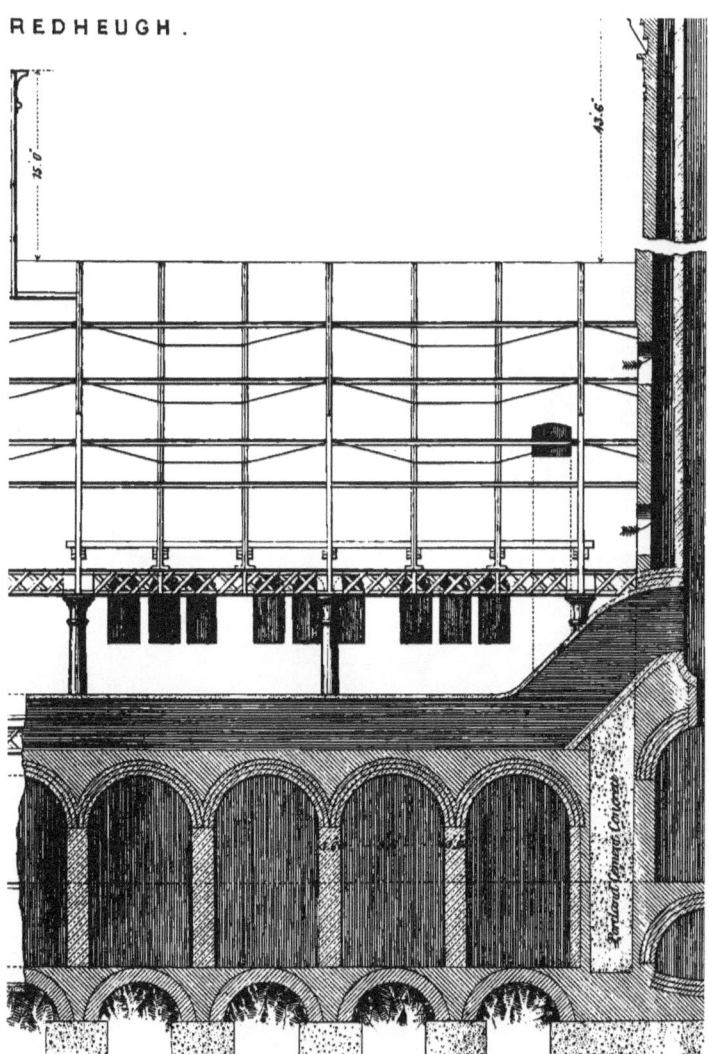

Feet.

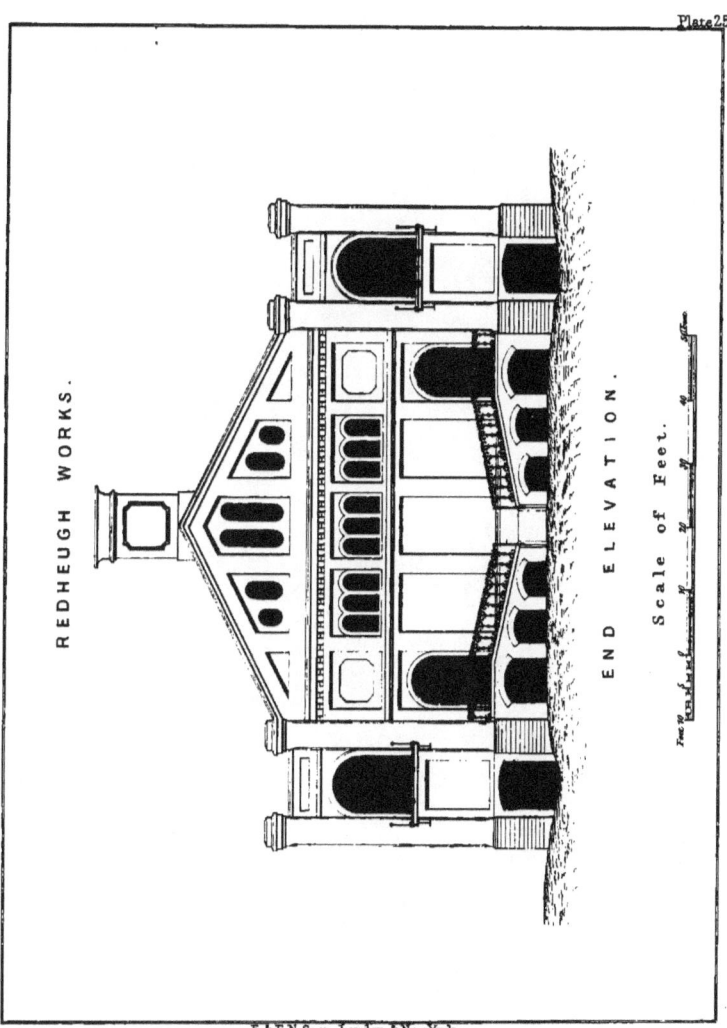

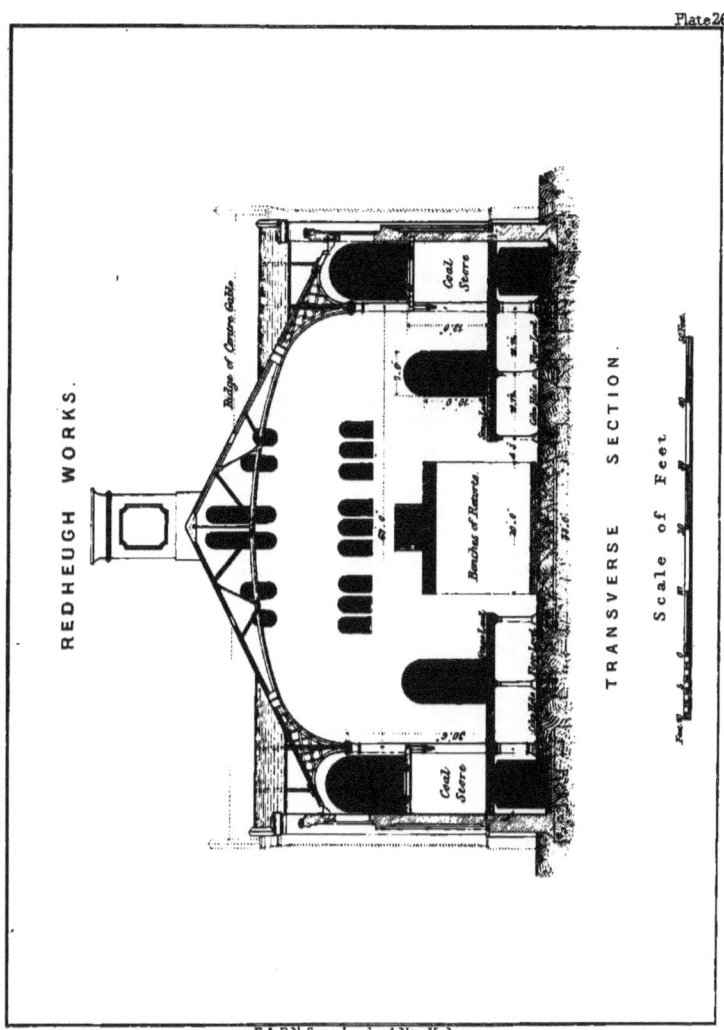

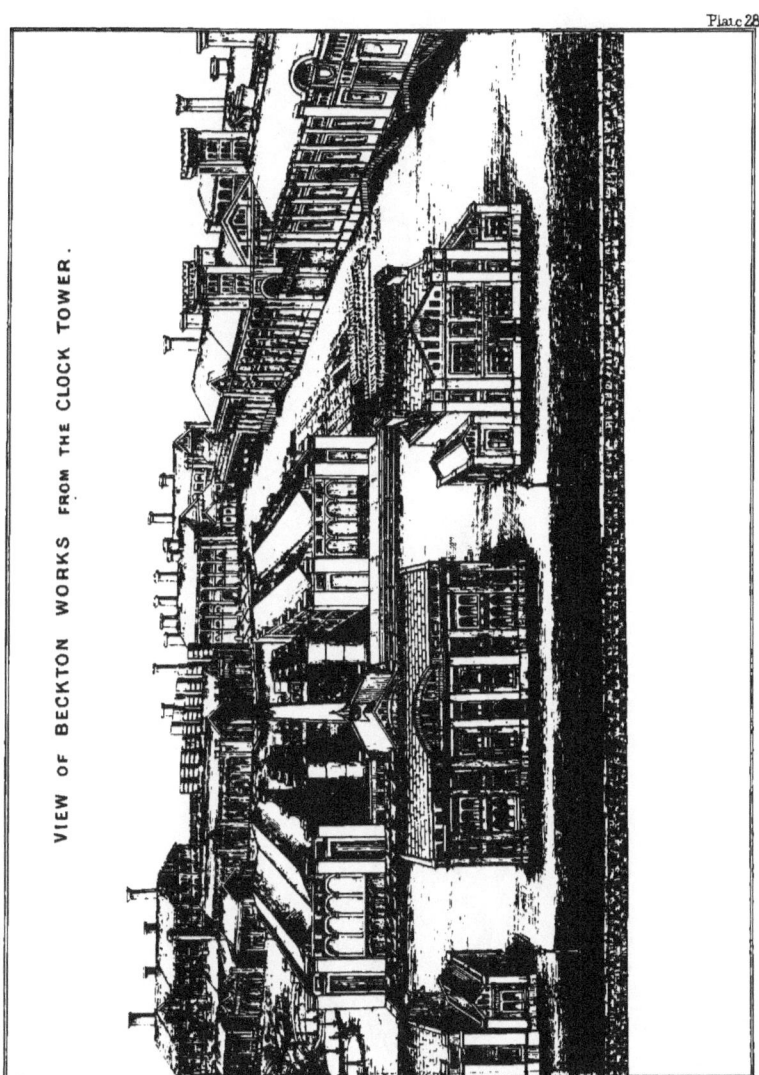

Plate 28.

VIEW OF BECKTON WORKS FROM THE CLOCK TOWER.

F. & F. N Spon London & New York.

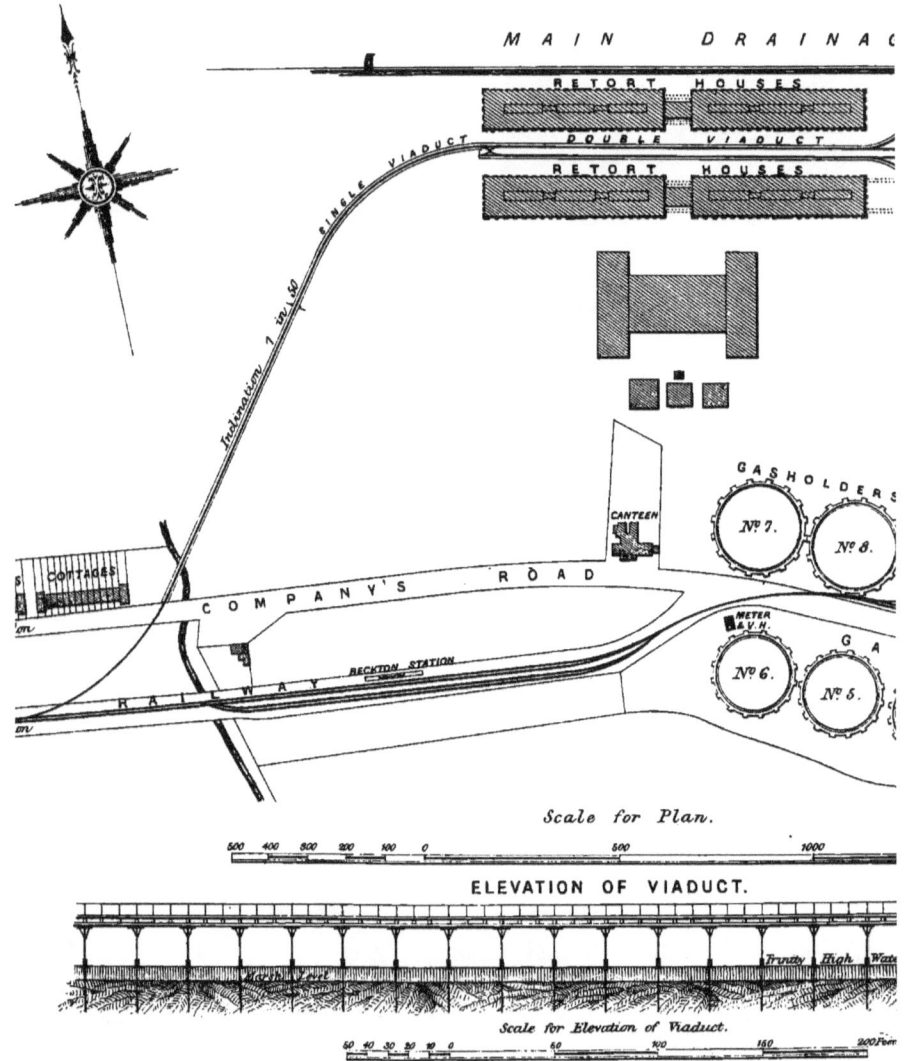

BECKTON WORKS.

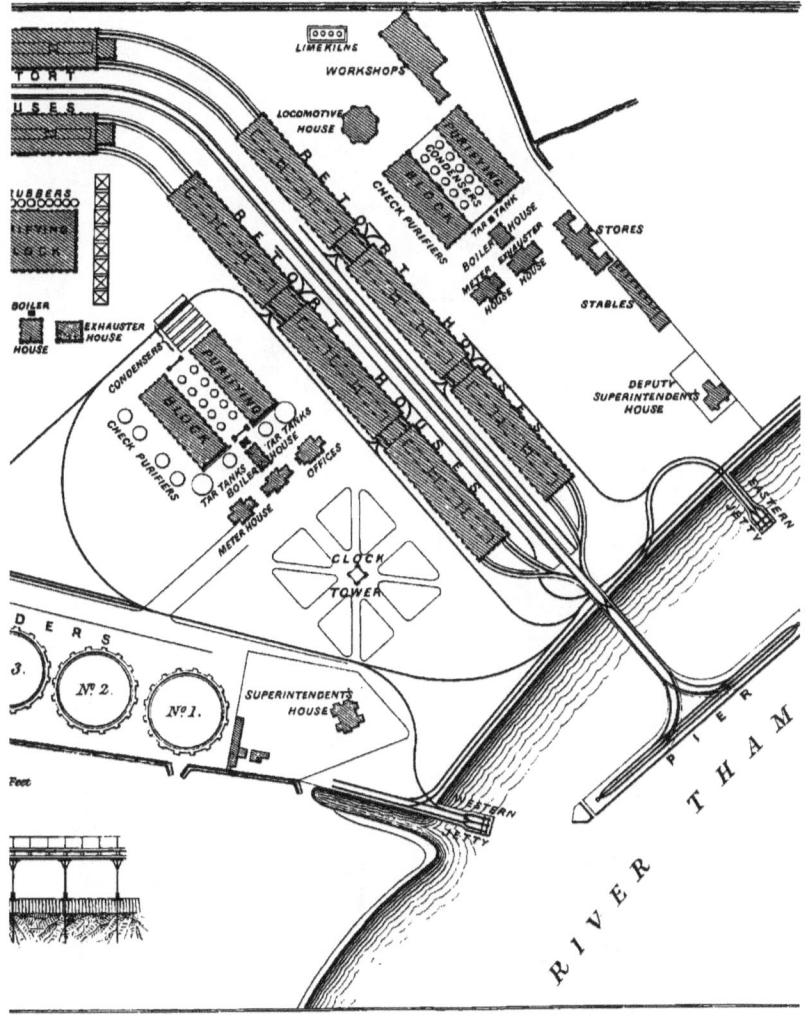

LIST OF PLATES.

Plate. 1.—Retort Settings of a Works of the First Magnitude.
- „ 2.—Setting of One Iron Retort.
- „ 2.—Anderson's Setting of Three Retorts.
- „ 3.—Setting of Two Retorts. Setting of Three Retorts.
- „ 4.—Setting of Five Retorts.
- „ 5.—Hutchinson's Setting of Five Retorts.
- „ 6.—Cathels's Setting of Seven Retorts.
- „ 7.—Skoines's Setting of Six Retorts.
- „ 8.—Methven's Setting of Ten Retorts.
- „ 9.—Tower Scrubber.
- „ 10.—Two Tower Scrubbers.
- „ 11.—Imperial Gasholder.
- „ 12.—Gasholder in Cast-Iron Tank, Elevation and Plan.
- „ 13.—Telescope Holder.
- „ 14.—Part of Plan of Gasholder at Croydon.
- „ 15.—Half-section of Tank of Redheugh Works.
- „ 16.—Plan of Purifiers.
- „ 17.—Elevation of Purifiers.
- „ 18.—Station Governor.
- „ 19.—Station Meter.
- „ 20.—Plan of Works of Three Million Feet.
- „ 21.—Plan of Works of Sixty Million Feet.
- „ 22.—Board of Trade Plan.
- „ 23.—Elevation of the Retort House of the Redheugh Works.
- „ 24.—Part of Longitudinal Section of the same.
- „ 25.—End Elevation of the same.
- „ 26.—Transverse Section of the same.
- „ 27.—View of the Beckton Works from the Thames.
- „ 28.—View of the Beckton Works from the Clock Tower.
- „ 29.—Plan of the Beckton Works.

ADVERTISEMENTS.

ESTABLISHED 1844.

London, 1851.
New York, 1853.
Paris, 1855.
London, 1862.

Dublin, 1865.

THE SIX MEDALS AWARDED
TO THOMAS GLOVER FOR HIS
PATENT DRY GAS-METERS;

Paris, 1867.

The latter being the Highest Medal awarded for Dry Gas-Meters by the Imperial Commissioners for the Universal Exhibition, Paris, 1867.

THOMAS GLOVER & CO.,
DRY GAS-METER MANUFACTURERS,
214 TO 222, ST. JOHN STREET, CLERKENWELL GREEN,
LONDON, E.C.

THOMAS GLOVER & CO.'S PATENT DRY GAS-METERS,

- 1st. Are a Remedy for all the Defects of Wet Meters;
- 2nd. Are Suitable for all Climates, whether Hot or Cold;
- 3rd. Incur no Loss of Gas by Evaporation;
- 4th. Cannot become Fixed by Frost, however severe;
- 5th. Are the most Accurate and Unvarying Measurers of Gas;
- 6th. Prevent Jumping or unexpected Extinction of the Lights;
- 7th. May be Fixed either Above or Below the Level of the Lights;
- 8th. Cannot be Tampered with without visibly Damaging the Outer Case;
- 9th. Will Last much Longer than Wet Meters;
- 10th. Will not Cost more than One-Half for Repair that Wet or Water Meters do; Are upheld for Five Years without Charge.

a

ADVERTISEMENTS.

WET AND DRY GAS METERS.

WET METERS in Cast-iron Cases.

STATION METERS,
WITH PLANED JOINTS.

WILLIAM PARKINSON & CO.,

COTTAGE LANE, CITY ROAD,

LONDON.

GOVERNORS, PRESSURE REGISTERS, GAUGES, AND EXPERIMENTAL APPARATUS.

METERS FOR MEASURING WATER.

SMALL STATION METERS,
In Cylindrical Cases.

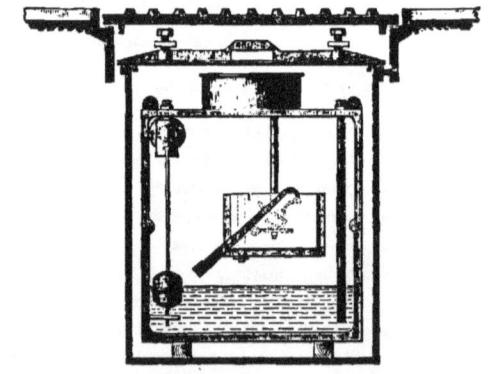

STREET LAMP METERS,
And all APPARATUS for the ADOPTION of the AVERAGE METER SYSTEM.

W. & B. COWAN,

ESTABLISHED 1827,

ORIGINAL LICENSEES AND MANUFACTURERS OF

WARNER & COWAN'S SELF-REGULATING GAS-METER

Remarks on some Points of Agreement and Difference in Meters that have been generally adopted for the Measurement of Gas.

WET METER (COMMON).	DRY METER (EVERY MAKE).	WET METER (COMPENSATING).	WET METER, Warner & Cowan's (SELF-REGULATING DRUM).
Variable measure, and affected by valvular (hood) action; **variation limited** to height of spout and range of float. Measures and valves (hoods) are constructed together, and form but parts of a simple wheel revolving on an axis; the variation of measurement arises from the variation of water-line. Measures constructed of incorrodible metal. Valves (hoods) formed by portions of the measures passing through the water in the meter. Action of the float and valve sensitive and limited.	Variable measure, and affected by valvular action. **Variation unlimited,** depending upon the condition of leather and valves. Measures and valves are separate and distinct parts of the Meter, and complicated by the mechanism necessary to transmit the motion of one part to the other. Measures formed of a porous leather. Valves formed by surfaces of metal rubbing over each other. Action of Meter sensitive from rubbing surfaces of valves. Allows gas to pass unregistered, but none of the others permit this.	Dependent on a true **as well as a fixed** water-line. Varies more or less, and to the extent of range of float, if it is adjusted to water-line. Measures and valves as in the common Meter in construction and action, but it is complicated by external arrangements, to ensure as far as possible a constant water-line, on which the accuracy of measurement is dependent. Valves (hoods) as in common Meter. Action of float and valve depends upon their position in the Meter.	Unvarying measure, and unaffected by valvular (hood) action. Measurement accurate through the whole range of float. Measuring-wheel constructed with an inner wheel by which the action of the Motor is neither impeded nor complicated, and which ensures **accurate measurement, with a varying water-line.** Measures incorrodible metal. Valves (hoods) as in common Meter. Range of float increased, and less liable to be affected by change of pressure or consumption.
The Range allowed by Sales of Gas Act necessary in these Meters.			No Range necessary.

W. & B. COWAN, WET AND DRY GAS-METER MANUFACTURERS,

Church Street, Milbank Street,	Fennel Street,	Buccleuch Street Works,
LONDON, S.W.	**MANCHESTER.**	**EDINBURGH.**

J. & J. BRADDOCK,

GAS ENGINEERS,

GLOBE METER WORKS, OLDHAM,

MANUFACTURERS OF

WET AND DRY GAS METERS,

OF THE HIGHEST EXCELLENCE ONLY.

Our Ordinary **IMPROVED CONSUMERS' GAS METERS** are good, plain, substantial, durable Meters, the action and registration most reliable, and require the least possible attention—once a quarter.

Our **PATENT COMPENSATING GAS METERS**, which are uniform with ordinary Meters, register with minute accuracy, and only require watering once, or at most twice, a year.

Licensees and Manufacturers of **METERS** with **Warner & Cowan's PATENT SELF-REGULATING DRUMS**, highly commended for accuracy of registration.

LAMP METERS and Cast-iron Boxes.

IMPROVED DRY GAS METERS, in best tinned Iron Cases only.

LARGE CONSUMERS' WET GAS METERS, in Round Cases, 100 lights and upwards, the most suitable for Public Buildings, Manufactories, &c.

GAS STATION METERS, in Round Cases, on Cast-iron Stands, with Byepass, Valves, and all Modern Appliances complete.

GAS STATION METERS, in Square Cases, of any magnitude, of appropriate design and substance, with Planed Joints, as first introduced by us, fitted complete with all Appliances requisite for the Supply, Level, and Overflow of Water, with substantial Gauges for showing the same and Gas Pressures at a glance. These Meters are highly appreciated for their extraordinary efficiency, stability, and completeness.

PATENT COMPENSATING GAS STATION GOVERNORS, that give a uniform pressure at all draughts of the main.

PATENT COMPENSATING GAS GOVERNORS, for Consumers.

IMPROVED WET DISTRICT GOVERNORS.

PRESSURE GAUGES, PRESSURE AND VACUUM REGISTERS, EXHAUST GOVERNORS, TEST HOLDERS, PUBLIC LAMPS, GAS COOKERS, ELBOW COCKS, MAIN COCKS, FERRULES, WATER COCKS, &c.

All Sizes of Consumers' Meters kept in Stock, and Orders almost invariably despatched on the day of receipt.

WORKS—WELLINGTON STREET, NEAR THE CENTRAL AND CLEGG STREET STATIONS, OLDHAM.

ADVERTISEMENTS.

Retort Settings complete.

H. SKOINES & CO.,
GAS ENGINEERS,

And Contractors for Setting Retorts, Erecting and Remodelling Gasworks of any extent at Home and Abroad.

Gasholders Manufactured of all Sizes.

The Trade and Shippers supplied.

THE IMPROVED GASWORKS,

FOR

LIGHTING VILLAGES,

MANSIONS,

RAILWAY STATIONS,

&c., &c.

Argyle Street, LONDON, W.C.

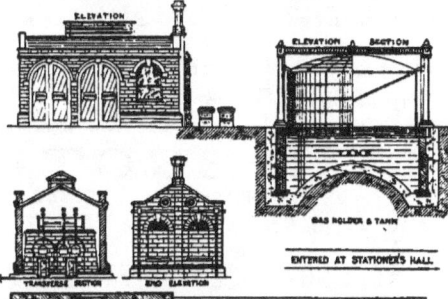

PLAN OF WORKS

ESTIMATES

FOR

SETTING RETORTS,

AND

SUPPLYING THE WHOLE

OR

ANY PORTION

OF THE

GAS APPARATUS,

AND

OTHER INFORMATION

SUPPLIED, WITH

PLANS & SPECIFICATIONS,

FREE OF COST.

The following PRICES of APPARATUS are inclusive of Chimney, Retorts, and Setting, Gasholder, and all Carriage and Fixing, likewise a complete Set of Tools and full Instructions, within a distance of 150 miles from London; or the complete Apparatus delivered free on board in the Thames, Mersey, Southampton, or Hull.

No. of Lights.	Retorts.	Condensers.	Purifiers.	Gas Holders.	Size of Pipes.	Weight.	Price.
	in. in. ft.		ft. ft. in.	ft. in. ft.	cub. ft.	in.	tons. cwt.	£
75	1—15 by 9 by 6	12 feet long	1—2 by 2 0	10 0 by 8 containing	650	3	4 5	100
100	1—15 „ 9 „ 6	24 „ „	1—2 „ 2 0	12 6 „ 8 „	1000	3	6 0	147
150	2—15 „ 9 „ 6	60 „ „	1—2 „ 3 0	14 0 „ 10 „	1500	3	7 10	168
200	2—15 „ 9 „ 6	80 „ „	2—3 „ 2 0	16 0 „ 10 „	2000	3	8 0	231
300	3—15 „ 9 „ 7	100 „ „	2—3 „ 2 0	20 0 „ 10 „	3100	3	10 5	294
500	4—15 „ 9 „ 7	100 „ „	2—6 „ 3 0	25 0 „ 10 „	5000	4	15 10	420
750	6—15 „ 9 „ 7	100 „ „	3—4 „ 2 6	35 0 „ 10 „	9600	5	22 0	546
1000	7—15 „ 9 „ 7	120 „ „	3—6 „ 3 0	40 0 „ 10 „	12,500	6	27 10	682

FOR RETORT SETTINGS, SEE PAGE 7.

ALEXANDER WRIGHT & CO.,
ENGINEERS,
AND MANUFACTURERS OF "DRY" AND "WET" GAS METERS
OF THE BEST CONSTRUCTION.

STATION METERS and STATION GOVERNORS for Regulating the Pressure and Supply of Gas.
STREET LAMPS, REGULATORS, LAMP METERS, and everything required for Public Lighting.
GAUGES for Indicating and Registering the Pressure of Gas.

TEST GASHOLDERS,
Of the Cubic Capacities of the Government Standards, constructed in accordance with the Regulations of the Standards Department of the Board of Trade.

TEST METERS,
Of ordinary construction, or identical with the STANDARDS *constructed by A. Wright & Co.* for the Standards Department of the BOARD OF TRADE, and all other Apparatus necessary for Testing Gas Meters, &c.

PHOTOMETERS,
And Apparatus of the most perfect description, for the Determination of the Value and Chemical Composition of Gas.

WROUGHT AND CAST IRON PIPES, &c.

PRICE LISTS ON APPLICATION.

CONTRACTORS FOR THE ERECTION OF GASWORKS,
55, 55a, and 56, Millbank Street, Westminster, London, S.W.

KÖRTING BROTHERS,
STEAM JET ENGINEERS,
7 and 17, LANCASTER AVENUE, MANCHESTER.

E. KÖRTING'S PATENT STEAM-JET GAS-EXHAUSTER,
IMPROVED CLELAND'S PATENT.

Small Cost — Compactness — Perfectly Self-Acting — Self-Regulating — Self-Cleansing — No Steam Engine — No Attention — No Extra Room required — No Wear and Tear — No Noise — No Oscillation in Vacuum or Back Pressure. **UPWARDS of 200 in USE,** supplied to

	Cub. feet per hour.		Cub. ft. per hr.		Cub. ft. per hr.
Chartered Gas Co., Beckton	100,000	Bury Corporation	40,000	*Leyland and Farrington ditto	9,000
Ditto, ditto	60,000	Ratcliffe Gas Works (London)	40,000	Dartford Gas Co.	7,000
*Ditto, Silvertown	21,000			*Horsham ditto	5,000
Ditto, ditto	2,000	Bolton Corporation	35,000	Cornholme ditto	3,500
Birmingham ditto, Adderley Street	100,000	Ipswich Gas Co.	25,000	Glasgow Corporation	3,500
		*†Liverpool ditto	25,000	*Nantwich Gas Co.	3,500
Ditto, Saltley	100,000	*†Ditto	25,000	Phœnix ditto	3,500
*London Gas Light Co.	60,000	*†Ditto	25,000	†Waterside Gas Works (Todmorden	3,500
Cheltenham Gas Co.	60,000	Longton Gas Co.	21,000		
Ditto	60,000	*†Lincoln ditto	18,000		
Exeter Gas Co.	50,000	*†Gloucester ditto	15,000	Enniskillen Gas Co.	2,000
South Metropolitan ditto	50,000	*Guildford ditto	15,000	F. W. Grafton & Co., Accrington	2,000
Staleybridge ditto	50,000	Kendal ditto	12,000		
*Wakefield ditto	50,000	Kirkintilloch ditto	12,000	Hanna, Donald, & Wilson, London	2,000
Brentford ditto	40,000	*Epsom and Ewell ditto	9,000		
Burnley Corporation	40,000	*Harrogate ditto	9,000	*Sandwich Gas Co.	1,000
Ditto	30,000	Hertford ditto	9,000		

Those marked thus (*) have the Steam Scrubber working in connection with the Steam-Jet Exhauster.
Those marked thus (†) have also the Slow Speed Condenser.

E. KÖRTING'S PATENT STEAM JET REVIVIFYING BLOWERS FOR THE PURIFYING MATERIAL.

Sole Licensees for **W. Cleland's Patent Steam Scrubber**, combining Small Cost and thorough Efficiency, with absence of all mechanical apparatus and necessity of renewing the Scrubbing Material.

H. SKOINES & CO.,

GAS ENGINEERS and CONTRACTORS for SETTING RETORTS,

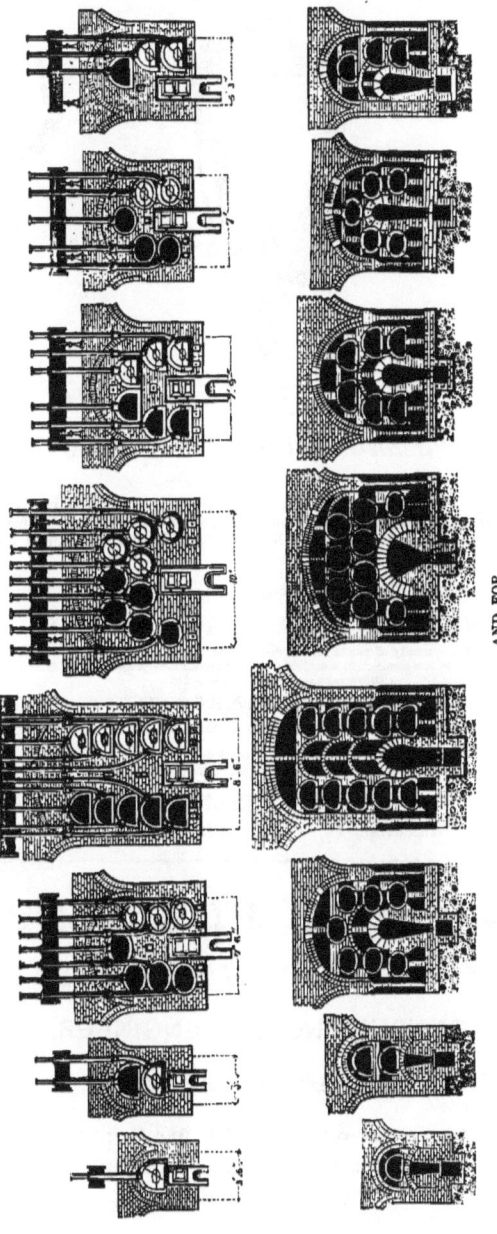

AND FOR SUPPLYING AND ERECTING GASWORKS AND GAS APPARATUS OF ALL DESCRIPTIONS,

ARGYLE STREET, KING'S CROSS, LONDON, W.C.

FOR PRICES OF GASWORKS, SEE PAGE 5.

THE WORKING DRAWINGS FOR THE ABOVE CAN BE HAD ON APPLICATION.

S. LEONI & CO.,
GAS ENGINEERS AND MANUFACTURERS TO THE TRADE.

Messrs. S. L. & Co. have been awarded the ONLY GOLD MEDAL which was presented by the Right Honble. the Earl of Derby, President of the Society for the Promotion of Scientific Industry, for their Novelty, Design, Workmanship, and Utility; also SILVER MEDAL, 1877, at South Shields Exhibition.

SOLE LICENSEES AND AGENTS IN GREAT BRITAIN FOR
GIROUD'S PATENT RHEOMETER;

A New Instrument, which regulates the quantity of gas to be consumed, notwithstanding variations of pressure of gas or alterations of the burners, and gives a much better light with a smaller consumption. They have already placed nearly 10,000, which are giving everywhere great satisfaction both to Gas Companies and Parishes where they have been adopted.

THERE COULD NOT BE A BETTER INSTRUMENT IN CONNECTION WITH THE AVERAGE METER SYSTEM.

A SAMPLE RHEOMETER, WITH SOCKET BURNER AND GLYCERINE COMPLETE, SENT CARRIAGE PAID ON RECEIPT OF 5s. IN STAMPS.

For Copies of Reports and References respecting the Rheometer, also for Pamphlets on the advantages of Cooking by Gas, Illustrations, Price Lists, &c., &c., apply at

Works and Offices, 54 to 66, ST. PAUL'S STREET, NEW NORTH ROAD, LONDON, N.;
Warehouse, 74, STRAND, ADELPHI, W.C.

WILLIAM RICHARDS,
INVENTOR OF THE DRY METER,
CONSULTING GAS ENGINEER,
11, ST. MICHAEL'S ROAD, STOCKWELL, LONDON,

May be Consulted on all matters connected with Gas Engineering, or on Instruments of Precision in connection therewith.

JAMES McKELVIE,
CANNEL COAL MERCHANT,
HAYMARKET, EDINBURGH.

ESTABLISHED 1840.

AGENT FOR THE NITSHILL COMPANY'S DUKE OF HAMILTON'S LESMAHAGOW CANNEL COAL.

PRICES AND ANALYSES OF ALL THE PRINCIPAL SCOTCH CANNEL COALS WILL BE FORWARDED ON APPLICATION.

JAMES MILNE & SON,
Gas Engineers, Gas Meter, Gas Apparatus, and Gas Fittings Manufacturers,
EDINBURGH;
2, KING EDWARD STREET, and NEWGATE STREET, LONDON;

METER WORKS IN LONDON—2, CROSS STREET, WILDERNESS ROW, E.C.

Station Meters, Governors, Consumers' Meters, Gas Lustres, Chandeliers, Brackets, etc., and every Description of Gas Fittings and Gas Apparatus.

ALBERT GAS COAL.

THE DERBYSHIRE SILKSTONE COAL COMPANY, LIMITED,
Can offer a GAS COAL of SUPERIOR QUALITY, delivered at any Station in England and Wales.

Purified Gas per Ton of Coal, in Cubic Feet (average) .. .. 10,775.   Weight of Coke in lbs. per Ton of Coal .. .. 1465.

Analysis and Prices on application to Mr. EDMUND TAYLOR, Secretary,

ALBERT COLLIERY, NEWBOLD, NEAR CHESTERFIELD.

8vo, cloth, 12s. 6d.

ANALYSIS, TECHNICAL VALUATION, PURIFICATION, AND USE OF COAL GAS.
BY THE REV. W. R. BOWDITCH, M.A.
WITH WOOD ENGRAVINGS.

Condensation of Gas — Purification of Gas — Light — Measuring — Place of Testing Gas — Test Candles — The Standard for Measuring Gas-light — Test Burners — Testing Gas for Sulphur — Testing Gas for Ammonia — Condensation by Bromine — Gravimetric Method of taking Specific Gravity of Gas — Carburetting or Naphthalizing Gas — Acetylene — Explosions of Gas — Gnawing of Gas-pipes by Rats — Pressure as related to Public Lighting, &c.

"While the Gas Question is claiming the attention of the inhabitants of London and many of the large provincial towns, Mr. Bowditch's book must prove useful. The whole subject of the production, purification, and use of gas, is treated in a perfectly exact, yet at the same time in a clear and concise manner. The gas manager will find in its pages a careful examination by a man of science and great experience of all the methods which have been introduced for removing the non-illuminating and offensive principles from coal gas. The use of the photometer in all its varieties is described, &c., with an important chapter on testing gas."—ATHENÆUM.

LONDON: E. & F. N. SPON, 46, CHARING CROSS.   NEW YORK: 446, BROOME STREET.

SPONS' DICTIONARY OF ENGINEERING,

CIVIL, MECHANICAL, MILITARY, AND NAVAL,

WITH TECHNICAL TERMS IN

French, German, Italian, and Spanish.

Super-royal 8vo, containing **3132** *Printed Pages, and* **7414** *Engravings.*

CAN BE HAD IN THE FOLLOWING BINDINGS:

	£	s.	d.
In 8 Divisions, Cloth	5	8	0
Ditto, French Morocco	7	4	0
Complete in 3 Vols., Cloth, Marbled Edges	5	5	0
Bound in a superior manner, 3 Vols., Half-morocco, Top Edge Gilt	6	12	0

3 DICTIONARY OF ENGINEERING is so arranged that particular branches of Civil, Mechanical, and Military Engi[n] an be referred to alphabetically. The subjects are treated in a thoroughly practical manner, and the majo[r] such length as to form complete treatises. Among the articles so treated may be instanced:—

[TURAL] INSTRUMENTS. 68 Illustrations.
[I]C SIGNS, with Examples of their use in [n]g Engineering Formulæ.
[Metalli]c) employed in the Useful Arts.
[M]ETERS.
[I]TT.—With a lengthened description of [su]ch Katrine Waterworks Aqueduct. 30 [tr]ations.
for Ships of War, with the various ex[perime]nts to determine the strength of Iron. 42 Illustrations.
[N] WELLS (boring and sinking of).—[me]ns amongst others a description of the [sin]king of an Artesian Well at Buite-aux[-Bois], by M. Dru. 60 Illustrations.
[NG].
[A]ND AXLE-BOXES FOR RAILWAY CAR[S]. 27 Illustrations.
[M]ACHINERY.—30 Illustrations.
[AR]REL. BARRACKS.
[..—T]he art of Barring a River or other [wa]tercourse, in order to facilitate Navigation [irr]igation, especially adapted to moun[tain]s and hot countries. 15 Illustrations.
[Y.—T]he various Galvanic Batteries.
[Galvanic] Batteries for Ore crushing, Iron Ships' [bott]ies. 46 Illustrations.
[Ch]arge).—How to construct to sound any [appen]ded note.
[A]ND BELTING.—43 Illustrations.
[—]89 Pages, and 140 Illustrations.
[MA]KING MACHINES.—80 Illustrations.
[G A]ND BLASTING.—83 Pages, and 128 [illustr]ations.
[Railway).—44 Pages, 67 Illustrations.
[MAK]ING MACHINES, and the Art of Brick[ma]k[in]g. 36 Illustrations.
[.—T]imber, Stone, and Iron. 214 Pages, 54 Illustrations.
[.—]MAKING MACHINE.
(Iron and Hemp), with Kirkaldy's ex[perim]ents on the strength of Welded Joints, [Por]tsmouth's experiments on the strength [of Iro]n and Hemp Cables.
[.].
[r]s.—31 Illustrations.
[LINING].—62 printed Pages, and 125 [Illust]rations.

COTTON MACHINERY.—32 Illustrations.
DAMMING.—42 Pages, and 75 Illustrations.
DETAILS OF ENGINES.—101 Illustrations.
DIVING BELL AND DIVING APPARATUS.
DOCKS AND HARBOURS.—40 Pages, and 65 Illustrations.
ELECTRO-METALLURGY.
ENGINES (Varieties of).—32 Pages, 64 Illustrations.
FAN.—With description of Struvé, Lemielle, Nasmyth, and other large Fans; Guibal's, for ventilating Collieries. 24 Illustrations.
FIRE-ARMS.—61 Illustrations.
FLAX MACHINERY.—20 Engravings of the latest and most approved Machines in use.
FORGING.—41 Illustrations.
FOUNDING AND CASTING.—39 Pages, 64 Illustrations.
GAS MANUFACTURE.—30 Illustrations.
GEARING.—80 Illustrations.
GLASS MACHINERY.—Machinery used in the Manufacture of Glass, particularly Plate Glass. 61 Illustrations.
GUNPOWDER, manufacture of.
GUN MACHINERY. — Machinery used in the rifling and manufacture of Guns.
HAND TOOLS.—425 Illustrations.
HARBOUR.—20 Illustrations.
HINGING.—46 Illustrations.
HYDRAULICS, AND HYDRAULIC MACHINES.—204 Pages, and 301 Illustrations.
ICE-MAKING MACHINERY.—10 Illustrations.
INDIA-RUBBER MANUFACTURE.
INDICATOR.
IRON.—115 Pages, and 435 Illustrations.
IRON SHIPBUILDING.—60 Illustrations.
IRRIGATION.—52 Illustrations.
JOINTS.—137 Illustrations.
KILNS.—52 Illustrations.
KNITTING MACHINERY.—50 Illustrations.
LIFTS, HOISTS, AND ELEVATORS.—34 Pages, and 103 Illustrations.
LIGHTS, BUOYS, AND BEACONS, with an account of the Wolf Rock Lighthouse. 24 Pages, 90 Illustrations.
LIMES, MORTARS, AND CEMENTS.
LOCKS AND LOCK-GATES.—36 Illustrations.
LOCOMOTIVE.—53 Illustrations.

MACHINE TOOLS.—30 Pages, and 80 [Illustra]tions.
MARINE ENGINE.—71 Illustrations.
MATERIALS OF CONSTRUCTION (Strength [of]) Pages, 84 Illustrations.
MECHANICAL MOVEMENTS.—31 Pages, a[nd] Illustrations.
METALLURGY.—212 Pages, 518 Illustra[tions].
METER.—Gas and Water Meters. 22 [Illustra]tions.
METRIC SYSTEM.
MILLS. — Comprising Wind, Oil, and [Water] Mills. 27 Pages, 92 Illustrations.
OBLIQUE ARCH.—21 Illustrations.
ORES.—25 Pages, 74 Illustrations.
PAPER MACHINERY.—27 Illustrations.
PILES AND PILE-DRIVING.—28 Pages, [Illustra]tions.
PUMPS AND PUMPING.
QUARRYING.
RAILWAYS AND RAILWAY ENGINEER[ING] Pages, 172 Illustrations.
RETAINING WALL.—25 Pages, 41 Illust[rations]
RIVERS AND RIVER ENGINEERING.
RIVETING.—24 Illustrations.
ROADS.
ROOFS. — Wood and Iron. 36 Pag[es] Illustrations.
ROPE-MAKING MACHINERY.—43 Illustra[tions]
SCAFFOLDING.
SCREW ENGINES.—27 Illustrations.
SCREW-MAKING MACHINE, for making Screws.
SIGNALS.—79 Illustrations.
SILVER.—27 Pages, 35 Illustrations.
STATIONARY ENGINES.—43 Illustrations
STAVE AND CASK-MAKING MACHINERY.
STEEL (Manufacture of).—50 Illustratio[ns]
SUGAR-MAKING MACHINERY.—40 Illust[rations]
SURVEYING AND SURVEYING INSTRUME[NTS] Pages, 72 Illustrations.
TELEGRAPHY. — Method of Constructi[on,] Laying out Lines, Manufacture and [Laying] of Cables, &c. 26 Pages, 64 Illustra[tions]
VENTILATION AND WARMING. — 27 Pa[ges,] Illustrations.
WATER-WORKS.—52 Illustrations.
WOOD-WORKING MACHINERY.—62 Illust[rations]

London: E. & F. N. SPON, 46, CHARING CROSS.
New York: 446, BROOME STREET.

www.ingramcontent.com/pod-product-compliance
Lightning Source LLC
Chambersburg PA
CBHW022116300426
44117CB00007B/739